PROFESSIONAL

Windows® Phone 7 Application Development

PROFESSIONAL WINDOWS® PHONE 7 APPLICATION DEVELOPMENT

INTRODUCTION ... xxv

CHAPTER 1 Taking the Metro with Windows Phone ... 1
CHAPTER 2 Free Transport with Visual Studio 2010 and Expression Blend 4 ... 15
CHAPTER 3 Designing Layouts Using Red Threads ... 27
CHAPTER 4 Adding Motion ... 63
CHAPTER 5 Orientation and Overlays ... 95
CHAPTER 6 Navigation ... 123
CHAPTER 7 Application Tiles and Notification ... 151
CHAPTER 8 Tasks ... 173
CHAPTER 9 Touch Input ... 195
CHAPTER 10 Shake, Rattle, and Vibrate ... 221
CHAPTER 11 Who Said That? ... 241
CHAPTER 12 Where Am I? Finding Your Way ... 279
CHAPTER 13 Connectivity and the Web ... 309
CHAPTER 14 Consuming the Cloud ... 349
CHAPTER 15 Data Visualization ... 385
CHAPTER 16 Storing and Synchronizing Data ... 417
CHAPTER 17 Frameworks ... 459
CHAPTER 18 Security ... 491
CHAPTER 19 Gaming with XNA ... 515
CHAPTER 20 Where to Next? ... 541

INDEX ... 571

PROFESSIONAL

Windows® Phone 7 Application Development

BUILDING APPLICATIONS AND GAMES USING VISUAL STUDIO, SILVERLIGHT®, AND XNA®

Nick Randolph
Christopher Fairbairn

Wiley Publishing, Inc.

Professional Windows® Phone 7 Application Development: Building Applications and Games Using Visual Studio, Silverlight®, and XNA®

Published by
Wiley Publishing, Inc.
10475 Crosspoint Boulevard
Indianapolis, IN 46256
www.wiley.com

Published simultaneously in Canada

ISBN: 978-0-470-89166-7

ISBN: 978-1-118-01534-6 (ebk)

ISBN: 978-1-118-01393-9 (ebk)

ISBN: 978-1-118-01394-6 (ebk)

Manufactured in the United States of America

10 9 8 7 6 5 4 3 2 1

For general information on our other products and services please contact our Customer Care Department within the United States at (877) 762-2974, outside the United States at (317) 572-3993 or fax (317) 572-4002.

Wiley also publishes its books in a variety of electronic formats. Some content that appears in print may not be available in electronic books.

Library of Congress Control Number: 2010932454

To the love of my life, Cynthia.

—Nick Randolph

To my princess forever after, Michele.

—Christopher Fairbairn

ABOUT THE AUTHORS

NICK RANDOLPH currently runs Built to Roam, which builds rich mobile applications for Windows Phone. Previously, Nick was co-founder and Development Manager at nsquared Solutions, where he led a team of developers to build inspirational software using next-wave technologies. Prior to nsquared, Nick was the lead developer at Intilecta Corporation, where he was integrally involved in designing and building their application framework.

After graduating with a combined Engineering (IT)/Commerce degree, Nick went on to be nominated as a Microsoft MVP in recognition of his work with the .NET developer community, and his interest and knowledge on developing for the Windows Mobile platform. He is still an active contributor in the device application development space via his blog at `http://nicksnettravels.builttoroam.com`.

Nick has been invited to present at a variety of events including TechEd Australia and NZ, MEDC, and Code camp. He has authored three other books covering Visual Studio, the latest being Professional Visual Studio 2010, and has helped judge the world-wide finals for the Imagine Cup for five years.

CHRISTOPHER FAIRBAIRN currently works as a lead developer for ARANZ Medical Limited developing mobile applications (and hardware) for wound care applications within the medical space. Prior to this, Christopher was Technical Architect for Blackbay working on the technical framework behind their flagship mobile freight logistics product, Delivery Connect. He has developed for a wide range of mobile platforms including iPhone, Windows Mobile, and Palm OS.

Christopher is currently a Microsoft MVP in recognition of his work within the Windows Mobile developer community. He has also presented at a variety of events throughout New Zealand, mostly organized through the thriving New Zealand MS Communities organization (formally NZ .NET User Groups Society). Whenever possible, Christopher maintains a blog at `www.christec.co.nz/blog/`.

ABOUT THE TECHNICAL EDITOR

SCOTT SPRADLIN has been programming professionally for 30 years and developing solutions in .NET since the early beta releases. He's always been fascinated by the ability to take code portable in handheld devices. As a Microsoft MVP, he is an advocate for the developer community at large and is on the board of INETA North America. You can find Scott on Twitter at `www.twitter.com/scotts` or his blog `http://geekswithblogs.net/sspradlin`.

CREDITS

ACQUISITIONS EDITOR
Paul Reese

SENIOR PROJECT EDITOR
Adaobi Obi Tulton

TECHNICAL EDITOR
Scott Spradlin

SENIOR PRODUCTION EDITOR
Debra Banninger

COPY EDITOR
Cate Caffrey

EDITORIAL DIRECTOR
Robyn B. Siesky

EDITORIAL MANAGER
Mary Beth Wakefield

FREELANCER EDITORIAL MANAGER
Rosemarie Graham

ASSOCIATE DIRECTOR OF MARKETING
David Mayhew

PRODUCTION MANAGER
Tim Tate

VICE PRESIDENT AND EXECUTIVE GROUP PUBLISHER
Richard Swadley

VICE PRESIDENT AND EXECUTIVE PUBLISHER
Barry Pruett

ASSOCIATE PUBLISHER
Jim Minatel

PROJECT COORDINATOR, COVER
Lynsey Stanford

PROOFREADER
Nancy Carrasco

INDEXER
Robert Swanson

COVER DESIGNER
Michael E. Trent

COVER IMAGE
© Thorsten Rust

ACKNOWLEDGMENTS

I'VE ALWAYS BEEN A MOBILE ENTHUSIAST, so when the opportunity came about to write one of the first books to be published on Windows Phone, I jumped at it, not quite envisaging how much of a challenge it would be. In the months since the early versions of the developer tools were made available, a great deal changed, often causing whole sections of the book to be completely rewritten. I can't say how grateful I am to my beautiful Cynthia. Without your unquestioning support and understanding, I would have never been able to complete this book.

The team at Wrox has been amazing to work with. Their attention to detail and ability to address even the most challenging aspects of this book have been a lifesaver — especially Paul Reese, Adaobi Obi Tulton, and the editors who worked on producing this book.

I also extend a big thank you to the Windows Phone team at Microsoft — especially Peter Torr and Anand Iyer, whose willingness to answer my questions saved hours of experimentation, and Dave Glover (DPE Australia), who provided an invaluable sounding board for issues and ideas I came across while writing.

Lastly, I thank my co-author Christopher for sharing his knowledge of multiple mobile platforms to make sure we addressed issues that were relevant to developers wanting to build applications for Windows Phone.

—Nick Randolph

I WOULD LIKE TO THANK MY CO-AUTHOR Nick for giving me the opportunity to collaborate on writing this book with him. His passion and in-depth knowledge of multiple modern technologies made coming up to speed with Windows Phone 7 a breeze. Nick was also very accommodating when it came to adapting to challenges that came up along the way, such as a magnitude 7.1 earthquake that occurred within my hometown of Christchurch, New Zealand near the end of the process.

Closer to home, without the patience, reassurance and support of my partner, Michele, I would never have made it to the end.

Lastly I would like to thank the team at Wrox who has been extremely easy to work with. Their clear explanation of expectations and quality of feedback has been truly awe-inspiring. Thank you to Paul Reese, Adaobi Obi Tulton, and the various editors who contributed to making this book what it is today.

—Christopher Fairbairn

CONTENTS

INTRODUCTION *xxv*

CHAPTER 1: TAKING THE METRO WITH WINDOWS PHONE 1

Minimum Specifications 2
Chassis Design 3
Screen Resolution 5
Metro Design Language 5
Principles 7
User Experience 8
Start and Lock Screens 9
Hubs 10
Developer Landscape 12
Summary 13

CHAPTER 2: FREE TRANSPORT WITH VISUAL STUDIO 2010 AND EXPRESSION BLEND 4 15

Visual Studio 2010 Express for Windows Phone 16
Expression Blend 4 21
Windows Phone Emulator 24
Summary 26

CHAPTER 3: DESIGNING LAYOUTS USING RED THREADS 27

Red Threads 27
Controls 29
Standard Controls 34
Layout 50
Resources and Styles 54
Themes 59
Summary 61

CHAPTER 4: ADDING MOTION 63

Visual State Management 63
Behaviors 71
Animation 79

Template Transitions 80
State Transitions 83
Panoramic and Pivot Controls 88
Summary 94

CHAPTER 5: ORIENTATION AND OVERLAYS 95

Device Orientation 95
Orientation Detection 96
Orientation Changes 99
Orientation Strategies 99
Soft Input Panel (SIP) 109
Application Bar 113
Icon Buttons 114
Menu Items 118
Opacity 119
StateChanged Event 120
System Tray 120
Summary 121

CHAPTER 6: NAVIGATION 123

Page Layout and Architecture 123
Navigation 126
Fragments and QueryString 131
UriMapping 132
Go Back 133
GoBack and CanGoBack 135
Animation 136
Wizards 139
Background Processing 140
Eligible for Termination 141
Scenarios 142
Saving State 146
Obscured 148
Summary 149

CHAPTER 7: APPLICATION TILES AND NOTIFICATION 151

Application Tile 151
Push Notifications 155
Priority 161
Tile Notifications 163

Toast Notifications 164
Raw Notifications 165
Examples 166
Errors 169
Summary **171**

CHAPTER 8: TASKS **173**

Windows Phone Tasks **173**
Where Did My Application Go? 175
Camera and Photos 176
Phone and SMS 179
E-Mail 184
Launchers 188
Extras **192**
Summary **194**

CHAPTER 9: TOUCH INPUT **195**

User Experience **195**
Guidelines 196
Touch Events **200**
Single Touch 201
Double-Tap 203
Multi-Touch 214
Summary **220**

CHAPTER 10: SHAKE, RATTLE, AND VIBRATE **221**

Accelerometer **221**
Working with the Emulator 226
Reactive Extensions for .NET 238
Vibration **239**
Summary **240**

CHAPTER 11: WHO SAID THAT? **241**

Media Playback **241**
MediaElement 241
SoundEffects with XNA 257
Microsoft Translator 267
Audio Recording **270**
Playback 272
Saving 272

Music and Video Hub 274
FM Tuner 276
Summary 278

CHAPTER 12: WHERE AM I? FINDING YOUR WAY 279

Geo-Location 279
GeoCoordinateWatcher 280
IGeoPositionWatcher 284
Bing Maps 295
Map Design 296
Map Credentials 297
Points of Interest and Lines 298
Events 302
Bing Maps Web Services 304
Summary 307

CHAPTER 13: CONNECTIVITY AND THE WEB 309

Connected Status 309
Network Availability 310
Service Reachability 310
Emulator Testing 312
Connectivity 315
WebBrowser Control 320
MultiScaleImage 326
Authentication 328
Windows Live ID 329
Summary 347

CHAPTER 14: CONSUMING THE CLOUD 349

HTTP Request 349
WebClient 350
HttpWebRequest 355
Credentials 359
Compression 360
WCF/ASMX Services 365
Service Configuration 365
Add Service Reference 367
Service Implementation and Execution 368
Custom Headers 371
Credentials 371

WCF Data Services 372
OData with WCF Data Services 372
JSON 379
Summary 383

CHAPTER 15: DATA VISUALIZATION 385

Data Binding 385
DataContext 386
BindingMode 388
Value Converters 392
Designing with Data 394
Sample Data 394
Design-Time Data 402
MVVM Light Toolkit 411
Element and Resource Binding 413
Summary 415

CHAPTER 16: STORING AND SYNCHRONIZING DATA 417

Isolated Storage 417
ApplicationSettings 418
IsolatedStorageFileStream 421
Data-Caching 422
Object Cache 422
Persistent Storage 430
Synchronization 442
Summary 458

CHAPTER 17: FRAMEWORKS 459

Managed Extensibility Framework 459
Import and Export 460
ImportMany 463
Application Composition 467
Microsoft Silverlight Analytics Framework 467
Testing 471
Unit Testing 471
Emulator Automation 486
Summary 490

CHAPTER 18: SECURITY **491**

On the Device **491**
Device Security 492
Device Management 492
Data Encryption 493
Over the Wire **496**
Transport 496
Authentication 497
Summary **514**

CHAPTER 19: GAMING WITH XNA **515**

Getting Started **515**
Game Loop 521
Game Life Cycle 522
Rendering **524**
Content 524
Sprites 525
Movement 526
Text and Fonts 527
Input **530**
Accelerometer 530
Touch 531
Keyboard 532
3D Rendering **532**
3D Model 533
Color and Lighting 535
Primitives 535
Textures 538
Summary **539**

CHAPTER 20: WHERE TO NEXT? **541**

Device Debugging **541**
Registering for Development 541
Debugging Applications 543
Deploying Applications 544
Third-Party Components **544**
Silverlight Toolkit 545
Database 546

Application Migration **547**
User Interface 547
Services and Connectivity 548
Data 548
Device Capabilities 548
Background Processing 549
User Interface Performance **549**
Performance Counters 549
Redraw Regions 551
Caching 551
External Systems **552**
Proxy Service (Exchange) 552
Shared Key Signatures (Windows Azure) 554
Publishing **563**
Application and Start Icons 564
Splash Screen 565
Capabilities 565
Trial Mode and Marketplace 567
Summary **569**

INDEX ***571***

INTRODUCTION

WINDOWS PHONE IS THE LATEST mobile platform from Microsoft. It brings with it a host of new features and services that make it one of the richest mobile platforms on the market. Applications and games for Windows Phone are developed using Visual Studio and designed with Expression Blend. The combined power of these tools will make Windows Phone the easiest mobile platform to develop for.

Professional Windows Phone 7 Application Development takes you on a journey through every major aspect of this new mobile platform. It will show you how to build applications and games that make use of the Windows Phone hardware and software services. You'll also learn how to connect your application to services running in the cloud.

Each topic provides sample code that you can use to walk through and adapt to help you learn more about the Windows Phone development platform.

WHO THIS BOOK IS FOR

Professional Windows Phone 7 Application Development was written to accommodate developers new to building mobile applications, as well as those who have built mobile applications for other platforms such as Windows Mobile, Android, or the iPhone.

In order to get the most out of this book, it is recommended that you have at least a good understanding of C# and the.NET Framework. Prior knowledge of Silverlight or WPF can assist with following some of the examples in the second half of the book.

WHAT THIS BOOK COVERS

Applications and games for Windows Phone have a unique set of requirements and challenges that need to be addressed. Through the course of this book, you will learn about the design philosophy that went into creating the Windows Phone operating system and how you can apply it to what you develop. You'll learn how to interact with various aspects of the Windows Phone platform, as well as how to structure your application so as to provide a consistent and reliable user experience.

This book does not seek to provide thorough coverage of either Silverlight or the .NET Framework. Instead, it aims to give you the background on building applications specifically for Windows Phone.

HOW THIS BOOK IS STRUCTURED

Professional Windows Phone 7 Application Development has been structured to assist you in getting up and building applications as quickly as possible. The chapters at the beginning of the book are designed to give you an understanding of the tools, technologies, and design guidelines for building applications for Windows Phone. Subsequent chapters cover working with both the hardware and software services available on the device. The last set of chapters covers topics such as connectivity to the Web, working with data, application frameworks, and security; all are slightly more advanced topics but are still equally important when building a successful Windows Phone application.

- **Chapter 1: Taking the Metro with Windows Phone** — In Chapter 1 you'll be introduced to the Windows Phone platform and the design philosophy that went into the user experience.
- **Chapter 2: Free Transport with Visual Studio 2010 and Expression Blend 4** — Windows Phone development is done using Visual Studio and Expression Blend. In Chapter 4 you will learn how to get started with these tools and how to use the built-in emulator to debug and test your applications.
- **Chapter 3: Designing Layouts Using Red Threads** — The small screen size of mobile devices makes it essential to get the layout of your application right. Chapter 3 covers how you can leverage the power of Silverlight to allow you to rapidly build rich user interfaces.
- **Chapter 4: Adding Motion** — In Chapter 4 you'll learn how to make your application more dynamic. You'll also see how you can use Expression Blend to create storyboards and state transitions.
- **Chapter 5: Orientation and Overlays** — Windows Phone supports several different orientations that you can elect to make use of within your application. Chapter 5 will show you how to handle changes in the orientation of your application and respond when your application is obscured by overlays such as an incoming phone call.
- **Chapter 6: Navigation** — One of the most unique features of Windows Phone is the application life-cycle model that governs both inter- and intra-application navigation. In Chapter 6 you'll see how to navigate between pages and what to do when your application goes into the background.
- **Chapter 7: Application Tiles and Notification** — Chapter 7 covers how your application can integrate with the Windows Phone Start screen. You'll learn how you can customize the tile that appears on the Start, as well as how you can use notifications to update the tile or notify the user of important events.
- **Chapter 8: Tasks** — What sets mobile applications apart from desktop or web applications is their ability to integrate with the device capabilities. In Chapter 8 you'll see how you can send SMS messages, initiate phone calls, and integrate with the Pictures hub.
- **Chapter 9: Touch Input** — Windows Phone has been designed to be operated through the use of touch gestures. Chapter 9 covers how you can extend your application to respond to many of the standard touch gestures such as Flick, Pan, Pinch, and Stretch.

- **Chapter 10: Shake, Rattle, and Vibrate** — Chapter 10 covers the use of the built-in accelerometer and how you can use it within your application to extend the user experience. You'll also learn how you can simulate the accelerometer within the Windows Phone emulator.
- **Chapter 11: Who Said That?** — Speech and sound are important senses when it comes to our daily life. In Chapter 11 you'll learn how you can play and record sounds from within your application.
- **Chapter 12: Where Am I? Finding Your Way** — Windows Phone includes a sophisticated Location Service that is able to integrate GPS, Cellular, and Wi-Fi information. In Chapter 12 you'll see how easy it is to build a location-aware application using these services.
- **Chapter 13: Connectivity and the Web** — Connecting to the Web in order to send or receive data is an important consideration for your application. Chapter 13 covers how to make your application network-aware and working with the `WebBrowser` control in order to display both local and remote HTML data.
- **Chapter 14: Consuming the Cloud** — Chapter 14 continues the discussion on how to connect to services across the Web, providing examples and strategies for optimizing your application for Windows Phone.
- **Chapter 15: Data Visualization** — In Chapter 15 you'll learn how you can take advantage of the rich data-binding capabilities of Silverlight in building your Windows Phone application.
- **Chapter 16: Storing and Synchronizing Data** — Working with data becomes complex when you need to integrate with existing backend systems. In Chapter 16 you'll learn how to save data in Isolated Storage and synchronize with WCF Data Services.
- **Chapter 17: Frameworks** — In Chapter 17 you'll learn about some of the existing frameworks that you can plug into your application to assist with application structure, usage tracking, and testing.
- **Chapter 18: Security** — Any mobile application that captures or displays data is a potential security risk. In Chapter 18 you'll learn how to use encryption and authentication to improve the security of your application and the data it works with.
- **Chapter 19: Gaming with XNA** — In addition to building applications and games using Silverlight, you can also use the XNA Framework to develop for Windows Phone. Chapter 19 provides a walk-through of some of the important features of this framework.
- **Chapter 20: Where to Next?** — The last chapter, Chapter 20, covers some additional points that you should consider while developing for Windows Phone. This includes some steps you should take to prepare your application for publishing on the Windows Phone Marketplace.

WHAT YOU NEED TO USE THIS BOOK

To use this book effectively, you'll need to download and install the Windows Phone developer tools. This is covered in Chapter 2, which discusses Visual Studio, Expression Blend, and the Windows Phone emulator in more detail.

This book doesn't cover the multitude of features of Visual Studio, other than in passing, when illustrating how to use a particular feature related to Windows Phone development. For more detailed information on Visual Studio, you might consider another Wrox title: *Professional Visual Studio 2010* by Nick Randolph, David Gardner, Chris Anderson, and Michael Minutillo (Wrox, 2010).

In some chapters there are references to third-party tools that may assist you in building your Windows Phone application. These are not necessary to understand the concepts discussed but are included to help you develop for Windows Phone.

CONVENTIONS

To help you get the most from the text and keep track of what's happening, we've used a number of conventions throughout the book.

Boxes with a warning icon like this one hold important, not-to-be forgotten information that is directly relevant to the surrounding text.

Notes provide additional, ancillary information that is helpful, but somewhat outside of the current presentation of information.

HEADING

Boxes with headings explain important points in more detail or give further explanatory material.

As for styles in the text:

- We *highlight* new terms and important words when we introduce them.
- We show keyboard strokes like this: Ctrl+A.
- We show URLs and code within the text like so: `persistence.properties`.
- We present code in two different ways:

```
We use a monofont type with no highlighting for most code examples.
We use bold to emphasize code that's particularly important in the present context.
```

SOURCE CODE

As you work through the examples in this book, you may choose either to type in all the code manually or to use the source code files that accompany the book. All of the source code used in this book is available for download at www.wrox.com. You will find that the code snippets from the source code are accompanied by a download icon and note indicating the name of the program so you know it's available for download and can easily locate it in the download file. Once at the site, simply locate the book's title (either by using the Search box or by using one of the Title lists) and click the Download Code link on the book's detail page to obtain all the source code for the book.

Because many books have similar titles, you may find it easiest to search by ISBN; this book's ISBN is 978-0-470-89166-7.

Once you download the code, just decompress it with your favorite compression tool. Alternately, you can go to the main Wrox code download page at www.wrox.com/dynamic/books/download.aspx to see the code available for this book and all other Wrox books.

ERRATA

We make every effort to ensure that there are no errors in the text or in the code. However, no one is perfect, and mistakes do occur. If you find an error in one of our books, like a spelling mistake or faulty piece of code, we would be very grateful for your feedback. By sending in errata you may save another reader hours of frustration, and at the same time you will be helping us provide even higher quality information.

To find the errata page for this book, go to www.wrox.com and locate the title using the Search box or one of the Title lists. Then, on the Book Details page, click the Book Errata link. On this page you can view all errata that has been submitted for this book and posted by Wrox editors. A complete book list including links to each book's errata is also available at www.wrox.com/misc-pages/booklist.shtml.

If you don't spot "your" error on the Book Errata page, go to www.wrox.com/contact/techsupport.shtml and complete the form there to send us the error you have found. We'll check the information and, if appropriate, post a message to the book's Errata page and fix the problem in subsequent editions of the book.

P2P.WROX.COM

For author and peer discussion, join the P2P forums at http://p2p.wrox.com. The forums are a Web-based system for you to post messages relating to Wrox books and related technologies and interact with other readers and technology users. The forums offer a subscription feature to e-mail you topics of interest of your choosing when new posts are made to the forums. Wrox authors, editors, other industry experts, and your fellow readers are present on these forums.

At `http://p2p.wrox.com` you will find several different forums that will help you not only as you read this book, but also as you develop your own applications. To join the forums, just follow these steps:

1. Go to `p2p.wrox.com` and click the Register link.
2. Read the terms of use and click Agree.
3. Complete the required information to join as well as any optional information you wish to provide and click Submit.
4. You will receive an e-mail with information describing how to verify your account and complete the joining process.

You can read messages in the forums without joining P2P, but in order to post your own messages, you must join.

Once you join, you can post new messages and respond to messages other users post. You can read messages at any time on the Web. If you would like to have new messages from a particular forum e-mailed to you, click the "Subscribe to This Forum" icon by the forum name in the Forum listing.

For more information about how to use the Wrox P2P, be sure to read the P2P FAQs for answers to questions about how the forum software works as well as many common questions specific to P2P and Wrox books. To read the FAQs, click the FAQ link on any P2P page.

1

Taking the Metro with Windows Phone

WHAT'S IN THIS CHAPTER

- How Windows Phone has changed Microsoft's approach to the mobile industry
- What the Metro Design Language is and how it came about
- An overview of the Start and Lock Screens and how they help users access information on the phone
- Why the use of Hubs creates a more connected user experience
- What it means to be a Windows Phone developer

Microsoft has been building mobile devices for well over 10 years, starting with a variety of Windows CE-based devices, such as the Handheld PC and the Palm-size PC, first released in 1996. Beginning around 2000, these disparate operating systems began converging into what became Windows Mobile, based on the principle of delivering a PC to your pocket. New features were predominately driven by enterprise needs such as device management and security. This eventually worked to the detriment of the platform as it didn't appeal to the average consumer. Devices were more robust than sexy, and the user interface mirrored that of the desktop, even having a Start menu, rather than providing an experience.

Throughout this chapter, and in other parts of this book, there will be references to both Windows Mobile *and* Windows Phone. *This is intentional, and they are not the same thing.* Windows Mobile *refers to the previous mobile operating system from Microsoft that at the time of writing is Windows Mobile 6.5.3.* Windows Phone *refers to Microsoft's latest offering in the mobile space and starts with Windows Phone 7.*

In February 2010 at the Mobile World Congress, Microsoft unveiled the Windows Phone 7 series, a new-look mobile operating system featuring hard edges, full bleed pages, and sharp typography. Code-named *Metro*, the new look-and-feel is more akin to an immersive experience than to a collection of mobile applications. It appears that Microsoft has taken several leaves out of the playbooks of other mobile platforms, while still innovating and delivering on their own set of values and practices.

Before launching into how you can get started building applications for Windows Phone, it is important to understand the Metro user experience. This will help you build applications that not only run on a Windows Phone but also integrate into the mobile experience.

MINIMUM SPECIFICATIONS

Traditionally, Windows Mobile has had the stigma attached to it that it is slow and unreliable. This had little to do with the underlying operating system, but rather the other stakeholders involved in getting a device into market. Manufacturers, telecommunication companies, application developers, and other third parties all contribute to what comes prepackaged on a device. Each one of these parties builds or adds features that they think will benefit the user. Unfortunately, quite often these features either degrade the overall experience — for example, hogging precious device resources — or don't play well with other aspects of the phone. This has led to an overall negative impression of the Windows Mobile platform as a whole.

Microsoft took the opportunity with Windows Phone to restructure the ecosystem in which devices operate. Although they haven't been so arrogant as to come out with the *Microsoft Phone*, which many were anticipating, they have put some checks and balances in place to ensure that users receive an amazing experience, and, furthermore, that this experience is uniform throughout the phone and for the duration of the phone's life.

It all starts with the hardware. Previously Microsoft has been overly optimistic in specifying the minimum specifications for Windows Mobile. This resulted in many devices that were woefully underpowered, and although this kept the price point low, the devices were frustratingly slow and unresponsive to use. Going forward, Microsoft has defined a much higher set of minimum hardware requirements for Windows Phone, which includes a 1-GHz processor and support for graphics hardware acceleration. When you look at the frameworks that are to be used to develop applications and games for this platform, it is very evident why such high hardware specifications are required.

In addition to having graphics acceleration, Windows Phone devices will appear to be highly responsive because of the use of capacitance screens. This, in turn, lends itself to supporting multi-touch. The net effect is that users will interact with a Windows Phone device using gestures such as tap, pinch, and swipe with their fingers, rather than the more traditional mechanism of using a stylus.

There is currently no intention to support a non-touch-screen Windows Phone device. However, device manufacturers will still be able to differentiate their devices through different device ergonomics and the optional inclusion of a hardware keyboard. A hardware keyboard will complement the Windows Phone experience, making it easier to enter text rapidly. This is particularly useful for e-mail, messaging, and annotating documents on the road.

Chassis Design

So far you know that hardware manufacturers can optionally include a keyboard, but what else can they modify? In the past this was quite an open-ended discussion as manufacturers could build a device to meet a certain price point. For example, for high-end devices, they could include GPS, an accelerometer, and a high-resolution camera; a low-end device may only have a T9 keypad and no camera at all. Even the number and layout of physical hardware buttons could change between devices. Of course, all these options come at a cost, and the first place this took hold was with developers. When building applications, developers were seldom able to rely on a particular hardware feature being present. Instead, you would typically either query an API to determine if hardware existed, or simply attempt to address the hardware. Failure or an exception would indicate the lack of supported hardware.

After the application had been developed, the problem was transposed to the end users. They would see a product advertised as being compatible with Windows Mobile and purchase it, only to discover that it required hardware that they didn't have. No two Windows Mobile devices were alike. When it came to Marketplace for Windows Mobile, Microsoft acknowledged this issue, and as part of application submission, developers had to indicate what device capabilities their application required. The Marketplace client running on the device would then restrict the list of applications to only those that matched the device capabilities.

For Windows Phone, Microsoft has taken the proactive position of enforcing a set of requirements around device capabilities. This has been achieved by taking the traditional minimum hardware specifications and turning them into what Microsoft calls a *chassis design*. This specifies the external buttons, and in some cases their location, and the inclusion of particular hardware features such as Wi-Fi, GPS, accelerometer, compass, camera, light and proximity sensors, and the ability to vibrate. A device that doesn't include all of the features dictated by a chassis design cannot be called a Windows Phone.

On the front-facing side of a Windows Phone there will be three buttons: Back, Start, and Search. There will also be dedicated camera, power, and volume controls. Figure 1-1 shows an example device in both portrait and landscape, illustrating the relative positions of the three front-facing hardware buttons.

FIGURE 1-1

It's important to note that these hardware buttons have a dedicated purpose. Unlike in Windows Mobile, where the buttons could be assigned by the user to different functions, and then applications could elect to override all or some of the buttons, on a Windows Phone the buttons have a sole purpose. This reinforces the overall user experience through a consistent interface.

The one exception to this rule is the Back button. Within your application you are able to control the navigation sequence. This means that you are able to intercept and handle the Back button. However, it is important to remember that the purpose of this button is to navigate back to whence the user came. For example, if they click to delete an item from a list and the application displays a confirmation prompt, the Back button should dismiss this prompt without deleting the item. Similarly, if the user has clicked an item in a list and gone through to a Details view, the Back button should navigate the user back to the list of items.

If you ensure that you correctly handle the Back button, there should be little need for your application to include navigation controls within the context of your application. Forward navigation is typically done through interaction with content; Back navigation is instigated from the Back button, and exiting your application is little more than pressing the Back button when on the first page of the application.

You can think of every application you open as being placed on a stack. When you hit the Back button and have not purposely handled it within your application, the application is popped off the stack and the previous one is displayed. This analogy works well as Windows Phone will automatically close your application when it goes out of focus by being popped off the application stack.

Similar to other mobile platforms, Windows Phone has a dedicated "I'm lost, take me to a known location" button. As it's a Microsoft platform, this button is logically called the *Start button* and takes the user back to the Start experience. As you will learn, the Start is an area on the device that contains a personalized set of tiles that reflect what's important to the user. It also acts as a launching point for accessing areas of the device and applications that the user may have installed.

The last of the buttons on the front of the device is the Search button, also known as the *Bing button*. Pressing the Search button launches a context-sensitive search. For example, if you are looking through your contacts, pressing the Search button will filter your contacts based on the search criteria. If there is no appropriate search context, pressing the Search button will launch Bing Search, allowing you to search over Web content, images, and maps. At this stage it is not possible to integrate the Search button into your application, so tapping it within any third-party application will launch Bing Search.

The inclusion of Wi-Fi seems like an obvious requirement, but with the advent of 3G+ networks that are continuing to get cheaper, it would have been an easy cost saving for manufacturers to omit the Wi-Fi stack. In the early days of Windows Mobile, before Microsoft tightened security, it used to be possible to synchronize your contracts, calendar, and e-mail with Outlook by connecting to ActiveSync through a Wi-Fi network. This capability is returning with the ability to synchronize across your home Wi-Fi network to your Zune desktop experience.

Location is definitely one of the hip new fads being talked about across the software development community. Software that is aware of the user's location means that it can locate information and people nearby. Of course, there are all manner of privacy issues to navigate, but it is important that

Windows Phone can provide location information. This topic will be covered in detail in the context of the location services offered by the platform in Chapter 12, but it's enough to say that having a GPS is essential in order to accurately geolocate the user.

One thing that you will notice about Windows Phone is that it is the first mobile offering from Microsoft that has been designed with a consumer rather than enterprise or business user focus. Previously, Windows Mobile was more tailored for the mobile worker, with support for enterprise features such as device deployment and management at the expense of a consistent set of hardware capabilities. As a consumer device Windows Phone will offer a minimum of a 5-megapixel camera with integrated flash. Windows Phones will also include light and proximity sensors that will be used to enhance the user experience.

In building your application, you need to be very aware of the experience you are constructing for the user. Where you would have once provided simple on-screen feedback, you can now use more complex animation and sounds and even have the device vibrate. You should use all visual and hardware effects sparingly as it is easy to overwhelm the user and drain the phone's battery in the process.

Screen Resolution

Dealing with differing hardware was just one of the challenges faced by developers working with the Windows Mobile platform. In seeking to be a platform that could be tailored to a wide variety of user scenarios, Windows Mobile 6.5 supported six touch-screen and five non-touch-screen resolutions. As you have already learned, there won't be support for non-touch screen in Windows Phone, but what's even more exciting for developers is that with Windows Phone there will only be two different screen resolutions that you need to accommodate.

The initial platform will be released with WVGA (480 × 800) resolution. A second resolution, HVGA (320 × 480), will follow sometime in the future.

Whenever you see a Windows Phone being demonstrated, it is likely that it will be in Portrait mode. In fact, if you look at some of the core areas of Windows Phone, such as Start, they only support being displayed in Portrait mode. However, this doesn't prevent you from taking advantage of running in Landscape if that's more suitable for your application. In fact, the best applications are those that allow either orientation, reorganizing the layout in order to make best use of the available screen real estate. Windows Phone also provides the necessary extension points for your application to handle the change in device orientation during operation, such as sliding out a physical keyboard on a device that has one.

METRO DESIGN LANGUAGE

Before getting much further into the ins and outs of Windows Phone, it's worth taking a step backward and looking at the approach that Microsoft took in designing the user experience. Rather than simply refining what they already had, it was time to make a clean break and come up with a revolutionary design. The outcome of this process was not only a unique mobile interface but also a design language that you can, and should, adopt as part of building your application.

As part of going forward, it's important to look back, even if only for a moment to wave as it disappears into the sunset. In the case of Windows Phone, you can see from Figure 1-2 the legacy of Windows Mobile from Pocket PC 2003 SE, through Windows Mobile 5.0, 6.1.4, and finally 6.5.3. As you can see, each has been an incremental approach with little to excite the user, save the more touch-friendly home screen and controls in Windows Mobile 6.5.3.

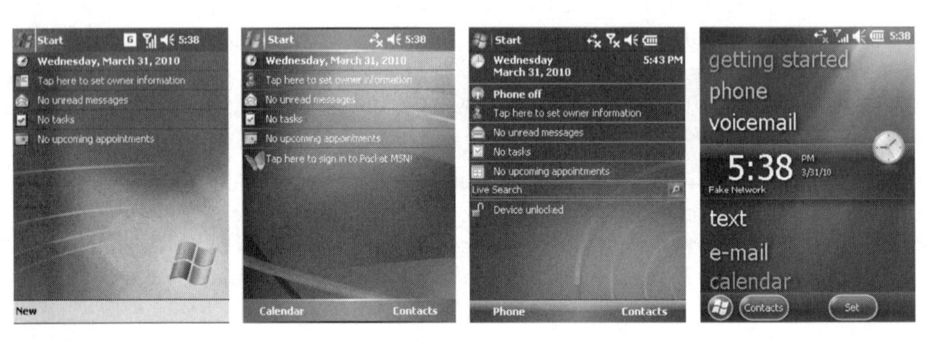

FIGURE 1-2

What's important to recognize is that although the Windows Mobile user interface appears somewhat dated now, there are some important concepts embraced within the layouts — for example how relevant information could easily be accessed right from the home screen, and that applications were only a click or two away via the Start menu.

The user interface on nearly all the current-generation smartphones is geared around applications. The underlying operating system typically handles standard information types such as e-mail, calendar, and contacts, but it is left to individual applications to handle other types of data (e.g., updates from your favorite social networking site need to have an application running on the device). Unfortunately, these applications are often built in isolation and don't interact with each other or even integrate into the phone experience. In building Windows Phone, it was important for Microsoft to build an immersive user experience, rather than a set of disjointed applications. This needed to encompass all the features that make up Windows Phone, as well as the ability for third-party applications to be built to integrate into the same experience.

It was decided that Windows Phone should have a fresh start and not just the new Start that you'll see when you first unlock a Windows Phone. This should not just be about an ad hoc change to the way applications look, but a new language for communicating with users. What Microsoft came up with is what they refer to as the *Metro* design language, based on generations of refinement that have gone into the use of signposting, signaling, iconography, fonts, and layout in the transportation industry. The name *Metro* was chosen to reflect its heritage in the language used to efficiently guide people to their destinations.

If you examine signs used around and on buses, trains, and other forms of transportation, and at stations, airports, and other transportation hubs, you can clearly see a set of principles that govern them. Most of them have graphics that are instructional, yet minimal in design. Furthermore, the signs are typically universal and feature simple icons. The use of color is significant and plays an important role alongside weight and style when applied to text. Unlike in a lot of software systems, where it is almost an afterthought, the use of different typography in signs can dramatically affect the readability and thus the effectiveness of those signs.

Using transportation signage as a base, the Windows Phone team collaborated with other teams within Microsoft in order to flesh out the design language. For example, the use of motion and animation in the Xbox, the use of navigation via content on the Zune and Media Center, and the intimacy of the ZuneHD interface were all sources of inspiration for Metro.

The resulting Metro design language is about being modern with a clean and fast user experience. It's about delivering an experience that is in motion and all about the content. Being authentically digital in design and using clear typography are also important throughout the Windows Phone experience.

Principles

In order to build a consistent experience across the Windows Phone platform, it was important not only to have a concept of a design language, but also to have a set of principles that would govern all decisions made in constructing the user experience. The following are the Metro design principles:

- Clean, light, open, fast
- World class motion
- Integrated hardware and software
- Content, not chrome
- Soulful and alive

Before you learn more about the Metro experience itself, take a moment to review these principles and more importantly think about how they can affect the way you build the user experience for your application. The first principle talks about the experience being "Clean, light, open, and fast." Although this doesn't limit itself to certain aspects of the interface, you can apply this principle by making sure your application isn't cluttered with icons and images and that any action the user takes is fast and responsive.

Just because screens on mobile phones have gotten larger over the last couple of years, this doesn't mean that you can pack more information onto a single screen. Doing so makes the content hard to read and difficult to navigate. Build a clean and light interface that displays only the important and relevant information to the user. This will necessitate larger and clearer fonts, which is where rich typography becomes important. The Metro design language celebrates typography and provides you with some rich fonts out-of-the-box for making your applications clear and easy to read.

The introduction of touch screens brought with it a need for haptic, or at least responsive, feedback. If you imagine touching the water in a glass or on a pond, you can immediately see the water ripple out from where you first touched the water. Instantaneous feedback, such as the changing of background color or a slight vibration of the device, gives the impression that the application is alive and awaiting the user's instructions. You will notice that motion is built into the core platform. The Start is made up of tiles that dynamically update and respond to being touched. Transitions provide visual indicators to the user of not only where they are going, but where they have come from. The use of animation and other visual effects using motion is essential in building for Windows Phone. Within Windows Phone 7 the capacitive touch screen and other sensors enable gestures to control hardware accelerated animations and transitions which enhance the application. Hardware and software features blend together to provide a unified user experience.

Generations of applications across all manner of computer systems have adopted a windowing approach in which the border allows the user to control the window. On Windows Phone, there is no chrome! There is no border on the window, on controls, or even on content. Chrome takes up valuable screen real estate and for what, to illustrate that one piece of content has ended and the next has begun. If this is the case, use the content itself to indicate where those transitions are. For example, when you have adjacent images, instead of separating them by a few pixels, simply fade out the trailing edge of one image so that it's clear where the second image begins, as shown in Figure 1-3.

FIGURE 1-3

The first set of images in Figure 1-3 illustrates how images can literally be separated using empty space. Applying the content not chrome principle, this can be adapted to the second set of images, where the trailing edge, in this case the right edge, of the image is blurred or faded in order to make way for the next image.

At the end of the day, don't forget that you are building a software application for a mobile phone, which should be intimately attuned to its owner. Build applications that are soulful and alive, which offer a personalized and rewarding experience showing the information that matters most to the user with minimal excess. Bring the application to life by integrating the user experience with the rest of the device and making use of the platform's unique hardware capabilities to make it feel responsive to the user's gestures.

Applications should embody the three Red Threads of Windows Phone 7: Personal — your day, your way; Relevant — your people, your location; and Connected — your stuff, your peace of mind. We will investigate this theory further in Chapter 3.

User Experience

The Windows Phone user experience is the canonical representation of the Metro design language. However, having a design language isn't enough in itself to build a user experience. A user experience is governed by what you want to display and how users will interact with what is being displayed. For the Metro user experience, there were two guides that were used to direct the presentation of information: The user experience should focus on the individual and his or her tasks, and to help organize information and applications.

You have to ask the question, "Who was Windows Phone built for?" and, surprisingly, Microsoft has a remarkably well-defined answer. Windows Phone was designed for Life Maximizers and specifically Anna and Miles. *Life Maximizers* are defined as individuals who value the use of technology in their lives: They have both a busy personal life and professional life, yet they are more settled than seeking and interact with their respective families. To empathize with such individuals and to understand their needs and wants, the personas of Anna and Miles were created. They are a pair of married, 38-year-old Life Maximizers who demand the most from their devices. As you investigate further, you learn that Anna is a part-time PR professional and a busy mother to whom balancing friends, work, and family is essential. Meanwhile, Miles is growing his architectural business and thus needs information available wherever he finds himself. Although these personas were created for the purpose of creating the Windows Phone experience, it is important for you to consider them when building your application. What would Anna or Miles want or expect?

START AND LOCK SCREENS

Now that you have heard about the Metro design language, it's time for you to see it in action as you navigate through the Windows Phone experience. It seems natural to begin with Start. However, you'll soon realize that the first thing you see on a Windows Phone device is actually the Lock screen. This is the informational screen that appears when the device is locked to ensure both the privacy of your data as well as to prevent you from accidentally dialing someone while your phone is in your pocket. Figure 1-4 illustrates the Lock screen showing the current date and time in a beautiful, clear white font overlaid on one of the default images that are available on the device. Both the time-out, whether there is a password for unlocking the device, and the background image can all be customized via Settings ➪ lock & wallpaper.

FIGURE 1-4

Once you unlock the Windows Phone, you enter the Start screen. As you can see from Figure 1-5, the Start screen is made up of a series of square tiles (there are some built in tiles such as the tile for the Pictures hub that take up two squares). These aren't just static blocks representing links to applications, in other words, icons similar to application icons on other mobile platforms. The tiles, also known as *live tiles*, are just that — they're alive and will not only dynamically update, but they'll also respond to your touch. The information displayed on Start should be what is relevant and important to you.

When you simply tap a tile, it appears to enlarge slightly before transitioning to the appropriate window. The right image shows what happens when you tap and hold one of the tiles. In this case, the tile enlarges even more, going into a mode in which you can drag the tile around the screen. As you move the tile, the other tiles will move themselves to allow you to drop the tile where you want it positioned. When you are done, simply tap in an area where there are no other tiles. Alternatively, if you no longer want a

FIGURE 1-5

tile to be pinned to the Start area, you can tap the pin icon. This will remove that tile from the Start area.

It's all very well to have these dynamic, adjustable tiles, but what happens to all those applications that aren't immediately visible on Start? Well, there are a couple of places where you can go to look for them. In the left image of Figure 1-6, you can see a Start that contains a large number of tiles and that has been scrolled up to reveal additional tiles. While not immediately obvious, Start can contain as many or as few tiles as you want. It's worth adjusting these tiles so that those that are important to you appear at the top of Start. Hitting the right arrow icon takes you to an alphabetic list of the available applications, shown in the right image of Figure 1-6.

FIGURE 1-6

Whereas the left side of Start is designed for glance-and-go information, the applications list has been designed to be easy to navigate in order to get you where you need to go as quickly as possible.

HUBS

As you step out of Start, you will most likely end up in one of six hubs: People, Pictures, Music + Video, Games, Office, and Marketplace. Each of these hubs is structured around a Panoramic view in which the screen acts as a small portal into a larger display. The panoramas are designed to be cyclic. The content loops around so that if you keep scrolling either to the right or the left you will see each of the content groups available in the panorama. In Figure 1-7, you can see the People hub, which shows your contacts in groups that represent Recent, All, and What's New.

FIGURE 1-7

The intent is that through the People hub you have access to information about your contacts that you care the most about. Updates to your contacts are drawn from social networking sites so that you won't miss what your friends and family are up to.

In Figure 1-8, you can see the Pictures hub, providing you with access to not only the various galleries you have for photos, but again a What's New area that draws on photo information from both your phone and your online photo stores. You'll notice with the panoramas that they are a great demonstration of content, not chrome. There is little in the way of navigation controls, yet through clever positioning of text and the content, it is intuitive for the user to scroll to the left and right to explore the hubs.

FIGURE 1-8

The Music + Video hub is very similar to the Pictures hub in structure. Building on the success of the ZuneHD, Microsoft has made every Windows Phone a Zune, and as such, the Music + Video hub is where you go to organize and play music and video on the device.

Not only did Microsoft integrate the Zune experience for media playback, but they've also made the Windows Phone a gaming device. Through the Games hub, you can access games on the device. Your Xbox live account is also accessible via this hub, as shown in the focus area of Figure 1-9. This figure illustrates how the device acts as a view into the panoramic experience. The header "ames" indicates that there is content to the left, while the cut off "C" on the right of the screen indicates that there is more to the right as well.

FIGURE 1-9

With all the features designed for consumers you may be wondering whether Windows Phone will make a good corporate device. The answer is that it will, indeed. The Office hub, shown in Figure 1-10, is where you can go to access your documents both from the device and via SharePoint.

The last hub, but definitely not the least, is Marketplace. As you can imagine, this is where users will go to purchase and download applications that you have written. The Marketplace will be split into genres such as music, applications, and podcasts. It will also support searching for applications via categories, as well as notifications for when there are application updates.

FIGURE 1-10

The Metro user experience introduces several new ways to present and navigate information. The hub concept is designed to group related information together and make it easy to glance at the most relevant information, such as your friends' recent social networking updates. Through the use of panoramas and an area much larger than the actual screen, hubs allow the personas Anna and Miles to quickly access frequently used information and updates, while still providing access to a significant volume of details and features that may be used less frequently. Although you can build your own Panoramic views, you need to be careful that you don't overuse them in cases where a single view would be more appropriate.

DEVELOPER LANDSCAPE

Building simple applications for Windows Mobile has always been a trivial exercise. You could simply open up Visual Studio, create a new project, and hit Debug or Run in order to see your application run on either an emulator or a connected device. Unfortunately, over time, support within Visual Studio and the associated .NET Compact Framework didn't keep pace with expectations in the market for rich applications that both looked good and were responsive to the user. In order to build sophisticated applications for Windows Mobile, you had to hunt down information on frameworks such as DirectDraw or OpenGL in the hope that they would allow you to build an application with a rich user experience. Even then there were non-trivial issues to resolve as different manufacturers decided not to provide driver-level support for them, or introduced device-specific functionality, or worse yet, bugs.

One of the goals with Windows Phone was to address these issues and provide a platform upon which developers could build both applications and games. The time to get up-and-running should be minimal, and yet the tools and frameworks need to be sophisticated enough to handle the most complex of user interfaces. Figure 1-11 provides a thousand-foot view of the development ecosystem for Windows Phone developers.

	Tools and Support	Runtimes
Screen	Visual Studio & Expression Blend Windows Phone Emulator XNA Game Studio Samples & Documentation Guides & Community Packaging and Verification Tools	Silverligh & XNA Sensors, Media, Data, & Location Phone, Gamer Services, & Notifications .NET Framework Sandbox Windows Phone / XBox / Windows 7
	Portal Services	**Cloud Services**
Cloud	Registration & Marketplace Validation & MO/CC Billing Certification & Business Intelligence Publishing & Updated Management	Notifications & App Deployment Location Identity, Feeds, Social, and Maps Xbox Live Windows Azuret

FIGURE 1-11

Let's start in the top-left corner with Tools and Support. For anyone who is already using Visual Studio 2010, you'll be aware that there is no support for doing Windows Mobile development. This was primarily a decision as to where to invest resources by Microsoft, and I think everyone will agree that it was a decision well made. The tooling for Windows Phone throughout both Visual Studio 2010 and Expression Blend dramatically reduces the time to build applications. In addition, the fact that these two products share the same solution and project structure means that both developers and designers can work in harmony on the same project at the same time. These tools will be covered in more detail in the next chapter, which shows how easy it is to get started with building your first application and debugging it on the Windows Phone Emulator.

Looking to the right, you'll see the Runtimes square. One of the most significant aspects of Windows Phone is that it fulfills the Microsoft three-screens goal. As a developer you can build an application or game and have it target one or more of the screens running Windows Phone, Xbox, or Windows 7. Unlike the .NET Compact Framework, which provided a reduced feature set compared to the desktop, the Silverlight and XNA runtimes available within Windows Phone are virtually identical to those you are already familiar with. Although you need to make a decision as to whether you're going to use Silverlight or XNA, you can still access a wide range of functionality from across the device no matter what technology you build your application or game in.

In building applications for Windows Phone, you are likely to want to access some of the cloud-based services that Microsoft has to offer. These include a notification service, for when application data changes and you want to notify the user; a location service, which integrates the device capabilities such as GPS with online services for resolving Wi-Fi locations; and access to the user's Xbox Live information from within your application. Going forward, expect further support for building Windows Phone applications that integrate with other cloud offerings such as Bing Maps and Windows Azure.

The last quadrant refers to the Developer Portal Services, which covers all the online services through which you as a developer interact in order to have your application certified and published via Marketplace. As Microsoft wants to ensure the highest quality of applications in Marketplace, there will be a more rigorous process for developing for Windows Phone. The online portal will be the point of reference for all your applications and will, hopefully, be the place where you go to receive monies generated by your application.

SUMMARY

Throughout this chapter, you have learned about the Windows Phone user experience, its origin in the transportation industry, and the birth of the Metro design language. In the coming chapters, you will see this language and its associated principles in action as you discover more about what it takes to build a Windows Phone application.

Free Transport with Visual Studio 2010 and Expression Blend 4

WHAT'S IN THIS CHAPTER

- Where to get all the tools you need to start building amazing Windows Phone applications and games
- How to get started with Visual Studio Express for Windows Phone
- Working with Expression Blend 4 to Design Windows Phone applications
- Running applications in the Windows Phone Emulator

Developing for previous versions of Windows Mobile could be somewhat traumatic because there were many different technologies, frameworks, and tools that could be used, none of which made building rich user experiences easy. Windows Phone, however, supports just two development strategies: Silverlight, primarily for application development; and XNA, for games development. Silverlight is an event-driven system in which the layout is declaratively defined using XAML (eXtensible Application Markup Language), which can include styles, templates, states, and animation. XNA, on the other hand, is a game-loop-driven framework, specifically geared toward rendering two-dimensional (2D) and three-dimensional (3D) graphics.

In contrast to other mobile platforms, where the development tools are free, Windows Mobile has traditionally required at least Visual Studio Standard Edition. But Microsoft has announced that the development tools for Windows Phone will be free, in the form of Visual Studio 2010 Express for Windows Phone. This includes a device emulator that includes the Windows Phone operating system compiled for the x86 platform, allowing it to run on your computer taking advantage of any hardware acceleration available.

In this chapter, you will learn about the tools for building applications for Windows Phone, where to get them, and how to get started with your first application.

VISUAL STUDIO 2010 EXPRESS FOR WINDOWS PHONE

To get started building your first Windows Phone application, you need to download the appropriate tools. Microsoft has gone to great lengths to make this process as simple as possible. The Microsoft portal for Windows Phone development is available at `http://developer.windowsphone.com`. From this portal you will find all the tools you need to build great Windows Phone applications as well as links to additional documentation, blogs, and samples.

When you follow the link to download the developer tools, you will be prompted to download and install the tools via a single web installer. This will download and install the following components:

- Visual Studio 2010 Express for Windows Phone
- Windows Phone Emulator
- Silverlight for Windows Phone
- XNA 4.0 Game Studio

The only customization you can do during the installation process is to control the location to which the components get installed. Note that if you have a commercial edition of Visual Studio 2010, such as Visual Studio 2010 Professional, installed, the tools and templates for building Windows Phone applications will be made available within that edition of Visual Studio as well as the free Express edition installed as part of the Windows Phone Developer Tools.

Once installation is complete, it's time to get started writing your first Windows Phone application. From the Start menu, run "Microsoft Visual Studio 2010 Express for Windows Phone." Upon launch you will see the Start page, which has been customized to help support you as a Windows Phone developer. You can use the links on the Get Started tab to jump to information such as the Windows Phone class library, the UI design and implementation guide, and additional code samples. The Latest News tab is preconfigured with the RSS feed for the Windows Team, constrained to topics relating to Windows Phone development. This feed is worth keeping track of because it contains information on new tools and Marketplace updates.

To create your first application, click "New Project" from the Start page. Alternatively, you can always create a new project via the File menu. Both options will launch the "New Project" dialog, which contains templates for building Windows Phone applications. Figure 2-1 shows the New Project dialog with the "Silverlight for Windows Phone" tree item selected on the left-hand side of the dialog. In the middle area you see templates for creating a Windows Phone Application, a List Application, or a Class Library. In the right pane you see a description of each of these templates along with a preview of what the application would look like.

The other tree item worth noting is for "XNA Game Studio 4.0." Selecting this item reveals project templates for Windows Phone Game and Game Library, as well as templates for building Windows and Xbox games. One thing you will notice about these templates is that none of them has a preview for how the application will look. This is because XNA, unlike Silverlight, doesn't give you a designer upon which you can build your visual layout. As XNA is designed primarily for games development, it's assumed that most developers will want to compose their own experience from the ground up. Chapter 19 covers building a simple XNA game and discusses the differences between the technologies, which may be useful if you are unsure as to whether to select Silverlight or XNA to work with.

FIGURE 2-1

For the purpose of this example, select the "Silverlight for Windows Phone" tree item, followed by the "Windows Phone Application" template. Give your project and solution a name and click OK to proceed. This will create a new project consisting of a single page, MainPage, as shown in Figure 2-2. By default, Visual Studio 2010 Express for Windows Phone uses a vertical partition to allow you to visually design the portrait layout of your main page while also being able to read the corresponding XAML.

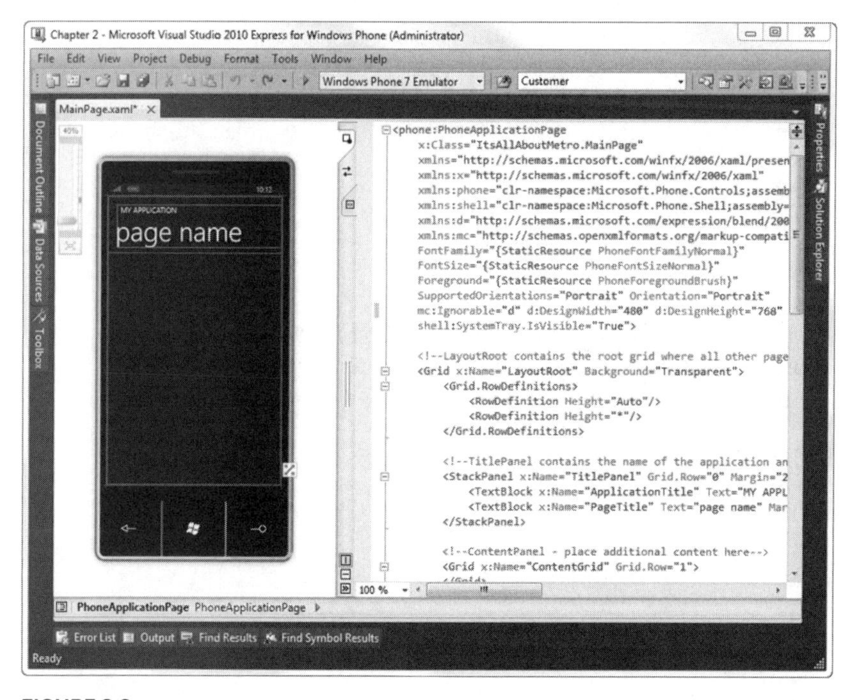

FIGURE 2-2

AUTOFORMATTING XAML

Unfortunately, the default configuration for Visual Studio 2010 Express for Windows Phone doesn't set up the XAML text editor to place each XML (eXtensible Markup Language) attribute on a new line. In the default vertical split view, this can make the XAML hard to read. Because this is an Express SKU of the Visual Studio 2010 product, the ability to configure the editor has been restricted, preventing you from changing the autoformatting settings via the Tools ➪ Options dialog. Normally you would set the Attribute Spacing property to Position each attribute on a separate thread on the Text Editor ➪ XAML ➪ Formatting ➪ Spacing tree node.

A work-around for this is to save the following snippet into a vssettings file and then import it into Visual Studio using the Tools ➪ Settings ➪ Import and Export Settings menu item:

Available for download on Wrox.com

```
<UserSettings>
  <ApplicationIdentity version="10.0"/>
  <ToolsOptions>
    <ToolsOptionsCategory name="TextEditor"
                          RegisteredName="TextEditor">
      <ToolsOptionsSubCategory name="XAML Specific"
                               RegisteredName="XAML Specific"
                               PackageName="Microsoft.VisualStudio.Xaml">
        <PropertyValue name="AutoReformatOnStartTag">true</PropertyValue>
        <PropertyValue name="AutoReformatOnEndTag">true</PropertyValue>
        <PropertyValue name="WrapTags">false</PropertyValue>
        <PropertyValue name="AutoReformatOnPaste">true</PropertyValue>
        <PropertyValue name="AttributeFormat">NewLine</PropertyValue>
        <PropertyValue name="KeepFirstAttributeOnSameLine">true
          </PropertyValue>
      </ToolsOptionsSubCategory>
    </ToolsOptionsCategory>
  </ToolsOptions>
</UserSettings>
```

Autoformatting.vssettings

Importing this set of settings will configure Visual Studio to place attributes on a new line. In order to reformat the XAML for the MainPage, place the cursor somewhere in the Editor window and press the key combination Ctrl+E, Ctrl+D.

Although you may find the split view convenient for working with both the design and XAML panes, quite often you will want to work with one view by itself. This is especially true if you are hand-editing the XAML content. The splitter bar between the two views has some useful buttons for toggling the panes on and off. In Figure 2-3, the "Collapse Pane" button (A) is used to hide the second pane. In the left image of Figure 2-3, the Design pane is on the left of the splitter bar, indicated by the tabs at the top of the splitter bar, so clicking the "Collapse Pane" button will collapse the XAML pane.

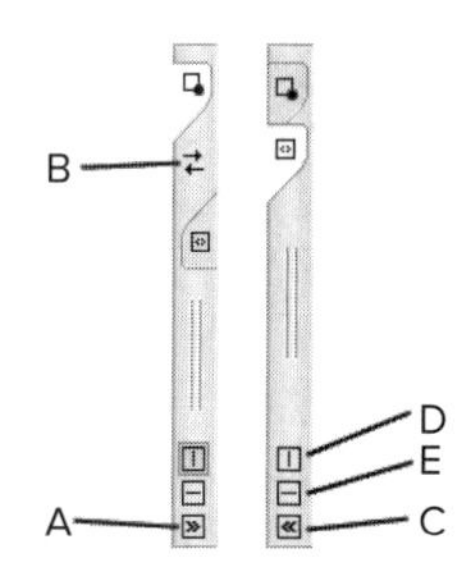

FIGURE 2-3

Instead, if you were to exchange the two panes by clicking the two opposing arrows (B) prior to using the "Collapse Pane" button, the XAML pane would remain visible. This is the state indicated by the right splitter bar, which indicates that only the XAML pane is visible as the XAML tab is selected and the opposing arrows are no longer visible. To expand the second pane, click the "Expand Pane" button (C). You can also change the split orientation from vertical (D) to horizontal (E) via the splitter bar.

Looking across to the Solution Explorer window, as shown in Figure 2-4, you can see the other files that were created as part of the Windows Phone Application template. In addition to MainPage.xaml, there is also App.xaml, which is used to define global styles as well as defining the entry point for your application. There are two PNG files that are used to define the application icon and the background of the Start tile for the application, StartScreenImage.jpg, which is a placeholder for the image to be displayed when your application is loading, and two XML files and AssemblyInfo .cs, which are used to define attributes for your application. These will all be covered in more detail when you look at the application properties.

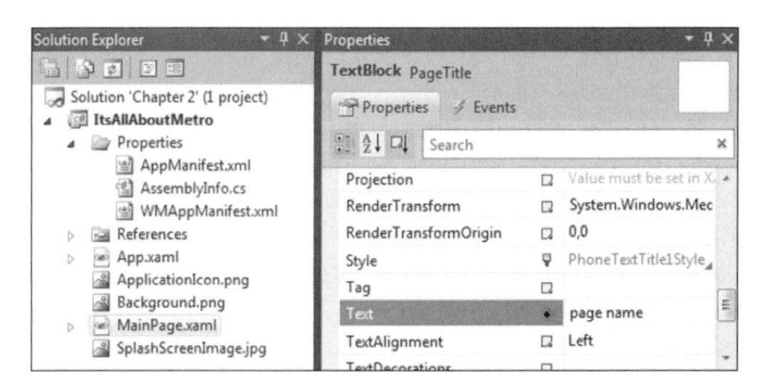

FIGURE 2-4

To the right of the Solution Explorer in Figure 2-4 is the Properties window. If you are familiar with Visual Studio, you will most likely have already seen the Properties window. In Visual Studio 2010, the Properties window has been upgraded to provide rich support for manipulating XAML attributes and data binding. In this case, the XAML element named `PageTitle` has been selected, and you can see a selection of the properties that you can modify on a `TextBlock`.

To get a feel for how easily you can change property values, select the `Text` property and change the value to `tasks`. Then select the other `TextBlock`, which currently says "MY APPLICATION," and change its `Text` property to `METRO TASKS`. You also want this application to support both portrait and landscape modes, so select the SupportedOrientations property of the PhoneApplicationPage and set it to `PortraitOrLandscape`.

FIGURE 2-5

You're going to start building up an interface through which you can query a list of tasks using a simple search filter. To do this, drag a textbox from the Toolbox window into the main area, and set its name to `SearchTextBox`. Then, using the Property window, reset the `HorizontalAlignment`, `Width`, and `Text` properties by clicking on the icon adjacent to the value textbox and selecting "Reset Value" from the dropdown menu, and set the `VerticalAlignment` to `Top`. Lastly, set the `Margin` property to `0,0,160,0` and the `Height` property to `70`. Follow the same process to add a button called SearchButton, except instead of resetting the `Text` property, change the `Content` property to `Search`, set the `HorizontalAlignment` to `Right`, the `Margin` to `0`, and leave the default `Width`. This should give you a layout similar to that shown in Figure 2-5.

Double-click the SearchButton to automatically wire up an event handler for the `Click` event. Alternatively, you can select the SearchButton and locate the `Click` event in the Event list on the Properties window. In the code window you can raise a dialog displaying the contents of the `SearchTextBox`:

```
private void SearchButton_Click(object sender, RoutedEventArgs e){
    MessageBox.Show(this.SearchTextBox.Text, "Search", MessageBoxButton.OK);
}
```

The two XML files and AssemblyInfo.cs that sit under the Properties node in the Solution Explorer hold all the application properties. If you double-click the Properties node, the Application Properties window will be displayed, as shown in Figure 2-6. An interesting point to note here is that in addition to specifying the usual properties such as "Assembly name" and "Default namespace," you can also define an application Title and Icon under "Deployment options," as well as a Title and Background image under "Tile options." In Chapter 7 you will learn how to work with Start Tiles, but essentially the "Tile options" configure the text and background image that will be displayed by default when the application is pinned to the Start area of the Windows Phone, whereas the "Deployment options" configure the text and icon that will be displayed in the applications list.

Figure 2-6 also shows the "Assembly Information" dialog that is displayed when you click on the "Assembly Information" button on the Application tab of the Properties window. Make sure that you change the copyright information to reflect the year that the application was developed and the company or entity that owns the copyright for the application, which will in most cases not be Microsoft IT.

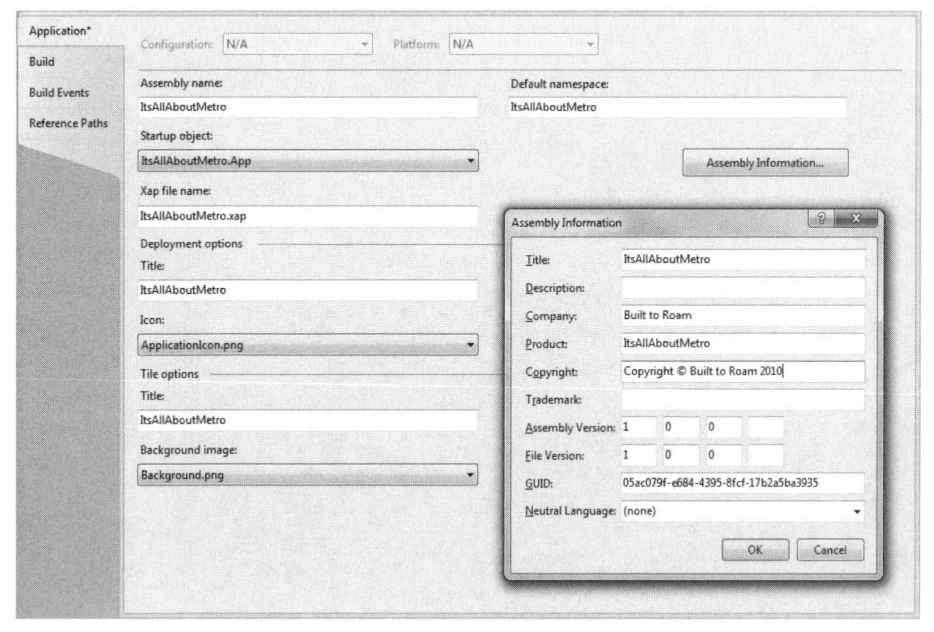

FIGURE 2-6

EXPRESSION BLEND 4

Although Visual Studio provides a basic design experience for laying out controls, it doesn't provide visual support for more complex design activities such as styles and animation. For these activities you'll want to use Expression Blend 4, which includes an add-in for building Windows Phone applications.

Expression Blend 4 can be downloaded from `http://expression.microsoft.com`. Follow the links to Downloads ➪ Blend + SketchFlow Downloads, and then you will need to download and install the following components:

- Expression Blend 4
- Expression Blend Add-in for Windows Phone
- Expression Blend Software Development Kit for Windows Phone

After completing the installation, run Expression Blend 4 (Blend) and open the solution file that you've been working on within Visual Studio. One of the goals behind the original creation of XAML for declarative markup of user interfaces was the ability for it to be read, manipulated, and persisted by a wide variety of development tools. A tool such as Visual Studio has been optimized for developers who write code, rather than designers who think and work in visual terms. On the other hand, Blend has been built from the ground up to focus on enabling designers to visually design applications. Both tools leverage the same solution, project, and file format, meaning that designers and developers can work in harmony, focusing on what they each do best.

As you can see from Figure 2-7, the layout of Expression Blend is significantly different from that of Visual Studio. What can be confusing to start with is that there is no Solution Explorer window. Instead there is a Projects window that by default is on the left of the screen. As with Visual Studio, each tool window can be detached or repositioned to suit your style.

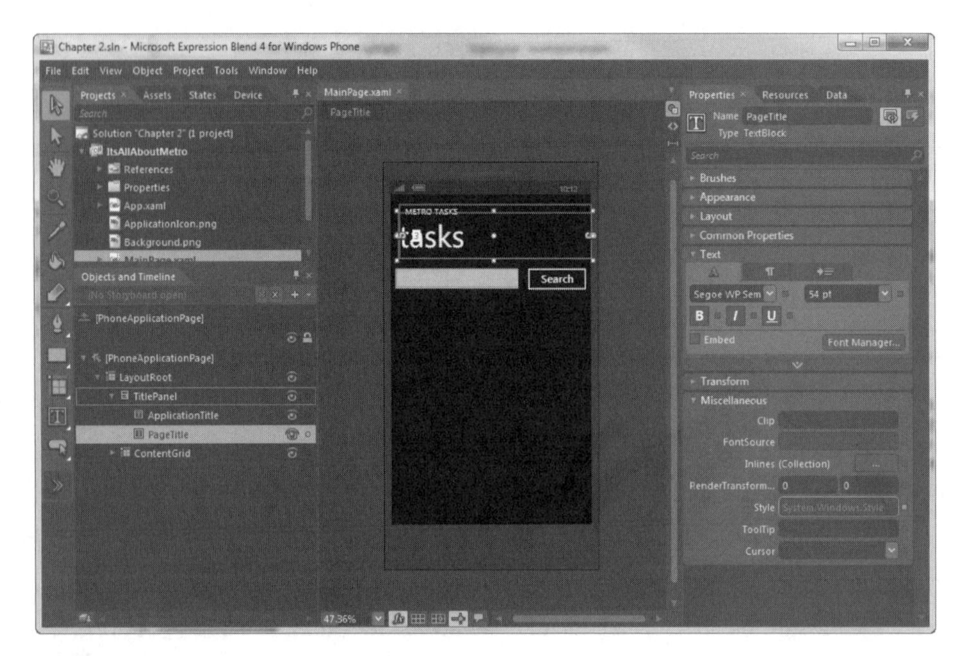

FIGURE 2-7

One unique feature of Blend is that you can define workspaces. A *workspace* is essentially a way of defining a window layout that is optimized for certain tasks. For example, if you occasionally need to hand-code some XAML, you may want to define a workspace called XAMLCoding where all the windows other than the main editor space are collapsed. To do this, simply click the pin icon on each of the windows, which will cause them to collapse when they lose focus. Once you have collapsed all the windows, select the Window ➪ Save as New Workspace menu item, and provide it a name. With the workspace saved, you can easily toggle between one of the predefined workspaces, such as Design or Animation, or your custom workspace.

MINIMIZING TOOL WINDOWS

There is actually a predefined shortcut for minimizing all the tool windows in Expression Blend. Pressing F4 will toggle all the tool windows between being minimized and the state defined by the current workspace. Another shortcut is F6, which will rotate between the workspaces you have defined.

To briefly demonstrate the animation capabilities of Expression Blend, modify the behavior of the Search button so that it will wiggle slightly when clicked. This can be done visually, without any code, by adding an animation storyboard that will be triggered when the button is pressed.

In the Objects and Timelines window, click the + button or the down-arrow, followed by New. This will create a new storyboard and prompt you to enter a name for it. Enter **SearchPressedAnimation** and click OK. You will see that a time line appears in the Objects and Timelines window and the main window shows a recording icon to indicate that a storyboard is currently being edited. Over the course of 0.5 second, you can make the SearchButton rotate –20 degrees. To do this, click at approximately the 0.5-second mark to move the yellow vertical line, which indicates the current point in the animation. Making sure that the `SearchButton` control is selected, set the rotation angle to –20 within the RenderTransform section of the Properties window, as shown in Figure 2-8.

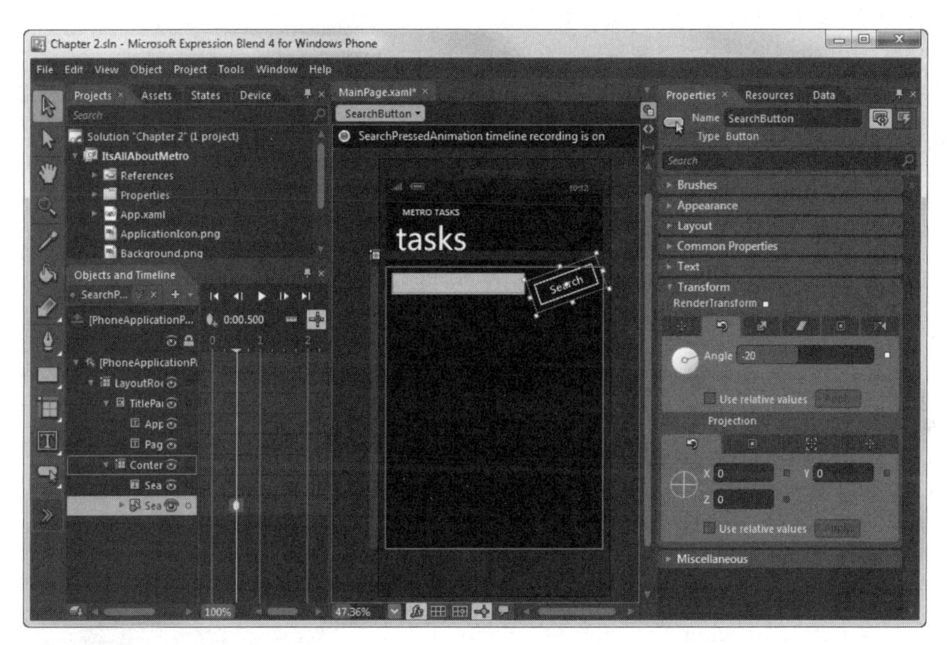

FIGURE 2-8

Next, click at the 1-second mark, and with the `SearchButton` still selected, change the angle to 20. If you press the Play button, shown above the time line in Figure 2-8, you can preview the animation of the button. With the storyboard complete, stop the storyboard editing by clicking the red button to the left of the storyboard name. It's important to remember at this point that although you have declared the animation and given it a name, you haven't linked it to any event, such as the user clicking the button or pressing a key. To do this, you'll want to flip back to Visual Studio by right-clicking the Solution node in the Projects window and selecting "Edit in Visual Studio (make sure you save changes before returning to Visual Studio)." When you return to Visual Studio, you may be prompted to reload MainPage.xaml because Visual Studio has detected that you have made changes and that it needs to reload the file. Allow Visual Studio to reload the file, and then navigate to the `Click` event handler that you created earlier for the `SearchButton`. Change the code to trigger the animation storyboard instead of displaying the message box:

```
private void SearchButton_Click(object sender, RoutedEventArgs e)
{
    this.SearchPressedAnimation.Begin();
}
```

You will notice that the animation created visually in Blend, and saved as XAML, is exposed to the developer as an object. Now that you have some basic functionality in your Windows Phone application, you will want to be able to run it in order to see how it looks and behaves. Both Visual Studio and Blend are capable of running your application in either the emulator that ships as part of the Windows Phone Developer Tools or a real device. In Visual Studio you can select whether to debug with the emulator or a device using the "Select Device" dropdown that has been added to the Standard toolbar, as shown in Figure 2-9.

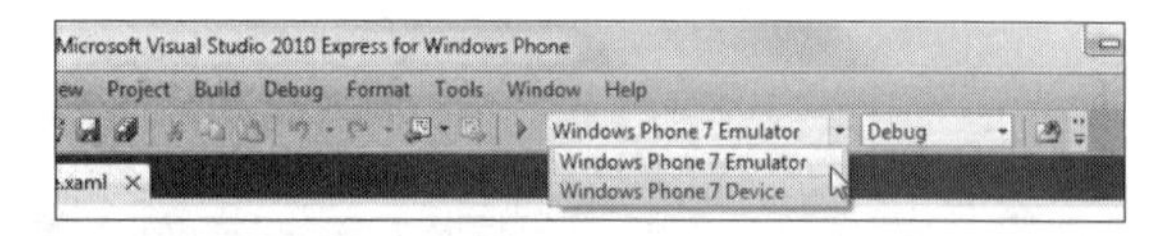

FIGURE 2-9

Alternatively, when you press F5 to run your application in Blend, you will be prompted to select whether to launch the application using the emulator or on a real device.

WINDOWS PHONE EMULATOR

The Windows Phone Developer Tools include a sophisticated emulator that can be used to debug and test your application without having to access a real device. The emulator is an x86 build of the Windows Phone operating system, which means that it can leverage the hardware acceleration available via the underlying desktop PC.

When you hit F5 in either Visual Studio or Blend, you will have the option of selecting to debug the application within the emulator. The first time you deploy an application using this option the emulator will load. This can take a minute or so as the virtual machine boots up. To save time while using the emulator, don't close it each time you finish debugging your application. Instead, leave the emulator open and simply restart your debugging session by pressing F5. This avoids the lengthy boot process as Visual Studio and Blend are able to reattach to the already opened emulator. Figure 2-10 shows the application you have been building, running in the emulator.

FIGURE 2-10

The Windows Phone emulator has been skinned to appear like a real device, rather than having the usual window chrome that allows you to resize or move the window. This can make moving or resizing the emulator difficult. However, there is an associated toolbar from which you can close or minimize the emulator. You can also move the emulator by dragging the toolbar around your desktop.

The third and fourth buttons on the toolbar allow you to rotate the emulator in an anti-clockwise or clockwise direction. For example, in Figure 2-11, you can see your application running in Landscape.

FIGURE 2-11

Note how the `SearchTextBox` automatically scales based on the new layout. Handling different screen orientations will be covered in more depth in Chapter 5.

The remaining buttons on the emulator toolbar allow you to autosize and adjust settings, respectively. The Settings dialog allows you to control the zoom level, in other words, the size, of the emulator. Rather than explicitly using the predefined zoom levels or specifying a custom level, it's usually easier to use the autosize option and allow the emulator to display at a size appropriate to your display.

As the emulator has been designed to replicate a real device, it also has three buttons on the front. These are, of course, the Back, Start, and Search buttons and they behave as you would expect on a real Windows Phone device. Of these buttons, the Back button is probably of the most interest because it can be used within your application to control the navigation between pages and dialogs.

The emulator can be used to see how your application behaves when it is pushed into the background and how it can integrate into the Start screen. You can also explore the emulator to learn more about how a real Windows Phone device will behave. In Figure 2-12, you can see the Start screen and Internet Explorer as they would appear on a real device.

FIGURE 2-12

The emulator can also connect to the Internet, meaning that you can use it to access remote sites or services during your development process. This is possible because the emulator makes use of the underlying computer's available networks. A less well-known fact is that `localhost` resolves to the underlying computer, rather than to the emulator. This means that during development you can easily expose a service running on your computer and have it consumed by a Windows Phone application running on the device.

One of the important things to remember when debugging your application in the Windows Phone emulator is that it is no substitute for testing on a real device. When you run your application on a real device, you will most likely notice that it performs differently. There are also some features, such as vibration, location and the accelerometer, that are not currently supported in the emulator.

SUMMARY

In this chapter, you have seen an overview of the tools that make up the Windows Phone developer experience. You have also seen how developers and designers can collaborate on a Windows Phone application by using tools such as Visual Studio 2010 Express for Windows Phone and Expression Blend 4, which are tailored to their specific needs yet can seamlessly interact at the project and solution file level.

The following chapters will cover many features of Windows Phone applications making use of both of these tools. Although you may feel more comfortable in one tool than the other, you should spend time working with both tools because they have their respective strengths when it comes to enabling you to build rich user experiences for Windows Phone.

3

Designing Layouts Using Red Threads

WHAT'S IN THIS CHAPTER

- What the Red Threads are and how they apply to building Windows Phone applications
- How to find and use the standard Windows Phone controls
- Adjusting the layout of your application
- How to apply styles, templates, and themes to your controls

One of the biggest changes in the last few years has been the advent of the declarative XAML (eXtensible Application Markup Language) — based user interface. Originally introduced as part of the Windows Presentation Foundation (WPF), XAML is now the markup language behind Silverlight both for the Desktop (via the Web) and Windows Phone. XAML not only includes a base set of controls, but also the ability to restyle and theme them. You can also create your own controls from scratch, and, as the rendering is vector-driven, you can be sure that your controls will scale as your application adapts to different screen resolutions and orientations.

As part of creating the Metro user experience, Microsoft kept to three design principles, which they refer to as *Red Threads*. In this chapter, you will learn about the default set of controls that ship with Silverlight, how to style them and use the Windows Phone themes, and how to use the three Red Threads to guide you as you build your user experience.

RED THREADS

So far you have learned about the Metro design language and how Windows Phone has been optimized for life maximizers, such as Anna and Miles. The challenge is to take this knowledge and use it to shape what, and how, you present information to the user. To assist with this, Microsoft came up with a concept that they refer to as *Red Threads*. A *Red Thread* is essentially a thread or theme that runs through the entire user experience guiding the presentation of information.

If you look up "Red Thread" on Wikipedia, one of the associated references is for "Red String," which refers to the wearing of a thin, red string in order to ward off misfortune.

In the case of Windows Phone, Microsoft derived three Red Threads that were applied through the creation of the base user experience. When you are building your application you too should consider the appropriateness of your design based on these threads:

- **Personal** — Your day, your way
- **Relevant** — Your people, your location
- **Connected** — Your stuff, your peace of mind

Now, if you have come from a background in building mobile applications, you may be thinking that these threads are either common sense (Personal and Relevant) or contrary to building mobile applications (Connected). I'd urge you to stay the course and at least consider the approach that Microsoft has offered.

Let's start with Personal. You may be thinking that this is a no-brainer, that you'd never design an application that isn't designed for the end user. Well, the reality is that we, as developers, often build applications that don't focus on the individual. There are a multitude of reasons for this, ranging from the lack of understanding about how our application will be used, to the need to meet corporate or other guidelines. The key to making your application Personal is to only present information that involves the user in some way. For example, if you are building an application that displayed outstanding tasks for a project, you'd only display tasks that involve the current user.

One of the side effects of making your application Personal is that you don't need to transport large quantities of data. In the case of the tasks application, if you didn't make it Personal, you'd have to synchronize all outstanding tasks for the project. However, as the user is really only going to be interested in his or her tasks, you can reduce the amount of data that needs to be synchronized and displayed.

The next Red Thread is to make your application Relevant. Again you might be thinking that you always make your application Relevant to the user. The question is how Relevant *do* you make it versus how Relevant *can* you make it? There are many different dimensions over which you can apply the concept of being Relevant. With the inclusion of a location service in Windows Phone, the obvious one is to filter data based on where the user is located. For example, you can use location to display only tasks that relate to the aspects of a project being carried out at, or near to the user's current location. Other dimensions that are less obvious are using the accelerometer to determine if the user is on the move, or date and time to filter information by an appropriate time window.

The last Red Thread refers to building an application that is Connected. Now, you may be thinking that Microsoft has lost the plot and that it would be suicide to build a mobile application that is a connected system, and you'd be correct in thinking this. However, that's not the intent of this Red Thread. Microsoft is under no illusions that applications for Windows Phone have to be able to operate in a disconnected or occasionally connected fashion. However, they also understand the power of cloud computing and the use of hosted services in order to extend the reach and power of mobile applications.

Instead of building a stand-alone application, the Connected thread encourages you to think about wider scenarios where the user is able to collaborate with other users, share information and knowledge between devices, and access their information wherever they are. Build your application to be connected to online services, store information in the cloud, or communicate with other users. Don't build your application to be always connected and totally reliant on having a network connection. Mobile networks are still a long way from being omnipresent, and even in areas which have good network coverage, there will still be cases — such as in planes, trains, and automobiles — in which the device will lose signal. Instead, build an application that functions independently of the network connection, synchronizing and updating in the background whenever a connection is available.

Building a compelling user experience is an art form in itself. You need to combine the ability to determine what information the user needs and wants, with some natural creativity to make the experience visually pleasing. By adhering to the Red Threads of making your experience Personal, Relevant, and Connected, you will be ensuring that you are only presenting information that the user wants to see and engage with. Throughout the remainder of this book, keep returning to these principles and thinking about how they apply to what you are building.

CONTROLS

Enough with the design lesson — you want to get in and see what you can build with Windows Phone! In the previous chapter, you saw how easy it is to use Visual Studio 2010 and Expression Blend 4 to build a Windows Phone application. The development story gets even better when you look at the array of controls that are available to you for composing your application.

Figure 3-1 shows the Toolbox from Visual Studio in the left image. A similar list of controls is available within the Assets window in Blend, as shown in the middle and right images. When working in Blend, for the most part you will only need the controls located under the Controls node. However, it's important to be aware of the full list, as shown in the right image of Figure 3-1, which can be accessed by expanding the Controls node and clicking on the All node.

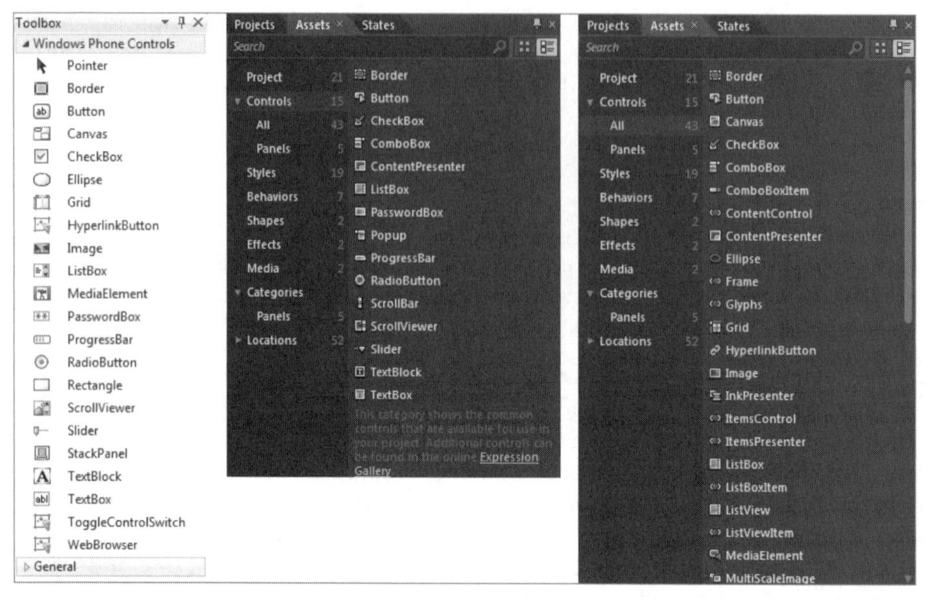

FIGURE 3-1

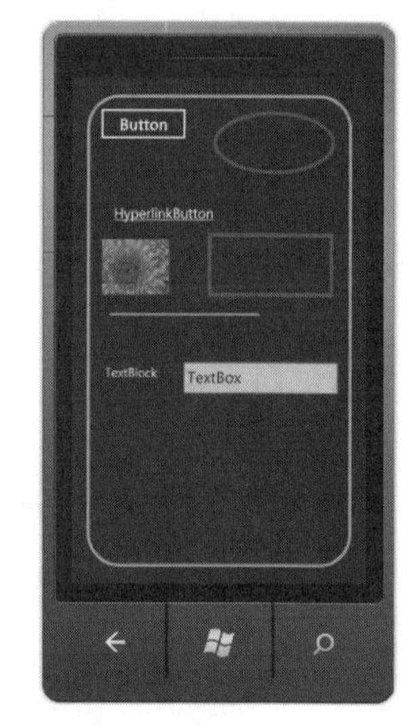

FIGURE 3-2

As noted in the previous chapter, any of the controls in the Toolbox or the Assets window can be dragged onto the design surface of your application. You can then set properties using the Properties window. Figure 3-2 illustrates the result of dragging a few of the common controls onto a page.

Before you go much further, you will need to take a bit of a side trip into the world of XAML. Every page or control that you build in Silverlight is typically represented by at least two files: a XAML file that defines the layout and a C# file that contains associated code such as event handlers (often referred to as the *code-behind file*). *XAML* originally stood for "eXtensible Avalon Markup Language," which is a reference to the original code-name, *Avalon*, for Windows Presentation Foundation (WPF).

XAML is an XML-based language created by Microsoft to allow the structure of a set of Common Language Runtime (CLR)–based objects and properties to be specified in a declarative manner. The intent was to make it easy for design tools to create layouts without suffering from the brittleness of parsing C# source code, as earlier technologies like Windows Forms required. This model has been so successful that other technologies such as Windows Workflow Foundation and Silverlight have adopted it. Recently Microsoft released documentation covering the technical specifications of XAML that is available from the Microsoft download center (`www.microsoft.com/downloads`). It is a common misconception to consider

XAML to be the graphical aspect of Silverlight or WPF, when in reality XAML is purely an XML-based file format for describing a desired hierarchy of CLR-based objects.

Although the designer experiences in Visual Studio 2010 and Expression Blend are approaching the point where you can do most of your application layout without having to worry about the XAML that is generated, it is still desirable to be able to read and write XAML. This will assist you in understanding how control layout works and enable you to manually code anything the designers aren't able to cope with. We'll start by working in Visual Studio and then go through more complex design tasks in Expression Blend later in the chapter.

Let's start by creating a new Windows Phone application project within Visual Studio named *RedThreadLayout* and taking a look at the XAML for the MainPage it creates. Once you create the project, you will see that the Designer view already contains two TextBlocks. These are nested within a StackPanel that is used to position the TextBlocks toward the top of the page. If you switch to the XAML Editor window, you will see that the XML tag hierarchy begins with a `PhoneApplicationPage` element, which is the class that MainPage inherits from. Each screen in a Windows Phone application inherits from `PhoneApplicationPage` and represents a page in the navigation system, which you'll learn about in Chapter 6.

```
<phone:PhoneApplicationPage
    x:Class="RedThreadLayout.MainPage"
    xmlns="http://schemas.microsoft.com/winfx/2006/xaml/presentation"
    xmlns:x="http://schemas.microsoft.com/winfx/2006/xaml"
    xmlns:phone="clr-namespace:Microsoft.Phone.Controls;assembly=Microsoft.Phone"
    xmlns:shell="clr-namespace:Microsoft.Phone.Shell;assembly=Microsoft.Phone"
    xmlns:d="http://schemas.microsoft.com/expression/blend/2008"
    xmlns:mc="http://schemas.openxmlformats.org/markup-compatibility/2006"
    FontFamily="{StaticResource PhoneFontFamilyNormal}"
    FontSize="{StaticResource PhoneFontSizeNormal}"
    Foreground="{StaticResource PhoneForegroundBrush}"
    SupportedOrientations="Portrait" Orientation="Portrait"
    mc:Ignorable="d" d:DesignWidth="480" d:DesignHeight="768"
    shell:SystemTray.IsVisible="True">

<!-- Page content goes here -->

</phone:PhoneApplicationPage>
```

The first attribute, `x:Class`, is the class declaration, in this case, `RedThreadLayout.MainPage`. The prefix, `RedThreadLayout`, is the namespace for the class. If you examine the MainPage.xaml.cs file, you will see that the namespace, class name, and base class match those in the XAML file:

```
namespace RedThreadLayout
{
    public partial class MainPage : PhoneApplicationPage
    {
        public MainPage()
        {
            InitializeComponent();
        }
    }
}
```

You will notice within the code-behind file that the class, `MainPage`, is declared as being partial. This is important because it enables the compiler to wire the XAML layout together with the code-behind file so that you can reference controls declared in the XAML from the code you write.

If you decide to rename the class, change the base class, or change the namespace of a Silverlight page, you'll need to ensure that the changes are done in both the XAML and CS files.

OK, so the compiler wires together the XAML and code-behind files. But what does this mean? When your application is compiled, the XAML file is processed, and an additional file is generated. In the case of `MainPage`, this file is called *MainPage.g.cs* and is located in the obj/Debug folder within your project. Since this folder is not visible by default, you will need to click the "Show All Files" icon in the Toolbar on the Solution Explorer window. When you open this file, you'll see that it declares several internal fields and an `InitializeComponent` method. In fact, there is one internal field for every named control within MainPage.xaml. The `InitializeComponent` method calls `LoadComponent` to parse the XAML file and generate the appropriate object hierarchy.

```
public partial class MainPage : Microsoft.Phone.Controls.PhoneApplicationPage {
    internal System.Windows.Controls.Grid LayoutRoot;
    ...
    private bool _contentLoaded;
    public void InitializeComponent() {
        if (_contentLoaded) {return; }
        _contentLoaded = true;
        System.Windows.Application.LoadComponent(this,
                       new System.Uri("/RedThreadLayout;component/MainPage.xaml",
                                      System.UriKind.Relative));
        this.LayoutRoot =
                 ((System.Windows.Controls.Grid)(this.FindName("LayoutRoot")));
        ...
        }
    }
```

Within the context of `MainPage` the `LoadComponent` method will create the page layout according to how it was created using the designer. After the controls have been created, `InitializeComponent` then calls `this.FindName` for each control on the page, assigning the returned value to the corresponding internal field. Actually, this isn't technically true; as the generated code will only locate controls that have an associated `x:Name` attribute within the XAML file. Returning to MainPage.xaml, find the element with the attribute `x:Name="ContentGrid"`:

```
<Grid x:Name="ContentGrid" Grid.Row="1">
</Grid>
```

If you look in MainPage.g.cs, you will see that there is an internal field called ContentGrid of type System.Windows.Controls.Grid and that there is a corresponding call to FindName to obtain this control once the XAML has been loaded. In MainPage.cs, navigate to the constructor and type **this.con**. You will see that the ContentGrid appears in the IntelliSense, as shown in Figure 3-3.

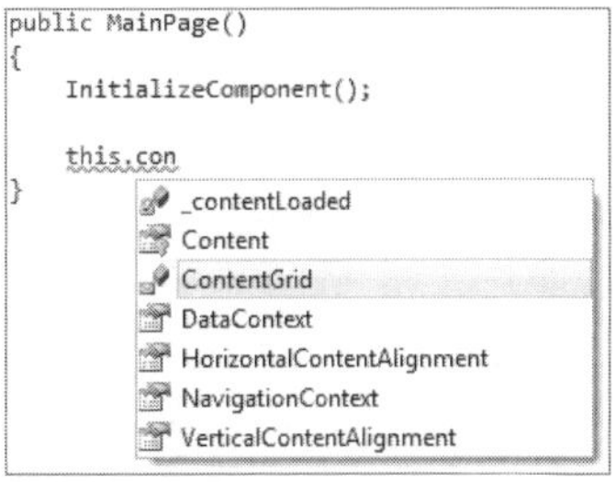

FIGURE 3-3

Now, go back to MainPage.xaml and remove x:Name="ContentGrid" from the Grid element. Save and rebuild your project. If you go to MainPage.g.cs, you'll see that the internal field is no longer there, and if you go back to MainPage.cs, you'll see that this control is no longer listed in the IntelliSense. While it can be convenient to name all your controls, note that there may be a small performance hit during initialization where each loaded control is linked to the internal backing field. On desktop systems, this is not likely to be a major issue, but on a mobile device, where performance can make a significant difference in the speed of the application and battery life, it's worth considering whether you need to be able to reference all of your controls. If not, remove the x:Name attribute for the controls you aren't going to be referencing. With more advanced concepts such as Data Binding and Model View View Model (MVVM), discussed in Chapter 15, you'll probably be quite surprised at how little you actually need to name controls.

So far you've seen how an element in XAML can represent an instance of a control. If you look closer at the XAML, you'll also note that XML attributes correspond to properties on the class. For example, in the following snippet, an instance of a TextBlock is being created with the Text property being set to "MY APPLICATION".

```
<TextBlock x:Name="ApplicationTitle" Text="MY APPLICATION"
           Style="{StaticResource PhoneTextNormalStyle}"/>
```

When this control is instantiated, the default class constructor is called, followed by the Text property being set to this value. Similarly, the Name and Style properties are also set. Note that the Style value uses curly braces, { }. These indicate that the value is not a simple text string and instead needs to be evaluated prior to the property being set. You'll learn more about using styles later in this chapter, and this syntax is also used in data binding, which is covered in Chapter 15. What's interesting is that this XAML can be rearranged by removing the Text attribute and instead specifying it via a nested element:

```
<TextBlock x:Name="textBlockPageTitle"
           Style="{StaticResource PhoneTextPageTitle1Style}">
    <TextBlock.Text>MY APPLICATION</TextBlock.Text>
</TextBlock>
```

The XAML parser knows to interpret the child element as a property setter because its name, TextBlock.Text, consists of the parent object's name and a property name separated by a period. In other words, the contents of the TextBlock.Text element will be used to set the Text property on the TextBlock. You might be wondering why there are two supported formats and which one you should use. It is recommended that, where possible, you set property values using the attribute format because it is more compact, yielding a smaller XAML file, and generally improves readability. However, there are circumstances in which the property value can't be represented as a simple string. As an example, the following XAML assigns multiple `RowDefinition` objects to the Grid's `RowDefinitions` property, which would be hard to specify as a string:

```
<Grid x:Name="LayoutRoot" Background="{StaticResource PhoneBackgroundBrush}">
    <Grid.RowDefinitions>
        <RowDefinition Height="Auto"/>
        <RowDefinition Height="*"/>
    </Grid.RowDefinitions>
    ...
</Grid>
```

Standard Controls

Let's take a closer look at some of the controls that are in the Toolbox. Figure 3-4 shows five of the controls that you will most likely be familiar with. Going from left to right, starting from the top row, the controls are the `Button`, `TextBox`, `PasswordBox`, `CheckBox` and `RadioButton`.

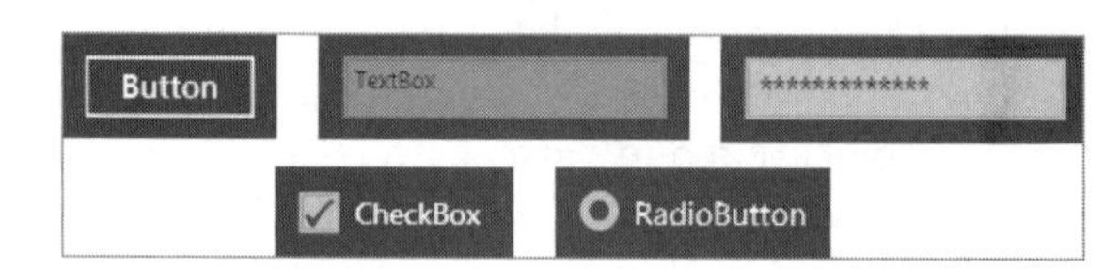

FIGURE 3-4

The one thing that you will notice about these controls is that they all have a black background and a white foreground. In actual fact, most of the controls are mostly transparent and will by default inherit the background of their parent control. You'll learn later on in this chapter how these colors are affected by styles and themes being applied.

When it comes to building a Windows Phone application, one of the most precious resources is the battery life of the device. It should come as no surprise that the more your application does, the more CPU (Central Processing Unit) cycles and hence battery life it uses up. Interestingly, another factor that can affect battery life is how much of the screen is illuminated. If you imagine the screen as a series of light bulbs that are turned on or off in order to display different shapes, then you can understand how a predominantly white control would consume more power than one that is predominantly black. With this in mind, it comes as no surprise that the default background color for a Windows Phone application is black, which, in turn, affects most of the controls that use a transparent background.

ContentProperty

Before you go through each of these controls, you need to go on a bit of a sidetrack to discuss the special `Content` property that certain controls have. Essentially, the `Content` property is a result of the control inheriting from `ContentControl`, rather than directly from `Control`. If you look at

Figure 3-5, you will see that the `Button` is a `ContentControl`, whereas the `TextBox` is not.

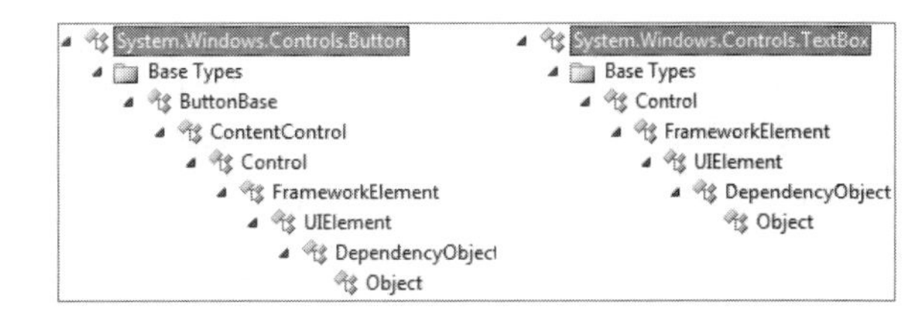

FIGURE 3-5

The purpose of the `Content` property exposed by the `ContentControl` is to allow the contents of the control to be either a simple value, such as a string, or a hierarchy of additional XAML. To try this out, place a button on the MainPage of your application. Go to the XAML view and locate the newly created button, and replace it with the following snippet:

```
<Button Height="115" HorizontalAlignment="Left" Margin="34,26,0,0" Name="button1"
        VerticalAlignment="Top" Width="330">
    <Canvas Margin="-25,-16,0,0" Width="330" Height="115">
        <TextBlock Canvas.Top="18" Width="330"
                   TextAlignment="Center" >Button</TextBlock>
        <TextBlock Text="Additional Button Text" TextAlignment="Center"
                   Canvas.Top="56" Width="330"/>
    </Canvas>
</Button>
```

Code snippet from MainPage.xaml

Here there is a canvas inside the button, and then within the canvas there are two `TextBlock` controls. If you view this in the Designer, you will see that there are now two lines of text contained within the `Button`. A control that derives from `ContentControl` can only have one child; as such, it is common to see this child be a container such as a `Canvas` or `Grid` that can then hold multiple controls. You'll also note that even though you are specifying the `Content` property, the `Canvas` has been specified as a direct descendant of the `Button`. This is a XAML shortcut. The previous snippet of XAML is equivalent to the following more verbose form:

```
<Button Height="115" HorizontalAlignment="Left" Margin="34,26,0,0"
        Name="button1" VerticalAlignment="Top" Width="330">
    <Button.Content>
        <Canvas Margin="-25,-16,0,0" Width="330" Height="115">
            <TextBlock Canvas.Top="18" Width="330"
                       TextAlignment="Center" >Button</TextBlock>
            <TextBlock Text="Additional Button Text" TextAlignment="Center"
                       Canvas.Top="56" Width="330"/>
        </Canvas>
    </Button.Content>
</Button>
```

We glossed over the fact that two `TextBlock` controls were added as descendants of the `Canvas`. This is because `Canvas` inherits from `Panel`, which has a `Children` property. Again, the XAML syntax here is shortened to remove the `Canvas.Children` element that you would expect to see. You'll learn more about using layout controls like `Canvas` in the next section.

Let's go back to examining the controls that are available to you. You've already seen the `Button` control in action, but it's worth noting that there is a property called `ClickMode` that can be used to determine when the `Click` event is raised. There are three options:

- **Release** — Triggers the `Click` event when the button is released.
- **Press** — Triggers the `Click` event when the button is pressed.
- **Hover** — Triggers the `Click` event when the mouse hovers over the control. There is no mouse on a Windows Phone, so avoid using this mode.

The `TextBox` and `PasswordBox` work as you'd imagine, although instead of a `Content` property they have `Text` and `Password` properties, respectively. If you want to detect when their values change, you can handle the `TextChanged` or `PasswordChanged` events. The `PasswordBox` has an additional property, `PasswordChar`, which determines the character that is displayed instead of the password.

The `CheckBox`, `RadioButton`, and `ToggleControlSwitch` controls all have an `IsChecked` property that determines the checked state. `RadioButtons` are grouped together by the `GroupName` property. By default, this property isn't specified, which means that all `RadioButtons` belong to the same group.

Figure 3-6 shows a TextBlock, Image and MediaElement, ProgressBar, and Slider in order from top to bottom. The `TextBlock` is not a `ContentControl` so, like the `TextBox` control, it has a `Text` property instead of a `Content` property. You can also control how the text is displayed by altering the `FontFamily`, `FontSize`, `FontStretch`, `FontStyle`, and `FontWeight` properties.

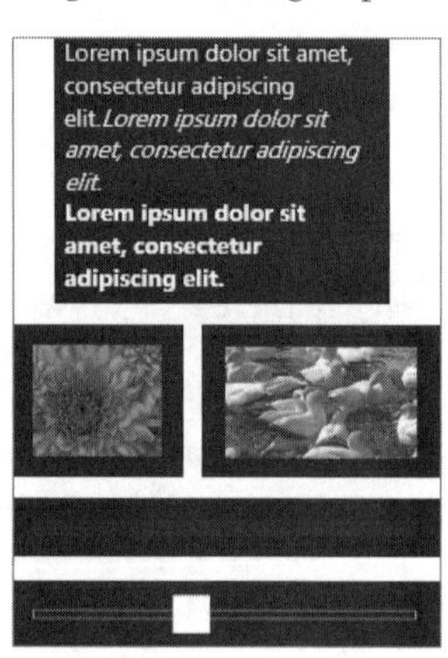

FIGURE 3-6

You may think that the `TextBlock` element is equivalent to a Windows Forms `Label` control; however, if you look at the text displayed in the single textblock of Figure 3-6, you will notice that there are three font styles in use: The first section is in a regular font; the next section, which continues the same paragraph, is in italics; and the remaining sentence begins on a new line and is in bold. The complex content of the textblock is created using the following code, which breaks up the text into chunks. Each chunk is called a *Run* and has an explicitly specified font, or inherits it from the parent `TextBlock`.

```
<TextBlock HorizontalAlignment="Left" VerticalAlignment="Top" TextWrapping="Wrap">
    Lorem ipsum dolor sit amet, consectetur adipiscing elit.
    <doc:Run FontStyle="Italic">Lorem ipsum dolor sit amet,
                                consectetur adipiscing
                                elit.</doc:Run><doc:LineBreak />
    <doc:Run FontWeight="Bold">Lorem ipsum dolor sit amet,
                               consectetur adipiscing
                               elit.</doc:Run>
</TextBlock>
```

But hang on, didn't we say that the `TextBlock` wasn't a `ContentControl`, in which case, it doesn't have a `Content` property? If this is the case, how can you nest text and other elements within the `TextBlock` element? The secret here is a little bit of magic called the `ContentProperty` attribute (this is a CLR attribute, rather than an XML attribute). This attribute can be applied to a class in order to specify which property should be set, based on the result of any XAML nested within the control. In the case of a ContentControl, the ContentProperty attribute specifies the Content property, while controls such as Grid or Canvas, which inherit from Panel, specify their Children property.

Interestingly, in the case of a TextBlock, the ContentProperty doesn't specify the Text property. Instead, it specifies a property called Inline, which is of type InlineCollection, which, as you can imagine, is a collection of objects of base type Inline. The Inline class is an abstract class, meaning it can't be instantiated directly, but it does contain the properties for setting font and color. Further investigation reveals that there are two classes that inherit from Inline, which are Run and LineBreak. So this means that you can add any number of Run and LineBreak elements within the TextBlock. Each one is added to an InlineCollection that is assigned to the Inline property of the TextBlock. The only missing piece of the puzzle is how the first line of text gets incorporated into the InlineCollection, as a string doesn't derive from the Inline base class. This isn't a great mystery as the InlineCollection supports an Add method that accepts a string as an argument, and the XAML parser is smart enough to use it.

ADD DOCUMENT NAMESPACE REFERENCE

In the TextBlock code snippet, you will see that the Run and LineBreak elements are prefixed with *doc*. This is a reference to the namespace System.Windows.Documents, which contains these classes. To use these classes within XAML, you need to specify the XML namespace within the root element of your page. This is similar to a C# using statement:

```
<phone:PhonweApplicationPage
       x:Class="RedThreadLayout.MainPage"
       xmlns:doc=
"clr-namespace:System.Windows.Documents;assembly=System
.Windows"
       ...
```

TextBlock, Run, and LineBreak

Let's explore how you can use a TextBlock with Run and LineBreak elements to build an RSS (Really Simple Syndication) reader. Note that while this example is for building an RSS viewer, if you have any text that contains markup for bold, italics, titles, and so on, you can use a similar process to view the content within a Windows Phone application. Later in this chapter, you'll actually see an alternative viewer that uses the WebBrowser control to view HTML markup.

To begin with, you're going to need to locate an RSS feed to work with. In this case you're going with the feed from the .NET Travels blog (http://nicksnettravels.builttoroam.com), and, for simplicity, you're going to use a snapshot of this feed, rather than dynamically loading it. To take a snapshot, use Internet Explorer to navigate to the RSS feed, click the "View feeds on this page" button in the toolbar (or press Alt+J), and select one of the RSS or Atom feeds. When the feed finishes loading, right-click anywhere on the page and select "View Source." This will open the XML source for the RSS feed in Notepad. Save this as **BlogPosts.xml** into the root folder of your Windows Phone project. In Visual Studio, right-click on the project node within Solution Explorer and select Add ➪ New Item. Select the "Resource File" template, and call the new item **RssResources.resx**, as shown in Figure 3-7.

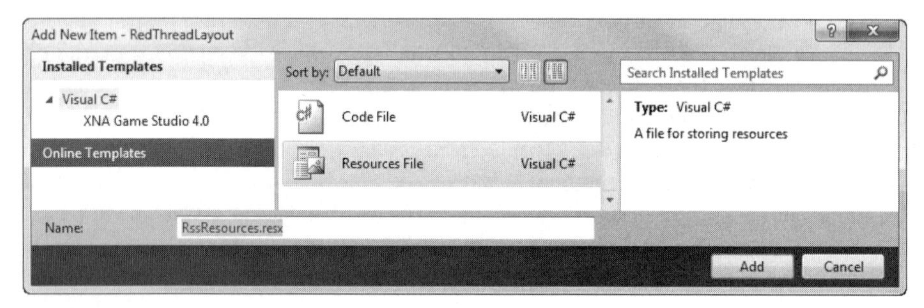

FIGURE 3-7

Click Add to create the new resource file, which will then be automatically opened in the resource file Editor window. Select Add Resource ➪ Add Existing File and select the BlogPosts.xml file you created earlier. This will automatically add the XML file to your project, as well as adding it to the RssResources resource file, as shown in Figure 3-8.

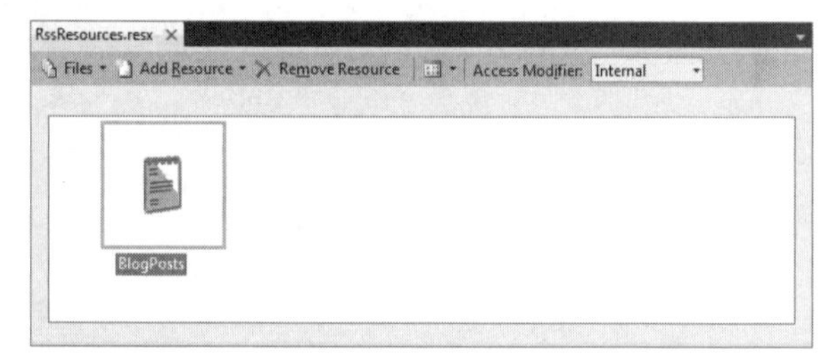

FIGURE 3-8

Before you write some code to parse the BlogPost XML file, you'll need somewhere to display the posts. Open up MainPage.XAML within the Visual Studio Designer, and replace any existing contents within the `ContentGrid` control with a `StackPanel` nested within a `ScrollViewer`. You can add controls, such as the `TextBlock`, to a `StackPanel`, and they will be stacked below each other going down the page. Once the end of the page is reached, the `ScrollViewer` will allow the user to scroll down in order to see controls that would otherwise be off the page. In the end the relevant part of XAML should look similar to the following snippet:

```
<Grid x:Name="ContentGrid" Grid.Row="1">
    <ScrollViewer Name="Scroller">
        <StackPanel Name="PostStack" Width="480"></StackPanel>
    </ScrollViewer>
</Grid>
```

Code snippet from MainPage.xaml

You're going to load the RSS feed during the `Loaded` event of the MainPage. To do this, open MainPage.xaml within the designer and select the PhoneApplicationPage node. Alternatively, you may find it easier to use the Document Outline window (View ➪ Other Windows ➪Document Outline menu item), as shown in the left image of Figure 3-9.

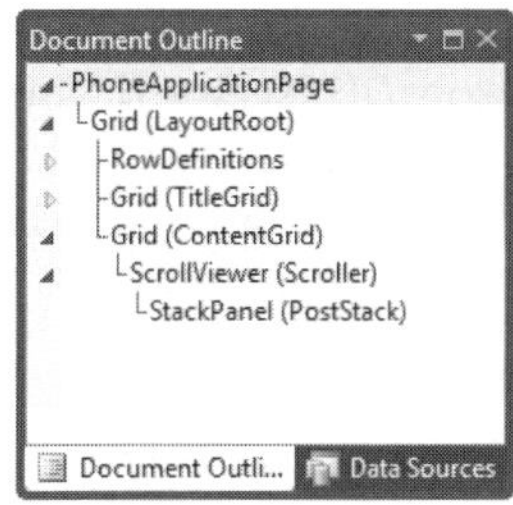

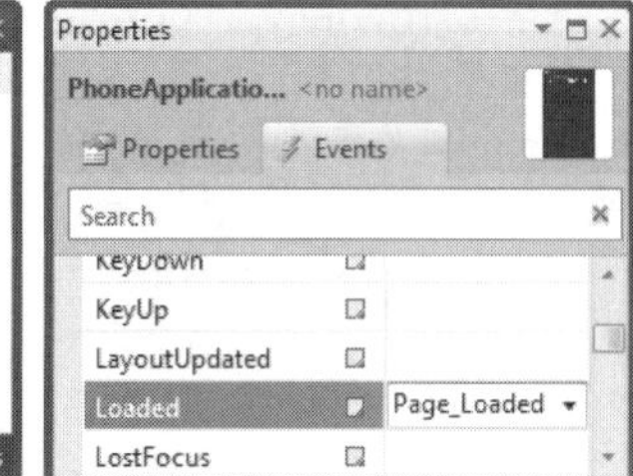

FIGURE 3-9

With the PhoneApplicationPage selected, open the Properties window, select the Events tab, and locate the `Loaded` event, as shown in the right image of Figure 3-9. Double-click in the cell next to this event to create and navigate to an event handler in the MainPage.xaml.cs file. The first thing you need to do is to load the contents of the BlogPosts XML file into memory so that you can work with it:

```
private void PhoneApplicationPage_Loaded(object sender, RoutedEventArgs e){
    // Load RSS into a feed
    var rss = RssResources.BlogPosts;
    using(var strm = new System.IO.StringReader(rss))
    using(var reader = System.Xml.XmlReader.Create(strm)){
        var feed = System.ServiceModel.Syndication.SyndicationFeed.Load(reader);
```

Rather than having to parse the XML file to extract blog posts, you can use the `SyndicationFeed` class located in the `System.ServiceModel.Syndication` namespace.

There is no Windows Phone–specific version of the `System.ServiceModel.Syndication` *assembly. However, you can add a reference to the desktop Silverlight 3 assembly in order to access the classes in this namespace* (`C:\Program Files\Microsoft SDKs\Silverlight\v3.0\Libraries\Client\System.ServiceModel.Syndication.dll`).

The `Load` method returns an instance of the `SyndicationFeed` class, which has an Items collection through which you can iterate to access the individual posts. You can then programmatically add a new `TextBlock` to the `StackPanel` to hold the first set of `Run` elements:

```
// Create the initial textblock
var txt = new TextBlock() { TextWrapping = TextWrapping.Wrap };
this.PostStack.Children.Add(txt);
foreach (var item in feed.Items){
    // Add a title
    txt.Inlines.Add(new Run() { FontWeight = FontWeights.Bold,
                                Text = item.Title.Text,
                                FontSize = txt.FontSize*2 });
    txt.Inlines.Add(new LineBreak());
    var itemText = item.Summary.Text;
```

Unfortunately, if you add the raw contents of the post to the `TextBlock`, you will end up seeing all the HTML tags that are usually scattered throughout the post. One option would be to blindly remove all these tags. However, this would yield a single block of text without images or formatting of the text. An alternative is to use a Regular Expression to locate each tag and perform some conditional logic depending on what tag it is.

```
// Regular expression to look for html tags
var tagFinder = new Regex("<(.|\n)+?>", RegexOptions.IgnoreCase);
```

Whenever an HTML tag is located, you will need to add the section of text between the start and end tags, to the current `TextBlock`. For example the following code extracts the text, including the start and end tags (e.g. "<b>Fred Blogs</b>"). It then trims the start and end tags by replacing them with an empty string, before adding the text to the TextBlock:

```
var text = source.Substring(startidx, endidx - startidx);
text = tagFinder.Replace(text, "");
txt.Inlines.Add(text);
```

If you locate an `<i>` or `<b>` tag, you can also apply italics and bold as part of adding the new `Run` to the `TextBlock`. To do this, you need to modify the `FontWeight` and `FontStyle` properties of the `Run` element you are creating:

```
txt.Inlines.Add(new Run(){
                         FontWeight = (isBold ? FontWeights.Bold :
                                                FontWeights.Normal),
                         FontStyle = (isItalics ? FontStyles.Italic :
                                                  FontStyles.Normal),
                         Text = text });
```

Alternatively, if you encounter an end of paragraph, `</p>`, or line feed, `</br>`, tag, you can add a `LineFeed` to the `TextBlock`:

```
txt.Inlines.Add(new LineBreak());
```

In order to display any images that are contained in the posts, they will have to be added directly to the `StackPanel`. At this point the previous `TextBlock` ends, an `Image` will be created and added to the `StackPanel`, and then a new `TextBlock` will be inserted to hold the next set of `Run` elements:

```
// Create an image and add to stackpanel
var img = new Image() { Source = new BitmapImage(new Uri(url)),
                        Width = this.PostStack.ActualWidth / 2,
                        Stretch = Stretch.UniformToFill };
this.PostStack.Children.Add(img);

// Create a new textblock and add to stackpanel
txt = new TextBlock() { TextWrapping = TextWrapping.Wrap };
this.PostStack.Children.Add(txt);
```

When you combine this logic you'll get a rough RSS reader capable of viewing most blogs and other RSS sources. The full source code for the `Loaded` event handler is as follows:

Available for download on Wrox.com

```
private void PhoneApplicationPage_Loaded(object sender, RoutedEventArgs e){
    // Load RSS into a feed
    var rss = RssResources.BlogPosts;
    using(var strm = new System.IO.StringReader(rss))
    using(var reader = System.Xml.XmlReader.Create(strm)){
        var feed = System.ServiceModel.Syndication.SyndicationFeed.Load(reader);

        // Regular expression to look for html tags
        var tagFinder = new Regex("<(.|\n)+?>",
                       System.Text.RegularExpressions.RegexOptions.IgnoreCase);

        // Create the initial textblock
        var txt = new TextBlock() { TextWrapping = TextWrapping.Wrap };
        this.PostStack.Children.Add(txt);
        foreach (var item in feed.Items){
            // Add a title
            txt.Inlines.Add(new Run() { FontWeight = FontWeights.Bold,
                                        Text = item.Title.Text,
                                        FontSize = txt.FontSize*2 });
            txt.Inlines.Add(new LineBreak());
            var itemText = item.Summary.Text;

            var match = tagFinder.Match(itemText);
            int startidx = 0, isBold = 0, isItalics = 0;
            while (match.Index >= 0 && match.Length > 0){
                // Extract the section of text up until the tag
                ExtractText(txt, itemText, startidx, match.Index,
                            tagFinder, isBold > 0, isItalics > 0);
                startidx = match.Index + match.Length;

                var isEndTag = match.Value.Contains("/");

                if (match.Value == "</p>" || match.Value == "<p />"
                                          || match.Value == "<br />"){
                    // Found the end of a paragraph, so add line break
                    txt.Inlines.Add(new LineBreak());
                }
                else if (match.Value=="<b>" || match.Value=="</b>"){
                    isBold += isEndTag ? -1 : 1;
                }
                else if (match.Value == "<i>" || match.Value == "</i>"
                                              || match.Value == "<em>"
                                              || match.Value == "</em>"){
                    isItalics += isEndTag ? -1 : 1;
                }
                else if (match.Value.Contains("<img")){
                    // Locate the url of the image
                    var idx = match.Value.IndexOf("src");
                    var url = match.Value.Substring(
```

```
                                                  idx + 5, match.Value.IndexOf("\"",
                                                  idx + 6) - (idx + 5));

                    // Create an image and add to stackpanel
                    var img = new Image() {
                                   Source = new BitmapImage(new Uri(url)) };
                    img.Width = this.PostStack.ActualWidth / 2;
                    img.Stretch = Stretch.UniformToFill;
                    this.PostStack.Children.Add(img);

                    // Create a new textblock and add to stackpanel
                    txt = new TextBlock() { TextWrapping = TextWrapping.Wrap };
                    this.PostStack.Children.Add(txt);
                }

                // Look for the next tag
                match = tagFinder.Match(itemText, match.Index + 1);

            }
            // Add the remaining text
            txt.Inlines.Add(itemText.Substring(startidx));

            // Add some space before the next post
            txt.Inlines.Add(new LineBreak());
            txt.Inlines.Add(new LineBreak());
        }
    }
}

private static void ExtractText(TextBlock txt, string source, int startidx,
                                int endidx, Regex tagFinder, bool isBold,
                                bool isItalics){
    var text = source.Substring(startidx, endidx - startidx);
    text = tagFinder.Replace(text, "");
    if (!isItalics && !isBold){
        txt.Inlines.Add(text);
    }
    else{
        txt.Inlines.Add(new Run(){
                    FontWeight = (isBold ? FontWeights.Bold : FontWeights.Normal),
                    FontStyle = (isItalics ? FontStyles.Italic : FontStyles.Normal),
                    Text = text
                });
    }
}
```

Code snippet from MainPage.xaml.cs

Figure 3-10 illustrates the RSS Reader in action.

Image, MediaElement, and ProgressBar

FIGURE 3-10

In creating your RSS reader, you were able to incorporate images through the dynamic creation of an `Image` element. This included setting the `Source` property to a `BitmapImage` wherein the constructor accepted a Uri pointing to the location of an image at a particular URL (Uniform Resource Locator). What if you wanted to reference an image that was packaged within your application? Well, you can do that by specifying a relative source. Start by adding an `Image` to the `Grid` named `LayoutRoot`. You can do this easily by selecting "Grid (LayoutRoot)" from the Document Outline window (Ctrl+Alt+T), then double-clicking the Image control from the Toolbox window. Once added, open the Properties window and reset the `HorizontalAlignment`, `Height`, `VerticalAlignment`, `Width`, and `Margin` properties. This is the equivalent of removing these attributes from the XAML. Also set the `Grid.RowSpan` to 2, so that the `Image` fills the entire page. Next, click on the ellipses in the cell next to the Source property. This will open the "Choose image" dialog, as shown in Figure 3-11. Click the Add button and browse to a picture you wish to use as a background. In this case, use Chrysanthemum.jpg and Desert.jpg, which are in the Sample Pictures folder on Windows 7.

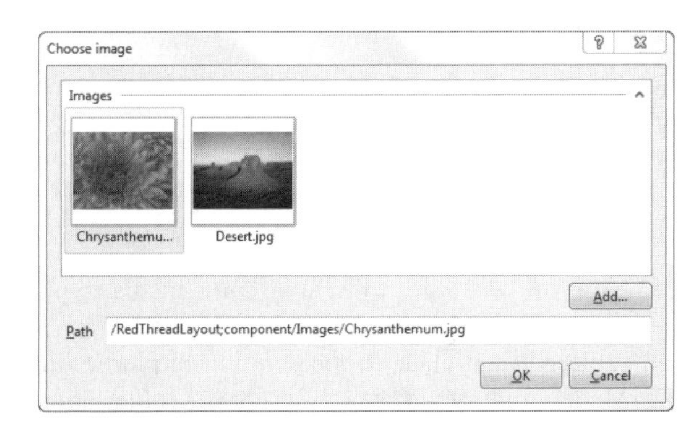

FIGURE 3-11

Notice how the Path is defined in Figure 3-11 for the selected image, which in this case is Chrysanthemum.jpg. It refers to the assembly, `RedThreadLayout`, and folder, Images, that contains the desired file. It also has *;component* specified in the middle. This is the format that you should use whenever you are referencing images or media that are compiled into your application using the Resource build action. When you select an image and click OK, you will notice that the image may appear within the Designer somewhat distorted. This is because the default `Stretch` property on the `Image` control is set to `Fill`. If you change this to `UniformToFill` or `Uniform`, you will see that the image is scaled uniformly, which will not distort your image. The final XAML for the image should look like the following:

Available for download on Wrox.com

```
<Image Grid.RowSpan="2"
       Name="image1"
       Stretch="UniformToFill"
       Source="/RedThreadLayout;component/Images/Chrysanthemum.jpg" />
```

Code snippet from MainPage.xaml

You may have noticed that when you added the image all your other content appeared to be missing. In actual fact, this is because the image you just added is sitting over the top of everything. You can resolve this by moving the Image XAML to be the first child after the `Grid.RowDefinitions`. *Alternatively, you can let Visual Studio do this for you by right-clicking on the image in the Designer and selecting Order ➪ Send to Back.*

Figure 3-6 contained both a MediaElement and a ProgressBar. You're going to add these to the bottom of the MainPage, so you may need to make space by reducing the height of the ScrollViewer that houses the RSS reader from earlier. Add both a MediaElement and a ProgressBar to the MainPage by selecting the ContentGrid and then either double-clicking or dragging the controls from the Toolbox onto the design surface. Once they have been added, you may want to lay them out and rename them. You should end up with XAML similar to the following:

Available for download on Wrox.com

```
<MediaElement Height="150"
              Margin="0,420,0,0"
              Name="SampleMedia"
              VerticalAlignment="Top"/>
<ProgressBar Height="10"
             Margin="0,600,0,0"
             Name="MediaProgress"
             VerticalAlignment="Top"/>
```

Code snippet from MainPage.xaml

Now you will want to wire up some media to play when the application starts, and you'll want the ProgressBar to stay in sync with the media. To begin with, you need to add some media into your project. Right-click on the Solution Explorer and select Add ➪ New Folder, name the new folder **Media**, and press Enter to accept the folder name. Now, right-click on the new folder, select Add ➪ Existing Item, and locate a video to add to the project. In this case, we've added Wildlife.wmv, which is a sample video that ships with Windows 7. Once this video has been added, select it and press F4 to view the Properties window. Make sure the Build Action is set to Content and that the "Copy to Output Directory" is set to Copy always.

Select the MediaElement control and find the Source property in the Properties window. Enter **Media\Wildlife.wmv** (if your video is called something different, then change the value entered here). Then locate and check the box against AutoPlay. You've just set up the MediaElement to load and play your video when the page has finished loading.

To keep the ProgressBar in sync with the video playback, you need to do two things. First, when the video is loaded, the `Maximum` property of the `ProgressBar` needs to be updated. Select the MediaElement and go to the Events tab of the Properties window. Double-click on the empty cell next to the `MediaOpened` event. Add code to the automatically generated event handler to update the `Maximum` property:

Available for download on Wrox.com

```
private void SampleMedia_MediaOpened(object sender, RoutedEventArgs e){
    this.MediaProgress.Maximum = this.SampleMedia.NaturalDuration.TimeSpan.Ticks;
}
```

Code snippet from MainPage.xaml.cs

The last thing you need to do is to connect the ProgressBar to the MediaElement so that as the `Position` property of the `MediaElement` changes, the `Value` property of the `ProgressBar` is updated to match. With Silverlight you can do this with zero lines of code using data binding. Start by locating the `Value` property of the `ProgressBar` in the Properties window and clicking on the small icon between the word *Value* and the textbox cell. This will display a dropdown list as shown in Figure 3-12.

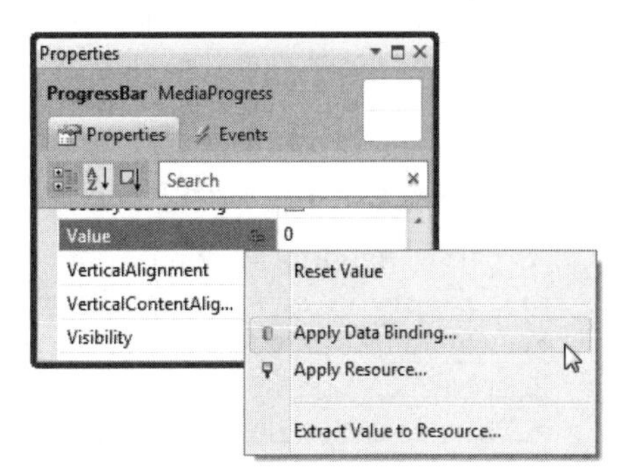

FIGURE 3-12

Selecting "Apply Data Binding" will cascade down a much more complex dropdown selector. When this dropdown initially displays, the Source tab is shown. In this case, select ElementName followed by SampleMedia (which is the name of the `MediaElement` control), then select the Path tab, as shown in Figure 3-13.

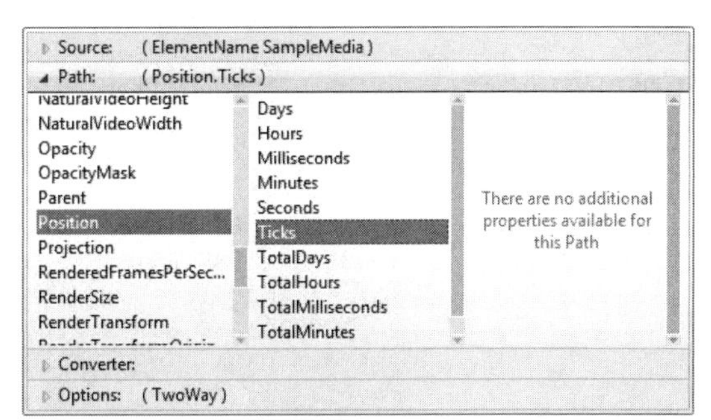

FIGURE 3-13

The Path tab shows all the properties of the `MediaElement`. In our case, you're interested in the `Position` of the video being played. When you select this property, you will see that you can then select which child property to bind to. In this case, you want to connect to the `Ticks` property, which is consistent with setting the `Maximum` value of the `PropertyBar` to being equal to the total ticks in the `NaturalDuration` of the `MediaElement` that you did in code earlier.

Let's add another Slider to our page to accept user input to control the position of the video in the MediaElement. Once you have added the Slider, select the `ValueChanged` event in the Properties window and double-click on the empty cell in order to create the event handler. All you have to do is update the `Position` property on the `MediaElement`. You also need to set the `Maximum` value for the `Slider` as you did previously for the `ProgressBar`:

Available for download on Wrox.com

```
private void MediaSlider_ValueChanged(object sender,
                                        RoutedPropertyChangedEventArgs<double> e){
    this.SampleMedia.Position = new TimeSpan((long)this.MediaSlider.Value);
}

private void SampleMedia_MediaOpened(object sender, RoutedEventArgs e){
```

```
        this.MediaProgress.Maximum = this.SampleMedia.NaturalDuration.TimeSpan.Ticks;
        this.MediaSlider.Maximum = this.SampleMedia.NaturalDuration.TimeSpan.Ticks;
    }
```

Code snippet from MainPage.xaml.cs

You can now run your application and see the ProgressBar update as the media plays. When you change the position of the Slider, you will see that both the MediaElement and ProgressBar are updated to reflect the new position in the media playback.

Ellipse, Rectangle, and Path

As part of building an application, you may need to draw shapes either as a diagram or just to add to the aesthetics of your application. From the Toolbox you have two base shapes, the Rectangle and the Ellipse, which you can use to draw some primitive shapes. You also have the ability to create Paths, which are essentially a freeform way of drawing on your application canvas. The following XAML describes an `Ellipse`, a `Rectangle`, and a `Path`:

```
<Ellipse Height="100" HorizontalAlignment="Left" Margin="200,21,0,0"
         Name="ellipse1" Stroke="#FF962E2E" StrokeThickness="6"
         VerticalAlignment="Top" Width="200" StrokeStartLineCap="Flat" />
<Rectangle Height="100" HorizontalAlignment="Left" Margin="191,214,0,0"
         Name="rectangle1" Stroke="#FF962E2E" StrokeThickness="6"
         VerticalAlignment="Top" Width="206" />
<Path Data="M58,649 C52.929039,631.43274 50.789375,628.06335 59,610
            C61.93206,603.54944 62.495438,601.50458 68,596 C73.749435,590.25055
            78.888641,589.87183 87,588 C132.55473,577.48737 119.31644,586.39557
            151,626 C163.11189,641.13989 161.29994,642 185,642 C199.38113,642
            205.88039,630.64954 216,618 C239.76944,588.28821 236.71027,578.05463
            288,582 C306.44083,583.41852 313.61087,594.12445 322,613
            C329.44449,629.75012 329.8949,635.21021 320,655 C301.07568,692.84863
            286.45102,691.37256 239,700 C203.02031,706.54175 164.19038,704.34009
            128,694 C108.45914,688.41687 94.84214,685.87372 80,674 C69.716911,665.7735
            63.758663,658.16559 57,646"
         Fill="#FFFF8080" Height="123.513" Margin="53.51,0,93.485,32.297"
         StrokeStartLineCap="Flat" Stretch="Fill" StrokeEndLineCap="Flat"
         Stroke="#FF962E2E" StrokeThickness="5" StrokeMiterLimit="10"
         StrokeLineJoin="Miter" UseLayoutRounding="False"
         VerticalAlignment="Bottom"/>
```

Code snippet from ControlsPage.xaml

This XAML creates the shapes shown in Figure 3-14. What's interesting in this case is that the shape specified by the `Data` attribute on the `Path` element is completely freeform. This doesn't have to be the case, however, and by modifying the contents of the `Data` string different arcs can be chosen to join the individual points along the path.

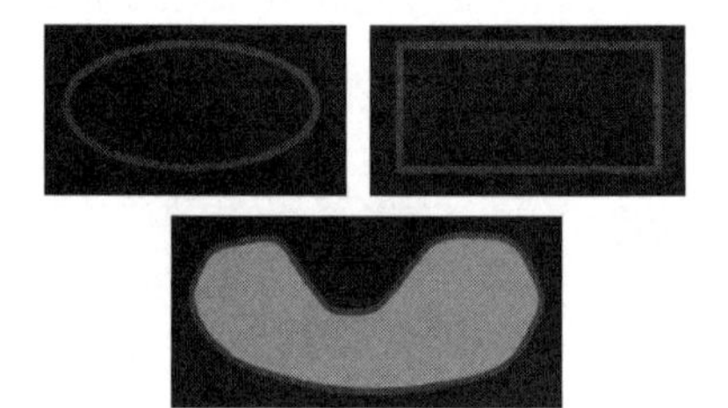

FIGURE 3-14

Unfortunately, although Visual Studio will display `Path` elements on the design surface, there is no designer support for *creating* them. To do this, you need to switch across to Blend. Within Blend there are two tools that allow you to add `Path` elements to your page. From the Toolbar, right-click the Pen icon. This will expand the dropdown tool selector shown in Figure 3-15.

FIGURE 3-15

The distinction between a Pen and a Pencil doesn't really convey the difference between these two tools. If you select the Pencil, you can literally draw freeform on the page. The corresponding XAML will be a large array of points representing the shape you have drawn. On the other hand, the Pen draws line or arc segments. In Figure 3-16 you can see two paths: the one on the left is being drawn using the Pencil, the one on the right with the Pen.

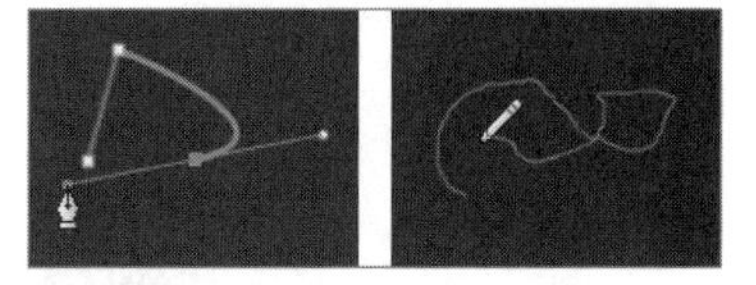

FIGURE 3-16

Taking a look at the XAML for two paths created using the Pencil and Pen tools, you'll see that there are a large number of values stored against a single `Data` property:

```
<Path Data="M292.5,557.5 C311.16061,563.49805 317.82581,568.15363 339,565"
      HorizontalAlignment="Right" Height="14.516" Margin="0,0,80,167.984"
      Stretch="Fill" Stroke="#FF962E2E" StrokeThickness="6" UseLayoutRounding="False"
      VerticalAlignment="Bottom" Width="52.5"/>
<Path Data="M226.5,556 L252,484 C252,484 396,497.5 345,527.5"
      HorizontalAlignment="Right" Height="78" Margin="0,0,66.068,181" Stretch="Fill"
      Stroke="#FF962E2E" StrokeThickness="6" UseLayoutRounding="False"
      VerticalAlignment="Bottom" Width="135.432"/>
```

The easiest way to think about a path is as a series of line segments. Even a continuous path drawn with the Pencil is made up of a series of small, curved line segments. The drawing of a path commences from the coordinates (0,0), and each set of coordinates in the `Data` property represents the next coordinate to travel to. The route taken from one coordinate to the next is determined by the type of line segment to be drawn. In the previous code snippet, the `Data` property of both `Path` elements starts with the character *M*, followed by a coordinate. The *M* stands for "Move" and will move the drawing cursor to the next coordinate set without drawing a line. Using *M* at the beginning of the `Path` essentially shifts the starting point from which the `Path` will be drawn. However, *M* can be used anywhere within a `Path` if you don't want to connect two points. Imagine that you are drawing a single line but lifting the pencil off the screen midway through drawing.

After the start point (i.e., the *M* followed by a coordinate) in the first `Path`, the next character is a *C* followed by three coordinates. This represents a "cubic Bézier curve" and requires three points to determine the tangent to the curve (the first two coordinates) and the end point of the curve. In the second `Path`, the next character is an *L* followed by a single coordinate. This represents a straight `Line` segment to the end point. This path then continues with a cubic Bézier curve. Table 3-1 summarizes the different line segments you can use and any parameters required to fully specify how the segment should be drawn.

TABLE 3-1: Path Segments

SYNTAX	DESCRIPTION
`M x,y`	**Move** — Moves the drawing cursor to the specified point.
`L x,y`	**Line** — Connects a straight Line to the specified end point.
`H x`	**Horizontal Line** — Connects a Horizontal line to the point with the corresponding X coordinate.
`V y`	**Vertical Line** — Connects a Vertical line to the point with the corresponding Y coordinate.
`C c1 c2 x,y`	**Cubic Bézier Curve** — Connects a Cubic Bézier curve with a tangent specified by the line segment from coordinate c1 to coordinate c2, to the specified end point.
`Q c1 x,y`	**Quadratic Bézier Curve** — Connects a Quadratic Bézier curve using the coordinate c1 to the specified end point.
`S c2 x,y`	**Smooth Cubic Bézier Curve** — Connects a cubic Bézier curve using the last coordinate (e.g., c1) from the previous curve and the coordinate c2, to the specified end point.
`T c2 x,y`	**Smooth Quadratic Bézier Curve** — Connects a quadratic Bézier curve using the last coordinate (e.g., c1) from the previous curve and the coordinate c2, to the specified end point.
`A sizex, sizey angle large dir x,y`	**Elliptical Arc** — Connects an elliptical Arc to the specified end point. `Size`, as an *x,y* pair, determines the *x* and *y* radius of the arc. `Angle` determines the rotation angle. `Large` determines if the arc should be the small (0) or large arc (1). `Dir` (direction) indicates whether to draw in increasing (1) or decreasing (0) angle.
`Z`	**Close** — Connects to the start point of the `Path`.

WebBrowser

In Blend if you look in the Assets window under the node Controls ➪ All, you will notice that there is a much larger list of controls than what you saw in Visual Studio. You won't go through each of these, but before you go onto the next section to look at layouts, let's take a quick look at the `WebBrowser` control. You heard earlier that there was an alternative to parsing and then rendering a blog post using the `TextBlock`. This alternative involves using the `WebBrowser` control to display the contents of the blog post. To get started, return to Visual Studio and remove the controls that you added previously to display the blog post. These were the `ScrollViewer` and the nested `StackPanel`. You also need to remove the code within the `Page_Loaded` method; otherwise, your application will no longer compile. Next, locate the `ContentGrid`, and add a `Button` and

`WebBrowser` control by dragging them from the Toolbox window. This will add them to the `ContentGrid` with some default properties. Resize, rename, and tweak the layout of these controls. The resulting XAML should look similar to the following:

Available for download on Wrox.com

```
<Button Content="Load in WebBrowser Control"
        Name="LoadInWebBrowser" VerticalAlignment="Top" />
<phone:WebBrowser Margin="0,80,0,240"
        Name="PostBrowser" Visibility="Collapsed" />
```

Code snippet from MainPage.xaml

To create an event handler for the `Button`, instead of having to open the `Properties` window and locate the `Click` event, you can use the XAML IntelliSense to help you. Place your cursor at the end of the XAML that declares the `Button`, immediately after the last set of quotes. Add a space, and then start typing the word **Click**. You will see the IntelliSense appear as in the first image of Figure 3-17. Press Tab to autocomplete the word **Click**. Autocomplete will not only complete the word *Click*, but it will also add `=""` and immediately display a new set of IntelliSense options, as shown in the second image of Figure 3-17. Use the down-arrow on your keyboard to select "<New Event Handler>" and again press Tab. This will create a new event handler in the code-behind file, as well as entering the new name into the XAML.

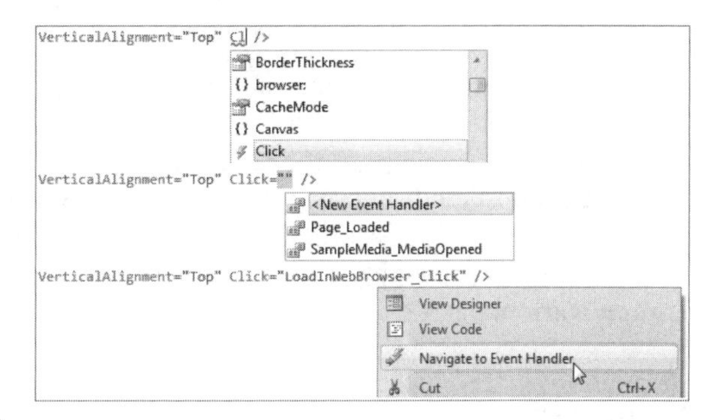

FIGURE 3-17

When you right-click on the name of the event handler in the XAML, shown in the last image of Figure 3-17, you can select "Navigate to Event Handler" to go directly to the method in the code-behind file.

There isn't a keyboard shortcut defined for the "Navigate to Event Handler," but you can easily create your own. Open the Options dialog from the Tools menu, then locate the Keyboard subnode under Environment. Enter **NavigateToEventHandler** *in the Search box, and then select EditorContextMenus.XAMLEditor.NavigatetoEventHandler from the commands list. Go to the "Press shortcut keys" textbox, and enter the key combination you want to assign, followed by clicking the Assign button. You can now use that key combination to navigate to the event handler from the XAML.*

Now, add the following code to load the posts out of the resource file, display the `WebBrowser` control, and tell it to display the content of the RSS feed (which is done by supplying the HTML to be rendered to the `NavigateToString` method):

```
private void LoadInWebBrowser_Click(object sender, RoutedEventArgs e){
    var rss = RssResources.BlogPosts;
    this.PostBrowser.Visibility = System.Windows.Visibility.Visible;
    this.PostBrowser.NavigateToString(rss);
}
```

Code snippet from MainPage.xaml.cs

When you run this and click the "Load in WebBrowser Control" button, you will see that the `WebBrowser` control is displayed and it is filled with the contents of the RSS feed, as shown in Figure 3-18.

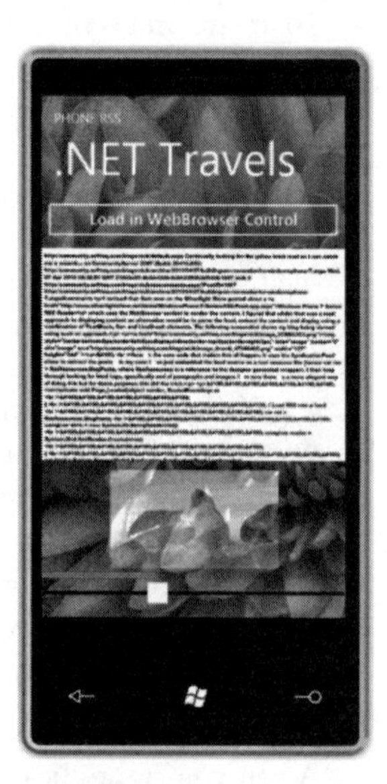

FIGURE 3-18

As this is simply displaying the entire RSS feed the user experience isn't great. Remember the Red Threads discussed previously? Displaying a single post at a time with subtle navigation features between posts will improve the experience. Since individual posts are HTML, which the WebBrowser will correctly format and lay out, this should be a reasonably straightforward feature for you to add to the application.

LAYOUT

In the previous section you saw a number of controls that are available for you to build your Windows Phone application. So far you've been simply adding these to your page and explicitly positioning and resizing them to look good. What you may not have appreciated is that you've been typically adding your controls inside of an existing instance of the `Grid` control. A grid can be used to automatically position and size controls within the available space, but before you learn about this and other layout containers that assist you in laying out your application, let's take a look at a simpler case of using the `Border` control to add a border around another control. While not technically a layout container, as it doesn't inherit from `Panel` and only supports a single child control, you will learn about it here as you typically would use it to wrap one of the other layout controls.

Switch over to working with Expression Blend as it makes designing the layout of your controls and pages much easier. To add a border to the `ContentGrid`, expand out the nodes in the Objects and Timelines window in Blend until you have the ContentGrid selected. Right-click and select Group Into ⇨ Border, and you're done! Well, you've wrapped the `ContentGrid` with a default `Border`.

Now, make sure you have the newly created Border selected and open the Properties window. Find and adjust the following properties:

Brushes ➪ BorderBrush ➪ Select Solid color brush ➪ #63AB0000

Appearance ➪ BorderThickness ➪ 10 (all sides)

Appearance ➪ CornerRadius ➪ 20

Once you have set these, you should see a thick, deep red border around all the content on the screen. The border itself is semitransparent, allowing you to still see the background image coming through, and the XAML for the border should look similar to the following. The `Grid` element has been included here to illustrate that the `Border` wraps around this element.

```
<Border Grid.Row="1" BorderBrush="#63AB0000"
        BorderThickness="10" CornerRadius="20">
    <Grid x:Name="ContentGrid">
        ...
    </Grid>
</Border>
```

Earlier in this chapter when you were displaying blog posts, you added a `ScrollViewer` followed by a `StackPanel`. The `StackPanel` is a simple layout control that stacks controls in either the horizontal or vertical direction depending on what its `Orientation` property is set to. As you add even more controls to the `StackPanel` it will eventually grow beyond the dimensions of the screen. In this case, you can place it within a `ScrollViewer` so that the user can scroll the `StackPanel` to see the rest of the content. The built-in smarts within the `ScrollViewer` update the scrollbars to indicate how large the underlying content is, and thus the current vertical and horizontal position within the content.

If you use a `StackPanel` *with the* `Orientation` *property set to* `Horizontal`, *the contents of the* `StackPanel` *will appear as a single row of controls without any wrapping. If this is the case, you can display the horizontal scrollbars by setting the* `HorizontalScrollbarVisibility` *on the* `ScrollViewer` *to either* `Auto` *or* `Visible`.

You can now move on to the more sophisticated layout panels, the `Canvas` and the `Grid`. The `Canvas` is actually a very simple layout panel to work with, particularly if you come from Windows Forms development, where you are familiar with defining the exact location and size of controls on a form. Unfortunately, the `Canvas` can also be the most limiting when it comes to taking advantage of some of the auto-layout features offered by Silverlight. To illustrate how a `Canvas` works, return to Visual Studio and create a new PhoneApplicationPage within your application by right-clicking on Solution Explorer and selecting Add ➪ New Item, then selecting the "Windows Phone Portrait Page" template. Specify a name, in this case **LayoutPage.xaml**, and click OK.

Into the `ContentGrid` of the newly created page, add a `Canvas` and remove all the default attributes added to the `Canvas` XAML element. This will set your `Canvas` to take up the entire `ContentGrid` area. Now drag any control you wish from the Toolbox onto the Canvas. You will see that the

XAML created for each control uses two special `Canvas.Top` and `Canvas.Left` attributes to specify where the controls are located on the `Canvas`. This is illustrated in the following XAML for a `Button` and `Image` that have been added to a `Canvas`:

```
<Canvas>
    <Button Canvas.Left="61" Canvas.Top="49"
            Content="Button" Height="70" Width="160" />
    <Image Canvas.Left="64" Canvas.Top="159" Height="150"
           Stretch="Fill" Width="200" />
</Canvas>
```

Code snippet from LayoutPage.xaml

`Canvas.Top` and `Canvas.Left` are known as *attached properties* because they are in effect attached to the nested controls by the parent, in this case, the `Canvas` control. You will notice that the attribute names use a similar syntax as demonstrated previously for setting complex properties via nested XML elements. At run time the `Canvas` is responsible for using these properties to determine where the nested controls will be displayed. As you can see, the `Canvas` has its strengths in being able to specify precisely where a control is located, but this is also generally a weakness as it means the layout won't automatically adapt in scenarios such as the screen orientation changing.

ABSOLUTE POSITIONING

Unlike controls used in building a Windows Forms application, WPF and Silverlight controls don't have properties that specify their location. The rationale behind this is that it's not up to the control to specify its location but rather the job of its parent container to determine where the control is displayed. In essence, attached properties are a way for the parent control to track properties that pertain to individual child controls. For example, the `Canvas` control tracks `Top` and `Left` properties for controls that are nested within it. These properties are set as attributes of the nested control using the `Canvas.Top` and `Canvas.Left` syntax, indicating that they are attached properties.

Probably one of the more sophisticated layout controls is the `Grid`. You've seen this in action a couple of times so far, but in each case, we have glossed over how the `Grid` works. As you would imagine, a `Grid` can be made up of one or more rows and one or more columns. By default, there are one of each, and if you don't specify which row or column your controls belong to, they are all assumed to belong in the first row and column and the controls appear stacked on top of each other. Create a `Grid` on the same page as the `Canvas` by firstly reducing the height of the `Canvas` to half the screen and then dragging a `Grid` across onto the `ContentGrid` from the Toolbox. Again, remove the default attributes except the `Margin`, and then resize the `Grid` so that it takes up the remaining space on the page. When the `Grid` is selected, you should see a border appear on the top and left sides of the `Grid`, as shown in Figure 3-19. These borders are used to show where the rows and columns are defined.

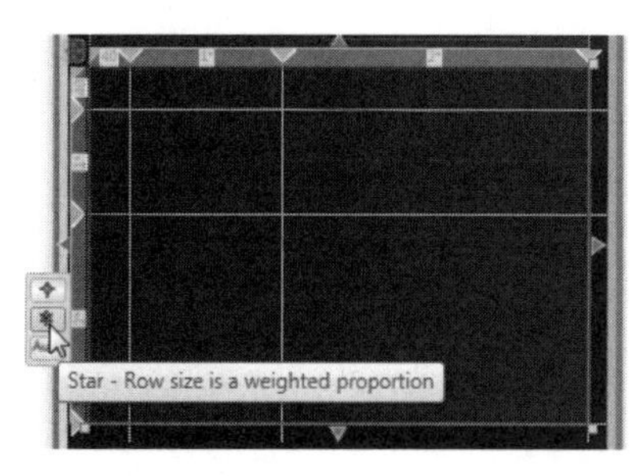

FIGURE 3-19

In Figure 3-19 there are four columns defined: one close to the left edge of the grid with an absolute width of 40, two proportional columns with a ratio of 1:2, and finally a column that is set to Auto, which currently has a width of 0 and is only identifiable by the column marker along the top edge of the grid. You can add rows or columns by simply clicking in either the left or top borders, respectively. Once it is created, if you hover your mouse over the row or column, you will see a small floating toolbar similar to that shown at the side of Figure 3-19. The three icons indicate how the row or column will be sized, fixed or absolute, proportional, or Auto. As you would imagine, a fixed row or column will remain the same size regardless of other rows or columns or its content. Auto-sized rows and columns resize in order to fit the content nested within that row or column. Since you have not added content to the grid, the Auto-sized rows and columns currently have a size of 0, rendering then invisible. Proportional rows and columns determine their size based on any remaining height or width available to the grid. For example, after the size of all the fixed and Auto columns is determined, if there is 150 remaining and there are two proportional columns with widths set to 1* and 2*, respectively, the calculated widths will be 50 and 100. Note that the use of the * after the number indicates that these widths are proportional. This `Grid` layout in XAML is the following:

Available for download on Wrox.com

```
<Grid Margin="0,300,0,0" Name="grid1">
    <Grid.RowDefinitions>
        <RowDefinition Height="40"/>
        <RowDefinition Height="*"/>
        <RowDefinition Height="2*"/>
        <RowDefinition Height="Auto"/>
    </Grid.RowDefinitions>
    <Grid.ColumnDefinitions>
        <ColumnDefinition Width="40"/>
        <ColumnDefinition Width="*"/>
        <ColumnDefinition Width="2*"/>
        <ColumnDefinition Width="Auto"/>
    </Grid.ColumnDefinitions>
    ...
</Grid>
```

Code snippet from LayoutPage.xaml

When you add controls to a `Grid`, you will again see the use of attached properties to determine which row and/or column each control is placed in. If either the row or the column is omitted, the control is assumed to be in the first row or column. Add a `Button` to the last row and column of the `Grid`. The XAML for the `Button` should be similar to the following:

```
<Button Content="Button" Grid.Column="3" Grid.Row="3" Height="70" Width="160" />
```

Note how the attached properties `Grid.Row` and `Grid.Column` are used to specify which row and column the control is in. In order to get the `Button` into this row and column, you may initially have to drop the `Button` into one of the other cells and then modify these properties by hand. You will notice that when you do add the `Button` to the last row and column, they both resize to accommodate the height and width of the button. This is because this row and column are set to Auto size. Also notice how the proportional rows and columns reduce in size but remain in the same proportion to each other.

Placing a control into a cell within a `Grid` does not mean that the control will always stretch to cover the entire cell space. There are several properties you can set on a control to determine how it makes use of the space allocated to it by its parent. As an example, the following XAML adds an `Ellipse` to the third row and column:

```
<Ellipse Grid.Column="2" Grid.Row="2"
         Margin="22,22,22,22" Name="ellipse1" Stroke="Black" StrokeThickness="1" />
```

The `Ellipse` is specified with an exterior `Margin` of `22` on all four sides. This space will be subtracted from the size of the `Grid` cell to determine the final size and placement of the `Ellipse`. An alternative would be to explicitly specify the height and width of the `Ellipse`, as in the following example:

```
<Ellipse Grid.Column="2" Grid.Row="2"
         Margin="22,22,0,0" Name="ellipse1" Stroke="Black" StrokeThickness="1"
         HorizontalAlignment="Left" Width="163"
         VerticalAlignment="Top" Height="121" />
```

There are some points to note here as to how this will be rendered. First, the right and bottom parts of the `Margin` are now `0`, which means that the `Ellipse` may consume the space between 22 on the left and top, and the right and bottom edges of the cell. You may wonder what will happen when the space allocated to the `Ellipse` is larger than that specified by its explicit `Width` and `Height` and `Margins`. The answer is to look at the `HorizontalAlignment` and `VerticalAlignment` properties, which are set to `Left` and `Top`, respectively. This means that the ellipse is anchored to always sit 22 from the left and top edges of the cell. If you were to set these properties to `Center` and the ellipse's width and height were less than the space available, you would see the ellipse float in the center between 22 and the far edge of the cell. Figure 3-20 shows the ellipse consuming all the horizontal space but floating between the margin and the edge of the cell vertically.

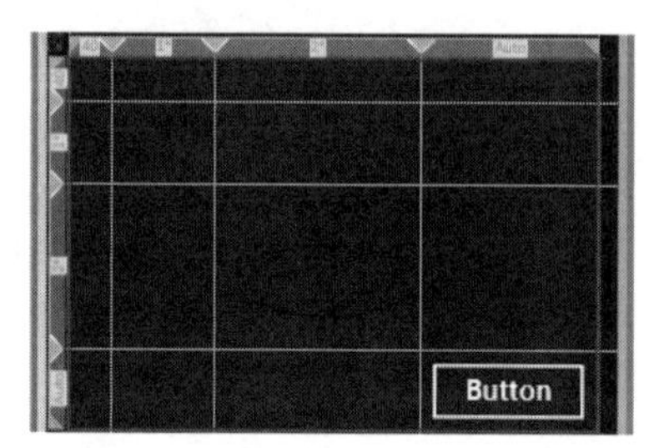

FIGURE 3-20

RESOURCES AND STYLES

Up until now you may be thinking that this whole Silverlight thing hasn't really offered much more than what you could have done with Windows Forms. In this section, you'll learn how to use resources and styles to adjust the way a control looks in ways simply impossible with Windows Forms.

When you created the new PhoneApplicationPage in the previous section, you may have noticed that the `Style` property of the `ApplicationTitle TextBlock` was set using curly braces. This essentially means that instead of using a hard-coded value, an expression is evaluated at run time and the property is set to the calculated value. In this case, the `Style` property is being set using a `StaticResource` called `PhoneTextNormalStyle`:

```
<TextBlock x:Name="ApplicationTitle" Text="MY APPLICATION"
           Style="{StaticResource PhoneTextNormalStyle}"/>
```

In WPF there are two types of resources that can be used throughout your application: `StaticResource` and `DynamicResource`. Essentially the difference is that `StaticResources` are completely evaluated when the XAML is initially loaded, whereas a `DynamicResource` is evaluated when an instance of the object is requested. In Silverlight, and hence for Windows Phone development, you only need to worry about `StaticResources`.

When the XAML is loaded, `StaticResources` are resolved by walking up the XAML tree looking for the resource in any resource dictionary found along the way. If it cannot be found anywhere within the current page, the application-wide resource dictionary, located in the App.xaml file, is consulted. In this case the `PhoneTextNormalStyle` resource is actually located in a global resource dictionary that is available to all Windows Phone applications.

A *resource dictionary* is simply a collection of named .NET objects that are readably available for reuse within an application. In the following snippet you can see two resources defined in the application-wide resource dictionary: firstly a `Color` and a `SolidColorBrush`. The `x:Key` attribute on each XAML element declares the key needed when accessing the resource from code. You'll also notice that the `SolidColorBrush` resource actually references the `Color` resource. The order of these two resources is important. If the `SolidColorBrush` were defined first, an exception would be thrown when the parser attempted to locate a resource that hasn't yet been added to the resources dictionary.

Available for download on Wrox.com

```
<Application.Resources>
    <Color x:Key="ButtonBackgroundColor">#FFFF5C5C</Color>
    <SolidColorBrush x:Key="ButtonBackgroundBrush"
                     Color="{StaticResource ButtonBackgroundColor}"/>
</Application.Resources>
```

Code snippet from App.xaml

If you look at the `TextBlocks` that are created by default when you create a new PhoneApplicationPage, you will see that they both have their `Style` property set via an attribute. The interesting thing to note here is that a `Style` resource can be used to set a number of properties on any control that it is applied to. The `Style` property, which is defined on the `FrameworkElement` class (which is one of the parent classes for Silverlight controls), is used as a way of grouping together a set of property setters. Instead of manually applying the same set of property values to multiple controls to keep a consistent look-and-feel, you can define a single `Style` object and then apply it to multiple controls by specifying their `Style` property. The following style, this time declared as a resource within a page, sets the `FontFamily` and `FontSize` properties.

```
<phone:PhoneApplicationPage.Resources>
    <ResourceDictionary>
        <Style x:Key="MyBaseStyle" TargetType="TextBlock">
            <Setter Property="FontFamily"
                    Value="{StaticResource PhoneFontFamilySemiLight}"/>
            <Setter Property="FontSize"
                    Value="{StaticResource PhoneFontSizeLarge}"/>
        </Style>
    </ResourceDictionary>
</phone:PhoneApplicationPage.Resources>
```

Looking at the XAML, you will also note a `TargetType` attribute defined that determines the type of controls the `Style` can be applied to. In this case, the `Style` is defined for use with all `TextBlock` controls. `Styles` can also inherit from other `Styles`. For example, you may have a `Style` that defines `FontFamily` and `Foreground`. Then you may inherit from that to define a `Style` where you specify the `Foreground`. The following code snippet can be added to your page before the `LayoutRoot Grid`. It adds two `Style` resources to the scope of the `PhoneApplicationPage`.

Available for download on Wrox.com

```
<phone:PhoneApplicationPage.Resources>
        <Style x:Key="MyBaseStyle" TargetType="TextBlock">
            <Setter Property="FontFamily"
                    Value="{StaticResource PhoneFontFamilySemiLight}"/>
            <Setter Property="FontSize"
                    Value="{StaticResource PhoneFontSizeLarge}"/>
        </Style>
        <Style x:Key="MyTextBlockStyle"
               TargetType="TextBlock" BasedOn="{StaticResource MyBaseStyle}">
            <Setter Property="Foreground"
                    Value="{StaticResource PhoneForegroundBrush}"/>
        </Style>
</phone:PhoneApplicationPage.Resources>
```

Code snippet from LayoutPage.xaml

Control Templates

Earlier in this chapter you will have seen how a `ContentControl`, such as a `Button`, can have an arbitrarily complex XAML tree placed within its `Content` property. This is one way that you can change the look of individual controls. However, if you have to do this on every item within a list, or every button throughout an application, it can become quite repetitive to not only write but also maintain. Luckily, using a concept similar to Styles, you can create a control template that can then be applied across multiple controls.

Although you can code Control templates in either Visual Studio or Blend, you'll find that the designer support within Blend means that you can get away with writing significantly less code. Open Blend, and select the "Load in WebBrowser Control" button that you created earlier. You're going to give this button a rounded border to match the deep red border on the main page. Right-click the button and select Edit Template ➪ Edit a Copy. This will prompt you to name the new template — in this case, call it **BrowserButton**, and select where you want the template to reside. If you want to reuse the template across the whole application, the template should be placed in the App.xaml file. In this case, you only want to use it within this page. When you press OK, Blend will take the existing Button template, duplicate it so that there is a copy you can tweak, and then open the template so that you can make changes to it. In Figure 3-21 the newly created BrowserButton template is open for editing as it is listed at the top of the Objects and Timeline window. Beneath it are the contents of the template, which contains a Grid, followed by a Border, followed by a ContentControl.

FIGURE 3-21

In this case, you're going to make some simple adjustments to the `ButtonBackground Border` control to give the button a rounded border similar to the border you added around the content of the page. In the Objects and Timeline window, select the ButtonBackground node, then locate the BorderBrush within the Brushes section of the Properties window. Click the yellow square next to the BorderBrush color, and select Reset from the dropdown, as shown in Figure 3-22. The reason that the square was yellow was to indicate that this value was data-bound, and by resetting it you are now able to set the `BorderBrush` explicitly.

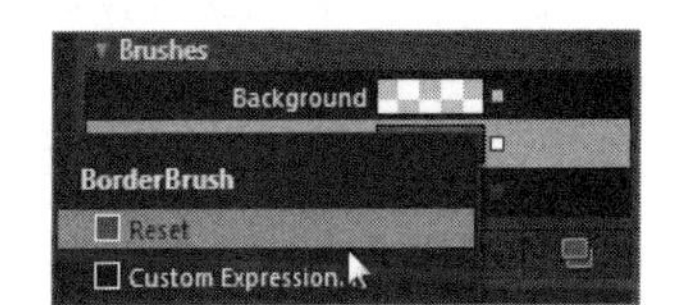

FIGURE 3-22

Change BorderBrush to be a Solid color brush, and then change the color to **#98AB0000** using the Color Editor. Next go to the Appearance section and reset the `BorderThickness` property. Set all sides of the `BorderThickness` to **10** and the `CornerRadius` to **20**. Since you are done editing the template, you can click on the LoadInWebBrowser label just underneath the MainPage.xaml tab at the top of the screen (shown in Figure 3-21) to exit Template Editing mode. If you go into the XAML view, you can see that creating this template has added a large and quite complex chunk of XAML to your MainPage.xaml. The following extract illustrates the main structure of the template:

```
<phone:PhoneApplicationPage.Resources>
    <Style x:Key="BrowserButton" TargetType="Button">
    ...
        <Setter Property="Template">
            <Setter.Value>
```

```
                <ControlTemplate TargetType="Button">
                    <Grid Background="Transparent">
                        ...
                        <Border x:Name="ButtonBackground"
                                Background="{TemplateBinding Background}"
                                CornerRadius="20"
                                Margin="{StaticResource PhoneTouchTargetOverhang}"
                                BorderBrush="#98AB0000" BorderThickness="10">
                            <ContentControl x:Name="foregroundContainer"
                 ContentTemplate="{TemplateBinding ContentTemplate}"
                 Content="{TemplateBinding Content}"
                 Foreground="{TemplateBinding Foreground}"
                 FontSize="{TemplateBinding FontSize}"
                 FontFamily="{TemplateBinding FontFamily}"
                 HorizontalAlignment="{TemplateBinding HorizontalContentAlignment}"
                Padding="{TemplateBinding Padding}" VerticalAlignment="{TemplateBinding
                 VerticalContentAlignment}"/>
                        </Border>
                    </Grid>
                </ControlTemplate>
            </Setter.Value>
        </Setter>
    </Style>
</phone:PhoneApplicationPage.Resources>
```

Code snippet from MainPage.xaml

What's interesting is that this template has been nested within a `Style` resource. To apply this template, you actually set the `Style` property, which indirectly configures the `Template` property of the `Button` control. If you look further down the MainPage.xaml file, you will see that Expression Blend has modified the `Button` XAML as follows:

```
<Button Content="Load in WebBrowser Control" x:Name="LoadInWebBrowser"
        VerticalAlignment="Top" Click="LoadInWebBrowser_Click"
        Style="{StaticResource BrowserButton}" />
```

Code snippet from MainPage.xaml

When this `Style` is applied to the `Button`, it sets several properties (which have been omitted for brevity in the above code snippet) and then sets the `Template` property. The `Template` property is set to the `ControlTemplate` to which you added a border. By changing the `Button`'s `ControlTemplate`, you have total control over the look-and-feel of the button.

If you only want to alter a single button, you can instead explicitly create a `ControlTemplate` and then set the `Template` property directly. For example, the following defines a `ControlTemplate` that simply has an ellipse in the middle of it:

```
<Buton x:Name="CustomButton">
  <Button.Template>
    <ControlTemplate x:Key="CustomTemplate" TargetType="Button">
        <Grid>
            <Ellipse Fill="#FFA3A3F9" Margin="170,8,159,8" Stroke="Black"/>
        </Grid>
    </ControlTemplate>
  </Button.Template>
</Button>
```

Via a few lines of XAML or a few mouse button clicks within Blend, Silverlight allows you to completely change the look-and-feel of any control. Try doing that within a Windows Forms project!

THEMES

In the previous section, you learned that there are some global resources that are available to all Windows Phone applications. Although it would appear that these resources are just the same as any resource you define, they're actually slightly different. These special resources obtain their values from the theme currently in use by the Windows Phone device. If you go to Settings ➪ Themes on Windows Phone, you can see that there are two Background colors — Dark and Light — and a number of Accent colors, as shown in Figure 3-23. Changing the theme will change the values assigned to the resources used by your application.

FIGURE 3-23

If you look at the controls available in the Toolbox within Visual Studio, you will notice that they adapt as the current theme alters details such as the foreground and background color. You don't end up with white text on a white background, for example. The standard controls achieve this by making use of standard resources such as `PhoneBackgroundBrush`. If you customize the style of a control, you should equally make sure that you use these resources where possible instead of hard-coding color and brush details.

One thing you should note is how your application will look when a different theme is selected. Figure 3-24 shows the application built within this chapter running with the Dark and Light themes, respectively. You'll notice that it is harder to read the text when running with the Light theme (where the text is in black) than with the Dark theme.

There are a couple of ways to address this issue. The first way is to detect which theme you are running when your application launches and then change the background of the page accordingly. For example, the following code detects if the `PhoneBackgroundBrush` is White, in which case it assumes that it is the Light theme:

Available for download on Wrox.com

```
var backgroundBrush = this.Resources["PhoneBackgroundBrush"] as SolidColorBrush;
if (backgroundBrush.Color == Colors.White){
    // Light Theme
}
else{
    // Dark Theme
}
```

Code snippet from MainPage.xaml.cs

The second way to address this issue is to modify the opacity level of the background image so that it allows some of the background color to come through. This way, when you're running the Dark theme, the black background will darken the background image, making it easier to see the white text. Alternatively, when you're running the Light theme, the light background will lighten the image, making it easier to see the black text. Figure 3-25 shows the Dark theme (left) and Light theme (right) with the opacity of the background image set to 50 percent.

FIGURE 3-24

FIGURE 3-25

SYSTEM RESOURCES

There are a number of resources that are defined for Windows Phone applications based on the current theme on the device. Figure 3-26 illustrates the list of available System Brushes, located in the Brush Resources tab of the color selector in the Properties window, and the list of available Style resources. Note that only the Style resources that are applicable to the currently selected item, in this case a TextBlock, will be visible. Both lists also include items defined in either the local or application-wide resource dictionaries.

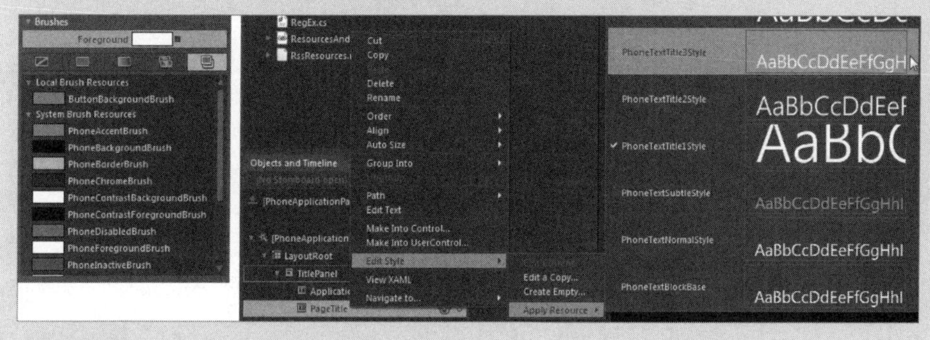

FIGURE 3-26

SUMMARY

In this chapter you've learned about Red Threads and how you can apply them when designing and building your application. You've also seen how to work with the set of controls that are available in the box with the Windows Phone development tools.

The power of the Windows Phone platform is in the richness provided through the use of Silverlight. If you want to modify or extend the default controls, you can override either the Style or Template of the control, or both. The use of Windows Phone theme resources means that you can adapt your application to fit in with the theme that the user sets for his or her phone.

Adding Motion

WHAT'S IN THIS CHAPTER

- Why motion is such an important consideration when building Windows Phone applications
- What visual states are and how to transition between them
- How animation can be used to transition between states
- Using the panoramic and pivot controls to build a consistent user experience

If you look at most computers, you'll see that they are designed for individuals sitting at desks. Mobile phones, however, are not most computers — they have become an integral part of our life in motion. Users have come to expect their mobile phones to keep up with where they are and what they are doing. They also expect an experience that matches the way they live through the creative use of motion within the applications.

Motion can be expressed through varying the visual state to reflect some form of interaction, through animation of controls or data, or through the smooth transition between pages and views. In this chapter, you will learn how to build and use visual states and animation, as well as how to build panoramic and pivot views as used by the Windows Phone platform.

VISUAL STATE MANAGEMENT

You may be wondering what state management has to do with adding motion and animation to your application. The truth is that animations are commonly used to transition between two distinct visual states of an application or control. In fact, animations are commonly described in terms of a start and an end state. Before you can construct an animation, you need to know how to define and use states.

This all sounds very theoretical, so let's talk about a practical example. Start off by creating a new project using the Windows Phone Application project template. Next, add a single button to the ContentGrid and set its x:Name to PressMeButton. Run the application and press the button. Although the button doesn't do anything, you should notice that the visual appearance — in other words, the *visual state* — of the button changes. This is shown in Figure 4-1, wherein the image on the left shows the button in the normal, that is, "unpressed," state, and the image on the right shows the button in the "pressed" state.

FIGURE 4-1

The Button control, like many controls, has several different visual states, and although some, such as the Normal and Pressed states, are mutually exclusive, others can be active at the same point in time. As an example, a Button can be in both the Pressed and Focused states concurrently.

The different states that a control can be in are defined programmatically within the class definition. As an example, the following is part of the Button class definition:

```
[TemplateVisualState(Name="Unfocused", GroupName="FocusStates"),
 TemplateVisualState(Name="Focused", GroupName="FocusStates"),
 TemplateVisualState(Name="Pressed", GroupName="CommonStates"),
 TemplateVisualState(Name="Disabled", GroupName="CommonStates"),
 TemplateVisualState(Name="Normal", GroupName="CommonStates"),
 TemplateVisualState(Name="MouseOver", GroupName="CommonStates")]
public class Button : ButtonBase
```

You will notice that each Button state is declared via a TemplateVisualState attribute. Individual states can also be grouped via use of the GroupName property. By placing the Normal and Pressed states in the same group, Silverlight knows that the application can only be in one of these states at a time. Similarly, Focused and Unfocused belong in a different group because a button with focus by definition cannot be unfocused.

BUTTON TEMPLATE VISUAL STATES

The list of available template states for a control such as the Button can be discovered several ways. One way is to use a tool such as RedGate's .NET Reflector (www.red-gate.com/products/reflector) to disassemble the class. To inspect the class declaration for the Button class, you would open the System.Windows .dll in .NET Reflector and select the Button class within the System.Windows .Controls namespace. Pressing the spacebar opens the Disassembler window, showing the class declaration seen earlier that includes the TemplateVisualState attributes.

The `TemplateVisualState` attribute simply documents the name of each available control state. It does not offer any definition of what visual change will occur when the state becomes active. This information is, instead, defined within the underlying control template for the `Button`. In other words when a user taps a button, the logic implementing the `Button` control will set the current state to Pressed. It's then up to the template to define how this change in state is conveyed visually.

Let's explore this a little further by making a copy of the default Button template and extending its visual style. Open your current solution in Expression Blend, and within the Objects and Timeline window, right-click on the PressMeButton you created earlier. Select Edit Template ➪ Edit a Copy, and specify the name *ExtendedButton* and where the copied template should be defined. In Figure 4-2 you can see that you're going to define the template at the Application level, as this means it can be easily reused across multiple pages within the application.

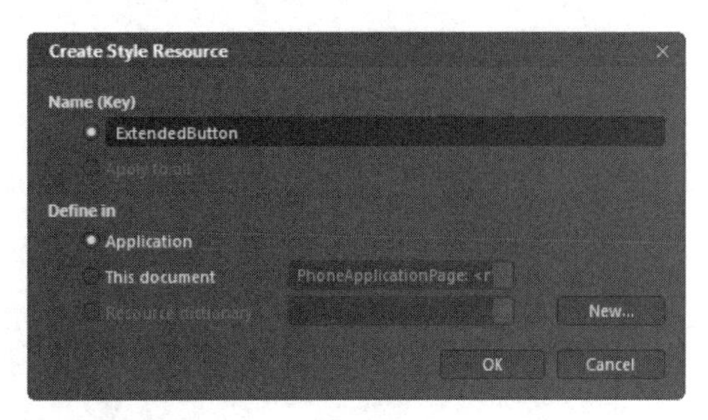

FIGURE 4-2

APPLICATION TEMPLATES

In Chapter 3, the BrowserButton template was defined within the page's XAML file. This restricted the definition to being usable only within that page. In order to enable a template to be reused across multiple pages, it can be defined at the Application level. The template, in this case, the ExtendedButton template, is a `Style` element within the Application.Resources section of the App.xaml file:

```
<Application.Resources>
    <Style x:Key="ExtendedButton" TargetType="Button">
    ...
    </Style>
</Application.Resources>
```

This template can then be used the same way as you would use any other static resource.

```
<Button . . . Style="{StaticResource ExtendedButton}" />
```

Once you click OK, the default Button template will be copied into the App.xaml file and immediately opened in Expression Blend so that you can make changes. Figure 4-3 shows the template editing experience with both the Objects and Timeline window and States window showing.

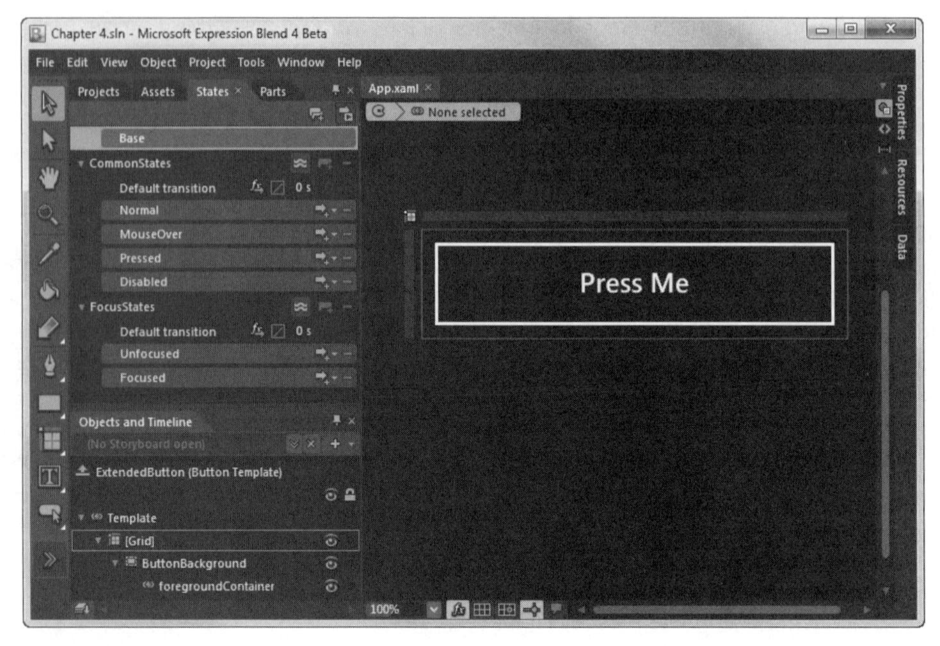

FIGURE 4-3

Within the States window, you can see the various states that are defined within our new ExtendedButton template. Perhaps the most important state is the Base state, which corresponds to the initial or unmodified state. Unless told to go to a particular state, the `Button` will load and remain in this state.

Earlier you saw how two distinct groups of states were defined using the `TemplateVisualState` attribute on the `Button` class. This grouping is also evident here with the various other states listed underneath the CommonStates and FocusStates parent nodes. Within these groups are the individual states. Recall that within each group only one state can be active at a time.

You're going to extend this template to include a new group of states that indicate if the button has ever been pressed. There are going to be two states — HasNotBeenPressed and HasBeenPressed — and you'll define them in a mutually exclusive group called *PressedStatus*. Start by clicking the "Add state group" button at the top of the States window, shown in Figure 4-4.

FIGURE 4-4

When you click this button, a new state group will be created at the bottom of the States window and will be selected so that you can immediately amend the name to **PressedStatus**. Next, click the "Add state" button located on the right side of the PressedStatus group node, shown in Figure 4-5.

FIGURE 4-5

The new state will appear under the PressedStatus node, and again will be in focus so that you can change the name to **HasNotBeenPressed**. You will also notice that there is a red dot appearing next to the state name. This indicates that you have entered State Recording mode. Across, in the main design area, you can confirm this from a red border being present, along with another red dot and a message that says "HasNotBeenPressed state recording is on," as shown in Figure 4-6. Repeat these steps to add a second state, and name it **HasBeenPressed**.

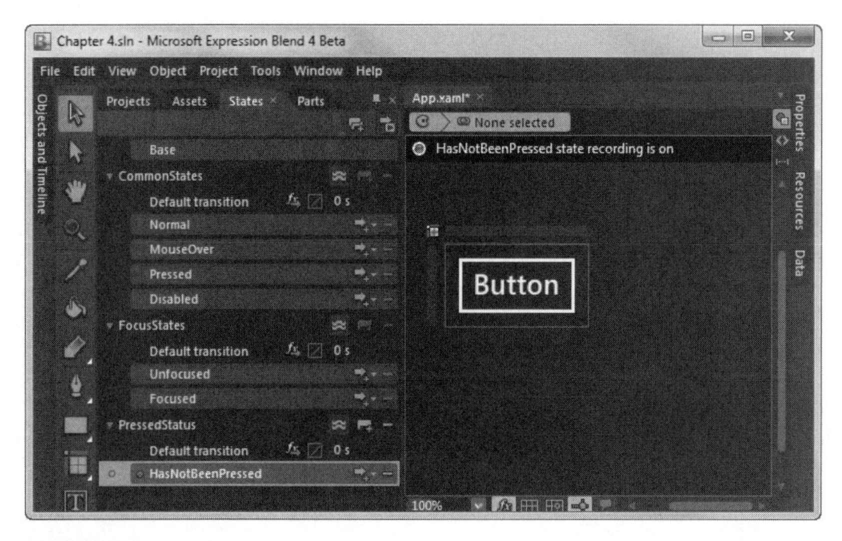

FIGURE 4-6

You should now see a message above the main design window that reads "HasBeenPressed state recording is on." So what does this actually mean?

One of the neat features of Expression Blend is the ability to design not just a static view of a page or control, but also all the slight variants required for the various states that it may have. When you enter state recording mode, Blend enables change tracking on the designer. Any changes you make are recorded into the specified state as a differential from the Base state.

You may think that all these indicators about state recording are overkill. Be warned: Accidentally leaving State Recording mode on will bite you at some point. Say you meant to alter the Base state, for example, to change the font, so that it is applied across all control states. Instead, you accidentally modify the font while in State Recording mode. Then, when you run your application, you get frustrated because you can't work out why the font hasn't changed. The font has changed but only in the state that you were recording when you made the change. There is no golden bullet to solve this problem — you just have to be aware of the red border when making changes — should it really be there?

To indicate that the button has been pressed, you're going to add a blue dot to the top-right corner of the button. The dot will only be visible when the button is in the HasBeenPressed state. However, instead of dynamically creating the dot when the button is pressed, you'll add the dot in the Base state for the button. Initially the dot will be hidden, and then when the button is clicked, the dot will be made visible.

To implement this, make sure that you have the Base state selected in the States window and that there is no red border around the design area. From the Objects and Timeline window, select the [Grid] node. Open the Assets window and navigate to Shapes, then double-click on Ellipse to create an instance of the `Ellipse` control within the `Grid`. In the Properties window, set both the Fill and Stroke to a Solid color brush with a color of **#FF0000FF** (Blue). Resize the Ellipse to a width and height of **10**. Set the HorizontalAlignment to Right and the VerticalAlignment to Top, and the Margin to 20 from the Top and Right sides. Finally, set the Visibility to Collapsed so that by default the dot is not visible.

From the States window, select the HasBeenPressed state. You will notice that the red dot and red border appear again to indicate that changes between the current and Base states are being recorded. At this stage, you're only going to change the `Visibility` property of the blue dot so whenever the HasBeenPressed state is active it will appear. To do this, select the Ellipse node from the Objects and Timeline window. Then, from the Appearance section of the Properties window, change the `Visibility` property to Visible. When you do this, you should see the `Visibility` property appear under the Ellipse node in the Objects and Timeline window to indicate that this property has been altered by this state. Figure 4-7 illustrates how the HasBeenPressed state should appear after making this change.

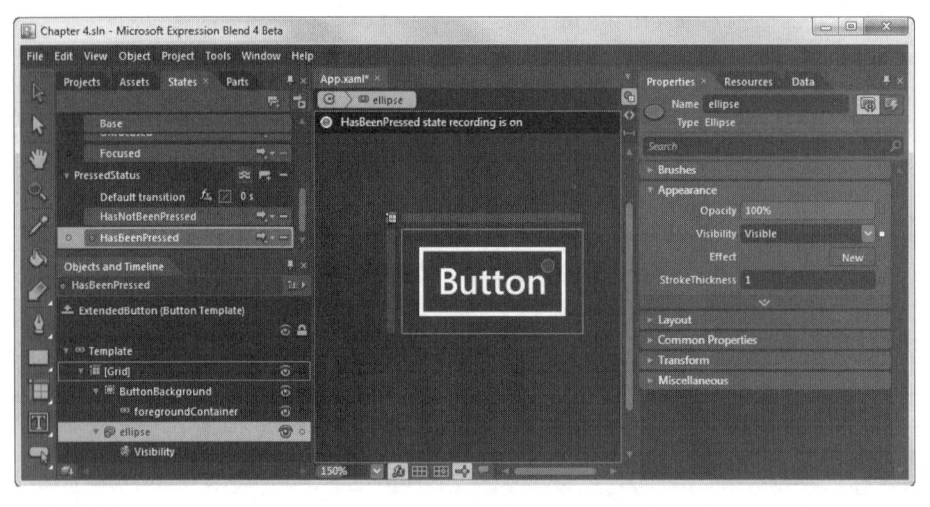

FIGURE 4-7

With the HasBeenPressed state defined, you can verify its behavior by clicking between the Base and HasBeenPressed states within the States window. You will see that the dot appears and disappears appropriately.

Let's take a quick look at the XAML that Blend generated for this new state. If you change to XAML view and locate the ExtendedButton template, there are a couple of points worth taking a look at. First, if you take a look at the `ControlTemplate`, you will see that the `Ellipse` you created is contained within the `Grid` as you would expect. The contents of the `ControlTemplate` element are effectively the Base state of the control:

```
<ControlTemplate TargetType="Button">
    ...
    <Grid Background="Transparent">
        <Border x:Name="ButtonBackground" ... >
            <ContentControl x:Name="foregroundContainer" ... />
        </Border>
        <Ellipse x:Name="ellipse" Fill="Blue" HorizontalAlignment="Right"
                Height="10" Width="10"
                Margin="0,20,20,0" Stroke="Blue" VerticalAlignment="Top"
                Visibility="Collapsed"/>
    </Grid>
</ControlTemplate>
```

You would have also noticed that there is a large chunk of XAML contained within a `VisualStateManager.VisualStateGroups` element. The `VisualStateManager` class is responsible for managing the states and transitions between them for controls within your application. It exposes a static property, `VisualStateGroups`, into which each of your state groups is added. States are subsequently added to those state groups. In the following snippet, you can see both the HasNotBeenPressed and HasBeenPressed states. The HasNotBeenPressed state doesn't have any children, which is to be expected since you didn't modify this state, making it identical to the Base state. On the other hand, the HasBeenPressed state contains a storyboard, and nested several levels further in is a `Visibility` element with a value of `Visible`. Essentially this storyboard defines that when the `Button` gets assigned to this state, the `Visibility` property of the `Ellipse` should be set to `Visible`. You'll learn more about storyboards and keyframes in conjunction with animation later in this chapter.

```
<VisualStateManager.VisualStateGroups>
    ...
    <VisualStateGroup x:Name="PressedStatus">
        <VisualState x:Name="HasNotBeenPressed"/>
        <VisualState x:Name="HasBeenPressed">
            <Storyboard>
                <ObjectAnimationUsingKeyFrames
                  Storyboard.TargetProperty="(UIElement.Visibility)"
                  Storyboard.TargetName="ellipse">
                    <DiscreteObjectKeyFrame KeyTime="0">
                        <DiscreteObjectKeyFrame.Value>
                            <Visibility>Visible</Visibility>
                        </DiscreteObjectKeyFrame.Value>
                    </DiscreteObjectKeyFrame>
                </ObjectAnimationUsingKeyFrames>
            </Storyboard>
        </VisualState>
    </VisualStateGroup>
</VisualStateManager.VisualStateGroups>
```

If you run the application at this stage you will notice that the blue ellipse is never shown. Although the HasBeenPressed state works at design time, you haven't specified when or how the `Button` should enter this state. To do this, open MainPage.xaml from the Projects window. Select the "Press Me" button, and from the Events tab of the Properties window, double-click on the empty cell beside the `Click` event. This will create an event handler for the `Button`'s `Click` event, which can then be specified as follows:

Available for download on Wrox.com

```
private void PressMeButton_Click(object sender, System.Windows.RoutedEventArgs e){
    VisualStateManager.GoToState(this.PressMeButton,"HasBeenPressed",true);
}
```

Code snippet from MainPage.xaml.cs

This code snippet uses the static `GoToState` method of the `VisualStateManager` to assign the `PressMeButton` the HasBeenPressed state whenever it is clicked. The last parameter indicates whether transitions should be used or not. You'll learn about this in the context of animations later in the chapter, so for the time being, set this parameter to `true` to enable transitions.

At this point you may be wondering why you went to the length of creating a HasNotBeenPressed state since it is no different from the Base state. Well, the dirty little secret of the `VisualStateManager` is that there is no way to reset or return a control to the Base state. If you think about it, this would actually be undesirable in most cases. For example, if you consider a `Button`'s Pressed and Focused states, you may remember that these belong to different state groups, which means that they can be applied independently. Thus it is possible for a button to be Pressed and Focused at the same time. This is typical as someone places his finger on the button, for example. However, when the user lifts his finger off the button, the control becomes unpressed (normal) yet still has focus. If lifting your finger off the button returned it to the Base state, it would no longer visually be focused. The recommended solution to this problem is for each VisualStateGroup to define a default state that is empty. In other words, like HasNotBeenPressed, it doesn't change any properties from the Base state.

In large applications with many visual states, it is worth settling on a naming convention for the default state. Since the names of your visual states have to be unique, it is worth combining the name of the template with the state group name. For example, Default_ExtendedButton_PressedStatus *immediately identifies this state as the default state for the PressedStatus group of the ExtendedButton template.*

When you want to reset out of a particular state without affecting other states, you tell the `VisualStateManager` to go to a groups default state. To demonstrate this, open the MainPage.xaml from the Projects window and drag another button onto the page. Modify this button to read "Reset Pressed Status," and from the Events tab, wire up a `Click` event handler with the following code:

Available for download on Wrox.com

```
private void ResetPressedStatus_Click(object sender, RoutedEventArgs e){
    VisualStateManager.GoToState(this.PressMeButton,"HasNotBeenPressed",true);
}
```

Code snippet from MainPage.xaml.cs

If you run up your application, you will see that when you press the "Press Me" button a blue dot appears. Regardless of whether you release your finger from the button, the blue dot will remain, indicating that it is still in the HasBeenPressed state. When you click the "Reset Pressed Status" button, the "Press Me" button is assigned to the HasNotBeenPressed state, which removes the blue dot by virtue of also reverting changes made by the HasBeenPressed state, which belongs to the same state group. The interesting thing here is that the `VisualStateManager` is intelligent enough to work out what properties need to be changed and what values they should have in order to revert changes made by the HasBeenPressed state despite the HasNotBeenPressed state being empty.

You've seen how you can use states to manage the appearance of a control. This technique can also be used to manage the layout of an entire page within your application. You essentially follow the same process of creating a VisualStateGroup, adding states, and then adding transitions to go between states. This will be covered in the next chapter, when you look at the different ways of handling device orientation.

BEHAVIORS

In the previous section, you saw how you could visually define additional visual states with the designer tools offered by Blend. Unfortunately, when you wanted the `Button` to enter one of these states, you needed to drop into code. It seems unnatural that you can do so much in XAML and yet the final step has to be done programmatically. The other, and probably more significant, issue is that this code had to be duplicated for each `Button` that uses the ExtendedButton template. Ideally you would want to simply apply the ExtendedButton template to a button and have the blue dot appear automatically as needed.

The good news is that a Silverlight feature called *Behaviors* enables you to do just that. Before you learn how to create your own custom Behaviors, you will learn see how a behavior called `GoToStateAction` can replace the manually written C# code from our previous application. First, remove the existing `Click` event handlers for the "Press Me" button from MainPage. (You need to remove the method both from the .cs file as well as from the `Click` event in the Properties window.)

The previous step can be done in either Visual Studio or Expression Blend. However, you will want to be working in Blend to edit the ExtendedButton template in App.xaml.

EDITING RESOURCES

When you open the App.xaml file, you may see a message saying "App.xaml cannot be edited in design view." In actual fact, this isn't the truth — it's just that it can't work out what you want to edit because App.xaml isn't a page of your application. Instead, the App.xaml file contains resources, so if you open the

continues

(continued)

Resources window, you will see a list of all the resources contained within App .xaml. Click on the icon next to the resource you want to edit, in this case, the `ExtendedButton`, to open it in the main editor, as shown in Figure 4-8.

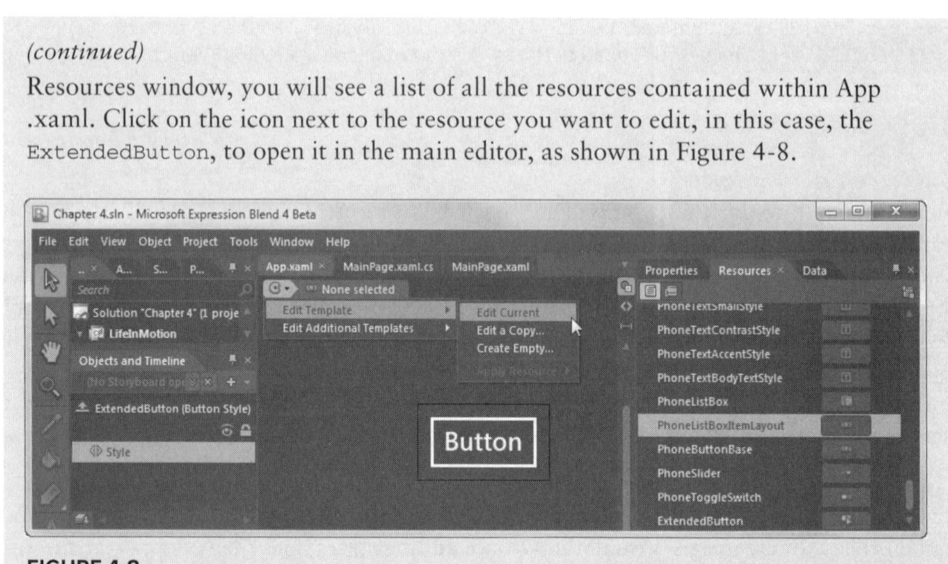

FIGURE 4-8

In this case, what is displayed is actually the Style itself. If you recall, the template actually gets nested within a `Style` resource. So, to edit the template, you have to select Edit Template ➪ Edit Current from the top of the screen.

With the `ExtendedButton` template open for editing, from the Behaviors node in the Assets window drag the `GoToStateAction` over to the [Grid] node in the Objects and Timeline window. Then, within the Properties window, make sure that MouseLeftButtonDown is selected in the EventName dropdown, and select the HasBeenPressed state for the StateName. Whereas the `Button` exposes a `Click` event, the `Grid` does not, exposing only the raw mouse events such as left button down and up. Your control should look similar to that shown in Figure 4-9.

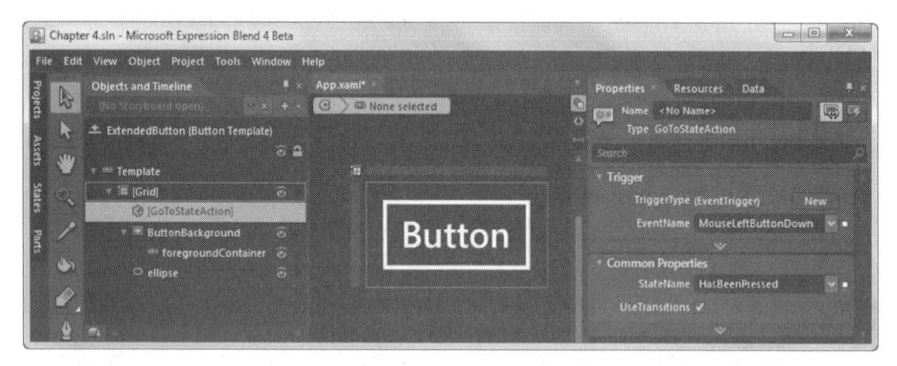

FIGURE 4-9

Rerun the application, and you should see that it behaves identically to the previous version. The `GoToStateAction` behavior listens for the event designated by the `EventName` property. When this event is raised, it makes a call to the `VisualStateManager` to go to the state defined by the `StateName` property.

You just saw how you can use a behavior as part of a control template to trigger a change in state based on an event. In this case, the `MouseLeftButtonDown` event triggered a change in the HasBeenPressed state. You can also attach behaviors directly to control events within your application. If you recall, you added a `Click` event handler to the "Reset Pressed Status" button to return the state of the "Press Me" button to the HasNotBeenPressed state. This can be replaced with a `GoToStateAction` behavior.

Open up MainPage from the Projects window, and remove the `Click` event handler. Then, drag the `GoToStateAction` from the Assets window onto the "Reset Pressed Status" button. This time, in addition to setting the `EventName` to `Click` and the StateName to HasNotBeenPressed, you also need to change the `TargetName` property to `PressMeButton`, as shown in Figure 4-10. By default, the `GoToStateAction` behavior targets, or changes, the state of the element it is attached to. In this case, you want the `Click` event on the "Reset Pressed Status" button to change the state of the button named *PressMeButton*. Changing the value of the `TargetName` property specifies this intent.

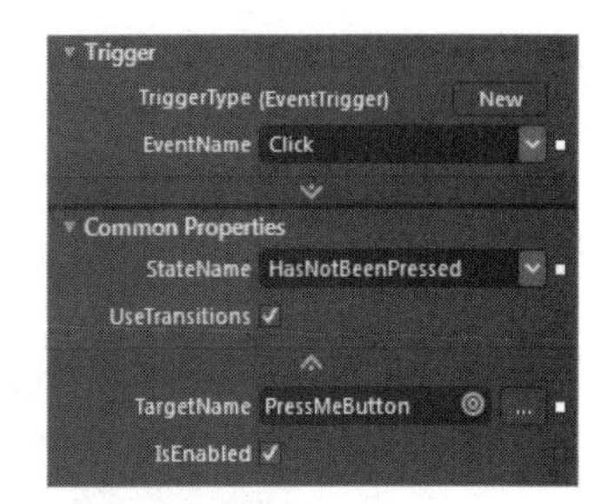

FIGURE 4-10

Behaviors enable you to add functionality to your Silverlight page or control without having to write code. The idea is that a developer can encapsulate a small amount of common behavior-based logic (such as changing the visual state in response to an event being raised) and package it in such a way that a designer can use Blend to configure and apply it to one or more XAML elements. As you've seen, this is done by dragging the behavior from the Assets window onto an element in the Objects and Timeline window. By setting various properties, the designer can define the behavior of the application without calling on a developer to write additional code.

From the developer perspective, there are three flavors of behaviors, as shown in Table 4-1.

TABLE 4-1: Behaviors

BASE CLASS	USAGE
`Behavior<T>`	This is the simplest form of behavior in that it only exposes `OnAttached` and `OnDetaching` methods that you can override. You'd typically use these to wire up event handlers to the `AssociatedObject` (the XAML element that the behavior is attached to).
`TriggerAction<T>`	One of the most common forms of behavior is the invocation of an action in response to an event. For example, you might want to navigate to a page when the user clicks a button. A `TriggerAction` allows the designer to specify which event on the associated control the behavior should observe. It then calls the overrideable `Invoke` method whenever the event triggers.
`TargettedTriggerAction<T>`	The last form of behavior is an extension of the `TriggerAction` that allows the designer to specify the target element. Within the `Invoke` method you can reference the `Target` element, which may not be the same element that the behavior is attached to.

To get a feel for how to write a behavior, you're going to add some additional visual effects to the buttons. The first effect will be to adjust the opacity of the background, toggling between a solid background and 50% opacity each time the button is pressed. Without behaviors this is something you would have done in code. In order to avoid duplication of code, you would have had to encapsulate this functionality in a Base class. If you then happen to want to take the same functionality and apply it to another type of control, you couldn't easily do so. Furthermore, you would have had to worry about wiring up event handlers for the `Button`'s `Click` event, and any code you wrote would have become mixed up with any application logic in the code-behind file of the page.

When you write behaviors, you'll want to be in Visual Studio because, although Blend provides great designer support for adding behaviors to an application's user experience, actually creating a behavior is a code-focused task better suited to Visual Studio. You can easily switch over using the "Edit in Visual Studio" menu item when you right-click on the Solution node in the Projects window in Blend.

Open your project in Visual Studio, then right-click on your project in Solution Explorer, and select Add ➪ New Item. Select the Class item template, name the file **ToggleOpacityAction.cs**, and click OK. This will create a new class in your project to contain your behavior's implementation. To make it into a behavior, you need to inherit from one of the three base classes. In this case, you'll go with `Behavior<T>`, which is the simplest of the three. This base class exposes two methods, `OnAttached` and `OnDetaching`, that you'll want to override in order to add functionality to your behavior, as per the following snippet:

Available for download on Wrox.com

```
public class ToggleOpacityAction : Behavior<Button>{
    protected override void OnAttached() {
        base.OnAttached();
    }
    protected override void OnDetaching() {
        base.OnDetaching();
    }
}
```

Code snippet from ToggleOpacityAction.cs

The other thing to note in this code snippet is that you will derive this behavior from `Behavior<Button>`; in other words, a behavior that can only be applied to `Button` controls. If you are building a behavior that has more generic applicability, you may want to use `UIElement`, `FrameworkElement`, or `Control` as your generic argument.

In the previous snippet the two method overrides didn't do anything other than call the base methods. To change the opacity, you're going to need to access the button that this behavior has been attached to. Luckily, all three behavior base classes expose an `AssociatedObject` property.

Through this property you can interact with the XAML element to which the designer associates the behavior, in this case, toggle the opacity level. There are a couple of things to note here. First, you want to toggle the opacity each time the button is clicked. This means that you'll need to add an event handler for the `Click` event, which, in turn, leads to the next issue, which is where to subscribe to this event. If you attempt to do this in the constructor of the behavior, you'll find that the `AssociatedObject` hasn't been set yet. Instead, you can use the `OnAttached` method, which gets invoked when the behavior is being attached to an element. With this in mind, add an event handler for the `Click` event. Don't forget to remove the event handler as part of the `OnDetaching` method, which will be called if the behavior is ever removed from the element.

```
protected override void OnAttached() {
    base.OnAttached();
    this.AssociatedObject.Click += AssociatedObject_Click;
}

protected override void OnDetaching() {
    base.OnDetaching();
    this.AssociatedObject.Click -= AssociatedObject_Click;
}

void AssociatedObject_Click(object sender, RoutedEventArgs e)
{ ... }
```

Code snippet from ToggleOpacityAction.cs

You're most of the way there now as the `AssociatedObject_Click` method will get invoked each time the button is clicked. What remains is to toggle the opacity. The following snippet shows an updated `AssociatedObject_Click` event handler, which toggles the `Opacity` of the button:

```
void AssociatedObject_Click(object sender, RoutedEventArgs e){
    if (this.AssociatedObject.Opacity == 1.0)
        this.AssociatedObject.Opacity = 0.5;
    else
        this.AssociatedObject.Opacity = 1.0;
}
```

Code snippet from ToggleOpacityAction.cs

When you return to Blend you should receive a prompt saying that the project has been modified and needs to be reloaded. Accept the new project modifications and force a rebuild by selecting the Project ➪ Rebuild Project menu item. If you now open the Assets window and expand out the Behavior tab, you will see your newly created `ToggleOpacityAction` behavior. Drag an instance of this behavior across to the PressMeButton node on the Objects and Timeline window. Repeat this last step for the ResetPressedStatus button.

When you run your application and then click on the "Press Me" button, you will see that the opacity of the button changes, as per Figure 4-11 (You may want to adjust the background color of the page or the button to clearly see this effect).

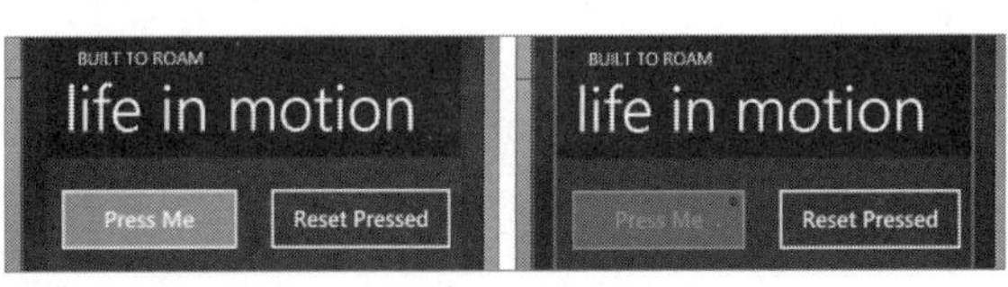

FIGURE 4-11

It is so common for behaviors to trap an event on the associated element and perform an action whenever the event triggers that Silverlight provides its own base class to help out with the implementation This is the purpose of the `TriggerAction<T>` base class. You can rewrite the `ToggleOpacityAction` to derive from `TriggerAction<T>` and in the process eliminate the lines of code that wire up the Click event handler. To do this, return to Visual Studio and open the ToggleDropShadowAction.cs file. Change the base class from `Behavior<Button>` to `TriggerAction<Button>` and remove the method overrides for `OnAttached` and `OnDetaching`. Instead, you'll override the `Invoke` method as in the following sample. (Note that this is all the code you now need in the `ToggleOpacityAction` behavior.)

Available for download on Wrox.com

```
public class ToggleOpacityAction : TriggerAction<Button>{
    protected override void Invoke(object parameter){
        if (this.AssociatedObject.Opacity == 1.0)
            this.AssociatedObject.Opacity = 0.5;
        else
            this.AssociatedObject.Opacity = 1.0;
    }
}
```

Code snippet from ToggleOpacityAction.cs

Rebuild and run your application to see this refactored behavior in action. In some cases, this can cause build or runtime errors; if you see this issue, then remove the instances of the `ToggleOpacityAction` from your page XAML and add them again from the Assets window in Blend. You will notice that if you select one of the instances of the `ToggleOpacityAction` in the Objects and Timeline window, there is an `EventName` property in the Properties window, which defaults to the `Click` event.

As another example of a custom behavior, let's modify the application so that after clicking the "Press Me" button the user can't immediately reset the state by pressing the "Reset Pressed Status" button. Instead, the reset button will be immediately disabled and only become re-enabled after 10 seconds has passed. Since this behavior may be useful in various situations and applications, it is a good candidate to be implemented as a behavior.

Since you want to capture an event from one control and then perform an action on another, the most suitable base class to use for this particular behavior is `TargetedTriggerAction<T>`. Return to Visual Studio and add a new class to your project called `DisableButtonAction`. Alter this class to inherit from `TargetedTriggerAction<Button>`. You'll also need to create a dependency property that Blend can display within the Properties window to allow the designer to specify the duration in milliseconds that you want the target control to be disabled for:

Available for download on Wrox.com

```
public class DisableButtonAction : TargetedTriggerAction<Button>{
    public int DisabledTimeout{
        get { return (int)GetValue(DisabledTimeoutProperty); }
        set { SetValue(DisabledTimeoutProperty, value); }
    }

    public static readonly DependencyProperty DisabledTimeoutProperty =
```

```
                  DependencyProperty.Register("DisabledTimeout", typeof(int),
                                                 typeof(DisableButtonAction),
                                                 new PropertyMetadata(0));
}
```

Code snippet from DisableButtonAction.cs

DEPENDENCY PROPERTIES

In Silverlight a dependency property extends the basic functionality of a .NET CLR property and is one of the foundations that underpins the data-binding system that you'll learn about in Chapter 15. Essentially, a dependency property can be set through styling, data binding, animation, and inheritance, and works well with visual editors like Visual Studio and Blend. Dependency properties can only be created on an object that inherits from `DependencyObject`, which exposes methods such as `GetValue` and `SetValue` that permit the storing of keyed values in a property store.

To create a dependency property, you start with a normal CLR property, but instead of using a backing field, you call `GetValue` and `SetValue` in the getter and setter, respectively. This is illustrated with the `DisabledTimeout` property. The calls to `GetValue` and `SetValue` reference a static read-only `DependencyProperty` field. The creation of the `DependencyProperty` registers the dependency property, along with information such as the type of the property and a default value.

Within the CLR property for a dependency property, you should never invoke any code other than `GetValue` and `SetValue`. The reason for this is that the Silverlight Framework can bypass the CLR property and access the underlying property store directly. If you place additional code in the CLR property, it will only get invoked when you call the CLR property from your code. A better alternative is to change the `DependencyProperty` instance to include an event callback for when the property changes. The following code illustrates this with a static `DisabledTimeoutChanged` method that gets invoked whenever the `DisabledTimeout` property on any `DisableButtonAction` instance changes:

```
public static readonly DependencyProperty DisabledTimeoutProperty =
         DependencyProperty.Register("DisabledTimeout", typeof(int),
                                     typeof(DisableButtonAction),
                                     new PropertyMetadata(0,
                                     DisabledTimeoutChanged));

private static void DisabledTimeoutChanged
                                   (DependencyObject d,
```

continues

(continued)

```
                                    DependencyPropertyChangedEventArgs
                                    e){
    var buttonAction = d as DisableButtonAction;
    var str = "Property " + e.Property + " changed from " +
              e.OldValue.ToString() + " to " + e.NewValue.ToString();
}
```

The first step in the callback method is to access the instance of the `DisableButtonAction` by casting the `DependencyObject` argument. The `DependencyPropertyChangedEventArgs` contains information about which property changed and what the old and new values are.

As you have inherited from `TargetedTriggerAction`, the `DisabledButtonAction` class already has a property for capturing which element the action should be applied on. This can be visually configured via the Properties window in Blend. From within the behavior you can access this element via the `Target` property:

```
private Timer timer;
public DisableButtonAction(){
    this.timer = new Timer(DisabledTimeoutComplete);
}

protected override void Invoke(object parameter){
    this.Target.IsEnabled = false;
    this.timer.Change(DisabledTimeout, DisabledTimeout);
}

private void DisabledTimeoutComplete(object state){
    // Disable the timer
    this.timer.Change(Timeout.Infinite, Timeout.Infinite);
    this.Dispatcher.BeginInvoke( ()=>this.Target.IsEnabled = true);
}
```

Code snippet from DisableButtonAction.cs

In this code snippet you can see that the `Invoke` method has been overridden in order to set the `IsEnabled` property on the `Target` element to `false`. The `Invoke` method is also responsible for starting the timer to make certain that the button is only disabled for the correct amount of time. Since the timer will trigger periodically, based on the `DisabledTimeout` value, it is important that the first step the `DisabledTimeoutComplete` method performs is to disable the timer by setting the time-out value to `Infinite`. It then returns the `IsEnabled` property of the `Target` element back to `true` to allow the user to interact with the button again.

BEWARE THE BACKGROUND THREAD

In this example, a timer was used to restore the `IsEnabled` state to `true` after a period of time. When the timer is triggered, it will invoke the callback method on a background thread. If within that callback method you wish to modify a property of a visual object (e.g., a button), you need to ensure that the code is being executed on the user interface (UI) thread. This is done by calling `BeginInvoke` on the `Dispatcher` object and passing in either a method to invoke or, as in this case, an anonymous method.

Another alternative here would have been to use the `DispatcherTimer`, which executes on the UI thread. This would eliminate the need to jump back to the UI thread. The important point to note here is that occasionally you will want to invoke actions on a background thread, such as loading data, so it is important to understand that you need to return to the UI thread in order to update any control properties.

With the class developed, save all changes and return to Blend. Rebuild your project and then locate the `DisableButtonAction` from under the Behavior node on the Assets window. Drag this behavior onto the PressMeButton and then go to the Properties window to configure the behavior. You will notice that Figure 4-12, which lists the properties for the `DisableButtonAction` behavior, is very similar to Figure 4-10, which lists the properties for the `GoToStateAction` behavior. This is because both behaviors inherit most of their properties from the `TargetedTriggerAction<T>` base class.

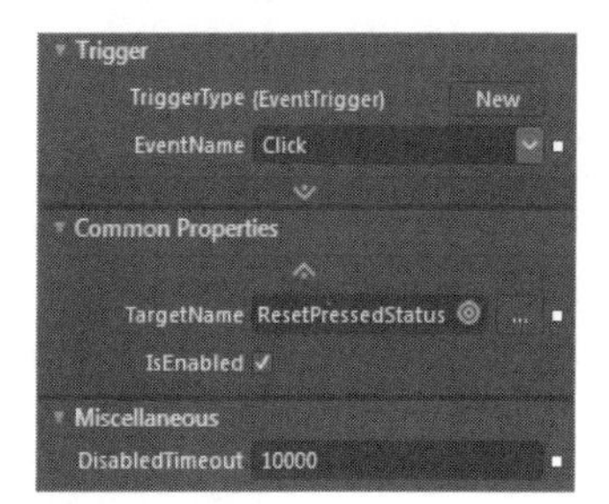

FIGURE 4-12

To disable the `ResetPressedStatus` button for 10 seconds, you need to set the `DisabledTimout` to 10,000, since this property represents the time-out in milliseconds. Now when you build and run your project, you should notice that the "Reset Pressed Status" button is disabled for 10 seconds after each click of the "Press Me" button. By encapsulating the logic into a custom behavior, you can implement the required logic once, yet easily reuse it within multiple situations. The ability to visually drag-and-drop the behavior onto a control and configure required properties within Blend means that this is something a designer will also feel comfortable doing.

ANIMATION

Adding states to your controls and pages is a great way to indicate to the user that information has changed or to provide feedback when they interact with the application. However, by themselves, changes in state can often appear jarring or jerky in nature. This can be solved by using animation to provide smooth transitions between distinct states. Animations can be used both within control templates and within your page to provide a rich, yet smooth, experience for the user.

Template Transitions

To start with, you're going to add an animation to a button template in order to provide more visual feedback when the button has been pressed. By default, the Pressed state of a button inverts the foreground and background colors to indicate that the button has been pressed. What you're going to do is to animate the button so that it expands and collapses in size over the space of a second when the Button first enters the Pressed state. Begin by adding a new button called **BounceButton** to the MainPage using the Blend design area. As you have done previously, assign the button a custom control template by right-clicking on the button and selecting Edit Template ➪ Edit a Copy. Give the new template a name, **BounceButtonTemplate**, and save it to the current document.

You can think of animations as being the transition between two distinct visual states. Within the States window, you will see that at the top of each state group there is a line entitled "Default transition." This property determines the default function and timing that are used when calculating the transitions between states within the state group. As in the left image of Figure 4-13, the standard values are 0 seconds with no easing function, which results in each state transition happening immediately.

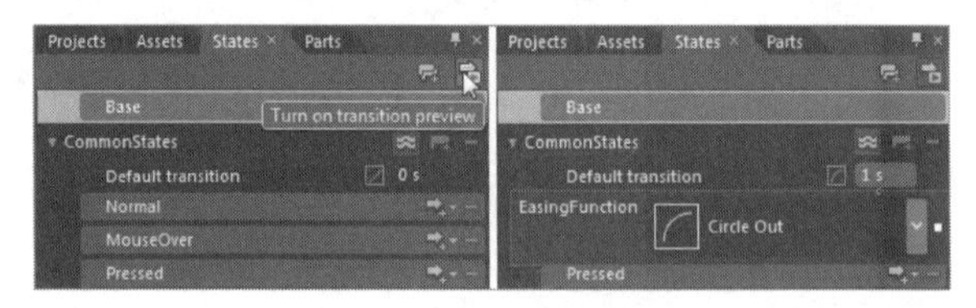

FIGURE 4-13

In the right image of Figure 4-13 the transition time has been increased to 1 second ("1 s") and the "Circle Out" easing function has been selected. Without an easing function defined, increasing the transition time to 1 second will result in a linear transition over that period. For example, if a change between two states required that the button be translated by 100 pixels over 1 second, a linear transition would result in the button moving 10 pixels every tenth of a second. By applying a "Circle Out" easing function, the transition starts off faster and then tapers off as the target value is reached toward the end of the transition. There are many other predefined easing functions, shown in Figure 4-14, that can be selected for the transition.

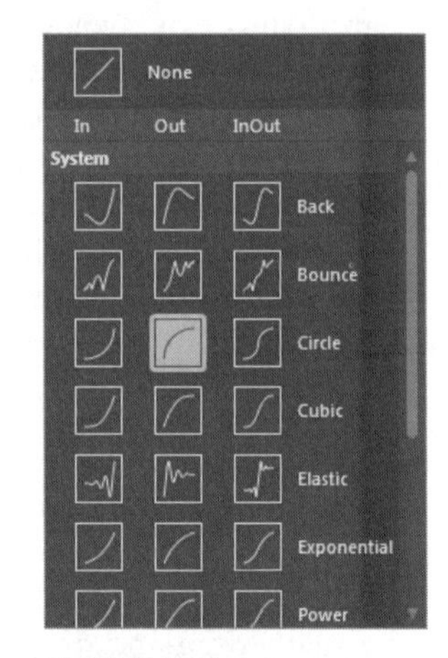

FIGURE 4-14

By itself, changing the default transition values won't give you the result you are after. In actual fact, since currently the only difference between the Normal and Pressed states is that the colors are inverted, the only change you will notice is that it takes a second for this color change to occur. In order to change the size of the button when it is in the Pressed state, enter State Recording mode by selecting the state within the States window. With the Grid node selected in the Objects and Timeline window, locate the Scale tab of the RenderTransform section of the Properties window (see Figure 4-15). Modify the X and Y scale properties from 1 to **1.15**. This will give the appearance that the button has grown by 15 percent.

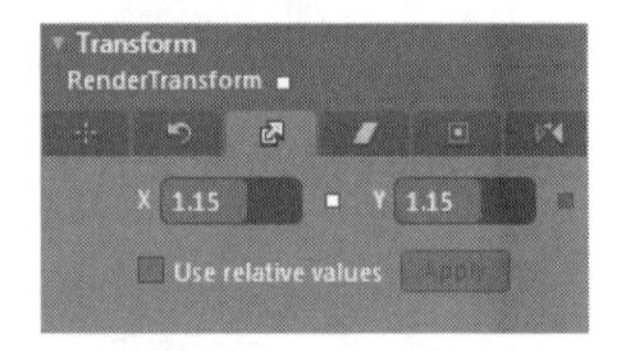

FIGURE 4-15

If you then select the Normal state within the States window, you should see that the button decreases in size over a period of 1 second. Similarly, if you again select the Pressed state, the button will grow over the same period of time.

Although being able to see the transitions between states within Blend can be useful, as the length and complexity of the transitions increase you may find that this feature can get in your way. Luckily, you can easily toggle transition previewing on and off by selecting the "Turn off transition preview" button at the top-right corner of the States window (shown in Figure 4-13).

If you go back to what you initially set out to do, you will notice that you haven't quite got the animation right yet. You've caused the button to smoothly grow in size when moving to the Pressed state. However, the actual requirements were for the button to expand and collapse back to its original size over the 1-second period. As this is more complex than a simple transition that moves a start value along a path to an end value, you will need to override the default transition. Before you do this, go back and set the Default transition to 0 seconds and undo the RenderTransform changes you made to the Pressed state. The easiest way to undo these changes is to select the Pressed state, locate the RenderTransform section on the Properties window, and reset the values by selecting the white square dot next to the section title and choosing "Reset."

To add a more specialized state transition, press the "Add transition" button on the same line as the Pressed state. This will display a dropdown displaying all possible transitions, as shown by Figure 4-16. Remember that you can only transition between states within the same state group, which is why there are no transitions listed to or from states such as Focused or Unfocused that belong in the FocusStates group.

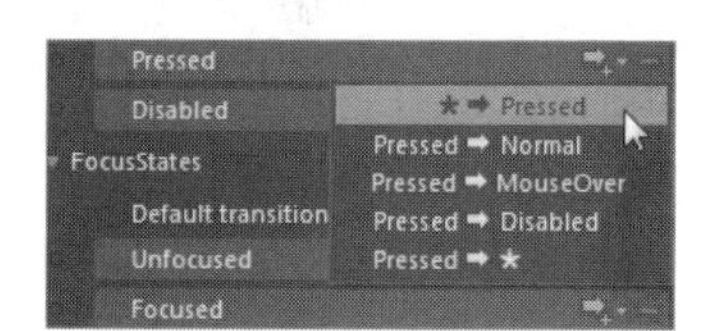

FIGURE 4-16

The first transition in the dropdown represents the transition from any state into the Pressed state. Actually, it represents the transition from all states where there isn't a custom transition defined. For example, if you were to define a special transition for Normal → Pressed, this would take precedence over any transition effects defined within the * → Pressed transition. Similarly, the last transition represents the transition to all other states where there isn't a custom transition defined. In this case, you want to define a transition that will be applied whenever the Pressed state is entered, regardless of what state the button was in previously. This matches the first transition, * → Pressed. When you select this transition, you will see that a new line is added to the States window and the time line region in the Objects and Timeline window is displayed. In Figure 4-17 you can see a yellow vertical line at the 0 marker on the time line. This represents the initial state of the transition, which by default will correlate to the current state that the button is in. At this moment the duration of the transition is 0 second ("0 s"), as indicated in the States window, and the transition doesn't modify any properties of the button. Of course, when the transition is complete, the button will take on the Pressed state, and any properties that don't match (e.g., the foreground and background colors) will be adjusted to match the new state.

FIGURE 4-17

Within Blend a transition is made up of a number of keyframes, each one representing property values at a given point in time within the transition. To make your button wobble in size, select the ButtonBackground node in the Objects and Timeline window, followed by clicking the "Record Keyframe" button (the oval icon toward the top of the Objects and Timeline window). This will create a new keyframe at the 0 marker. The new keyframe will represent the starting point of the transition, and thus you don't have to modify any values. Create another keyframe by first moving the yellow time marker to the 0:00.250 time marker and then clicking the "Record Keyframe" button.

You may find it difficult to select this point in time as the default snapping options define a resolution of 10 per second. This means that the time marker will snap to 1/10th of a second. You can either click the "Turn off Timeline snapping" button, manually enter the time by clicking the time itself and entering the value, or alter the Snapping resolution by clicking the "Snapping options" button and entering the resolution "20 per second."

With the yellow time marker at 0:00.250, open the Properties window and change the ScaleX and ScaleY properties to **1.15**, as you did previously with the Default transition. Your transition should look similar to Figure 4-18.

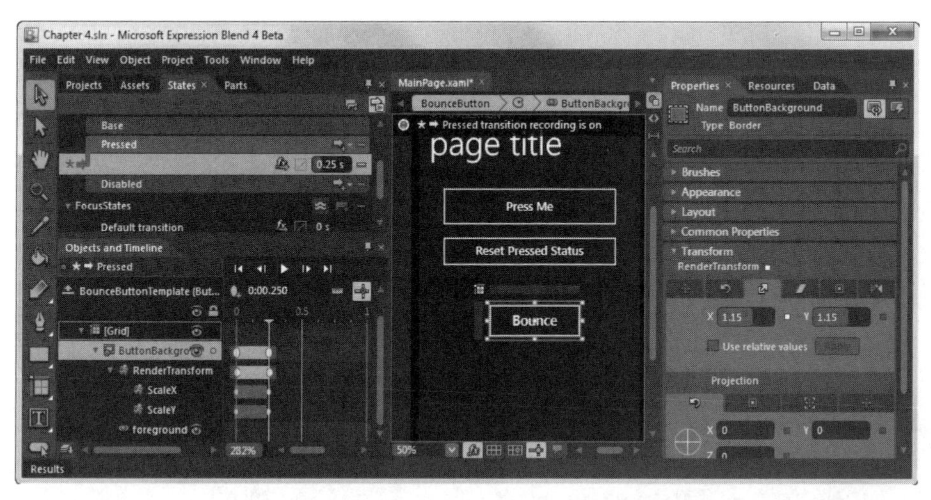

FIGURE 4-18

Go ahead and create two additional keyframes, the first at 0:00.750 and the second at 0:01.000. You don't need to modify any of the property values at 0:00.750 as you want the button to remain in the expanded state for half a second, but then at 0:01.000 you want to reset both ScaleX and ScaleY to 1 to return the button to its normal size. The final time line should be similar to that shown in Figure 4-19, showing four keyframes that change the `ScaleX` and `ScaleY RenderTransform` properties of the `ButtonBackground` over the duration of 1 second.

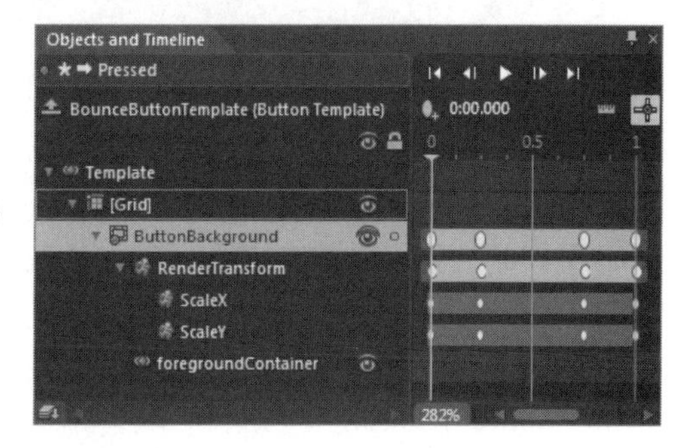

FIGURE 4-19

With this transition now complete, you can either preview the transition by changing between states in the States window as you have done in the past, or by pressing the Play button above the Timeline. This second technique is useful while tweaking the effect to your liking.

With the updated state transition, when you rerun the application and hold the button for more than a second, you will see the button enlarge, remain large for half a second, and then return to its original size. You'll notice that it's not until the button returns to its original size that the button colors invert as defined by the Pressed state, which is only entered once the transition completes.

State Transitions

So far you've been working with states in the context of a control template. You can also apply the concept of states to your whole page, making it very useful to manipulate multiple controls at the same time based on user interaction. In this section, you're going to add three buttons to the page called *B1* (Ball 1), *B2* (Ball 2), and *B3* (Ball 3). When the user clicks the Bounce button that you were working with earlier, these additional buttons will bounce off the bottom of the screen and back to a new location. To do this, you'll first need to add the three buttons to the page and name them appropriately. You'll then need to create a new state group within the MainPage, called *BounceStates*, and three states called *Bounce1*, *Bounce2*, and *Bounce3*, as shown in Figure 4-20.

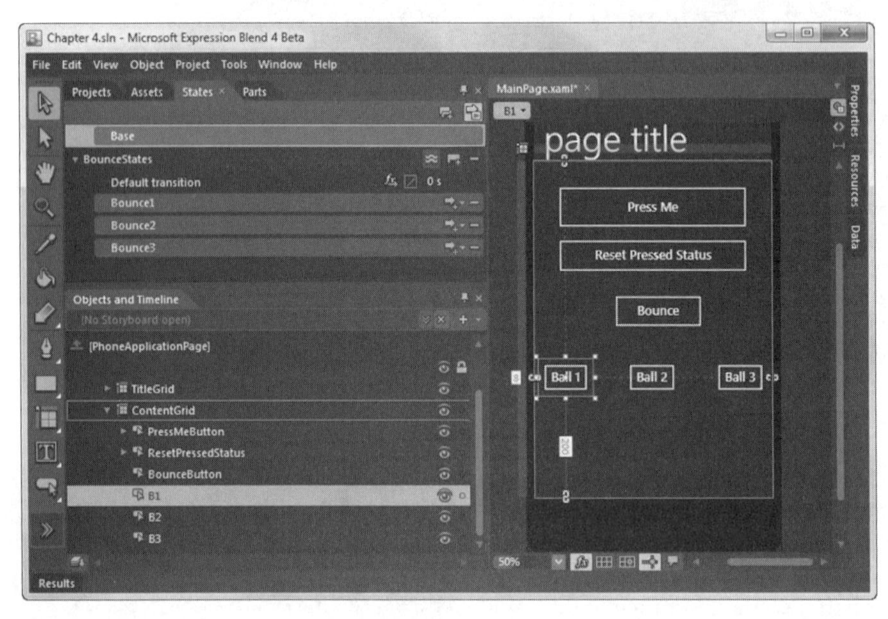

FIGURE 4-20

With the Base state selected, make sure the three buttons are in the correct position. Bounce1 will be the default state for this state group and thus will correspond to the Base state. Next, you need to modify the Bounce2 and Bounce3 states to reorder the buttons. For Bounce2, move B1 to where B2 was, B2 to where B3 was, and B3 to where B1 was. Similarly, for state Bounce3, move B1 to B3's original position, B2 to B1's original position and B3 to B2's original position. This should give you states similar to the following XAML:

```
<VisualStateManager.VisualStateGroups>
    <VisualStateGroup x:Name="BounceStates">
        <VisualState x:Name="Bounce1"/>
        <VisualState x:Name="Bounce2">
            <Storyboard>
                <DoubleAnimation Duration="0" To="170"
    Storyboard.TargetProperty="(UIElement.RenderTransform).
(CompositeTransform.TranslateX)"
    Storyboard.TargetName="B1" d:IsOptimized="True"/>
                <DoubleAnimation Duration="0" To="-340"
    Storyboard.TargetProperty="(UIElement.RenderTransform).
(CompositeTransform.TranslateX)"
    Storyboard.TargetName="B3" d:IsOptimized="True"/>
                <DoubleAnimation Duration="0" To="170"
    Storyboard.TargetProperty="(UIElement.RenderTransform).
(CompositeTransform.TranslateX)"
    Storyboard.TargetName="B2" d:IsOptimized="True"/>
            </Storyboard>
```

```
        </VisualState>
        <VisualState x:Name="Bounce3">
            <Storyboard>
                <DoubleAnimation Duration="0" To="-170"
    Storyboard.TargetProperty="(UIElement.RenderTransform).
(CompositeTransform.TranslateX)"
    Storyboard.TargetName="B2" d:IsOptimized="True"/>
                <DoubleAnimation Duration="0" To="-170"
    Storyboard.TargetProperty="(UIElement.RenderTransform).
(CompositeTransform.TranslateX)"
    Storyboard.TargetName="B3" d:IsOptimized="True"/>
                <DoubleAnimation Duration="0" To="340"
    Storyboard.TargetProperty="(UIElement.RenderTransform).
(CompositeTransform.TranslateX)"
    Storyboard.TargetName="B1" d:IsOptimized="True"/>
            </Storyboard>
        </VisualState>
    </VisualStateGroup>
</VisualStateManager.VisualStateGroups>
```

Code snippet from MainPage.xaml

When the user clicks the Bounce button, you want the Ball buttons to move in a cycle between the three states. However, at the moment, the transitions would result in the buttons moving horizontally into their new locations. To make them bounce off the bottom of the screen, you need to create a custom transition for moving between each of the states. Start with Bounce1, and select Add transition ⇨ Bounce1→ Bounce2 to specify how the page transitions from state Bounce1 into state Bounce2.

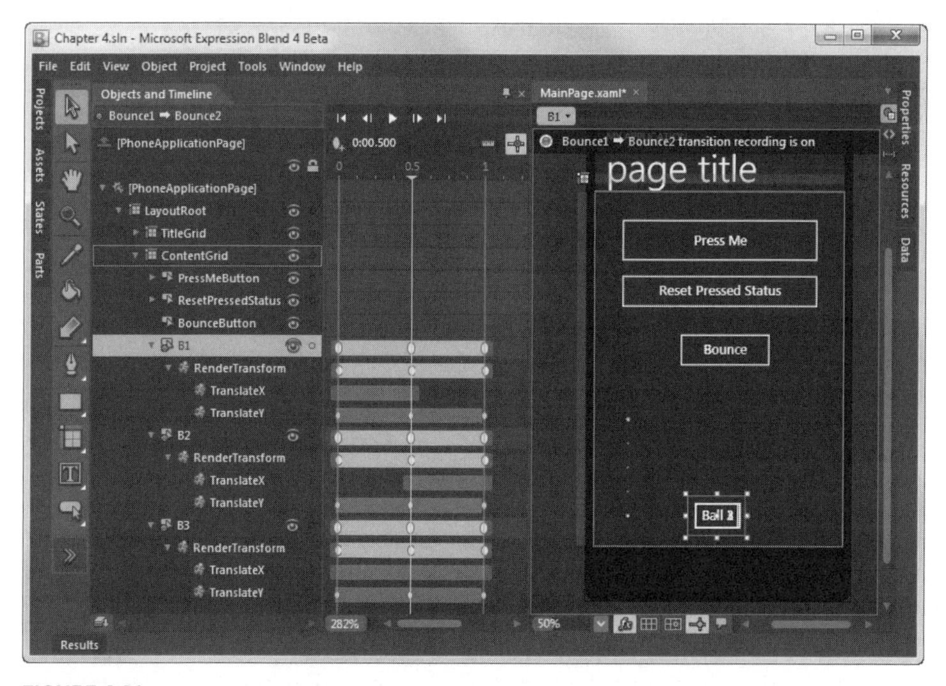

FIGURE 4-21

Within this transition add three keyframes for each of the three buttons. The initial keyframe will determine the starting position for the buttons; at the middle keyframe (i.e., 0:00.500, as shown in Figure 4-21), all three buttons should be moved toward the bottom of the screen and overlap one another; the last keyframe should have the buttons in their new locations. If you take a look at the XAML, you will see that it looks similar to the following:

Available for download on Wrox.com

```
<VisualStateGroup.Transitions>
    <VisualTransition From="Bounce1" GeneratedDuration="0" To="Bounce2">
        <Storyboard>
            <DoubleAnimation Duration="0:0:0.5" To="170"
    Storyboard.TargetProperty="(UIElement.RenderTransform).
(CompositeTransform.TranslateX)" Storyboard.TargetName="B1"/>
            <DoubleAnimation Duration="0:0:1" To="-340"
    Storyboard.TargetProperty="(UIElement.RenderTransform).
(CompositeTransform.TranslateX)" Storyboard.TargetName="B3"/>
            <DoubleAnimation Duration="0:0:0.5" To="170"
    Storyboard.TargetProperty="(UIElement.RenderTransform).
(CompositeTransform.TranslateX)" Storyboard.TargetName="B2" BeginTime="0:0:0.5"/>
            <DoubleAnimationUsingKeyFrames
    Storyboard.TargetProperty="(UIElement.RenderTransform).
(CompositeTransform.TranslateY)" Storyboard.TargetName="B3">
                <EasingDoubleKeyFrame KeyTime="0" Value="0"/>
                <EasingDoubleKeyFrame KeyTime="0:0:0.5" Value="180"/>
                <EasingDoubleKeyFrame KeyTime="0:0:1" Value="0"/>
            </DoubleAnimationUsingKeyFrames>
            <DoubleAnimationUsingKeyFrames
    Storyboard.TargetProperty="(UIElement.RenderTransform).
(CompositeTransform.TranslateY)" Storyboard.TargetName="B2">
                <EasingDoubleKeyFrame KeyTime="0" Value="0"/>
                <EasingDoubleKeyFrame KeyTime="0:0:0.5" Value="180"/>
                <EasingDoubleKeyFrame KeyTime="0:0:1" Value="0"/>
            </DoubleAnimationUsingKeyFrames>
            <DoubleAnimationUsingKeyFrames
    Storyboard.TargetProperty="(UIElement.RenderTransform).
(CompositeTransform.TranslateY)" Storyboard.TargetName="B1">
                <EasingDoubleKeyFrame KeyTime="0" Value="0"/>
                <EasingDoubleKeyFrame KeyTime="0:0:0.5" Value="180"/>
                <EasingDoubleKeyFrame KeyTime="0:0:1" Value="0"/>
            </DoubleAnimationUsingKeyFrames>
        </Storyboard>
    </VisualTransition>
</VisualStateGroup.Transitions>
```

Code snippet from MainPage.xaml

Notice here that since the X value for the buttons only transitions in one direction between two values, a `DoubleAnimation` has been used. However the Y value increases, to push the buttons to the bottom of the screen, then decreases, to pull the buttons back up to the original line. This requires the use of `DoubleAnimationUsingKeyFrames`. To complete this scenario, repeat the process to add transitions for Bounce2 → Bounce3 and Bounce3 → Bounce1.

The last thing you need to do is to initiate the next state transition in the pattern whenever the Bounce button is clicked. Select the Bounce button, open the Event tab of the Properties windows, and double-click the empty cell next to the `Click` event. The following code will move the buttons progressively between all three states:

```
int state=0;
private void BounceButton_Click(object sender, System.Windows.RoutedEventArgs e){
    switch(state){
        case 0:
            VisualStateManager.GoToState(this,"Bounce2",true);
            break;
        case 1:
            VisualStateManager.GoToState(this,"Bounce3",true);
            break;
        case 2:
            VisualStateManager.GoToState(this,"Bounce1",true);
            break;
    }
    state=(state+1)%3;
}
```

Code snippet from MainPage.xaml.cs

You can now run your application and click the Bounce button to see the other buttons bounce between the three states. Figure 4-22 shows a series of images taken of the buttons in motion between two of the states.

FIGURE 4-22

Most of the time when you are building an animation it is as a transition between two states. However, it is also possible to create stand-alone animations that aren't tied into the visual state management model. These can be invoked as a result of a user action or as a result of a background event to provide feedback to the user. Stand-alone animations are often referred to as *storyboards*, and within Blend they are created and edited within the Objects and Timeline window. To animate the Bounce button via a storyboard, you must first create a new one by clicking the plus button next to the field that currently reads "(No Storyboard open)" at the top of the Objects and Timeline window. This will prompt you for a name for the new storyboard, before opening it. You will notice that like when you are in State Recording mode, there is a red border around the main editor area. However, this time the text reads "Timeline recording is on." To exit Storyboard editing, click the cross next to the storyboard. In Figure 4-23, the storyboard is called `RotateBounceButton`.

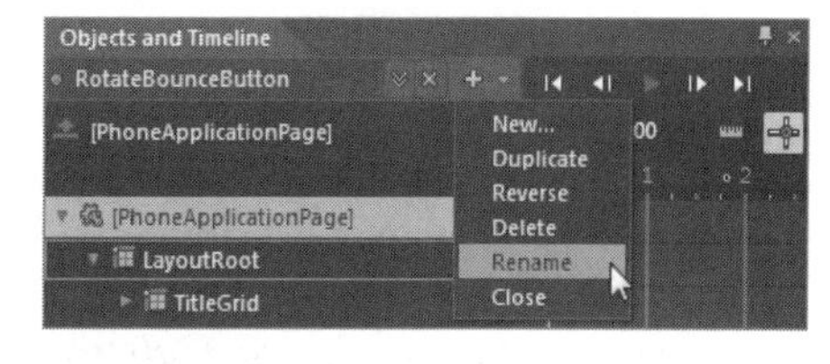

FIGURE 4-23

Figure 4-23 also shows the dropdown that you can use to duplicate, reverse, delete, or rename a storyboard. Quite a common task is to create a second storyboard that reverses the effects of a previously created storyboard. To do this, select the storyboard you want to reverse, duplicate the storyboard, and with the duplicate selected, select the reverse option.

With the `RotateBoundButton` storyboard selected, find and select the `BounceButton` control in the Objects and Timeline window. Create three new keyframes at 0:00.000, 0:01.000, and 0:02.000. At the 0:00.000 keyframe, make sure the Transform ➪ RenderTransform ➪ Angle property is set to 0, and the Brushes ➪ Background ➪ Solid Color Brush ➪ Color is set to #00FF0000. Then at 0:01.000, set the Color to #FFFF0000. Lastly, at 0:02.000, set the Color back to #00FF0000 and the Angle property to 360.

When you click the Play button in the Objects and Timeline window, you should see the Bounce button rotate through a full circle while changing color. The last thing to do is to wire up this storyboard to play when the user clicks the Bounce button. Here you can use a different predefined behavior called `ControlStoryboardAction`. Make sure you're not in Timeline Recording mode by clicking the "Close storyboard" button. Then, drag the `ControlStoryboardAction` behavior from the Assets window onto the BounceButton node in the Objects and Timeline window.

The new behavior will automatically be wired up to the `Click` event of the `BounceButton`. All you need to set is the `Storyboard` and the `ControlStoryboardOption` properties. Figure 4-24 shows these properties set to `RotateBounceButton`, which is the storyboard you just created, and `Play`, which is the default option and causes the storyboard to begin playing.

FIGURE 4-24

PANORAMIC AND PIVOT CONTROLS

One of the most revolutionary designs incorporated into the Windows Phone user interface is the concept of a *hub*. The continuous Panoramic view allows the user to flick between sets of related data, making it quick to locate what they are searching for. It also extends the effective size of the screen, allowing the presentation of more information. The other control that is unique to Windows Phone is the Pivot, which is typically used to provide alternative views across the same set of data. Each view has a title that appears at the top of the control, similar to the traditional tab control. The Pivot responds to the user tapping the title of the view they want to navigate to but also allows the user to pan/flick to the left or right to see the adjacent views. The Windows Phone developer tools ships with both Panorama and Pivot controls.

The Panorama control is just like any other control with the exception that it takes up the full screen. As such, you may want to add a new PhoneApplicationPage to your solution. Right-click on your Windows Phone project in the Solution Explorer again, and this time select Add ➪ New Item from the context menu. Select the "Windows Phone Portrait Page" template and give the new page a name, for example, *SampleHub*; then click Add.

As you now have two pages in your application, you'll also need a way to navigate from the MainPage to the SampleHub. Add a button to the MainPage; call it **GoToHubButton**; double-click the button to add an event handler; and then add the following code:

```
private void GoToHubButton_Click(object sender, System.Windows.RoutedEventArgs e){
    this.NavigationService.Navigate(new Uri("/SampleHub.xaml",UriKind.Relative));
}
```

Details of what the `Navigate` method does at this point are covered later in Chapter 6. It's enough to point out that it closes the MainPage and displays the SampleHub.

Let's jump over to Expression Blend and add the panorama to the SampleHub page. Before continuing, make sure you rebuild your solution in Blend to update the list of controls in the Assets window. Open the SampleHub from the Projects window. Start by removing all child elements in the `LayoutRoot Grid`. Then, drag the Panorama from the Assets window onto the page (you may need to reset the `Height`, `Width` and `Margin` properties to get the control to fill the page). The `Title` property of the `Panorama` is the large text that runs along the top of the hub. In the case of this example, we're building a hub for our outback adventures, so the Title is set to *Outback Adventure*.

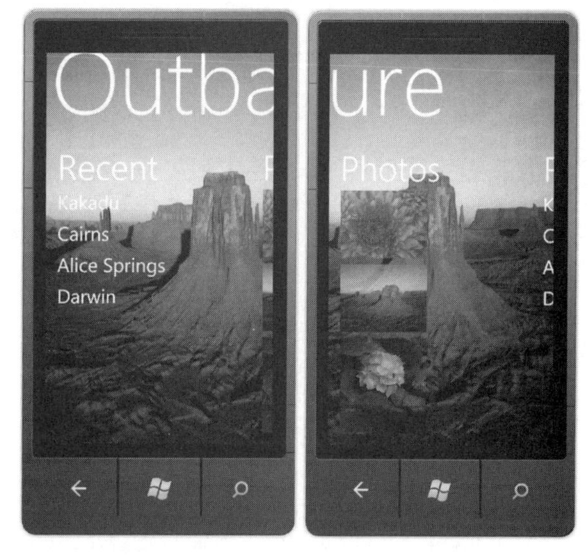

FIGURE 4-25

Working with the `Panorama` is relatively straightforward. You can have as many sections as you like, each one represented by a `PanoramaItem`, which can be dragged across from the Assets window. Each `PanoramaItem` has a `Header` property, which is the Header for the section. Figure 4-25 illustrates the two sections of the outback hub for "Recent trips" and Photos.

Within each `PanoramaItem`, you can place whatever controls you like in order to build up the layout you want. Here we're going to walk through creating a simple stack of text items and then how to create a list based on a set of sample data. Starting with the stack of text items, the following code illustrates the XAML that is created after adding a `StackPanel` and several `TextBlocks` to the first `PanoramaItem`:

```
<controls:Panorama Title="Outback Adventure">
    <controls:Panorama.Background>
       <ImageBrush Stretch="Fill" ImageSource="/panorama.jpg"/>
    </controls:Panorama.Background>
    <controls:PanoramaItem Header="Recent">
       <StackPanel CacheMode="BitmapCache">
            <TextBlock TextWrapping="Wrap" Text="Kakadu"
                       Style="{StaticResource PanoramaTextBlock}" />
            <TextBlock TextWrapping="Wrap" Text="Cairns"
                       Style="{StaticResource PanoramaTextBlock}" />
            <TextBlock TextWrapping="Wrap" Text="Alice Springs"
                       Style="{StaticResource PanoramaTextBlock}" />
            <TextBlock TextWrapping="Wrap" Text="Darwin"
```

```
                        Style="{StaticResource PanoramaTextBlock}" />
            </StackPanel>
        </controls:PanoramaItem>
    </controls:Panorama>
```

This makes use of the `PanoramaTextBlock` style that needs to be added to the Resources section of the page:

```
<phone:PhoneApplicationPage.Resources>
    <Style x:Key="PanoramaTextBlock" TargetType="TextBlock">
        <Setter Property="FontSize" Value="40"/>
        <Setter Property="Height" Value="64"/>
    </Style>
</phone:PhoneApplicationPage.Resources>
```

Creating the list of photos is slightly harder because you need to create a sample set of images to be displayed. However, start by creating another `PanoramaItem`, set the `Header` property to `Photos`, and add a `ListBox` to it. To create the sample images, open the Data window and click on the "Create sample data" button (second icon in from the right), followed by "New Sample Data." Give the sample data a name — **SamplePhotos** — select Define In ➪ This Document, and then click OK. In the Data window you will now see a tree similar to that shown in Figure 4-26 under the "This document" node. By default, in the Collection node there will be Property1 and Property2. Delete Property2 and rename Property1 to **Photo**.

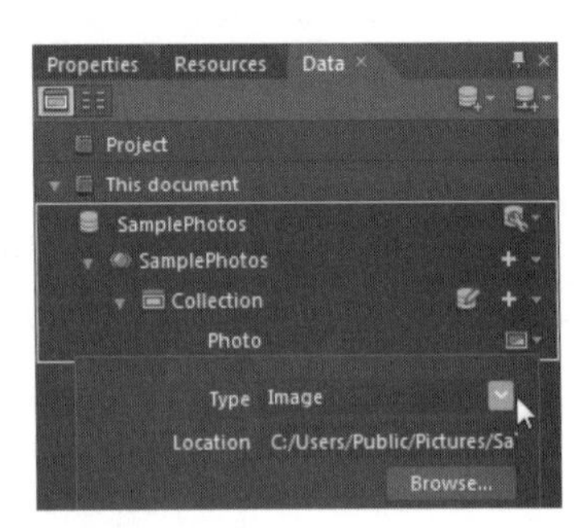

FIGURE 4-26

In Figure 4-26, if you look to the right side of the image in line with the word *Photo*, there is a dropdown that allows you to select the type of sample data that is created. Change this value to Image and select a location from where the sample data will be created (there is a Sample Pictures folder installed with Windows under your My Pictures folder).

Now that we've created the sample data, it's relatively easy to add them to the listbox you created earlier. Simply drag the Photo node from the Data window across onto the listbox in the Objects and Timeline window. This is all you need to do in order to get the data binding to work (this will be covered in more detail in Chapter 15).

The last thing to do is to style the listbox. This is again done using a `Style`, resulting in the following XAML for the Resources and Panorama control:

Available for download on Wrox.com

```
<phone:PhoneApplicationPage
    xmlns="http://schemas.microsoft.com/winfx/2006/xaml/presentation"
    xmlns:x="http://schemas.microsoft.com/winfx/2006/xaml"
    xmlns:phone="clr-namespace:Microsoft.Phone.Controls;assembly=Microsoft.Phone"
    xmlns:shell="clr-namespace:Microsoft.Phone.Shell;assembly=Microsoft.Phone"
    xmlns:d="http://schemas.microsoft.com/expression/blend/2008"
    xmlns:mc="http://schemas.openxmlformats.org/markup-compatibility/2006"
```

```
xmlns:SampleData="clr-namespace:Expression.Blend.SampleData.SamplePhotos"
xmlns:controls=
"clr-namespace:Microsoft.Phone.Controls;assembly=Microsoft.Phone.Controls"
x:Class="LifeInMotion.SampleHub"
SupportedOrientations="Portrait" Orientation="Portrait"
mc:Ignorable="d" d:DesignHeight="768" d:DesignWidth="480">
<phone:PhoneApplicationPage.Resources>
    <SampleData:SamplePhotos x:Key="SamplePhotos" d:IsDataSource="True"/>
    <Style x:Key="PanoramaTextBlock" TargetType="TextBlock">
        <Setter Property="FontSize" Value="40"/>
        <Setter Property="Height" Value="64"/>
    </Style>
    <Style x:Key="PanoramaImageListBox" TargetType="ListBox">
        <Setter Property="Padding" Value="1"/>
        <Setter Property="Background" Value="Transparent"/>
        <Setter Property="Foreground"
                Value="{StaticResource PhoneForegroundBrush}"/>
        <Setter Property="HorizontalAlignment" Value="Left"/>
        <Setter Property="VerticalAlignment" Value="Top"/>
        <Setter Property="HorizontalContentAlignment" Value="Left"/>
        <Setter Property="VerticalContentAlignment" Value="Top"/>
        <Setter Property="ScrollViewer.HorizontalScrollBarVisibility"
                Value="Disabled"/>
        <Setter Property="ScrollViewer.VerticalScrollBarVisibility"
                Value="Auto"/>
        <Setter Property="BorderBrush"
                Value="{StaticResource PhoneForegroundBrush}"/>
        <Setter Property="BorderThickness" Value="0"/>
        <Setter Property="ItemsPanel">
            <Setter.Value>
                <ItemsPanelTemplate>
                    <StackPanel HorizontalAlignment="Left"
                                VerticalAlignment="Top"
                                Orientation="Vertical" />
                </ItemsPanelTemplate>
            </Setter.Value>
        </Setter>
        <Setter Property="ItemContainerStyle">
            <Setter.Value>
                <Style TargetType="ListBoxItem">
                    <Setter Property="Template">
                        <Setter.Value>
                            <ControlTemplate TargetType="ListBoxItem">
                                <Image Width="185" Margin="0,0,12,12"
                                       Opacity="0.75" Source="{Binding Photo}"/>
                            </ControlTemplate>
                        </Setter.Value>
                    </Setter>
                </Style>
            </Setter.Value>
        </Setter>
    </Style>
</phone:PhoneApplicationPage.Resources>
```

```
        <Grid x:Name="LayoutRoot"
              Background="{StaticResource PhoneBackgroundBrush}"
              DataContext="{Binding Source={StaticResource SamplePhotos}}">
            <controls:Panorama Title="Outback Adventure">
                <controls:Panorama.Background>
                    <ImageBrush Stretch="Fill" ImageSource="/panorama.jpg"/>
                </controls:Panorama.Background>
                <controls:PanoramaItem Header="Recent">
                    <StackPanel CacheMode="BitmapCache">
                        <TextBlock TextWrapping="Wrap" Text="Kakadu"
                                   Style="{StaticResource PanoramaTextBlock}" />
                        <TextBlock TextWrapping="Wrap" Text="Cairns"
                                   Style="{StaticResource PanoramaTextBlock}" />
                        <TextBlock TextWrapping="Wrap" Text="Alice Springs"
                                   Style="{StaticResource PanoramaTextBlock}" />
                        <TextBlock TextWrapping="Wrap" Text="Darwin"
                                   Style="{StaticResource PanoramaTextBlock}" />
                    </StackPanel>
                </controls:PanoramaItem>
                <controls:PanoramaItem Header="Photos">
                    <ListBox ItemsSource="{Binding Collection}"
                             Style="{StaticResource PanoramaImageListBox}"
                             Height="500"/>
                </controls:PanoramaItem>
            </controls:Panorama>
        </Grid>
    </phone:PhoneApplicationPage>
```

Code snippet from SampleHub.xaml

The `Panorama` allows you to build experiences that are similar to what users will come to expect on their Windows Phone. However, you should be careful where you use it as it can easily make your application confusing. Typically you would only use a single `Panorama` within your application for this reason. You will probably use the `Panorama` at the beginning of your application as a landing page. However, you can use it elsewhere in your application if it suits the data that you're presenting to the user.

You typically use the `Pivot` to display multiple views over the same set of data. In this case we're going to display a list of adventure companies. We'll use the `Pivot` to allow the user to switch between three views of the data that filter on whether they have been used recently, and whether they operate in the summer or winter.

As the `Pivot` also takes up a full screen we'll again create a new `PhoneApplicationPage` within the application called SamplePivot.xaml and we'll add a button to the MainPage.xaml that navigates to this page.

```
private void GoToPivotButton_Click(object sender, System.Windows.RoutedEventArgs e){
    this.NavigationService.Navigate(new Uri("/SamplePivot.xaml",UriKind.Relative));
}
```

In the SamplePivot page create an instance of the Pivot by dragging it from the Assets window in Blend. Like with the Panorama you will probably have to reset the Height, Width, and

Margin properties so that the control takes up the full screen. For demonstrative purposes we're going to add three sets of sample data in the same way as you did earlier for the SampleHub. Each set of sample data will represent a subset of the full data set that would be in the application. For example, the RecentHolidays list (Figure 4-27) represents only the adventure companies that the user has recently purchased holidays from.

Add three PivotItem controls to the Pivot by dragging them from the Assets window onto the Pivot. For each set of sample data, drag the set onto a PivotItem. This will automatically create a listbox with the appropriate data bindings, as in the following code

FIGURE 4-27

```
<controls:Pivot Title="holidays">
    <controls:PivotItem Header="recent">
        <Grid>
            <ListBox ItemTemplate="{StaticResource
HolidaysItemTemplate}"
                     ItemsSource="{Binding RecentHolidays}"/>
        </Grid>
    </controls:PivotItem>
    <controls:PivotItem Header="summer">
        <Grid>
            <ListBox ItemTemplate="{StaticResource HolidaysItemTemplate}"
                     ItemsSource="{Binding SummerHolidays}"/>
        </Grid>
    </controls:PivotItem>
    <controls:PivotItem Header="winter">
        <Grid>
            <ListBox ItemTemplate="{StaticResource HolidaysItemTemplate}"
                     ItemsSource="{Binding WinterHolidays}"/>
        </Grid>
    </controls:PivotItem>
</controls:Pivot>
```

Code snippet from SamplePivot.xaml

In this case we've tidied up the code so that each listbox uses the sample ItemTemplate which is declared within the resource dictionary for the page.

```
<phone:PhoneApplicationPage.Resources>
    <SampleData:HolidayDataSource x:Key="HolidayDataSource" d:IsDataSource="True"/>
        <DataTemplate x:Key="HolidaysItemTemplate">
            <StackPanel>
            <TextBlock Text="{Binding Name}"
                       Style="{StaticResource PhoneTextExtraLargeStyle}"/>
            <TextBlock Text="{Binding Website}"
                       Style="{StaticResource PhoneTextSmallStyle}"/>
            <TextBlock Text="{Binding Description}" TextWrapping="Wrap"/>
        </StackPanel>
```

```
    </DataTemplate>
</phone:PhoneApplicationPage.Resources>
```

Code snippet from SamplePivot.xaml

The result is a Pivot that is made up of three views over the list of adventure companies, as shown in Figure 4-28.

FIGURE 4-28

SUMMARY

In this chapter, you have learned about visual states and how you can combine them with transitions and other forms of animation to extend your static user interface into one connected through motion. Building a user experience requires you to apply these capabilities to come up with rich and engaging layouts that are intuitive, easy to use, and immediately familiar to the user.

This is the last of the fundamental chapters that really form an overview of Silverlight. From here you'll learn about how you design applications that are specifically tailored to take advantage of device capabilities.

5

Orientation and Overlays

WHAT'S IN THIS CHAPTER

- How to detect and react to changes in device orientation
- Using states to define different layouts for each device orientation
- Working with the Soft Input Panel
- Creating Application Bar icons and menu items
- How to maximize screen real estate by hiding the System Tray

Imagine that you are viewing photos or scrolling through a table of data. Both of these actions are best suited to a landscape view. Alternatively, if you are scrolling through a list of contacts or viewing your schedule for today, you are most likely going to be in portrait view. Your Windows Phone application needs to know how the device is oriented so that it can optimize the user experience.

Windows Phone supports several overlays that can affect how your application is rendered. When you design your application you need to know how the Application Bar and the Soft Input Panel (SIP) will affect how users can interact with your application.

In this chapter, you will learn how to handle changes to the orientation of the device, how to control both the Application Bar and the SIP, and how you can run your application in Full-Screen mode in order to hide the System Tray.

DEVICE ORIENTATION

A large proportion of applications written for Windows Phone will be designed with an elongated, or as it's more commonly known, portrait screen in mind. This screen layout, where there is significantly more height than width, is great for displaying lists of information because you can fit quite a large number of items on the screen and the user can simply scroll the list up and down to see more items. However, there are scenarios in which it makes sense for the screen to be rotated into what's known as landscape orientation. In this case, the width of the screen is larger than the height. Landscape and Portrait are referred to as device orientations and can dynamically change through the lifetime of your application.

If you were simply to design your application with Portrait in mind, when the user switches to Landscape, only a limited portion of your application will be visible and there will be large areas of wasted space. For example, if you are presenting a list of items, in Portrait mode you might get 10 to a screen before the user has to scroll down to see more. When running in Landscape mode, this might be as low as two or three items, thus making it frustrating for the user.

Whereas the natural scrolling behavior for a Portrait screen is in the vertical plane (i.e., up and down), when you switch into Landscape mode it becomes more natural to scroll horizontally (i.e., left and right). This again highlights the need to think about how your application is going to be used when the phone is in different orientations and modes of use.

To add an additional level of complexity, Landscape, can appear in two ways depending on whether the device is rotated to the left or right, from being held in the Portrait orientation. Figure 5-1 illustrates, from left to right, the LandscapeLeft, PortraitUp, and LandscapeRight screen orientations.

FIGURE 5-1

Orientation Detection

As you can imagine, it's important that your application can detect which orientation the device is currently in. This is done by querying the `Orientation` property of your PhoneApplicationPage, which will return one of the values from the PageOrientation enumeration. You will notice that there is a PortraitDown enumeration value that isn't shown in Figure 5-1. This value is there for completion sake but represents an orientation not actually available on either device or emulator. It was found that the addition of this state meant that it was too easy for users to accidentally flip the screen when using their devices, so it was removed.

```
public enum PageOrientation {
        None = 0,
        Portrait = 1,
        Landscape = 2,
        PortraitUp = 5,
        PortraitDown = 9,
        LandscapeLeft = 18,
        LandscapeRight = 34
}
```

In most cases, you will want to query the `Orientation` property to determine whether the device is in Portrait or Landscape mode, not being concerned about differences between modes such as LandscapeLeft or LandscapeRight. This is easily done using a switch statement like the following:

```
var orientation = this.Orientation;
switch (orientation){
    case PageOrientation.Portrait:
    case PageOrientation.PortraitUp:
    case PageOrientation.PortraitDown:
        MessageBox.Show("Portrait");
        break;
    case PageOrientation.Landscape:
    case PageOrientation.LandscapeLeft:
    case PageOrientation.LandscapeRight:
        MessageBox.Show("Landscape");
        break;
}
```

In this snippet the orientation value is tested against all Portrait and Landscape permutations to work out which orientation the device is in. An alternative is to use a binary AND statement to test for specific orientation values. The `IsOrientation` method in the following snippet compares the value `orientation` with that of `testOrientation` by and'ing them and then testing if the result is greater than zero:

Available for download on Wrox.com

```
public static class Utilities{
    public static bool IsInLandscape(this PhoneApplicationPage page){
        return !page.IsInPortrait();
    }

    public static bool IsInPortrait(this PhoneApplicationPage page){
        return page.IsInOrientation(PageOrientation.Portrait |
                                    PageOrientation.PortraitUp |
                                    PageOrientation.PortraitDown);
    }

    public static bool IsInOrientation(this PhoneApplicationPage page,
                                       PageOrientation testOrientation){
        var orientation = page.Orientation;
        return page.Orientation.IsOrientation(testOrientation);
    }
```

```
        public static bool IsOrientation(this PageOrientation orientation,
                                         PageOrientation testOrientation){
            return (orientation & testOrientation) > 0;
        }
    }
```

Code snippet from Utilities.cs

Within your page, you can then reduce your logic down to a single conditional statement, for example:

Available for download on Wrox.com

```
if (this.IsInLandscape()){
    MessageBox.Show("Landscape");
}
else{
    MessageBox.Show("Portrait");
}
```

Code snippet from MainPage.xaml.cs

PAGEORIENTATION ENUMERATION

The values of the PageOrientation enumeration might appear unusual as they are neither incremental (i.e., 0, 1, 2, 3. . .) or flag values (i.e., 0, 1, 2, 4, 8. . .). However, if you look at the binary representation of the values you will notice a pattern:

000000	None = 0
000001	Portrait = 1
000010	Landscape = 2
000101	PortraitUp = 5
001001	PortraitDown = 9
010010	LandscapeLeft = 18
100010	LandscapeRight = 34

The important thing to note is that `Portrait`, `PortraitUp` and `PortraitDown` all have the lowest bit (working from the right) equal to 1. Conversely, `Landscape`, `LandscapeLeft` and `LandscapeRight` all have the second bit equal to 1. This property is used in the `IsOrientation` method to test for either `Landscape` or `Portrait`. For example if the `testOrientation` is `Landscape`, the `IsOrientation` method will return `true` when `orientation` equals `Landscape`, `LandscapeLeft` and `LandscapeRight`.

Orientation Changes

Now that you can determine what orientation the device is currently in, the next thing that you'll want to be able to do is detect when the user switches the device orientation. Remember that because the device has a built-in accelerometer, it will automatically attempt to switch orientation when the user rotates the device. Windows Phone exposes the changing of the device orientation through the `OrientationChanged` event. When the user changes the orientation of the device, the `OrientationChanged` event is raised.

You can attach to the orientation event in one of two ways. You can attach an event handler to the event itself, or you can override the base class method `OnOrientationChanged`. In Visual Studio, if you opt to attach an event handler, you can simply select the PhoneApplicationPage in the Document Outline window, then select the appropriate event from the Properties window. Double-clicking on the empty cell next to the event will create and wire up an event handler. For example, an event handler for the `OrientationChanged` event would look like the following:

```
private void PhoneApplicationPage_OrientationChanged(object sender,
                                                     OrientationChangedEventArgs e)
{ ... }
```

The other alternative is to override the base class method. In this case, the equivalent method is `OnOrientationChanged`, which you can override as follows:

```
protected override void OnOrientationChanged(OrientationChangedEventArgs e)
{
    base.OnOrientationChanged(e);
    ...
}
```

There is no hard-and-fast rule as to which alternative you should use. However, adding event handlers can be preferable as it is easier to see that there is code attached to an orientation event when using the designer. If you override the base class methods, there is no visual indication within the Visual Studio IDE that code is being executed when the orientation changes. You should, of course, be aware that you can wire up the event handler in code, but again this makes it harder to see which events are being handled in the designer.

Orientation Strategies

Upon detecting that the orientation of the device has changed, it is up to you to work out how your application should behave and best use the available screen real estate. There are a number of strategies that you can use, and although there is no right or wrong answer, it really comes down to what makes your application easier to use and what provides the best user experience.

Fixed Orientation

If you look at the Start, this is an area of the Windows Phone user experience that does not change regardless of device orientation. It has been specifically designed and optimized for a Portrait-based layout. In fact, quite a few areas of the Windows Phone user experience, such as hubs and pivots, are designed to work exclusively in Portrait. You can adopt this strategy for all or part of your application by setting the `SupportedOrientation` property on each page.

There are likely to be many Windows Phone devices that have keyboards, some of which will slide out in Portrait, some in Landscape. If you elect to fix the orientation of your application, you may alienate users who would like to be able to use their keyboards in order to enter text within your application.

This property doesn't reuse the PageOrientation enumeration; instead, it uses the much simpler SupportedPageOrientation enumeration:

```
public enum SupportedPageOrientation
{
    Portrait = 1,
    Landscape = 2,
    PortraitOrLandscape = 3
}
```

When you create a new application in Visual Studio based on the Windows Phone Application template, you will notice that the `SupportedOrientation` property is set in MainPage.xaml:

```
<phone:PhoneApplicationPage
    x:Class="ApplicationLayouts.MainPage"
    ...
    SupportedOrientations="Portrait">
```

You can also specify this property in code. For example the following code sets the `SupportedOrientations` to `Portrait` or `Landscape`. In other words it supports both orientations.

```
public MainPage(){
    InitializeComponent();

    SupportedOrientations = SupportedPageOrientation.Portrait |
                            SupportedPageOrientation.Landscape;
}
```

This could be simplified by using the PortraitOrLandscape enumeration value because this value matches the binary `Or` value of the Portrait and Landscape values. By changing the assigned value to either the Portrait or Landscape values, you can force the page to be displayed in a specific orientation no matter how the device is held.

An interesting artifact of dynamically setting the `SupportedOrientation` *value in your code is that you can force the application into a particular orientation. For example, if the user has changed the orientation of the device to Landscape but then attempts to do something that will only work well in Portrait, you can set the* `SupportedOrientation` *property to* `Portrait`*, and this will force the application back into Portrait orientation.*

Forcing the device into a fixed orientation is actually quite an effective strategy and highly recommended if you have content that can really only be presented in one orientation. It is also by far the easiest orientation-handling technique to implement as you just need to specify which orientation should be supported and design your application layout accordingly.

Auto-Layout

The first approach to altering your page layout to handle both Portrait and Landscape orientations is to allow the controls to automatically scale and position themselves. For example, if you align a control to the bottom of the screen, when you change between Portrait and Landscape, the control will automatically move to ensure that it remains in the same relative position to the bottom of the screen.

Before implementing this technique, you should always double-check that a better approach is not available. Taking this strategy can leave your application looking untidy, unfinished, or just poorly architected. Apart from the simplest of page layouts, the automatic resizing and positioning of controls can lead to poor usability. In most cases, you would be better off only supporting a single orientation than to try and do both using the ability to automatically adjust layout.

In Figure 5-2 there are five buttons, each carrying a label indicating the sides of the page that they are aligned with. As you can see, when the orientation changes from Portrait to Landscape, the positions of the controls vary accordingly.

For example, the button with the content, TopRight, is the same distance from the top and right sides of the page in both orientations. This is because it is VerticallyAligned to the top and HorizontallyAligned to the right. In the

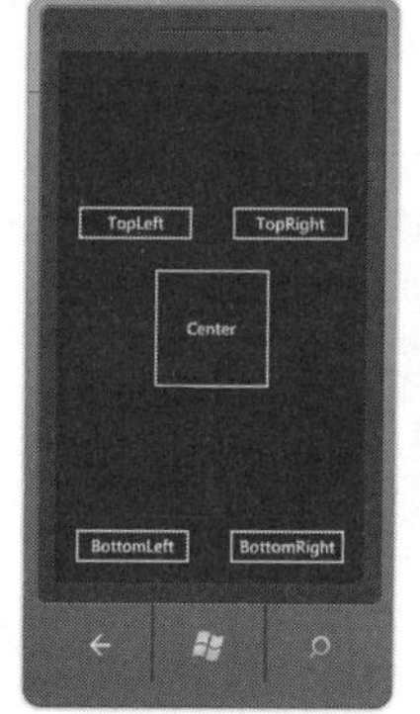

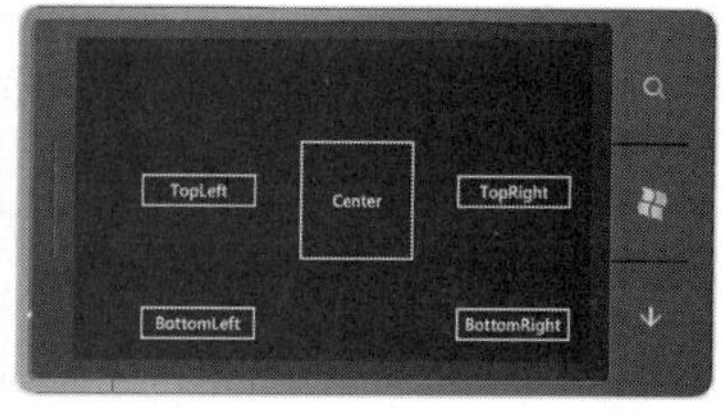

FIGURE 5-2

following code, you can see that the appropriate horizontal and vertical alignments have been set. You can also see that margins are used to determine the spacing between the button and the sides of the page.

Available for download on Wrox.com

```
<Grid x:Name="LayoutRoot" Background="Transparent">
    <Button Content="TopLeft"     HorizontalAlignment="Left"
                                  VerticalAlignment="Top"
                                  Margin="20,190,0,0" Width="200" />
    <Button Content="TopRight"    HorizontalAlignment="Right"
                                  VerticalAlignment="Top"
                                  Margin="0,190,20,0" Width="200" />
    <Button Content="BottomLeft"  HorizontalAlignment="Left"
                                  VerticalAlignment="Bottom"
                                  Margin="20,0,0,20"  Width="200" />
    <Button Content="BottomRight" HorizontalAlignment="Right"
                                  VerticalAlignment="Bottom"
                                  Margin="0,0,20,20"  Width="200" />
    <Button Content="Center" Margin="0" Height="200" Width="200" />
</Grid>
```

Code snippet from AutoLayoutPage.xaml

Another technique to automatically handle changes in screen orientation is to design for one orientation and configure scrollbars to allow the user to scroll through content if a change in orientation would obscure some content. When the device is in Portrait, all of the content might be in view, in which case there wouldn't be any scrollbars visible. Then, when the user changes the orientation of the device, the vertical scrollbar would appear because some of the elements are now off screen, allowing the user to scroll those into view. This is illustrated in Figure 5-3, where the screen on the left doesn't have a scrollbar, yet those on the right do (the scrollbar is only visible while scrolling), allowing the user to scroll in order to see the rest of the content.

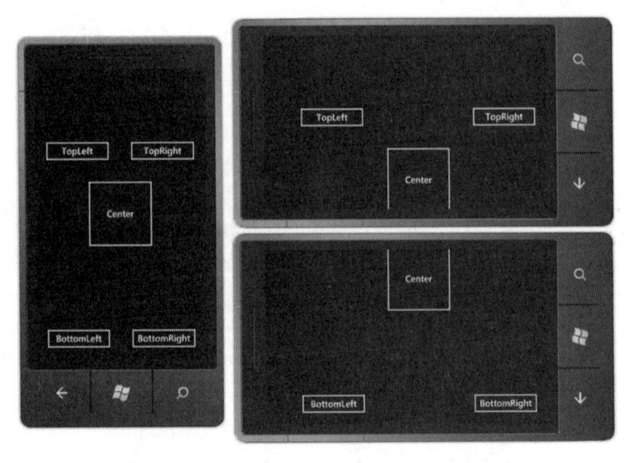

FIGURE 5-3

In order to achieve this layout, you can use a `ScrollViewer` and nest controls within it. The following code is similar to what you saw earlier, with the main difference being that the `Grid` is now nested within a `ScrollViewer`:

Available for download on Wrox.com

```
<Grid x:Name="LayoutRoot" Background="Transparent">
    <ScrollViewer HorizontalAlignment="Stretch" VerticalAlignment="Stretch"
                  VerticalScrollBarVisibility="Auto">
        <Grid Height="800">
```

```
            <Button Content="TopLeft"
                    HorizontalAlignment="Left"  VerticalAlignment="Top"
                    Margin="20,190,0,0" Width="200" />
            <Button Content="TopRight"
                    HorizontalAlignment="Right" VerticalAlignment="Top"
                    Margin="0,190,20,0" Width="200" />
            <Button Content="BottomLeft"
                    HorizontalAlignment="Left"  VerticalAlignment="Bottom"
                    Margin="20,0,0,20"  Width="200" />
            <Button Content="BottomRight"
                    HorizontalAlignment="Right" VerticalAlignment="Bottom"
                    Margin="0,0,20,20"  Width="200" />
            <Button Content="Center"
                    Margin="0" Height="200" Width="200" />
        </Grid>
    </ScrollViewer>
</Grid>
```

Code snippet from AutoLayoutWithScrollPage.xaml

Notice the `Grid` has its `Height` explicitly set to 800 pixels. Without this setting, the `Grid` would resize to fit the available space within the `ScrollViewer`. By explicitly specifying a height, the `Grid` will always be 800 pixels high. If this is taller than the height of the current screen orientation, the `ScrollViewer` will automatically add scrollbars. Since you have not explicitly set the `Width` of the `Grid`, it will continue to stretch or shrink to fit available space.

Manual Intervention

Since you are able to detect when the device has changed orientation, you can adjust the layout via code. This is more powerful than relying on automatic layout. In the following code, you'll see how you can change the position of the buttons so that they always remain in the same physical corner of the device (i.e., the buttons will remain in the same corner of the device regardless of the device orientation), as shown in Figure 5-4.

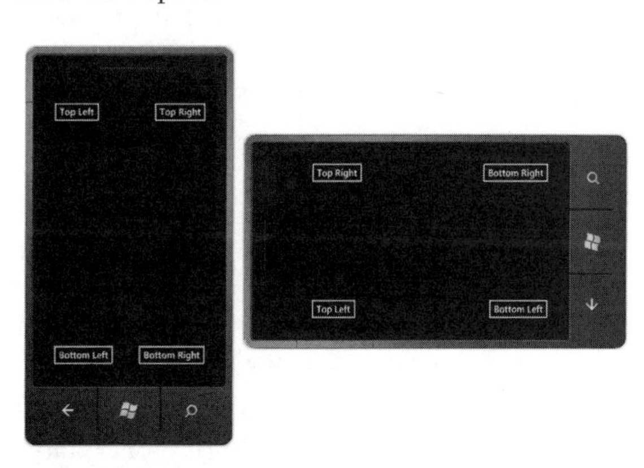

FIGURE 5-4

Available for download on Wrox.com

```
private void PhoneApplicationPage_OrientationChanged(object sender,
                                    OrientationChangedEventArgs e){
    if (this.IsInLandscape()){
        TopLeftButton.VerticalAlignment = VerticalAlignment.Bottom;
        TopLeftButton.Margin = new Thickness(20, 0, 0, 20);
        TopRightButton.HorizontalAlignment = HorizontalAlignment.Left;
        TopRightButton.Margin = new Thickness(20, 20, 0, 0);
```

```
            BottomLeftButton.HorizontalAlignment = HorizontalAlignment.Right;
            BottomLeftButton.Margin = new Thickness(0, 0, 20, 20);
            BottomRightButton.VerticalAlignment = VerticalAlignment.Top;
            BottomRightButton.Margin = new Thickness(0, 20, 20, 0);
        }
        else{
            TopLeftButton.VerticalAlignment = VerticalAlignment.Top;
            TopLeftButton.Margin = new Thickness(20, 20, 0, 0);
            TopRightButton.HorizontalAlignment = HorizontalAlignment.Right;
            TopRightButton.Margin = new Thickness(0, 20, 20, 0);
            BottomLeftButton.HorizontalAlignment = HorizontalAlignment.Left;
            BottomLeftButton.Margin = new Thickness(20, 0, 0, 20);
            BottomRightButton.VerticalAlignment = VerticalAlignment.Bottom;
            BottomRightButton.Margin = new Thickness(0, 0, 20, 20);
        }
    }
```

Code snippet from ManualLayoutPage.xaml.cs

The `Alignment` and `Margin` properties on the buttons are adjusted to reposition them in accordance with the new layout. Figure 5-4 shows how the buttons appear in Portrait and Landscape mode. Notice that the TopLeft button stays in the same location relative to the physical phone, rather than staying in the top-left corner of the screen relative to the user.

Although this approach for handling different device orientations can yield the best results, it requires a significant amount of code in order to update the layout for each orientation. It also makes it hard for the designer to make modifications to the layout. Rather than being specified in designer-friendly XAML, the layout is now part of the developer's code.

Changing States

Thinking about the situation, you may realize that changing the position of controls for each device orientation is essentially the same as having two visual states for the page, one for Portrait and one for Landscape. The advantage of this approach is that these states can be designed using state recording in Blend, and then when the device orientation changes, you just need to tell the `VisualStateManager` to update the state of the page.

To create two states for the page, open your solution in Blend, and in the States window create a new state group for your PhoneApplicationPage. Within this group create two states that will represent the Portrait and Landscape layouts. Since the Portrait state will be the same as the Base state, you don't need to modify that state. To alter the Landscape state, select the state in the States window to enable state recording. Rather than working with the designer skin in Portrait mode and trying to guess what it would look like in Landscape mode, you can use the Device window (accessible via the Window menu) to toggle the designer skin between Portrait and Landscape mode. Figure 5-5 shows the Blend Designer in Landscape mode with the Device window floating beneath the page layout.

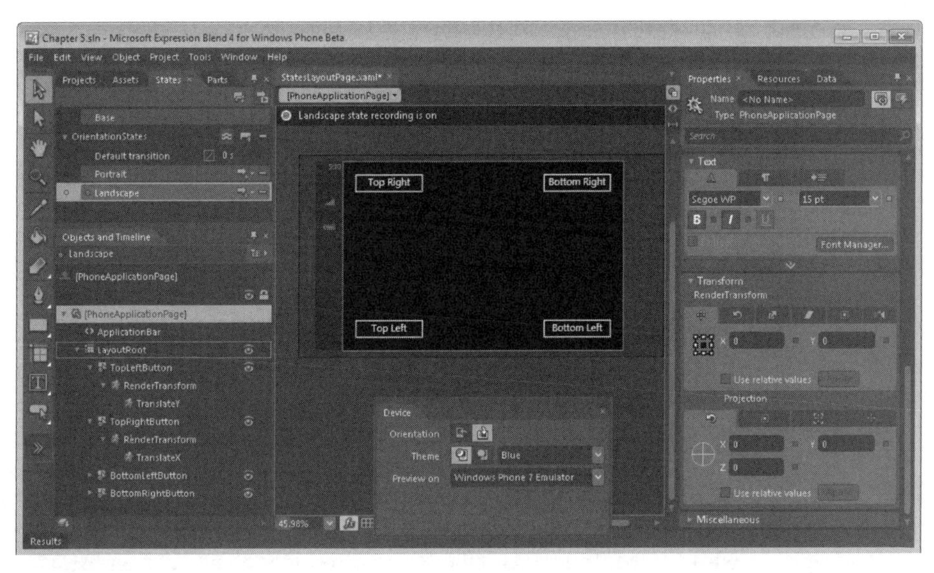

FIGURE 5-5

When you switch the design skin to Landscape mode, you may see a dialog pop up saying that you can't animate an automatically calculated property. You can ignore this warning because it is just referring to design time properties used by Blend and won't affect your application at run time.

The Devices Window in Expression Blend can be used to test how your layout will look under different device configurations. You have just seen how you can adjust the Orientation to see how changing between Portrait and Landscape can affect your page. You can also toggle between the Light and Dark theme, as well as changing the accent color. The Preview on drop down allows you to specify whether the application will be run on the emulator or on a real device (if one is connected) when you launch the application.

Once the Designer is in Landscape, rearrange your controls to how they should appear when the device is in Landscape mode. In this case, the controls have been moved into the correct location by adjusting the `TranslateY` or `TranslateX` properties. You can see the final layout of the buttons and the two states in Figure 5-5. In the Objects and Timeline window, you can see all the buttons and their corresponding changes.

With the two states defined, all the developer needs to do is to assign the page to a specific state whenever the device orientation changes. To do this, you need to trap the `OrientationChanged` event and direct the `VisualStateManager` to go to the new state:

Available for download on Wrox.com

```
private void PhoneApplicationPage_OrientationChanged(object sender,
                                        OrientationChangedEventArgs e){
    if (this.IsInLandscape()){
        VisualStateManager.GoToState(this, "Landscape", true);
    }
    else{
        VisualStateManager.GoToState(this, "Portrait", true);
    }
}
```

Code snippet from StatesLayoutPage.xaml

The `VisualStateManager` is instructed to assign the page to the appropriate state depending on what the current orientation of the device is. This technique, although more designer-friendly, still requires the developer to get involved. In the last chapter, you resolved this situation by using Behaviors. In particular, `GoToStateAction` was used to transition to a state whenever a particular event was raised. Although you could connect a `GoToStateAction` behavior to the `OrientationChanged` event, this would only go to a single state with no ability to make the state conditional on the current orientation.

To rectify this problem, you can create your own custom behavior that you can reuse among the pages of your application. Start by creating a new class in your application called `ChangeOrientationStateAction`. Modify this class so that it inherits from `TriggerAction<T>`, and add three dependency properties of type `string` that will be used to specify the names of states for each of the three possible orientations — LandscapeLeft, PortraitUp, and LandscapeDown:

Available for download on Wrox.com

```
public class ChangeOrientationStateAction:TriggerAction<PhoneApplicationPage>{
    public static readonly DependencyProperty LandscapeLeftStateNameProperty =
                    DependencyProperty.Register("LandscapeLeftStateName",
                                        typeof(string),
                                        typeof(ChangeOrientationStateAction),
                                        new PropertyMetadata("Landscape"));
    public string LandscapeLeftStateName{
        get { return (string)GetValue(LandscapeLeftStateNameProperty); }
        set { SetValue(LandscapeLeftStateNameProperty, value); }
    }

    public static readonly DependencyProperty LandscapeRightStateNameProperty =
                    DependencyProperty.Register("LandscapeRightStateName",
                                        typeof(string),
                                        typeof(ChangeOrientationStateAction),
                                        new PropertyMetadata("Landscape"));
    public string LandscapeRightStateName{
```

```
            get { return (string)GetValue(LandscapeRightStateNameProperty); }
            set { SetValue(LandscapeRightStateNameProperty, value); }
        }

        public static readonly DependencyProperty PortraitUpStateNameProperty =
                            DependencyProperty.Register("PortraitUpStateName",
                                                typeof(string),
                                                typeof(ChangeOrientationStateAction),
                                                new PropertyMetadata("Portrait"));
        public string PortraitUpStateName{
            get { return (string)GetValue(PortraitUpStateNameProperty); }
            set { SetValue(PortraitUpStateNameProperty, value); }
        }
    }
```

Code snippet from ChangeOrientationStateAction.cs

You will notice that the generic parameter for the `TriggerAction<T>` base class has been specified as `PhoneApplicationPage`, which is the class upon which the `OrientationChanged` event is defined. Since this is the event that this behavior is interested in, it makes sense for `ChangeOrientationStateAction` to be limited to being applied to instances of the `PhoneApplicationPage` class. When the event is triggered by the device changing orientation, the `VisualStateManager` needs to transition the page to the appropriate state. This is done in the `Invoke` method:

```
    protected override void Invoke(object parameter){
        var page = this.AssociatedObject;
        var newOrientation = (parameter as OrientationChangedEventArgs).Orientation;

        if (newOrientation.IsOrientation(PageOrientation.Portrait)){
            VisualStateManager.GoToState(page, this.PortraitUpStateName, true);
        }
        else{
            if (newOrientation.IsOrientation(PageOrientation.LandscapeLeft)){
                VisualStateManager.GoToState(page, this.LandscapeLeftStateName, true);
            }
            else{
                VisualStateManager.GoToState(page, this.LandscapeRightStateName, true);
            }
        }
    }
```

Code snippet from ChangeOrientationStateAction.cs

The parameter passed into the `Invoke` method is `OrientationChangedEventArgs`, from which you can determine the orientation that the screen is moving to. The new orientation is then used to determine which state to assign to the page.

After saving this behavior and rebuilding your project in Expression Blend, you will see that this behavior appears in the Assets window. You can then drag this across onto the [PhoneApplicationPage] node in the Objects and Timeline window. Assuming that you have states named *Landscape* and *Portrait*, you will only need to adjust the `EventName` property, setting it to `OrientationChanged`. When you run the application and change the device orientation, you will see the buttons change layout.

Smooth Transitions

The last thing to do in order to make the orientation change smooth is to add some animation to the changing of layout. This can be done by adding transitions to both the Portrait and Landscape states. Select the Landscape state in the States window, then open the Objects and Timeline window in Blend, and if the time line isn't visible, click the "Show Timeline" button. You will notice that the time line for this state is currently of zero length, which means that all the transitioning is done instantaneously. For each button there is a single keyframe that is at the 0-second mark, as shown in the left image of Figure 5-6. To make this into a smooth transition, drag each of the keyframes out to the 1-second mark, as shown in the second image of Figure 5-6.

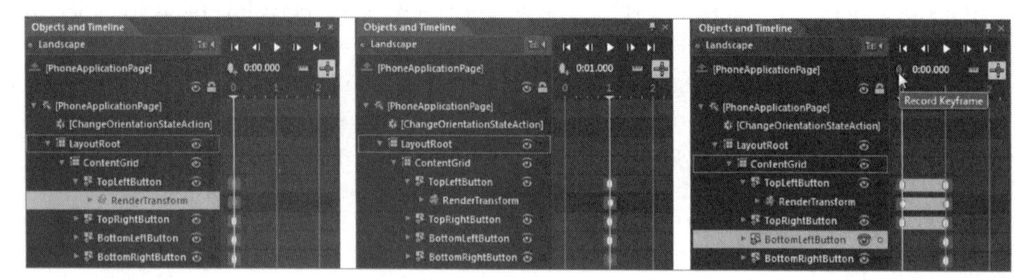

FIGURE 5-6

Last, move the yellow vertical time indicator back to 0, and in turn select each button followed by the "Record Keyframe" button, as shown in the third image of Figure 5-6. This will record a keyframe with the default button values, forming the starting state for the animation sequence. If you ran this now, what you would see is a smooth animated transition from Portrait to Landscape states but an instantaneous switch going the other way. Going the other way is similar, but you have to think in reverse.

Start by selecting the Portrait state in the States window and again opening up the Objects and Timeline window. As there isn't currently any time line for this state, move the yellow vertical time indicator to the 1-second mark, and for each button click the "Record Keyframe" button. This sets up the Base state for the animation. Next, move the time indicator to the 0-second mark, and again create a keyframe for each button. For each of the keyframes at the 0-second mark, go in and modify the location of the button by adjusting the `TranslateX` or `TranslateY` value. In Figure 5-7, you can see that the TopRightButton has been translated by –240 in the X direction, putting it in the old TopLeftButton position.

Once you have defined the translations for each button, your state transitions are complete. You should be able to run your application and see smooth animated transitions between the Portrait and Landscape orientations.

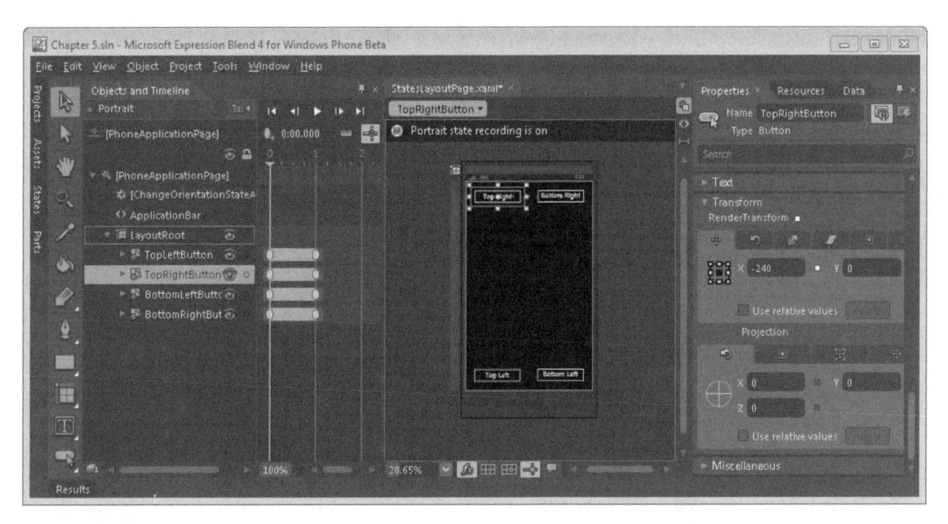

FIGURE 5-7

SOFT INPUT PANEL (SIP)

When you're designing your user experience, one of the things that you have to be aware of is how the user is going to input information. Wherever possible you want to reduce the amount of text that the user has to type. So, for example, rather than presenting a freeform text field, you can provide a list of the most frequently used responses. However, despite your best intentions, there will be cases in which the user will have to enter text. On a Windows Phone this is done via an on-screen keyboard, also known as the *Soft Input Panel* (SIP). Figure 5-8 shows the SIP that is displayed when you click into a text field within Internet Explorer.

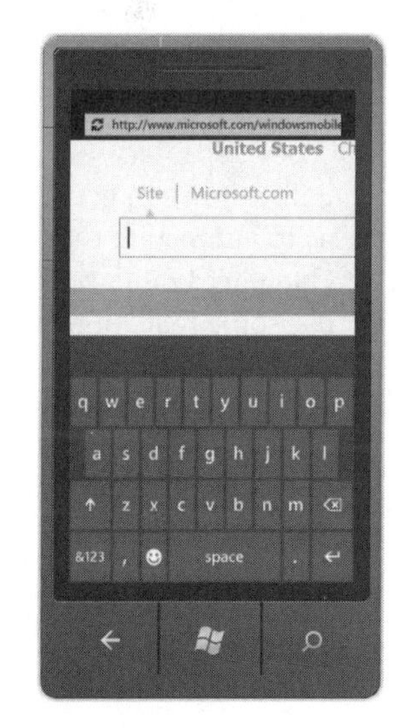

FIGURE 5-8

On devices where there is a physical keyboard, the SIP won't be displayed if the keyboard is accessible. What this means is that for devices where there is a keyboard that slides away, the SIP will be displayed when the keyboard is closed. When the keyboard is opened, the SIP will minimize.

The default behavior in the emulator when a TextBox gets focus is for the SIP to be displayed. While the SIP is displayed you cannot use the keyboard on the host computer to enter text into the emulator. Press the Pause/Break toggle between the keyboard on the host computer (which simulates a slide out keyboard from the point of view of the Windows Phone emulator) and the SIP.

There are some important issues to consider when working with the SIP. The first thing is that unlike previous versions of Windows Mobile where the SIP could be overridden by both application developers and hardware manufacturers, on a Windows Phone the SIP experience will always be the same. The other departure is that you no longer get to control the display or hiding of the SIP. Instead, the SIP is automatically displayed or hidden when a control that requires text input receives or loses focus.

The most commonly used text entry control is, of course, the `TextBox`, and although you can't override the Windows Phone SIP, you can specify the `InputScope` of the `TextBox`. You can think of `InputScope` as the context for which the user is entering text. For example, it might be a phone number, a URL, or a date. `InputScope` has been around in the full .NET Framework for a while but is seldom used. Most applications built for the desktop assume that the user will have access to a full QWERTY keyboard so it doesn't matter what they are entering. However, on a Windows Phone device, where there is only a limited amount of space on the screen, the `InputScope` becomes much more important. If the user is entering a phone number, it isn't necessary to display a full QWERTY keyboard; instead, a numeric pad with 0–9, #, and * would suffice. This would improve usability as the on-screen keys would be much larger and thus easier to find and tap.

To demonstrate the range of different `InputScope` values, you can iterate over all values of the InputScopeNameValue enumeration. In this example, you're going to add a listbox and a textbox to the MainPage. The listbox will display all the possible values of InputScopeNameValue enumeration, and when the user clicks on an item in the list, it will apply that `InputScope` to the `TextBox`. This, in turn, will determine the style of SIP that is displayed.

Before you can get started with the user interface, you need to add a helper method that will allow us to enumerate the values in the InputScopeNameValue enumeration. Unfortunately, the .NET Compact Framework doesn't support the `Enum.GetValues` method, so instead you will need to use some reflection to access a list of possible enumeration values. Add the following code to the `Utilities` class you created earlier:

Available for download on Wrox.com

```
public static KeyValuePair<string,T>[] EnumValues<T>(this Type enumerationType){
    var enumList = new List<KeyValuePair<string, T>>();
    if (enumerationType.BaseType == typeof(Enum)){
        var fields = enumerationType.GetFields(BindingFlags.Public |
                                               BindingFlags.Static);
        foreach (var field in fields){
            var enumValue = field.GetValue(null);
            if (((int)enumValue) > 0){
                enumList.Add(new KeyValuePair<string, T>(enumValue.ToString(),
                                                         (T)enumValue));
            }
        }
    }
    return enumList.OrderBy(kvp=>kvp.Key).ToArray();
}
```

Code snippet from Utilities.cs

This helper method will return an alphabetically ordered list of the different InputScopes that you can apply to a TextBox control. Put this into action by adding a ListBox and a TextBox to the top of your MainPage. The XAML for these controls should look similar to the following:

```
<ListBox Height="226" Margin="43,141,35,0" Name="InputScopeList"
         VerticalAlignment="Top" BorderThickness="2">
    <ListBox.ItemTemplate>
        <DataTemplate>
            <TextBlock Text="{Binding Key}"/>
        </DataTemplate>
    </ListBox.ItemTemplate>
</ListBox>
<TextBox Height="96" HorizontalAlignment="Left" Margin="43,392,0,0"
         Name="InputText" Text="TextBox" VerticalAlignment="Top" Width="402" />
```

Code snippet from MainPage.xaml

The ListBox control has an ItemTemplate property that is used to control what each item in the list looks like. Unlike more traditional list controls that may have only allowed a label and possibly an icon for each item, the Silverlight ListBox control allows you to completely design how each item looks. The ItemTemplate property accepts a single DataTemplate element, which, in turn, can contain a tree of visual elements. In this case, you're only styling each listbox item to contain a single TextBlock element. The text to be displayed in the textblock is specified using the data-binding notation. This will be covered in more detail in Chapter 15, but for now just accept that for each item being displayed in the list, the Key property will be displayed within the TextBlock of the ItemTemplate. The only thing left to do in order to display values in the ListBox is to set the ItemsSource property in the Loaded event handler for the page:

```
private void PhoneApplicationPage_Loaded(object sender, RoutedEventArgs e){
    this.InputScopeList.ItemsSource =
                    typeof(InputScopeNameValue).EnumValues<InputScopeNameValue>();
}
```

Code snippet from MainPage.xaml.cs

In order to change the InputScope of the TextBox, you will need to handle the SelectionChanged event of the ListBox. Select the ListBox, open the Event tab of the Properties window, and double-click on the empty cell next to the SelectionChanged event. The following code uses the SelectedItem property to create a new InputScope that is then applied to the TextBox:

```
private void InputScopeList_SelectionChanged(object sender,
                                             SelectionChangedEventArgs e){
    if (this.InputScopeList.SelectedItem == null) return;
    var selection =
       (KeyValuePair<string, InputScopeNameValue>) this.InputScopeList.SelectedItem;
    InputScope inputScope = new InputScope();
    inputScope.Names.Add(new InputScopeName() { NameValue = selection.Value });
```

```
        this.InputText.InputScope = inputScope;
    }
```

Code snippet from MainPage.xaml.cs

When you run your application, you will be able to select different `InputScopes` and see how the SIP varies. If you don't need to modify the `InputScope` of a `TextBlock`, you can simply include the `InputScope` in the XAML, as in the following example:

```
<TextBox Name="InputScopeText" InputScope="Chat"/>
```

There are eight different SIP configurations in total, which are listed in Table 5-1.

TABLE 5-1: SIP Layouts

SIP	SUMMARY
Default	QWERTY keyboard
Text	Default with additional screen for emoticons
E-Mail Address	Default with keys for @ symbol and .com
Phone Number	T9 text configuration (i.e., 12-key layout optimized for entering a phone number)
Web Address	Default with key for .com
Maps	Default with modified Enter key
Search	Default with keys for search and .com
SMS Address	Default with access to phone number layout

Figure 5-9 shows four of the different SIP configurations.

FIGURE 5-9

In the first image of Figure 5-9, you will notice that the two buttons TopLeft and TopRight are no longer on the screen. As the SIP expands, it adjusts the layout of the current page so that the textbox where text is being added is visible even if its default location would have caused it to be obscured by the SIP. What other controls are on the page, how the controls are aligned, and their visual state will determine whether they remain visible or not after the SIP has expanded. As you build your user experience, it is important to remember this and to thoroughly test each state of your application to make it easy for the user to see the active content even when the SIP is displayed.

As there are no events raised when the SIP is expanded or collapsed, you have to do some of this management yourself. The SIP is displayed when a `TextBox` *gets focus and is hidden when the* `TextBox` *loses focus. Unfortunately, there is no way to know definitively if the SIP is visible as there is no explicit event or property indicating the SIP state. Although you could do your own tracking based on* `TextBox` *focus, if the device has a physical keyboard, the SIP will not be displayed, making it impossible to know for certain whether the SIP is being displayed. You should also be aware that on devices that have a physical keyboard, the SIP will not be displayed when the keyboard is expanded.*

APPLICATION BAR

One of the guiding principles of building a user experience for a Windows Phone is to focus on building and displaying content, rather than chrome. However, there are times when you need to provide a mechanism for the user to perform actions that are either used infrequently, and thus don't necessitate dedicated screen space, or that can't be expressed in an intuitive way using content. For these scenarios there is the Application Bar.

The Application Bar resides on the same side of the screen as the three hardware buttons. Figure 5-10 shows the Application Bar, made up for one icon button and four menu items, displayed full in both Portrait and Landscape modes. Notice that in Landscape mode, the Application Bar still resides on the same side of the screen as the hardware buttons, and both the icon button and the text have been rotated so that they are readable.

FIGURE 5-10

Figure 5-10 illustrates the Application Bar in its expanded state. When it is minimized, only the icon buttons are visible. The ellipses on the right side (top side when in Landscape mode) are used to expand and collapse the Application Bar.

On a Windows Phone the Application Bar serves a similar purpose to more traditional menus. Unlike a menu system, however, there is no hierarchy of elements, and you are constrained by the number of icon buttons you can use to just four. While you can add more menu items, you should try to keep this to around five to prevent the user having to scroll through a list of items.

Icon Buttons

There is a trade-off between using icon buttons and menu items. On the one hand, icon buttons are smaller and will always be visible on the Application Bar, but on the flip side, an icon can be hard to interpret or misunderstood. You should ensure that the icons you select are meaningful. If they do a destructive task, such as deleting an item, you should always prompt the user to confirm that they wish to proceed. Since the icon buttons on the Application Bar remain visible whether the Application Bar is expanded or not, you should make the most frequently used tasks into icon buttons, with the remaining tasks menu items. When the Application Bar is expanded, in addition to displaying the menu items, the Text property of the icon button is displayed.

As the context of the application changes, you may want to dynamically adjust the icons that are visible. For example, in an e-mail application, you may have a "Create E-mail" icon that is always visible. Then, depending on whether an e-mail is selected, you can display icons for deleting or replying to the selected e-mail. The forward task may be something that gets added as a menu item as it is used less frequently.

Microsoft has come out with several recommendations around how to design icons that will work well within a Windows Phone application:

- The size of the icons in the Application Bar is 48 × 48 pixels.
- The circle that surrounds an icon is added by Windows Phone (so should not be included within your icon).
- To avoid overlapping the circle, your icon should appear within a 26 × 26 square in the center of the icon.
- The icon should have a transparent background with the actual icon being drawn in white. Depending on the theme that the user is running, the icon will be appropriately colored.

Figure 5-11 illustrates the basic process for creating an application icon. Start off with an image that is 48 × 48 pixels in size (the dotted square border in Figure 5-11 is there for illustrative purposes and shouldn't be added to your image). Add a circle that touches the edges of the image, followed by a 26 × 26 solid square in the center of the image. These are both guides to indicate where you should be drawing your icon. If your editing tool supports layers, then place these guides on a separate layer so that you can remove them once you've finished creating your icon.

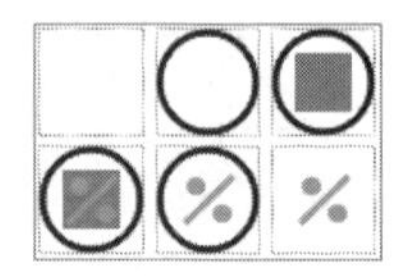

FIGURE 5-11

Add your icon using a white foreground (a colored percent sign has been added in Figure 5-11 for contrast but would be created in white when building a real icon). Then remove the guides, leaving your icon on a transparent background.

Icon Library

To give you a head start and to try to encourage developers to use a common set of icons, Microsoft has released a pack of icons that can be downloaded from `http://go.microsoft.com/fwlink/?LinkId=187311`. Figure 5-12 illustrates some of the icons that are available in the pack.

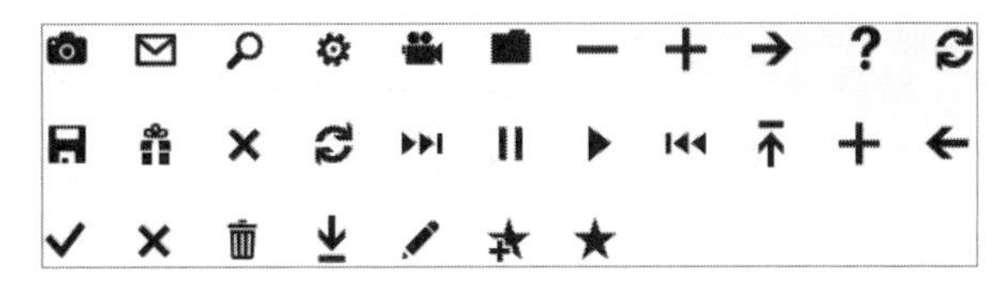

FIGURE 5-12

These icons are available for you to download and reuse within your application. This is particularly useful for common actions such as save, delete, search, and refresh where you can reuse the same icons that are used throughout the Windows Phone user interface.

Adding an Icon

Once you have created, or selected, the icon that you want to use within your application, the first thing you need to do is to include it in your solution. To keep the structure of your project tidy, you may want to create a subfolder called *icons* in which to place your icons. So that your icon can be used with your application you need to copy the icon into your project folder, or if you have created a subfolder, copy the icon into that folder. Within Visual Studio, use the Solution Explorer window to locate the icon, right-click on it, and select "Include In Project."

Your icon may not show up in Solution Explorer for a couple of reasons. First, you need to tell Solution Explorer to display all files, regardless of whether they are included in the project or not. Do this by selecting your project and clicking the "Show All Files" button in the icon bar of the Solution Explorer window. If Solution Explorer is already showing all files, your new icon may not be visible until Solution Explorer refreshes the files list. You can force an update by clicking the Refresh button in the icon bar of the Solution Explorer window.

By default, the icon will be included with a build action of Resource. You need to change the "Build Action" to Content and the "Copy to Output Directory" to "Copy always" in the Properties window.

Now you are ready to create your Application Bar and an icon to be displayed on it. There is only minimal designer support for creating Application Bar icons in Visual Studio, so it's easiest to manipulate the Application Bar in Blend. Open the PhoneApplicationPage that you want to add the Application Bar to and select the [PhoneApplicationPage] node in the Objects and Timeline window. From the Common Properties section of the Properties window, click the New button next to the ApplicationBar text. This will create a new ApplicationBar as shown in Figure 5-13.

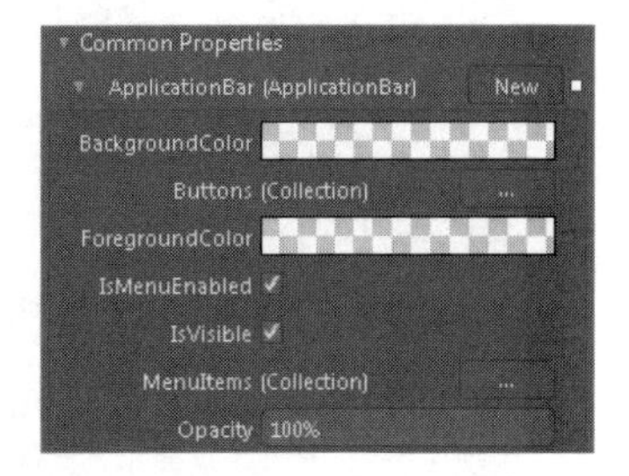

FIGURE 5-13

There are a number of properties that you can configure on the ApplicationBar that are listed in Table 5-2.

TABLE 5-2: ApplicationBar Properties

PROPERTY	DESCRIPTION
`BackgroundColor`	Sets the background color of the Application Bar.
`ForegroundColor`	Sets the foreground color of the Application Bar.
`IsMenuEnabled`	Whether or not the user can expand the Application Bar menu. If this property is set to false, clicking the ellipses will expand the Application Bar to show the text associated with the icon buttons but not the menu items. Use this as an alternative to adding and removing menu items.
`IsVisible`	By default, the Application Bar will be visible. Setting this property to `false` will hide both the icon bar and menu items.
`Opacity`	Controls the opacity of the Application Bar.

Clicking the ellipses next to the Buttons (Collection) text in Figure 5-13 will display the collection editor for adding, modifying, and removing ApplicationBar buttons. The collection editor is actually a generic editor that Blend uses to modify collections of any type. As such, in order to add an icon button to the ApplicationBar you have to specify what Type of item to add to the collection. Click the "Add another item" button, at the bottom of Figure 5-14, and locate the ApplicationBarIconButton class (you may have to check the "Show all assemblies" checkbox).

When you click OK, an instance of the ApplicationBarIconButton class is added to the collection (left pane of Figure 5-15), and the properties of the new item are displayed in the right pane.

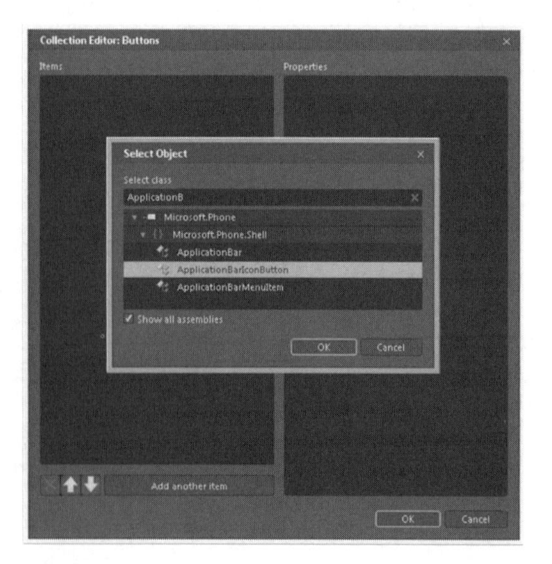

FIGURE 5-14

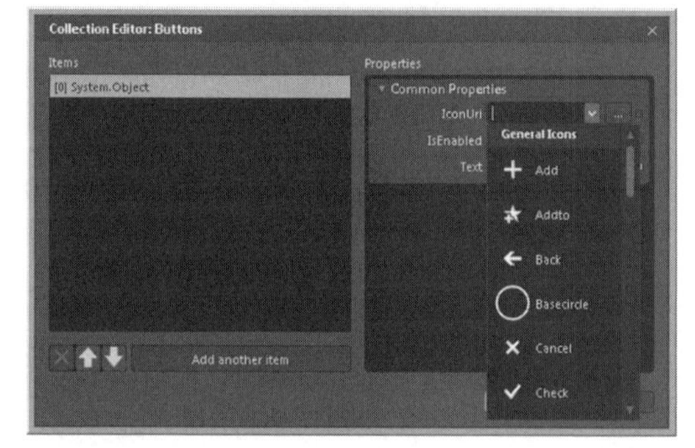

FIGURE 5-15

Table 5-3 lists the properties that can be configured on each icon button.

TABLE 5-3: ApplicationBarIconButton Properties

PROPERTY	DESCRIPTION
`IconUri`	The URI of the icon to be displayed. You can either select one of the predefined icons (these ship as part of the Windows Phone developer tools and will be automatically copied into your application if selected), or you can enter, or browse to, the URI of an icon you wish to use within your application.
`IsEnabled`	Whether the icon is enabled or not. Setting `IsEnabled` to `false` does not remove the icon from the Application Bar. Instead, it appears dimmed to indicate that it is not enabled.
`Text`	This is the text that will be displayed below the icon when the Application Bar is expanded. The text will always be displayed in lowercase for consistency across Windows Phone.

After adding and configuring the icon button you will notice that the icon appears at the bottom of the page in the designer (see Figure 5-16). If you select the icon button you will notice that the layout changes to expand the ApplicationBar, as shown in the right image of Figure 5-16.

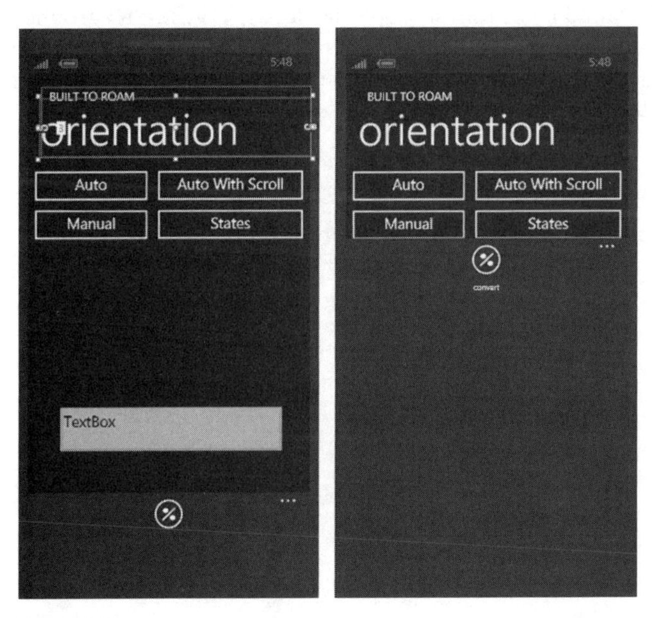

FIGURE 5-16

The following snippet contains the XAML for declaring an ApplicationBarIconButton and assigning it to a new instance of the ApplicationBar.

```
<phone:PhoneApplicationPage
    x:Class="ApplicationLayouts.AppBarPage"
    xmlns:phone="clr-namespace:Microsoft.Phone.Controls;assembly=Microsoft.Phone"
    xmlns:shell="clr-namespace:Microsoft.Phone.Shell;assembly=Microsoft.Phone"
    ... >
    <phone:PhoneApplicationPage.ApplicationBar>
        <shell:ApplicationBar>
            <shell:ApplicationBarIconButton Text="convert"
                                            IconUri="/icons/Percent.png"
                                            IsEnabled="True"
                                       Click="ApplicationBarIconButton_Click" />
        </shell:ApplicationBar>
    </phone:PhoneApplicationPage.ApplicationBar>
```

Here you only add a single `ApplicationBarIconButton`, but you can add more icons either through the collection editor, or by manually adding more `ApplicationBarIconButton` elements under the `shell:ApplicationBar` element. This snippet also includes an event handler for the Click event.

When you set the background and foreground colors on the Application Bar, you are overriding the default colors specified by the theme that the user has selected for his or her device. There is nothing to say that you can't do this, but you should be aware that you should only do this in order to adapt the Application Bar to your application style and branding. Where possible, it is recommended that you allow the user to specify the theme of his or her device and for your application to adapt to that theme.

As you would imagine, setting the background color for the Application Bar sets the background for both the icon bar and the Menu Items list when it is expanded. One thing to be aware of is that by setting the foreground color, you are not only setting the color of the text displayed for the menu items, but you are also changing the color that will be used to colorize any Application Bar icons.

Menu Items

Adding menu items is done the same way as adding icons to the Application Bar. Instead of adding an `ApplicationBarIconButton` to the Buttons collection, you add `ApplicationBarMenuItem` elements to the MenuItems collection. This is shown in the following snippet that adds a Clear menu item to the Application Bar:

Available for download on Wrox.com

```
<phone:PhoneApplicationPage.ApplicationBar>
    <shell:ApplicationBar IsMenuEnabled="True">
        <shell:ApplicationBar.MenuItems>
            <shell:ApplicationBarMenuItem Text="Clear"
                                          x:Name="ClearMenuItem"
                                          Click="ClearMenuItem_Click"/>
        </shell:ApplicationBar.MenuItems>
        <shell:ApplicationBarIconButton Text="convert"
                                            IconUri="/icons/Percent.png"
                                            IsEnabled="True"
                                       Click="ApplicationBarIconButton_Click" />
    </shell:ApplicationBar>
</phone:PhoneApplicationPage.ApplicationBar>
```

Code snippet from AppBarPage.xaml

As you can see, the `ApplicationBarMenuItem` has a `Text` property that defines the text that will be displayed for the menu item in the Application Bar. To avoid the menu item running over the edge of the screen, Microsoft recommends a maximum length of between 14 and 20 characters. As mentioned earlier, you should avoid having more than five menu items where possible to avoid the need for the user to scroll the Menu Items list. The last thing to be aware of is that the text of the menu item will always be converted to lowercase. This, again, is to preserve a consistent experience across the Windows Phone.

The `ApplicationBarMenuItem`, like the `ApplicationBarIconButton`, exposes an `IsEnabled` property that can be set via code or in XAML. This determines whether the menu item is enabled. When not enabled, the menu item appears dimmed to indicate that it is not operational.

Opacity

One of the unusual things about the Application Bar is that you can set the opacity level. Intuitively this doesn't make much sense since the Application Bar resides at the bottom of the screen, effectively reducing the screen size that you have to work within your application. This is the case if you leave the Opacity at its default value of 1. However, as soon as the Opacity is less than 1, the Application Bar sits over the page, allowing content to be rendered beneath it and showing through. Figure 5-17 illustrates the Application Bar with Opacities of 1, 0.5, and 0.

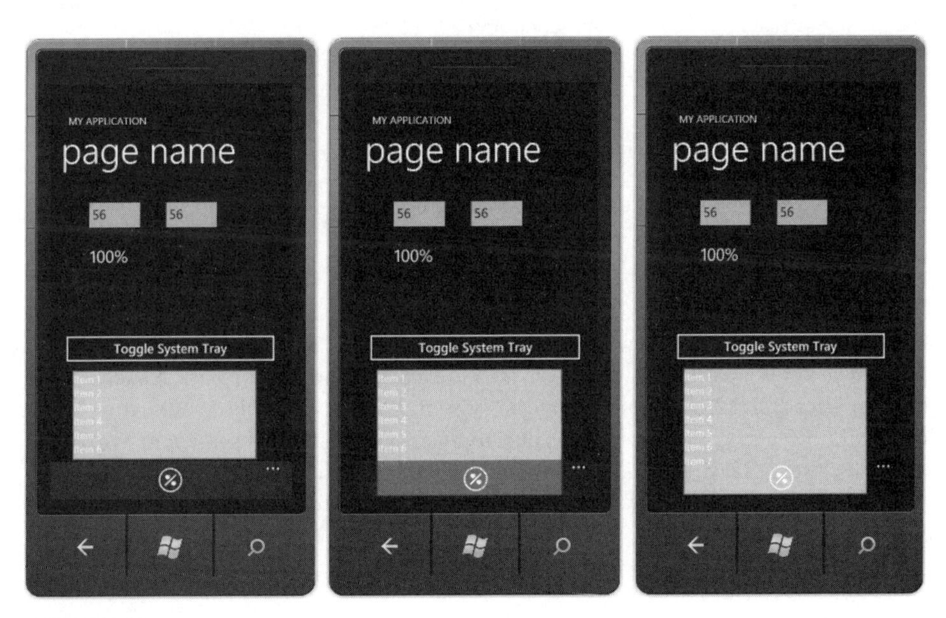

FIGURE 5-17

As you can see from Figure 5-17, as you adjust the Opacity, you can start to see the content appear through the Application Bar. However, you can interact with that content. In this case, this would obscure the listbox that is displayed at the bottom of the screen, making it hard to select some items.

StateChanged Event

The `ApplicationBar` exposes a `StateChanged` event that you can handle in order to detect when the Menu Items list is being displayed. You may want to adjust content layout or the menu items based on the current context. To attach an event handler, simply modify the XAML to include the name of the event handler you want to be invoked when the state of the `ApplicationBar` changes:

```
<phone:PhoneApplicationPage.ApplicationBar>
    <shell:ApplicationBar StateChanged="ApplicationBar_StateChanged">
    ...
    </shell:ApplicationBar>
</phone:PhoneApplicationPage.ApplicationBar>
```

Code snippet from AppBarPage.xaml

In this case, you're going to adjust the Opacity of the `ApplicationBar` dynamically, setting it to 1 when the Menu Items list is expanded (i.e., when the `IsMenuVisible` property is set to `true`), and restoring it back to its previous value when the Menu Items list is collapsed again:

```
private double PreviousOpacity { get; set; }
private void ApplicationBar_StateChanged(object sender,
                                        ApplicationBarStateChangedEventArgs e){
    var opacity = this.ApplicationBar.Opacity;
    this.ApplicationBar.Opacity = e.IsMenuVisible ? 1 : PreviousOpacity;
    this.PreviousOpacity = opacity;
}
```

Code snippet from AppBarPage.xaml.cs

SYSTEM TRAY

Windows Phone applications typically have a small amount of screen real estate dedicated at the top of the screen to what is known as the System Tray. This is an area where information such as battery level and connection strength can be displayed by the system. It is not possible to add icons or text to this area. You can, however, control whether your application runs in full screen, or whether it leaves space for the System Tray. This is done by setting the IsVisible property on the static SystemTray class. When you create a new PhoneApplicationPage the template sets this property to True by default.

```
<phone:PhoneApplicationPage
    x:Class="ApplicationLayouts.AppBarPage"
    xmlns:shell="clr-namespace:Microsoft.Phone.Shell;assembly=Microsoft.Phone"
    ...
    shell:SystemTray.IsVisible="True">
```

If you don't want the System Tray to be displayed you can either set this property to False, or you can modify this value in code.

```
private void SystemTrayButton_Click(object sender, RoutedEventArgs e){
    SystemTray.IsVisible = !SystemTray.IsVisible;
}
```

Figure 5-18 illustrates a page with the IsVisible property set to `False` (left image) and `True` (right image). Note that the System Tray does take up space that would otherwise be allocated to your application. If you attach an event handler to the SizeChanged event of the PhoneApplicationPage you can see how the size changed when the System Tray is visible or not. The SizeChanged event is also triggered if you adjust the Opacity of the Application Bar from 1 to a value less than 1. When the Opacity is set to 1 the Application Bar, like the System Tray, takes up space that the application would otherwise have for displaying content.

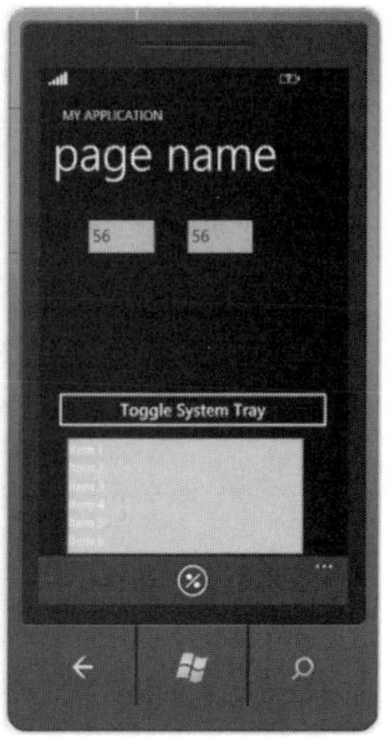

FIGURE 5-18

While your first instinct might be to set the IsVisible property to `false` *to maximize the amount of space for your application to display content, you have to be aware that this can have an adverse effect on your application. The System Tray is there to indicate connectivity, battery level and other important information to the user. If you do not display the System Tray you are not only preventing the user accessing this information, you are also breaking consistency with other applications and the Metro experience as a whole. There may be some pages within your application where you want to maximize screen real estate (such as viewing photos or a map) but in most cases you should display the System Tray where possible.*

SUMMARY

In this chapter, you have seen how to extend your Windows Phone application to handle different orientations. You also saw that different overlays such as the SIP and the Application Bar can affect the layout and usability of your application. It is important to consider the effects of both orientation and overlays when architecting the visual layout of your application.

Navigation

WHAT'S IN THIS CHAPTER

- How Windows Phone applications navigate between pages
- Handling the Back button to cancel operations within a page
- Using animations to transition between pages
- How the Windows Phone execution model works
- Handling to save persistent and transient data when the application goes into the background

Unless you're building a very simple application, your user experience will consist of multiple pages or views. As part of designing your application you will need to determine what transitions are possible and how you are going to navigate between different areas of your application.

You will also need to ensure that your application plays well with the rest of the Windows Phone. This includes integrating with the Back button, to ensure that users can cancel dialogs or navigate within the application. In this chapter, you will learn how to control navigation into, through, and out of your application. You will also learn how to persist the current state of your application so that the user isn't aware if your application has to be restarted.

PAGE LAYOUT AND ARCHITECTURE

Up until now everything that you've been doing has been on a single `PhoneApplicationPage`. In a real application it is likely that you will want to have multiple pages within your application through which the user will navigate. Designing applications for a Windows Phone is much harder than designing for the larger screens found on laptops or desktop computers. Where you might have had a single screen on a desktop application, you may need multiple pages in a Windows Phone application. This adds an extra dimension of complexity as it's very easy for users to get lost as they go between pages.

If you think about the design of a Windows Phone page, there are several options as to how you lay out the page (see Table 6-1). How you lay out the page will have some bearing on how you navigate to and from the page. You should be mindful about how users will use their Windows Phones and the navigation experience they will be familiar with from using the devices.

TABLE 6-1: Page Layouts

LAYOUT	DESCRIPTION
Panoramic	The Panoramic view, also known as a *hub*, is designed as a central point from which users can access information pertaining to a particular topic. For example, the Picture hub allows users to jump to different folders and view the current album. You should remember that the Windows Phone screen acts as a view port into the panorama. This means that at any stage, the user will only ever see a section of the panorama that's the width of the Windows Phone screen. You need to factor that into the design of the panorama so that users are always aware that they can (and perhaps be teased into doing so) scroll to the left and right. If you want to use a panorama in your application, you should consider using it as the first or home screen of your application.
Pivot	Not to be confused with the Panoramic view, the Pivot also encourages scrolling to the left and right. However, the Pivot is closer to the more traditional tabbed view in which each tab represents a subview that would be displayed when the tab is selected. There are no limits to the number of, or the content within, the views of a Pivot. This control is most commonly used to present different views of the same data. For example, the Pictures page uses a Pivot to allow the user to view all pictures, sort by date, or view their favorite pictures. In your application you should consider using a Pivot where you have long lists for which you want to provide alternative orderings (e.g., alphabetic, historic, most used), or for data for which it is possible to provide alternative visual representations.
List	If you have more items to display than will fit on a single page, it's quite useful to use a vertically scrolling list. You may want to consider whether you should make the list single orientation or allow the user to rotate into Landscape mode and enable horizontal scrolling. When users are scrolling horizontally, you may have more space to provide details on the items that they are scrolling through, so this layout is definitely worth considering. For large lists you may want to consider providing the user with ways to filter the list. This may be with a search box, or by grouping the items in some manner. You may also want to consider paging the list so that the user can easily scroll down a page at a time.

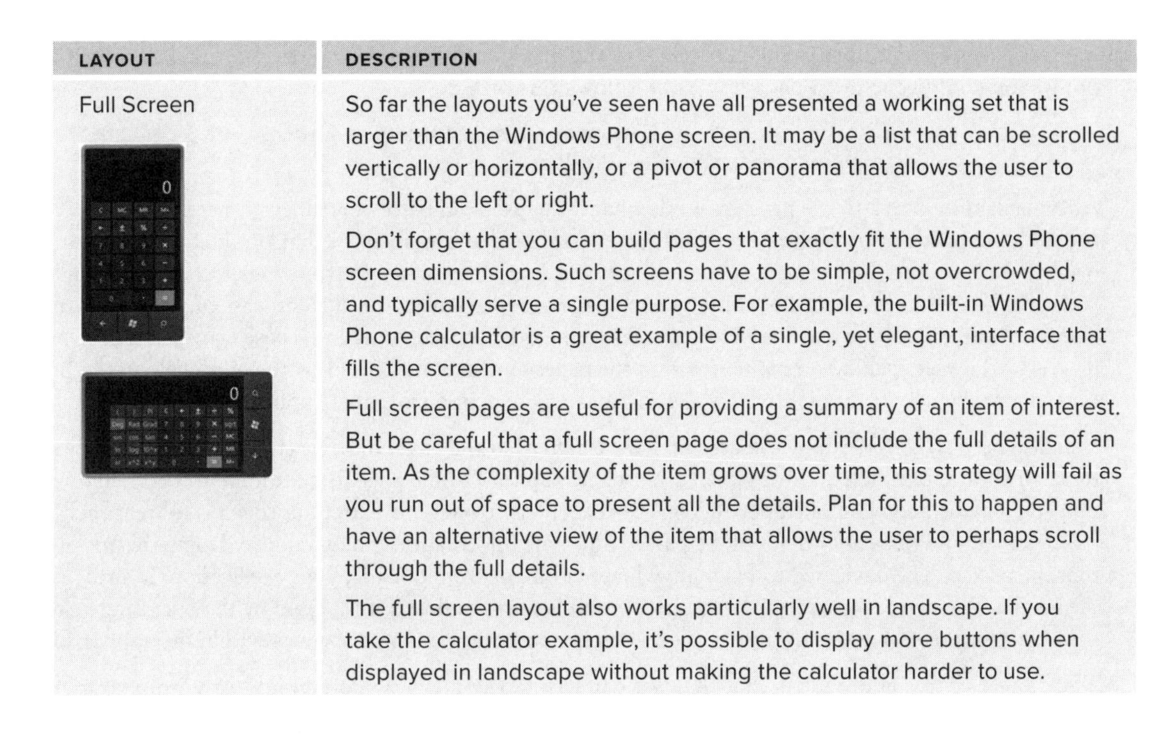

LAYOUT	DESCRIPTION
Full Screen	So far the layouts you've seen have all presented a working set that is larger than the Windows Phone screen. It may be a list that can be scrolled vertically or horizontally, or a pivot or panorama that allows the user to scroll to the left or right. Don't forget that you can build pages that exactly fit the Windows Phone screen dimensions. Such screens have to be simple, not overcrowded, and typically serve a single purpose. For example, the built-in Windows Phone calculator is a great example of a single, yet elegant, interface that fills the screen. Full screen pages are useful for providing a summary of an item of interest. But be careful that a full screen page does not include the full details of an item. As the complexity of the item grows over time, this strategy will fail as you run out of space to present all the details. Plan for this to happen and have an alternative view of the item that allows the user to perhaps scroll through the full details. The full screen layout also works particularly well in landscape. If you take the calculator example, it's possible to display more buttons when displayed in landscape without making the calculator harder to use.

In addition to understanding different screen layouts, you also need to consider how your application is pieced together. Typical rich client applications on the desktop have a central window from which dialogs are displayed — for example, Word 2010 in the left image of Figure 6-1. Alternatively, websites usually have chrome that includes branding and menus in the header, and a variety of useful links in the footer, as in the right image of Figure 6-1.

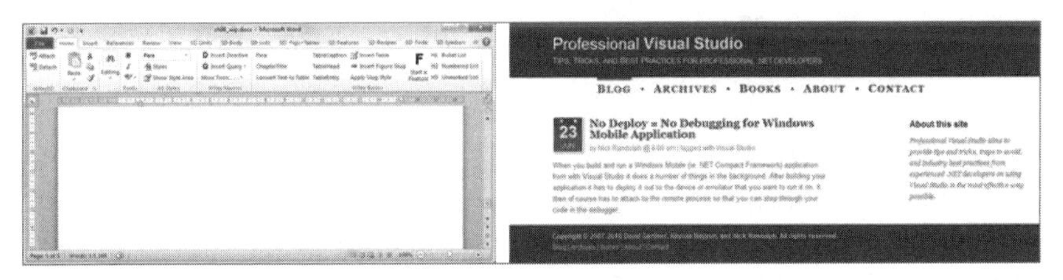

FIGURE 6-1

Pages within your Windows Phone application should be chrome-free. This means no header, no footer, and only minimal branding. Remember that one of the guiding principles of the Metro design language is "Content, Not Chrome." With this in mind, you have to think carefully about how the user is going to step through your application; how you're going to build a consistent, predictable, and intuitive navigation sequence; and how you're going to minimize the risk of the user getting lost within your application.

Let's take an example application, based on the AdventureWorks concept of a bike retailer. If you take a data-first approach, then there is a natural hierarchy of customers, who place orders, which are made up of order lines, which correlate to the product ordered. Using this scenario as a starting point, you could generate a page flow that follows this order:

List of Customer → Customer → List of Orders → Order (including order lines) → Product

After navigating down to the product level, what if the user wants to view all customers that have ordered that product, or wants to view all current orders that are open for that product, or even go back to the customer list? Instead of assuming that a user will always go from page to page in a single predefined sequence, you need to think of each page as representing the current state of your application and then map out all the reasonable transitions to the other states, or pages. These transitions may be driven by the user clicking on content within the page; they may be caused by the user clicking the Back button; or they may be a result of the user invoking some action from the Application Bar.

You also need to consider how the user may have arrived at the page. In some cases, you may want to go through your application and list which other pages can lead to the page in question. However, in doing this you may be adding implicit dependencies between those pages. A better approach is to treat each page as a stand-alone entity that will be passed any state information it may require. Doing this means that the page can be navigated to from anywhere within, or from outside, your application. It also means that you can easily persist the state of the application so that if it is closed by the operating system, the next time the user opens the application you can return her to where she was within the application.

The danger of this approach is that the user can easily become bewildered and lost within your application. In theory, the user could navigate to any page from any other page. As there isn't a single path through the application, users may get confused as to how to navigate to parts of the application they have not been to before, or may not discover a portion of the application because it isn't easily accessible. A strategy to address this is to have one, or in some cases, multiple, home pages. For small applications, a single home page, perhaps using a Panoramic view, that can easily be returned to from anywhere within the application, can serve as a launching pad. In the case of large applications, it may be necessary to split the home page into multiple pages, all acting as launching areas for particular sections of the application. You, of course, need to map how the user can go between these areas without getting lost.

NAVIGATION

Applications built using Windows Presentation Foundation (WPF) are designed in a more traditional manner using a *window* as the root container for the application. On the other hand, Silverlight was initially designed to run in the context of a Web browser so there was no concept of a *window*. Instead, the root container for a Silverlight application was a *UserControl*, and all controls the developer added, removed, showed, or hid were constrained to this single rectangular region:

SILVERLIGHT

```
<UserControl x:Class="SilverlightApplication.MainPage"
    xmlns="http://schemas.microsoft.com/winfx/2006/xaml/presentation"
    xmlns:x="http://schemas.microsoft.com/winfx/2006/xaml">
    <Grid x:Name="LayoutRoot">
        ...
    </Grid>
</UserControl>
```

It quickly became evident that, like the Web, a Silverlight application needed a navigation system to allow developers to create discrete screens and then have the user navigate between them. This led to the introduction of a frame control that could be inserted onto the user control and have different pages loaded within it:

SILVERLIGHT

```
<UserControl x:Class="SilverlightApplication.MainPage"
    xmlns="http://schemas.microsoft.com/winfx/2006/xaml/presentation"
    xmlns:sdk="http://schemas.microsoft.com/winfx/2006/xaml/presentation/sdk"
    xmlns:x="http://schemas.microsoft.com/winfx/2006/xaml">
    <Grid x:Name="LayoutRoot">
        <sdk:Frame x:Name="ContentFrame" Source="/Views/Home.xaml" />
    </Grid>
</UserControl>
```

Each page could then be designed and implemented in relative isolation:

SILVERLIGHT

```
<sdk:Page x:Class="SilverlightApplication.Home"
    xmlns="http://schemas.microsoft.com/winfx/2006/xaml/presentation"
    xmlns:x="http://schemas.microsoft.com/winfx/2006/xaml"
    xmlns:sdk="http://schemas.microsoft.com/winfx/2006/xaml/presentation/sdk"
    Title="Home">
    <Grid x:Name="LayoutRoot">
        ...
    </Grid>
</sdk:Page>
```

One of the advantages of this approach was that developers could customize the surrounds of the frame to give all pages within their applications a common header or footer. As you've learned previously, having such chrome in a Windows Phone application is not advisable, so the root container for a Windows Phone application is simply the frame control, into which pages are loaded.

In the applications you have built so far, you have mostly been working with a single `PhoneApplicationPage`, typically called `MainPage`. Within a Windows Phone application, a `PhoneApplicationPage` represents a page that is to be hosted within a `PhoneApplicationFrame`. If you examine the WMAppManifest.xml file, you will notice that there is a `DefaultTask` element that sets the default page for the application to open to be `MainPage.xaml`:

```
<Deployment
    xmlns="http://schemas.microsoft.com/windowsphone/2009/deployment"
    AppPlatformVersion="7.0">
    <App xmlns="" ProductID="{cefb062b-2e5d-4ea9-befd-efcddefe448d}"
         Title="Navigation" RuntimeType="Silverlight" Version="1.0.0.0"
         Genre="apps.normal"  Author="Navigation author"
         Description="Sample description" Publisher="Navigation">
        ...
        <Tasks>
            <DefaultTask  Name ="_default" NavigationPage="MainPage.xaml"/>
        </Tasks>
```

When the application loads, a PhoneApplicationFrame is created. This frame is then told to navigate to the MainPage. While the page is being loaded and navigated to, the splash screen is displayed. When navigation is complete the Navigated event is raised, at which time the PhoneApplicationFrame is set as the RootVisual for the application. The application is then ready to accept user interaction. Figure 6-2 illustrates this process, highlighting that the splash screen will be visible while the first page is loaded and navigated to.

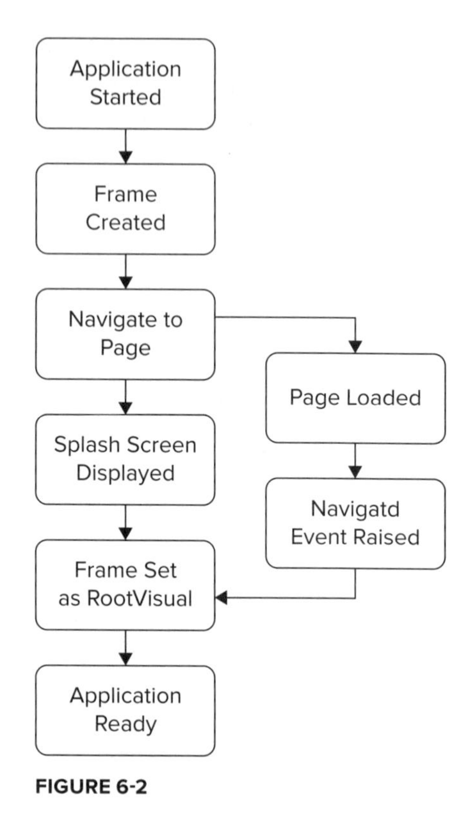

FIGURE 6-2

While the splash screen will be displayed, indicating to the user that the application is being loaded, it is advisable to keep the start time as short as possible for your application. There are specific policies in the documentation for the Windows Phone Marketplace that limit the length of time an application can take to load. It is advisable to consult this documentation (available from `http://developer.windowsphone.com`*) prior to submitting your application to the Marketplace.*

Once the application is running, the MainPage can be replaced with a different page by setting the frame's Source property to a relative URI that references the page you want to open:

Available for download on Wrox.com

```
private void ChangeSourceButton_Click(object sender, RoutedEventArgs e){
    var frame = Application.Current.RootVisual as PhoneApplicationFrame;
    frame.Source = new Uri("/SecondPage.xaml",UriKind.Relative);
}
```

Code snippet from MainPage.xaml.cs

Behind the scenes, setting the `Source` property causes the `Navigate` method on an instance of the `NavigationService` class to be called. In a Windows Phone application, when the frame is instantiated it creates an instance of the `NavigationService`. This instance is a singleton that is used throughout the application in order to journal all navigation actions between pages. When the `Navigate` method is called, the `NavigationService` is responsible for locating, loading, journaling, and raising appropriate events at various points while the content of the old page is removed and the new page is displayed. Instead of setting the `Source` property, the frame also provides a `Navigate` method, which is simply a pass through to the `Navigate` method on the `NavigationService`. Instances of the `PhoneApplicationPage` have a read-only `NavigationService` property that will return the instance of the `NavigationService` class, on which you can call `Navigate` directly:

Available for download on Wrox.com

```
private void NavigateViaFrameButton_Click(object sender, RoutedEventArgs e){
    var frame = Application.Current.RootVisual as PhoneApplicationFrame;
    frame.Navigate(new Uri("/SecondPage.xaml", UriKind.Relative));
}

private void NavigateViaPageButton_Click(object sender, RoutedEventArgs e){
    this.NavigationService.Navigate(new Uri("/SecondPage.xaml", UriKind.Relative));
}
```

Code snippet from MainPage.xaml.cs

When the `Navigate` method is called, there are several events that are raised to indicate when different stages of the navigation process are complete. Figure 6-3 illustrates that the first event to be raised is the `Navigating` event. Unlike the other navigation events, this event carries a `NavigationCancelEventArgs` argument that inherits from `CancelEventArgs`. The significance of this is that in the event handler you can set the `Cancel` property of this argument to `true` and the navigation will be canceled. The other events are `Navigated`, to indicate successful navigation; `NavigationStopped`, to indicate that the previous navigation was stopped (either by another navigation action or a call to `StopLoading`); and `NavigationFailed`, to indicate an exception or error in navigating.

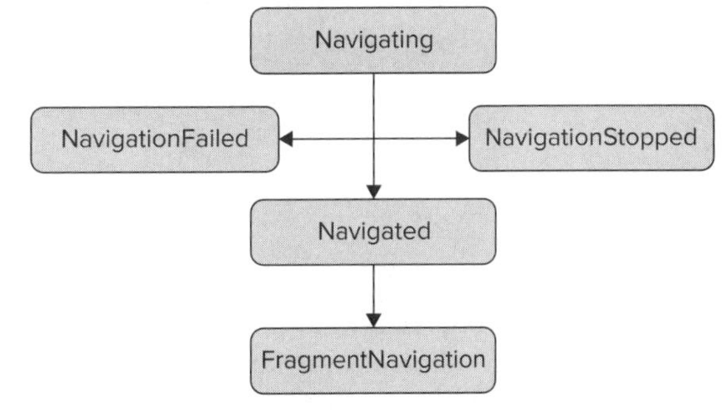

FIGURE 6-3

Navigation events are exposed in a number of ways within a Windows Phone application. On both the `NavigationService` itself and the `PhoneApplicationFrame` you can add event handlers for the appropriate event. For example, if you wanted to detect when a navigation action was triggered, you could add an event handler for the `Navigating` event on the `PhoneApplicationFrame` instance:

Available for download on Wrox.com

```
public App(){
    UnhandledException += Application_UnhandledException;
    InitializeComponent();
    InitializePhoneApplication();

    RootFrame.Navigating += App_Navigating;
}

void App_Navigating(object sender, NavigatingCancelEventArgs e){
    Debug.WriteLine("Navigating to " + e.Uri);
}
```

Code snippet from App.xaml.cs

The `PhoneApplicationPage` doesn't expose the navigating events directly. Instead, you need to override one of the virtual methods exposed by the `PhoneApplicationPage` base class:

- `OnNavigatingFrom` — Method invoked when the user is about to navigate away from this page
- `OnNavigatedFrom` — Method invoked when the page has been navigated away from
- `OnNavigatedTo` — Method invoked when this page has been navigated to
- `OnFragmentNavigation` — Method invoked to handle the navigation to a URI fragment once a page has been navigated to

For example, if you wanted to handle the navigation away from a page, you would override the `OnNavigatingFrom` method of the `PhoneApplicationPage`. In this case, the navigation will be canceled if the user decides not to confirm the navigation away from the page:

Available for download on Wrox.com

```
protected override void OnNavigatingFrom(NavigatingCancelEventArgs e){
    if (MessageBox.Show("Are you sure?", "Really?",
                        MessageBoxButton.OKCancel) != MessageBoxResult.OK){
        e.Cancel = true;
    }
    base.OnNavigatingFrom(e);
}
```

Code snippet from MainPage.xaml.cs

You may be wondering what the relative merits are between attaching event handlers to the `NavigationService` or overriding the navigation methods on a `PhoneApplicationPage` are. The simple answer is that from a functionality or performance point of view, there is minimal difference. However, if you attach event handlers to the `NavigationService` from within a page, you need to ensure that you detach the event handlers when the page is navigated away from. Otherwise you will end up with a buildup of navigation event handlers as more pages are opened. If, instead, you override the navigation methods within the PhoneApplicationPage, this is all handled for you behind the scenes. On the flip side, attaching to the events exposed by the NavigationService object can easily hook an event handler up for navigation events across all pages within your application with a single modification to your application's source code.

Fragments and QueryString

There is one additional event, FragmentNavigation, that may be fired once a page has been navigated to. This event is raised when a fragment is added to the page being navigated to. For example, in the following snippet, the string HelloWorld is added after a hash sign to the /SecondPage.xaml Uri that is being navigated to:

Available for download on Wrox.com

```
private void NavigateWithFragmentButton_Click(object sender, RoutedEventArgs e){
    this.NavigationService.Navigate(new Uri("/SecondPage.xaml#HelloWorld",
                                            UriKind.Relative));
}
```

Code snippet from MainPage.xml.cs

The way that the NavigationService processes the Uri is to, first navigate to the page (i.e., SecondPage) and then raise the FragmentNavigation event so that the fragment can be handled by the application after the page has loaded. In this case, in SecondPage you would override the OnFragmentNavigation method and then be able to extract the fragment from the supplied argument:

Available for download on Wrox.com

```
protected override void OnFragmentNavigation(FragmentNavigationEventArgs e){
    base.OnFragmentNavigation(e);

    MessageBox.Show("Fragment: " + e.Fragment);
}
```

Code snippet from SecondPage.xaml.cs

Specifying a URI fragment is great for static intra-page navigation — for example, specifying which section of a Panorama or Pivot control should be scrolled into view when a heading is tapped. However, it isn't good for passing more than one piece of information into the page being navigated to. Take the scenario in which you want to pass in a customer ID and the ID of the product they are interested in. In this case, there is another alternative, which is to add a query string to the end of the Uri:

Available for download on Wrox.com

```
private void NavigateWithQueryButton_Click(object sender, RoutedEventArgs e){
    this.NavigationService.Navigate(new Uri(
                                "/SecondPage.xaml?CustomerId=1234&Product=555",
                                UriKind.Relative));
}
```

Code snippet from MainPage.xaml.cs

Within the page being navigated to, you can access the NavigationContext property to extract the name–value pairs that correspond to the query string:

Available for download on Wrox.com

```
protected override void OnNavigatedTo(NavigationEventArgs e){
    base.OnNavigatedTo(e);

    foreach (var item in NavigationContext.QueryString){
        MessageBox.Show("Query String [" + item.Key + "] = " + item.Value);
    }
}
```

Code snippet from SecondPage.xaml.cs

One of the difficulties with both fragments and query strings is that they are a simple string or pairs of name–values. This can pose a problem if the data you want to pass between pages is more complex or slow to fetch. An alternative to trying to serialize these objects into a string is to make them available at an application level. For example, you might create a simple singleton key-based dictionary where objects can be cached.

You need to be careful, though, because this introduces a stateful dependency between pages. Each page assumes that the dictionary has been pre-populated with the required entries. This will in time make your application more brittle as you try to manage those dependencies. A compromise would be to have an application-wide store and to pass the keys into the page using the query string. The page should attempt to access the object from the store, but in the case that the object doesn't exist, it should handle re-creating the object from its source.

UriMapping

As your application grows you will probably have to use folders and perhaps subfolders in order to keep your project structure easy to follow. Navigating between pages can then become brittle as you have to ensure that you update all the navigation URIs whenever you move a page. The navigation system within a Windows Phone application allows you to use URI mapping to convert a short URI into the full path of the XAML page within your project.

Here is a simple example. In this case, you want to add URI overloads to ensure that you always end up at the SecondPage. In the App.xaml.cs file, update the constructor to add two URI mappings:

Available for download on Wrox.com

```
public App(){
    ...
    var mapper = new UriMapper();
    mapper.UriMappings.Add(CreateUriMapping("2", "/SecondPage.xaml"));
    mapper.UriMappings.Add(CreateUriMapping("Two","/SecondPage.xaml"));
    RootFrame.UriMapper = mapper;
```

```
    }

    private UriMapping CreateUriMapping(string uriAsString, string mappedUriAsString){
        return new UriMapping() {
                        Uri = new Uri(uriAsString, UriKind.Relative),
                        MappedUri = new Uri(mappedUriAsString, UriKind.Relative) };
    }
```

Code snippet from App.xaml.cs

The two `UriMapping` instances created in this snippet map `"2"` and `"Two"` to SecondPage.xaml. In the application you can now navigate to this page using any of the following navigation statements:

```
this.NavigationService.Navigate(new Uri("2", UriKind.Relative));
this.NavigationService.Navigate(new Uri("Two", UriKind.Relative));
this.NavigationService.Navigate(new Uri("SecondPage.xaml",
                                        UriKind.Relative));
```

But what happens if you want to do something more complex, like map any URI that starts with *Wizard* to the appropriate page within the WizardPages folder. This can be done using placeholders:

Available for download on Wrox.com

```
mapper.UriMappings.Add(CreateUriMapping(
                   "Wizard{page}",
                   "/WizardPages/WizardPage{page}.xaml"));
```

Code snippet from App.xaml.cs

In this case, any URI matching the pattern `"Wizard{page}"` will be mapped to `"/WizardPages/WizardPage{page}.xaml"`. You can think of `{page}` as a temporary variable that extracts content from the first URI and adds it into the mapped URI. For example, `"Wizard2"` would map to `"/WizardPages/WizardPage2.xaml"`. As such, you could navigate to WizardPage2.xaml in the WizardPages folder with the following:

```
this.NavigationService.Navigate(new Uri("Wizard2", UriKind.Relative));
```

UriMapping is a concept that is much more useful in desktop Silverlight applications, where it is primarily used to provide human and SEO (Search Engine Optimisation) friendly URIs to allow deep linking to pages within a Silverlight application. However, you can also use this concept within your Windows Phone application to manage page hierarchies.

Go Back

One of the hardest things to do when building a mobile application is to prevent the user from getting lost within your application. For this reason there is a dedicated hardware Back button on the front of all Windows Phone devices. Your challenge is to ensure that when the user presses the Back button your application behaves sensibly and that the user doesn't lose or change any data unexpectedly.

It would appear that the functionality of the Back button is relatively straightforward in that it should take the user back to where he was previously. However, on closer inspection, there are several different scenarios that you should consider when working with the Back button:

- **Previous Application** — When the user clicks the Back button, the current application is closed and the previous application is displayed. This is expected behavior when the user is on the first page that was loaded in the application.
- **Previous Page** — When the user clicks the Back button, the current page is hidden and the previous page is displayed. If you think of your application as pages that get added to a stack (the Back-stack) as they are displayed, then the Back button is the equivalent of popping pages off the stack. When the stack is empty, the user would expect the Back button to navigate back to the previous application.
- **Dialogs** — If the application prompts the user for some input, the Back button should be intercepted and should act as a way to cancel the prompt. This action should always be non-destructive and should cancel the current action.
- **Wizards** — If the user navigates through a set of pages that could be considered a wizard — for example, to set up account information — the Back button should not go back to previous steps within the wizard. A typical scenario would be:

  Page A ⇨ Page B ⇨ Wizard Page 1 ⇨ Wizard Page 2 ⇨ Wizard Last Page ⇨ Page B.

  Once the user returns to Page B, pressing the Back button should return the user to Page A, not to the last page in the wizard.
- **Animated Transitions** — The default transition during page navigation is an abrupt switch. If you want to build animated transitions between pages, you need to override the default Back button behavior so that you can insert your animation prior to the new page being displayed.

The first two of these navigation scenarios are already implemented by the `NavigationService`. When the user navigates through a series of pages, each page is added to a journal. If the user then clicks the Back button, the previous page in the journal will be displayed. Once the journal is empty, the current application will be closed, revealing the previous application.

The Back button is the only one of the hardware buttons that you, as a developer, can interact with on a Windows Phone device. By providing an event handler for the `BackKeyPress` event, or by overriding the `OnBackKeyPress` method, you can intercept the action of the Back button and control the behavior within your application.

In the following code, a custom dialog, which is essentially just a set of controls nested within a `Border` control, is presented when the user hits the DoSomethingButton:

Available for download on Wrox.com

```
private void DoSomethingButton_Click(object sender, RoutedEventArgs e){
    this.ProceedPopup.IsOpen = true;
}
```

Code snippet from MainPage.xaml.cs

When the user presses the Back button, the OnBackKeyPress method is invoked. If the ProceedDialog is visible, it is hidden and the back operation is canceled by setting the Cancel property of the CancelEventArgs argument. Clearly, if the user presses the Back button a second time, the back operation won't be canceled, navigating back to the previous page within the application.

Available for download on Wrox.com

```
protected override void OnBackKeyPress(CancelEventArgs e){
    base.OnBackKeyPress(e);

    if (this.ProceedPopup.IsOpen){
        this.ProceedPopup.IsOpen = false;
        e.Cancel = true;
    }
}
```

Code snippet from MainPage.xaml.cs

One thing to be aware of is that when the user presses the Back button the Navigating event on the NavigationService also gets raised. This event exposes a NavigationCancelEventArgs which can also be used to cancel the navigation. It is recommended that if you want to intercept the Back button, you should handle the BackKey event within the PhoneApplicationPage, rather than intercepting the Navigation event.

Occasionally you will want to prompt the user to confirm an action or for the user to make a choice as to what happens. In these cases, it's quite common to display a message box using the MessageBox class. Using the MessageBox class is not advisable for two reasons: First, it's not possible to style the dialog to make it appear as part of your application. Second, you can't use the Back button to cancel the dialog. The only way the Message Box dialog can be dismissed is by the user clicking on one of the buttons on the dialog. If you display a dialog to the user, you should ensure that the Back button can be used to dismiss the dialog.

GoBack and CanGoBack

In some cases, you may want to move back to the previous page within your application as a result of an action other than pressing the hardware Back button. To do this, you use the GoBack method on either the PhoneApplicationFrame or the NavigationService:

Available for download on Wrox.com

```
private void BackIfWeCanButton_Click(object sender, RoutedEventArgs e){
    if (this.NavigationService.CanGoBack){
        this.NavigationService.GoBack();
    }
}
```

Code snippet from MainPage.xaml.cs

As shown in this code snippet, there is a read-only property CanGoBack on both the PhoneApplicationFrame and the NavigationService that indicates whether the application can go back. You can think of this property as indicating whether there are pages remaining in the Back-stack. If there are no more pages left on the Back-stack, you can't go back beyond the current page. Note that it's when the `CanGoBack` property returns `false` that pressing the hardware Back button will close the current application, showing the previous application that the user had open.

If you explore the PhoneApplicationFrame or NavigationService class, you will see that there is also a CanGoForward property and a GoForward method. The former always returns `false`, and calling the GoForward method will throw an exception. These members are present for API level compatibility with Silverlight on the desktop. They don't equate to anything within a Windows Phone application, so they should never be used.

Animation

When you're structuring your application, one of the first things you have to consider is the order in which users will navigate through your application. To build a quality Windows Phone application, you not only need to think about the order that pages are displayed in, but you also need to create smooth transitions. The basic steps for creating an animated transition between two pages are as follows:

1. Intercept any action indicating the user is navigating away from the current page.
2. Start an animation storyboard that will hide the current page.
3. Navigate to the next page.
4. Intercept the navigation to the new page.
5. Start an animation storyboard to show the new page.

Hiding the Current Page

One way to intercept when a user is about to navigate away from the current page would be to simply replace the `Navigate` method call with a call to start the animation. The downside is that you may have to do this in multiple places within a single page if the user has multiple ways to cause navigation to another page. An alternative would be to override the `OnNavigationFrom` method and cancel the navigation:

```
private void AnimatedNavigationButton_Click(object sender, RoutedEventArgs e){
            this.NavigationService.Navigate(new Uri("/ThirdPage.xaml",
                                                UriKind.Relative));
}

protected override void OnNavigatingFrom(NavigatingCancelEventArgs e){
    base.OnNavigatingFrom(e);

    e.Cancel = true;
}
```

Code snippet from SecondPage.xaml.cs

All you've done so far is to cancel every navigation attempt. What you need to do is to invoke the animation storyboard to hide the current page. The following code invokes the `Begin` method on a storyboard called `HidePage` that will be used to hide the current view:

Available for download on Wrox.com

```
protected override void OnNavigatingFrom(NavigatingCancelEventArgs e){
    base.OnNavigatingFrom(e);

    e.Cancel = true;
    this.HidePage.Begin();
}
```

Code snippet from SecondPage.xaml.cs

Of course, canceling every page navigation won't allow the user to move away from the current page once the hide animation has completed hiding the current page. What you're going to do is cache the URI that the user was going to navigate to and then when the animation is over, call the `Navigate` method again. This time the navigation will proceed successfully.

Available for download on Wrox.com

```
public Uri UriToNavigateTo { get; set; }
protected override void OnNavigatingFrom(NavigatingCancelEventArgs e){
    base.OnNavigatingFrom(e);

    if (UriToNavigateTo == null){
        e.Cancel = true;
        UriToNavigateTo = e.Uri;
        this.HidePage.Begin();
    }
    else{
        UriToNavigateTo = null;
    }
}

private void HidePage_Completed(object sender, EventArgs e){
    this.NavigationService.Navigate(UriToNavigateTo);
}
```

Code snippet from SecondPage.xaml.cs

Displaying the New Page

Once the previous page has been hidden and the new page loaded, a similar process can be followed to animate its display. You need to override the `OnNavigatedTo` method on the new page and begin an animation storyboard that will show the page contents:

Available for download on Wrox.com

```
protected override void OnNavigatedTo(NavigationEventArgs e){
    base.OnNavigatedTo(e);

    this.DisplayPage.Begin();
}
```

Code snippet from ThirdPage.xaml.cs

Animation Storyboards

The animation to hide the current page might look like this:

Available for download on Wrox.com

```
<phone:PhoneApplicationPage
    x:Class="Navigation.SecondPage"
    ...
    x:Name="phoneApplicationPage">
    <phone:PhoneApplicationPage.Resources>
        <Storyboard x:Name="HidePage" Completed="HidePage_Completed">
            <DoubleAnimationUsingKeyFrames
Storyboard.TargetProperty="(UIElement.RenderTransform).
(CompositeTransform.TranslateX)" Storyboard.TargetName="phoneApplicationPage">
                <EasingDoubleKeyFrame KeyTime="0" Value="0"/>
                <EasingDoubleKeyFrame KeyTime="0:0:1" Value="-480"/>
            </DoubleAnimationUsingKeyFrames>
            <DoubleAnimationUsingKeyFrames
Storyboard.TargetProperty="(UIElement.RenderTransform).
(CompositeTransform.TranslateY)" Storyboard.TargetName="phoneApplicationPage">
                <EasingDoubleKeyFrame KeyTime="0" Value="0"/>
                <EasingDoubleKeyFrame KeyTime="0:0:1" Value="-800"/>
            </DoubleAnimationUsingKeyFrames>
        </Storyboard>
    </phone:PhoneApplicationPage.Resources>
        <phone:PhoneApplicationPage.RenderTransform>
        <CompositeTransform/>
    </phone:PhoneApplicationPage.RenderTransform>
```

Code snippet from SecondPage.xaml

This XAML would appear before the `LayoutRoot Grid` within your `PhoneApplicationPage`. When invoked, this storyboard translates the whole page from its current position to 480 to the left and 800 above the top of the screen (pushing the page out the top-left corner) over the period of 1 second.

In contrast, the animation to show the new page might look like this:

Available for download on Wrox.com

```
<phone:PhoneApplicationPage
    x:Class="Navigation.ThirdPage"
    ...
    x:Name="phoneApplicationPage">
    <phone:PhoneApplicationPage.Resources>
        <Storyboard x:Name="DisplayPage">
            <DoubleAnimationUsingKeyFrames
Storyboard.TargetProperty="(UIElement.RenderTransform).
(CompositeTransform.ScaleX)" Storyboard.TargetName="phoneApplicationPage">
                <EasingDoubleKeyFrame KeyTime="0" Value="0"/>
                <EasingDoubleKeyFrame KeyTime="0:0:1" Value="1"/>
            </DoubleAnimationUsingKeyFrames>
            <DoubleAnimationUsingKeyFrames
Storyboard.TargetProperty="(UIElement.RenderTransform).
(CompositeTransform.ScaleY)" Storyboard.TargetName="phoneApplicationPage">
                <EasingDoubleKeyFrame KeyTime="0" Value="0"/>
```

```
                <EasingDoubleKeyFrame KeyTime="0:0:1" Value="1"/>
            </DoubleAnimationUsingKeyFrames>
            <DoubleAnimationUsingKeyFrames
Storyboard.TargetProperty="(UIElement.RenderTransform).
(CompositeTransform.Rotation)" Storyboard.TargetName="phoneApplicationPage">
                <EasingDoubleKeyFrame KeyTime="0" Value="-720"/>
                <EasingDoubleKeyFrame KeyTime="0:0:1" Value="0"/>
            </DoubleAnimationUsingKeyFrames>
        </Storyboard>
    </phone:PhoneApplicationPage.Resources>
        <phone:PhoneApplicationPage.RenderTransform>
        <CompositeTransform/>
    </phone:PhoneApplicationPage.RenderTransform>
```

Code snippet from ThirdPage.xaml

This animation changes the scale of the page from 0 (i.e., a height and width of 0) to 1, again over the course of a second. It also rotates the page as it is scaled to fit the screen. This will give you a sequence similar to that shown in Figure 6-4.

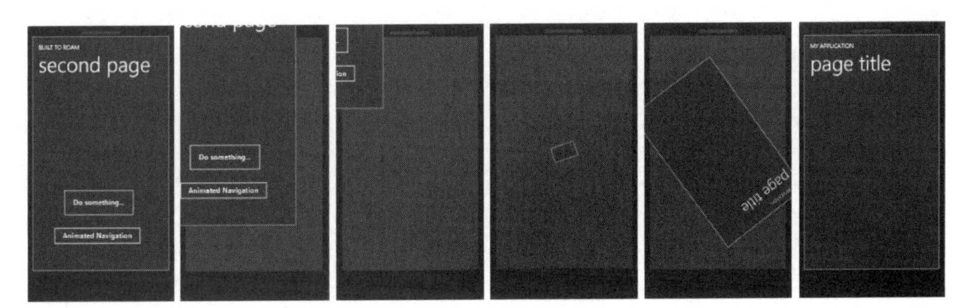

FIGURE 6-4

Wizards

In Silverlight for Windows Phone you don't have the ability to add or remove entries in the stack of pages that the Back button uses (i.e., the navigation journal). For the most part, this isn't a major issue since in most cases having the user return to the page he or she came from is expected behavior. However, there are some cases in which you will have to come up with work-arounds to compensate for this missing feature. One such scenario is where you want to present a series of pages that constitute a wizard, for example, for the user to enter account information or complete a questionnaire. In these cases, once the wizard is complete, the user shouldn't be stepped back through the pages of the wizard.

Without support for being able to remove items from the navigation journal, the only way to control the order in which the user can step back through the application is to control what gets added to the journal in the first place. Instead of navigating to a new page, which will add that page to the journal, you can place the content of that page into a `UserControl` and just display the `UserControl`. The important thing to remember is that when you use the `NavigationService` to navigate between pages, you are adding items to the journal, implicitly allowing the user to navigate back through the stack of pages.

BACKGROUND PROCESSING

One of the most radical changes in the Windows Phone is that only one application is executing at any given time. This is a significant departure from previous versions of Windows Mobile and many other mobile platforms that allow multiple applications to run concurrently. Windows Phone goes one step further, however — rather than simply closing an application when another is opened, Windows Phone may keep applications open in the background. This way, if the user presses the Back button, the application can quickly be returned to the foreground without having to be completely restarted. However, in developing your application you should work on the assumption that whenever your application is pushed into the background it may be closed by the system.

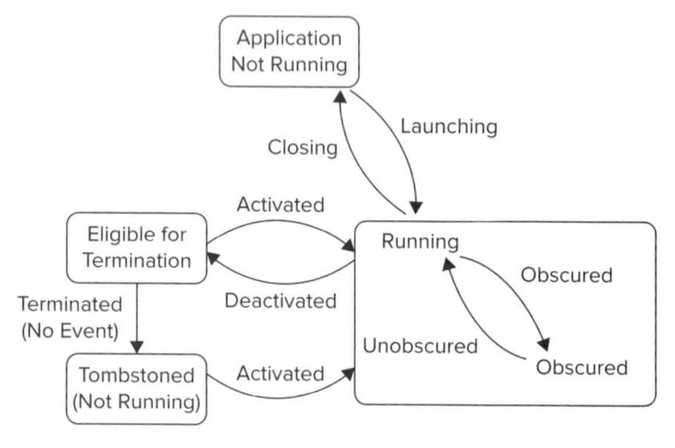

FIGURE 6-5

A Windows Phone application can be in one of four states, as shown in Figure 6-5: Not Running, Running, Eligible for Termination, and Tombstoned.

When your application is launched, it goes from the Not Running state to the Running state. There is a `Launching` event on the `App` class that is raised when the application first launches, and there is a `Closing` event which is raised when the application is closed by the user. When the user presses the Back button on the first page of the application, the application will be closed. The `Closing` event will be raised, the application will be terminated and the previous application the user was in will be displayed. Since the `Closing` event is only ever raised when the application is closed intentionally, you only need to save information that needs to be available the next time the user launches the application, in other words: a persistent application state.

If the user presses the Start button while inside your application, the application will be deactivated and marked as Eligible For Termination. From this state, your application can either be activated, if the user presses the Back button to return to your application, or it may silently be exited (i.e. Tombstoned) if the system needs to free up resources.

Figure 6-5 also illustrates that while in the Running state an application can be *Obscured*. This happens if a screen overlay is displayed over the top of the application, obscuring part of the screen. For example, in Figure 6-6, you can see that the Call in Progress dialog obscures the Calculator application that is running.

FIGURE 6-6

In the left image the Call Details view masks the entire application. However, clicking anywhere below the "End Call" button will minimize the Call Details view to the "Call in Progress" bar at the top of the second image. The application that was being masked is now accessible.

Eligible for Termination

When the user presses the Start button, in order to launch another application, the `Deactivated` event will be raised. It's important for your application to handle this event to stop any background processing and save any state that is in memory. Your application may continue to run in the background, in which case, if the user presses the Back button to return to your application, the `Activated` event will be raised.

Alternatively, if the system decides to terminate your application, your application will be closed, and marked as Tombstoned. If this happens the `Closing` event is not raised, so there is no way for you to detect that your application is being terminated. Further, if your application is activated from the Tombstoned state (for example if the user presses the Back button to return to your application), a new instance of your application will be executed but only the `Activated` event will be raised. The `Launching` event is not raised when the application is being restored from the tombstone state.

A Windows Phone application has both `Deactivated` and `Activated` events, and the default Windows Phone application template includes event handlers for these events, as well as the Launching and Closing events, in the App.xaml.cs file.

```
// Code to execute when the application is launching (eg, from Start)
// This code will not execute when the application is reactivated
private void Application_Launching(object sender, LaunchingEventArgs e){ }

// Code to execute when the application is activated (brought to foreground)
// This code will not execute when the application is first launched
private void Application_Activated(object sender, ActivatedEventArgs e){ }

// Code to execute when the application is deactivated (sent to background)
// This code will not execute when the application is closing
private void Application_Deactivated(object sender, DeactivatedEventArgs e){ }

// Code to execute when the application is closing (eg, user hit Back)
// This code will not execute when the application is deactivated
private void Application_Closing(object sender, ClosingEventArgs e){ }
```

These event handlers are wired up in the App.xaml file.

```
<Application x:Class="Navigation.App"
    ...>
    <Application.ApplicationLifetimeObjects>
        <shell:PhoneApplicationService
            Launching="Application_Launching" Closing="Application_Closing"
            Activated="Application_Activated"
            Deactivated="Application_Deactivated"/>
    </Application.ApplicationLifetimeObjects>
</Application>
```

Scenarios

In Figure 6-5 you saw that there are a number of states for your application. What is probably not clear is what happens when your application enters and leaves the Eligible for Termination or Tombstone states. To clarify this, let's look at a number of scenarios and the events that are raised.

Before you do this, you should be aware of two fundamental principles that govern the behavior of Windows Phone applications.

1. **Consistent Start** — This means that whenever the user goes to the Start screen and launches an application, the application should launch as if it wasn't already running. If the application was running in the background (which is the only other alternative if the user was at the Start menu) then it would be terminated and a new instance of the application would be started.

2. **No Phantom Back-stack** — The introduction of a Back button introduces a world of complexity around keeping applications running and providing a consistent user experience. In an ideal (i.e. unlimited device resources) world every application opened by the user would remain open. The user could then click Back to go back through all the applications they have opened.

   Unfortunately this is not the case and as device resources are very limited (in order to provide an optimal user experience for the application being run) it may be necessary to terminate applications when they are not in use. This introduces a challenge as to how to provide a consistent user experience around the Back button. On one hand you have the scenario that if the application is still running, the Back button will navigate back to it; on the other hand if the application has been terminated, the Back button won't show the application because it is no longer running. This is what is referred to as a phantom Back-stack.

   To resolve this dilemma Windows Phone behaves as follows:

   - When the user clicks the Back button on the home/first page of an application, that application will be closed and the Start page will be displayed.
   - A subsequent click on the Back button will take the user to the previous application they had opened:
     - If the application is still running, it will simply be displayed, showing the same page that was visible when the user pressed the Start button.
     - If the application is not running, the application will be executed and immediately navigated to the same page that was visible when the user pressed the Start button.
     - In both cases the Back-stack within the application is preserved so that subsequent presses of the Back button will navigate through the pages within the Application.

- If the user reaches the home/first page of the previous application and presses the Back button, it too will be closed, and the application before that will be displayed. Note that the Start screen is not displayed again.
- Eventually, after enough presses of the Back button, all applications that had been opened will have been re-activated and subsequently closed. The Back-stack will be empty and the Start screen will be visible.

Application Launched From Start

This is the scenario so far:

- **Start** — User selects Application1
- **Application1** — Executed
- **Application1** — App: Constructor is called
- **Application1** — App: Launching event raised
- **Application1** — Page1: NavigatedTo event raised
- **Application1** — Page1: Displayed
- **Application1** — User triggers navigation to Page2 in application
- **Application1** — Page1: NavigatedFrom event raised
- **Application1** — Page2: NavigatedTo event raised
- User presses Start Button
- **Application1** — Page2: NavigatedFrom event raised
- **Application1** — App: Deactivated event
- **Start** — . . .

Second Launch

Assume that Application1 hasn't been Tombstoned, so it is still running in the background.

- **Start** — User selects Application1
- **Application1 (previous instance)** — Terminated — No event raised
- **Application1** — Executed
- **Application1** — App: Constructor is called
- **Application1** — App: Launching event raised
- **Application1** — Page1: NavigatedTo event raised
- **Application1** — Page1: Displayed
- **Application1** — User clicks Back button
- **Start** — Displayed

- **Start** — User clicks Back button
- **Start** — Displayed (nothing on Back-stack)

The important thing to remember here is that in order for there to be a Consistent Start experience the first instance of Application1 is terminated and that the second instance is started as if there had been no other instance running.

Back: No Tombstone

Assume that Application1 hasn't been Tombstoned, so it is still running in the background.

- **Start** — User clicks Back button
- **Application1** — App: Activated event raised
- **Application1** — Page2: NavigatedTo event raised
- **Application1** — Page2: Displayed
- **Application1** — Page2: User clicks Back button
- **Application1** — Page2: NavigatedFrom event raised
- **Application1** — Page1: NavigatedTo event raised
- **Application1** — Page1: Displayed

Back: Tombstoned

Now, assume that Application1 has been Tombstoned, so it is no longer running in the background.

- **Start** — User clicks Back button
- **Application1** — Executed
- **Application1** — App: Constructor is called
- **Application1** — App: Activated event
- **Application1** — Page2: NavigatedTo event raised
- **Application1** — Page2: Displayed
- **Application1** — Page2: User clicks Back button
- **Application1** — Page2: NavigatedFrom event raised
- **Application1** — Page1: NavigatedTo event raised
- **Application1** — Page1: Displayed

Two Applications

Let's extend the original scenario by introducing a second application:

- **Start** — User selects Application1
- **Application1** — Executed

- **Application1** — App: Constructor is called
- **Application1** — App: Launching event raised
- **Application1** — Page1: NavigatedTo event raised
- **Application1** — User triggers navigation to Page2 in application
- **Application1** — Page1: NavigatedFrom event raised
- **Application1** — Page2: NavigatedTo event raised
- User presses Start Button
- **Application1** — Page2: NavigatedFrom event raised
- **Application1** — App: Deactivated event
- **Start** — User selects Application2
- **Application2** — Executed
- **Application2** — App: Constructor is called
- **Application2** — App: Launching event raised
- **Application2** — Page1: NavigatedTo event raised

Back: Two Applications

Assume that Application1 has been Tombstoned, so it is no longer running in the background.

- **Application2** — User clicks Back button
- **Application2** — Page1: NavigatedFrom event raised
- **Application2** — App: Closed event raised
- **Application2** — Exited
- **Start** — Displayed
- **Start** — User clicks Back button
- **Application1** — Executed
- **Application1** — App: Constructor is called
- **Application1** — App: Activated event raised
- **Application1** — Page2: NavigatedTo event raised
- **Application1** — Page2: Displayed
- **Application1** — Page2: User clicks Back button
- **Application1** — Page2: NavigatedFrom event raised
- **Application1** — Page1: NavigatedTo event raised
- **Application1** — Page1: Displayed

One thing you will notice about these scenarios is that they all end essentially the same way: with the user navigating back through the pages of Application1. There are some additional scenarios that pertain to Windows Phone tasks such as Launchers and Choosers. These are discussed in Chapter 8.

The second thing you should have observed is that when the user returns to Application1, if Page2 was previously displayed, then Page2 will be displayed. This happens regardless of whether the application was Tombstoned or not. There is also no requirement for you to navigate to Page2 if the application restarts. If the application is restarted as a result of being Tombstoned, the system will automatically navigate to Page2 instead of the default page. It's also important to note that Page1 does not get navigated to unless the user presses the Back button again to navigate from Page2 to Page1. This is particularly important if you are initializing data or application state within the first page of the application. You should always ensure pages can be loaded independently of any previous page to ensure the application will be correctly reloaded in the case that it has been Tombstoned. Further if you are initializing data in the `Launching` event of the application you should ensure the data is loaded within the `Activated` event.

Saving State

When your application is deactivated or closed you will need to save information so that the user can resume what they were doing or working on the next time they run the application. As mentioned earlier, if the application is closed, you only need to save persistent state information. However, when the application is deactivated, you not only need to save the state that needs to be persisted between application sessions, you also need to save the state of the current application. This can be referred to as the transient state of the application.

Page State

Let's start with saving the transient state of the current page. In the scenarios presented earlier whenever the user presses the Back button, the Start button, or navigates to a new page, the `NavigatedFrom` event is raised on the current page. At this point you can save the current state of the page to the `State` property which exposes an `IDictionary<object,string>`.

Available for download on Wrox.com

```
protected override void OnNavigatedFrom(NavigationEventArgs e){
    this.State["Text1"] = this.PageText1.Text;
    this.State["Text2"] = this.PageText2.Text;
    this.State["Text3"] = this.PageText3.Text;

    base.OnNavigatedFrom(e);
}
```

Code snippet from ThirdPage.xaml.cs

When the application returns to the foreground, or the user navigates back to the page within the application, the `NavigatedTo` event is raised. Again you can access the State property to retrieve any information that was saved. Remember that when the application is launched from the Start screen any transient state will be lost, which means the dictionary will be empty.

```
protected override void OnNavigatedTo(NavigationEventArgs e){
    base.OnNavigatedTo(e);

    this.PageText1.Text = this.State.SafeDictionaryValue("Text1") + "";
    this.PageText2.Text = this.State.SafeDictionaryValue("Text2") + "";
    this.PageText3.Text = this.State.SafeDictionaryValue("Text3") + "";
}
```

Code snippet from ThirdPage.xaml.cs

In this code snippet `SafeDictionaryValue` is a helper method used to retrieve saved values if they exist.

```
public static class Utilities{
    public static TValue SafeDictionaryValue<TKey,TValue>(
                                this IDictionary<TKey,TValue> dictionary, TKey key){
        TValue value;
        if (dictionary.TryGetValue(key, out value)){
            return value;
        }
        return default(TValue);
    }
}
```

Code snippet from Utilities.cs

Application State

In addition to a `State` property on the `PhoneApplicationPage` class there is also an application-wide `State` property that exists on the `PhoneApplicationService` class. The following code illustrates saving and extracting a Customer object from the application-wide `State` dictionary.

```
public class Customer
{
    public string Name { get; set; }
    public string PhoneNumber { get; set; }
}

CurrentCustomer = new Customer() {
                          Name = "Nick Randolph",
                          PhoneNumber = "+1 425 001 0001"
                      };

// Save current customer
PhoneApplicationService.Current.State["CurrentCustomer"] = CurrentCustomer;

// Restore current customer
CurrentCustomer = PhoneApplicationService.Current.State.SafeDictionaryValue
                                               ("CurrentCustomer") as Customer;
```

Code snippet from ThirdPage.xaml.cs

This snippet illustrates saving a `Customer` object into the `State` dictionary. Both the page and application level `State` dictionaries can be used to save any Type of object, so long as it is Serializable. You have to be careful that when you restore the saved state that you correctly cast the returned object.

Obscured

While your application is in the Running state, it may get interrupted by other events on the device. These may be an incoming phone call, a system notification such as an alarm, or even a toast notification from a different application. In each of these cases, part of the screen is obscured by the notification that appears on the screen. If the user is in the middle of doing something, you may want to pause anything running within the application while they process the notification. Take, for example, an action game — if you don't pause the game while the notification is on screen, their character might get killed because they couldn't see the enemy on the screen. Alternatively, if you are playing back audio or video, you may want to pause playback while their attention is focused elsewhere.

The PhoneApplicationFrame exposes both Obscured and Unobscured events. You can attach event handlers to suspend anything that is running when the application is obscured.

Available for download on Wrox.com

```
// Constructor
public App(){
    ...
    RootFrame.Obscured += RootFrame_Obscured;
    RootFrame.Unobscured += RootFrame_Unobscured;
}

void RootFrame_Obscured(object sender, ObscuredEventArgs e){
    // Handle obscured event
}

void RootFrame_Unobscured(object sender, EventArgs e){
    // Handle unobscured event
}
```

Code snippet from App.xaml.cs

Lock Screen

You'll notice that the second parameter in the Obscured event handler is of Type ObscuredEventArgs. This parameter contains an additional property, IsLocked, which indicates if the Lock screen is visible. The behavior of the Lock screen varies depending on whether your application has requested the ability to continue running under the Lock screen. If your application needs to run under the Lock screen (for example you might be tracking the route that the user takes when they go for a run), you first need to request the user's permission for the application to continue to run under the Lock screen. This is documented in the Windows Phone Application Certification Requirements document (available at `http://developer.windowsphone.com`).

Once you have requested the user's permission you can indicate to the operating system whether it should detect a lack of activity. By setting the `ApplicationIdleDetectionMode` to `Disabled` you are effectively telling the operating system not to monitor the application for activity. The default setting is `Enabled` which means the operating system will detect if the application is idling and will deactivate it when the Lock screen is activated.

```
private void RequestLockScreenButton_Click(object sender, RoutedEventArgs e){
    if (MessageBox.Show("Can I keep running under the Lock screen?",
                        "Lock Screen",
                        MessageBoxButton.OKCancel) == MessageBoxResult.OK){
        PhoneApplicationService.Current.ApplicationIdleDetectionMode =
                       IdleDetectionMode.Disabled;
    }
    else{
        PhoneApplicationService.Current.ApplicationIdleDetectionMode =
                       IdleDetectionMode.Enabled;
    }
}
```

SUMMARY

In this chapter you learned about the navigation service that underpins the Windows Phone application model. It breaks an application up into pages that are loaded within a frame. This model makes building an application simple as you create each page in isolation and then work out how to transition between them. If each page is self-contained, it is easy to launch the application and go straight to a particular page. This is particularly useful when your application is restored after being terminated after being pushed into the background.

7

Application Tiles and Notification

WHAT'S IN THIS CHAPTER

- How to add, and subsequently update, an application tile on the Start screen
- What the Push Notification service is and how to connect to it
- Understanding the different notification types
- Generating different types of push notifications

Part of building a great Windows Phone application is making it feel as though it is part of the users' mobile experience. As a developer, you can take advantage of various integration points that Windows Phone offers to make your application appear more seamless with the device. These include being able to pin the application to the Start area as a tile and using Push Notifications to send updates and alerts.

In this chapter, you will learn that there are three types of Push Notification that you can make use of in order to ensure that users have the most up-to-date information at their fingertips.

APPLICATION TILE

The Start is one of the most important aspects of the Metro user experience on a Windows Phone. It's displayed when the phone is unlocked and when the user presses the Start button on the front of the device. It's the single element of the user interface that is immediately accessible no matter what application the user is in or what they are currently doing. The Start is a launch screen for applications, but more importantly, it is alive with information.

On other mobile platforms, and even desktop platforms such as Windows 7, it is possible to pin an application to the home area or to the top of a programs list. This does little more than make the application more accessible, as it still requires the user to go into the application to see whether there have been any updates or alerts they should be aware of. Within Windows Phone, *Application Tiles*, also known as *Live Tiles*, appear on the Start to not only provide a link to the application, but also to inform the user of any changes that have occurred since the application was last run.

The description of Application Tiles may remind you of a feature within Windows Vista and Windows 7 called *desktop gadgets*, which represent slices of the Web that remain on the desktop. Gadgets can present interactive UIs and periodically update with data from the Web. Common uses of desktop gadgets are for news and weather updates. A similar concept was introduced in Windows Mobile 6.5 with widgets, but these were more akin to applications running on the device. Widgets aren't able to update the home screen with updates and were introduced as more of a lightweight application development framework. Application Tiles represent a middle-of-the-road approach. Although they integrate with the Start screen and provide a lightweight approach to having multiple applications asynchronously display update information, they have minimal interaction capabilities apart from launching the associated application.

An Application Tile has three components:

- **Background Image** — This is a 173 × 173 image that will make up the background of the Application Tile that will appear on the Start.
- **Title** — The Title is the text that will appear on the tile. This is usually a shortened application name, or represents the type of information that the application features. For example, "contacts" or "photos."
- **Count** — This represents the number of updates or changes relating to the application. In the case of e-mail, this would be the number of unread items, but it could represent the number of changes waiting to be synchronized, the number of bills to be paid, or the number of tasks to be completed. The important thing is to make this number intuitive to the user in the context of your application.

FIGURE 7-1

When coming up with a design for the Application Tile of your application, you need to consider the placement of both the Title and the Count. Figure 7-1 illustrates an Application Tile designed for a real estate application. The application is used for searching for properties that are for sale. The first image of Figure 7-1 is the Background Image of the tile, which shows images of some of the recent properties that have been viewed. The second image shows the overlay, which will be applied to the tile with a Title of *Properties*, and the Count, which in this case indicates that there are eight new properties that have been listed that may be of interest. The overlay is applied to the Background Image to give the Application Tile that is displayed on the Start, as shown in the third image. Note that the background of the second image is only shown here so that you can see the Title; it is, in fact, transparent.

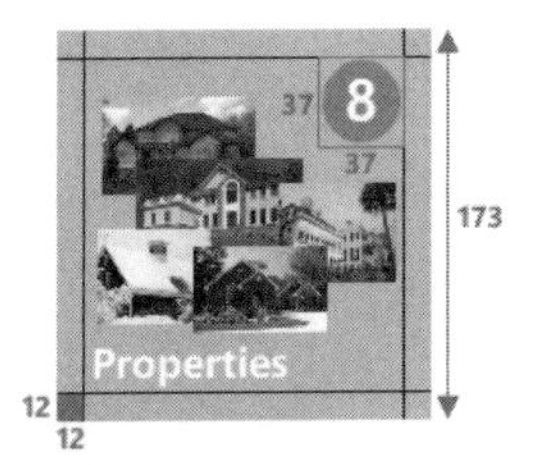

FIGURE 7-2

The most important thing to know when designing your Application Tile is the dimensions that you have to work with. Figure 7-2 shows a breakdown of the Application Tile, which is a square with the dimensions 173 × 173.

There is an internal border of 12, which is used to position both the Title and the Count. This border isn't drawn on the Application Title and is merely there as a guide so that you know where the Title and the Count will be displayed. The Count will be displayed within a colored circle with a diameter of 37.

When building an Application Tile, it can be quite useful to use a drawing package that supports multiple layers that can be toggled on or off. Figure 7-3 shows a sequence, going from left to right, of building up an Application Tile:

1. Starting with a background color or image of 173 × 173, add the border lines to a new layer inset by 12 from each side.
2. Next, on another layer, add a colored circle that touches the top and right borders with a diameter of 37.
3. On yet another layer, add text for the Title and the Count.
4. Finally, add any additional images to the original layer.
5. Now remove the layers that hold the border, circle, and text.

What you're left with is a base Application Tile that you can use in your application.

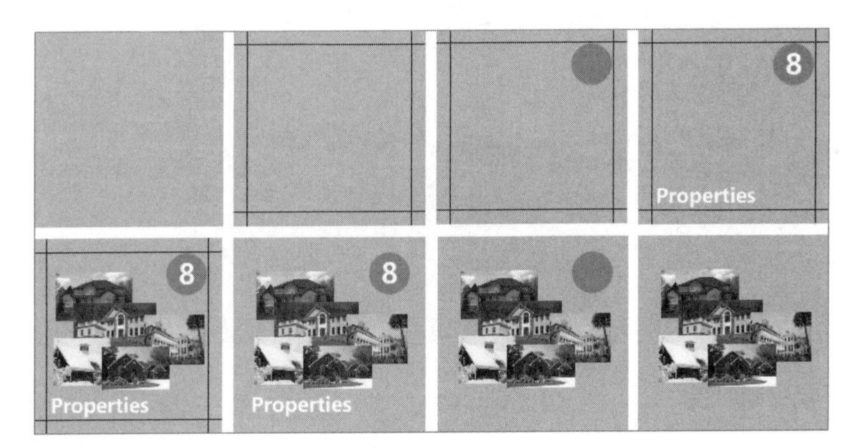

FIGURE 7-3

There are three ways to specify the Application Tile for your application. The first is to specify the Background Image in the Properties window of your Windows Phone application, as shown in Figure 7-4. You can also specify the Title that will appear in the overlay when your application is pinned to the Start. Note that unless a Tile Notification (discussed later in this chapter) is received, the colored circle will not be added to the Application Tile.

FIGURE 7-4

Changing the Background Image within the Properties window will set the `BackgroundImageURI` property in the WMAppManifest.xml file located within your application's project.

In order for your application to appear as a tile on the Start, the user has to choose the "pin to start" option from the applications list. The applications list can be found by scrolling the Start to the left, shown in the first two images of Figure 7-5. Tap and hold on the application you want to add to the Start. A context menu will appear from which you can select "pin to start" to add the application to the Start. The user can alternatively select "uninstall" from this menu to remove the application from the device.

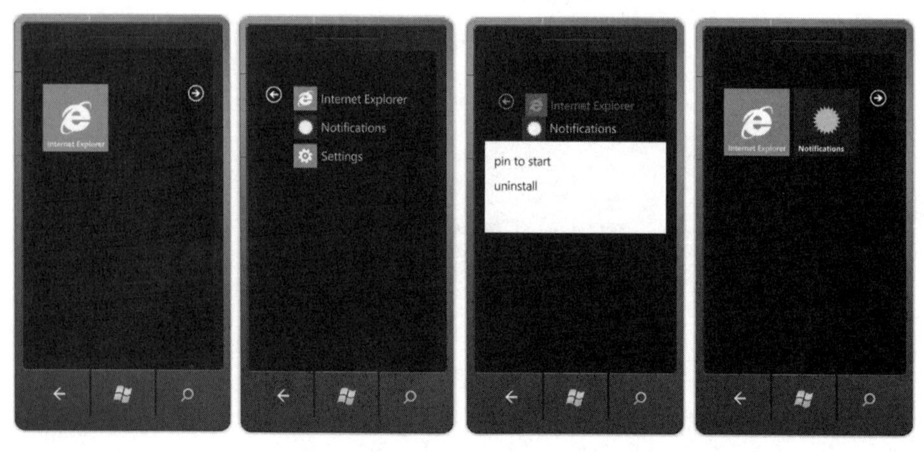

FIGURE 7-5

The second way to set the Application Tile is to establish a schedule for updating the Application Tile from images served up by a Web Server. You can think of the schedule as a timer that is set for an interval. When the timer goes off, the Application Tile is updated by downloading a new image from the specified remote URI. The following code establishes a schedule for updating the Application Tile. Once an hour for the next 5 hours, the Application Tile background will be updated:

Available for download on Wrox.com

```
private void ScheduleTileUpdate_Click(object sender, RoutedEventArgs e){
    var entrypoint = new ShellTileSchedule();
    entrypoint.RemoteImageUri =
 new Uri("http://www.builttoroam.com/books/devwp7/chapter7/notificationtile.png");
    entrypoint.Interval = UpdateInterval.EveryHour;
```

```
        entrypoint.Recurrence = UpdateRecurrence.Interval;
        entrypoint.StartTime = DateTime.Now;
        entrypoint.MaxUpdateCount= 5;
        entrypoint.Start();
    }
```

Code snippet from MainPage.xaml.cs

The Application Tile is updated by downloading the image from the URL specified by the `RemoteImageUri` property. In order to change the Application Tile, you simply need to change the image that is available for download at that URL.

The ShellTileSchedule doesn't rely on the application to be running in order for the Application Tile to be updated. It does however assume that the user has pinned the application to the Start in order for the Application Tile to be visible.

Updating the Application Tile based on some predefined schedule relies on the remote image being frequently updated. This strategy is most effective if the remote image is periodically updated, in which case you should seek to align the recurrence pattern of the Application Tile schedule with that of the remote image. The update process for Application Tiles has been well optimized for Windows Phone to ensure minimal impact on battery life. However, remember that each time the device has to download a remote image, it will be using processing and network bandwidth. These can affect battery life, so downloading an image that hasn't changed is somewhat wasteful.

The last way to specify the Application Tile is through the use of Tile Notifications. Tile Notifications are one of three types of notifications your application can make use of, discussed in the next section.

PUSH NOTIFICATIONS

There are two schools of thought when it comes to running applications in the background. Some platforms are based on the idea that a single application should be running at a time, limiting performance degradation when the user switches between applications. Other platforms allow multiple applications to run concurrently, based on the idea that the complexity of persisting the application state as discussed toward the end of the last chapter greatly outweighs the memory or CPU overhead of running multiple applications.

Windows Phone doesn't allow applications to run in the background, which is an issue if you want to periodically notify the user that data associated with your application has changed. Luckily, the platform supports three forms of Push Notifications through which you can provide updates to the user or your application directly, even when it is not running:

- **Tile** — Updates the Application Tile on the Start.
- **Toast** — Creates a toast pop-up that is displayed over the current screen.
- **Raw** — A notification intended to be processed by the application itself; it is transparent to the user, unless the application decides otherwise.

For each of these Push Notifications, the process is the same. There are three parties involved in Push Notifications — the Windows Phone application, the cloud-based Notification Service, and the Notification Source, as shown in Figure 7-6.

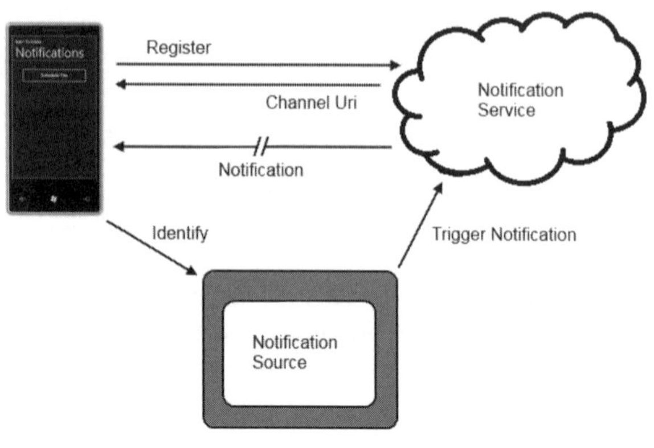

FIGURE 7-6

The Windows Phone application is the start of the notification work flow. It starts by registering itself with the cloud-based Notification Service. The Notification Service is hosted by Microsoft and is freely available for any Windows Phone application to take advantage of. Its purpose is to act as a message broker to transmit messages from the Notification Source to the Windows Phone. When the application registers with the Notification Service, it receives a Channel URI that uniquely identifies the combination of phone application, device, and channel name. You can think of the *channel name* as being a unique identifier defined by the application so that it can register for events from multiple Notification Sources, each of which will require a unique Channel URI from the Notification Service.

The Notification Source can be any application or service that is connected to the Internet and is capable of making an HTTP request. It may be a cloud-based service that is processing a large amount of work, it might be a desktop application that is running within an organization, or it might even be another Windows Phone application running on a different device.

In order to send a notification through to a specific Windows Phone device, the Notification Source needs a way to uniquely identify that device. This is done by the Windows Phone application directly identifying itself to the Notification Source by sending through its Channel URI. When the Notification Source wishes to send a notification, instead of having to locate the device, connect to the device, determine if the application is running, and so on, it can simply POST a message to the Channel URI (which forms part of the Microsoft-provided Notification Service). The Notification Service then handles the transmission of the notification to the application running on the device.

There are certain things you should note about the Notification Service: First, the channel that is established between the Notification Service and the application is quite *resilient*. This means that if device connectivity is lost, the channel will still operate once connectivity is restored. It's also *persistent*, which means that the application or device can restart and the channel will still operate. In fact, for some notification types, the application doesn't even need to be running for a notification to be received. The flipside of this is that the Notification Service does not guarantee delivery of a notification to the device. If you require guaranteed delivery or you want to get feedback that the notification has been received, you should implement a Web Service within your Notification Source that your Windows Phone application can call directly once the notification is received.

Regardless of which notification type you want to use within your application, the first step is always to register your application with the Notification Service. This is done by creating an instance of the `HttpNotificationChannel` with an appropriate channel name and then calling the `Open` method:

Available for download on Wrox.com

```
public const string channelName = "NotificationSample";
private HttpNotificationChannel Channel { get; set; }
private Uri ChannelUri { get; set; }

private void RegisterChannelButton_Click(object sender, RoutedEventArgs e){
    this.Channel = HttpNotificationChannel.Find(channelName);
    if (this.Channel == null){
        this.Channel = new HttpNotificationChannel(channelName,
                                                   "www.builttoroam.com");
        this.Channel.ChannelUriUpdated += httpChannel_ChannelUriUpdated;
        this.Channel.Open();
    }
    else{
        this.Channel.ChannelUriUpdated += httpChannel_ChannelUriUpdated;
        this.ChannelUri=this.Channel.ChannelUri;
    }

    this.Channel.HttpNotificationReceived +=
                                    httpChannel_HttpNotificationReceived;
    this.Channel.ShellToastNotificationReceived +=
                                    httpChannel_ShellNotificationReceived;
}
```

Code snippet from MainPage.xaml.cs

You will have to add a using statement for the Microsoft.Phone.Notification namespace to the top of the file in order to use the HttpNotificationChannel class.

The first line of the `RegisterChannelButton_Click` method attempts to locate an existing `HttpNotificationChannel` with the same channel name. The `Find` method will return the existing channel, along with its `ChannelUri`. Alternatively, if there is no existing channel, a new instance is created and opened. In both cases an event handler is attached to the ChannelUriUpdated event which gets raised when the channel has been successfully opened. There are a number of events defined on the `HttpNotificationChannel` class, as defined in Table 7-1.

TABLE 7-1: Notification Channel Events

EVENT	DESCRIPTION
`ChannelUriUpdated`	The `Open` method on the `HttpNotificationChannel` object is an asynchronous operation that will return immediately. When the channel has been successfully opened, the `ChannelUriUpdated` event will be raised so that you can extract the `ChannelUri` that will be forwarded to the Notification Source.
`ErrorOccurred`	If for some reason an error occurs with the use of the channel, an `ErrorOccurred` event will be raised.

(continues)

TABLE 7-1 *(continued)*

EVENT	DESCRIPTION
`ShellToastNotificationReceived`	The `ShellToastNotificationReceived` event is raised when the Windows Phone application is running and the Notification Source sends a Toast Notification. In this case, the toast is not displayed automatically to the user.
`HttpNotificationReceived`	If the application is running, the `HttpNotificationReceived` event will be raised when the Notification Source sends a Raw Notification.

When you trap the `ChannelUriUpdated` event, the URI of the notification channel is held in the `ChannelUri` property of the `NotificationChannelUriEventArgs` argument. As you can see in the following code, the `ChannelUri` is a standard HTTP URI with a unique identifier at the end:

```
void httpChannel_ChannelUriUpdated(object sender, NotificationChannelUriEventArgs e){
    this.ChannelUri = e.ChannelUri;
// Example ChannelUri: http://sn1.notify.live.net/throttledthirdparty/01.00/
AAHu1b30u2_2T6uCj-BDKLHVAgoOs1ADAgAAAAQOMDAwAAAAAAAAAAAAAAA
}
```

From the previous code snippets, you can see that there are two places where you need to extract the `ChannelUri`: when the channel exists and is returned from the `Find` method, and when the `ChannelUri` changes (i.e. when the channel is opened). In both of these cases, the phone application needs to send the URI directly to the Notification Source so that it can be used for Push Notifications targeted to your application.

In this case the Notification Source is going to be made up of two components, a WCF Service (which the Windows Phone application will communicate with to identify itself and its ChannelUri) and a simple WPF application that will be used to create and send the notifications. You'll start by creating the WCF Service. Add a new project based on the WCF Service Application template. Then update the default service to match the following code:

Available for download on Wrox.com

```
[ServiceContract]
public interface IChannelIdentification{
    [OperationContract]
    void Identify(string channelUri);

    [OperationContract]
    string RetrieveChannelUri();
}
```

Code snippet from IChannelIdentification.cs

```
public class ChannelIdentification : IChannelIdentification{
    private static string identifiedChannelUri { get; set; }
    public void Identify(string channelUri){
        identifiedChannelUri = channelUri;
    }

    public string RetrieveChannelUri(){
        return identifiedChannelUri;
    }
}
```

Code snippet from ChannelIdentification.svc.cs

As you can see this service has two methods, `Identify` and `RetrieveChannelUri`, which essentially set and retrieve the values of the static `identifiedChannelUri`. Clearly this is an overly simplified example that is only capable of identifying a single channel (in other words the Windows Phone application running on a single device). The Windows Phone application invokes the `Identify` service method, passing along the device's current `ChannelUri`:

Available for download on Wrox.com

```
private Uri channelUri;
private Uri ChannelUri {
    get{
        return channelUri;
    }
    set{
        channelUri = value;
        IdentifyWithNotificationSource(channelUri);
    }
}

void IdentifyWithNotificationSource(Uri channelUri){
    var client = new ChannelService.ChannelIdentificationClient();
    client.IdentifyCompleted += client_IdentifyCompleted;
    client.IdentifyAsync(channelUri.AbsoluteUri);
}

void client_IdentifyCompleted(object sender, AsyncCompletedEventArgs e){
    // Do nothing
}
```

Code snippet from MainPage.xaml.cs

Once you have created the WCF Service you need to return to your Windows Phone application and add a reference to the service. Right-click on the project node in Solution Explorer and select Add Service Reference. Browse to the service location, give it a name (in this case ChannelService) and click OK to add the service reference.

In the previous code you will also need to add a using statement to the System .ComponentModel namespace.

When the `ChannelUri` property is set, the `IdentifyWithNotificationSource` method is invoked. This, in turn, calls the `Identify` method on a WCF service in order to identify that the Windows Phone application is configured to accept notifications via the supplied channel URI.

The second part of the Notification Source, which will be used to send notifications to the Windows Phone application, is a simple Windows Presentation Foundation (WPF) application. Create another new project based on the WPF Application template, and add a service reference to the WCF Service you created earlier. The WPF application, which will be the source of all the Push Notifications, needs to call the `RetrieveChannelUri` method on the service to retrieve the current channel URI of the Windows Phone application:

Available for download on Wrox.com

```
private void FindChannelUriButton_Click(object sender, RoutedEventArgs e){
    var client = new ChannelService.ChannelIdentificationClient();
    this.ChannelUriLabel.Content = client.RetrieveChannelUri();
}
```

Code snippet from MainWindow.xaml.cs

Now that you have registered your Windows Phone application to receive Push Notifications, you can start sending notifications to the Notification Service. Let's quickly recap how the Windows Phone application, the Notification Service, and the Notification Source work by stepping through Figure 7-7.

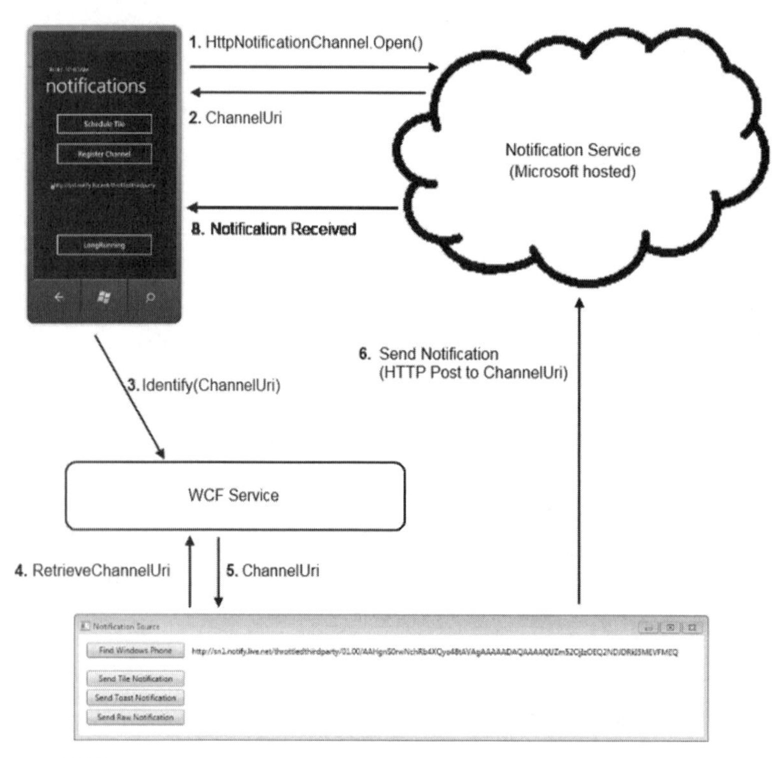

FIGURE 7-7

The sequence of events are as follows:

1. The Windows Phone application creates an instance of the HttpNotificationChannel and invokes the Open method to establish the notification channel.
2. As the Open method is asynchronous, at some point in the future the ChannelUriUpdated event is raised, exposing the ChannelUri.
3. The Windows Phone application invokes the Identify method on the WCF Service, passing the ChannelUri. The ChannelUri is recorded by the WCF Service.
4. The Notification Source application invokes the RetrieveChannelUri method on the WCF Service.
5. The WCF Service returns the ChannelUri recorded in step 3.
6. The Notification Source application sends a notification by doing a Http POST to the ChannelUri.
7. The notification is received by the Windows Phone application (or in the case of a Tile or Toast notification by the Windows Phone operating system).

Priority

As you've seen, there are three types of notifications, but there are also different priority levels. In your WPF application you can define an enumeration that lists the different notification types along with their priority levels. The numeric value is important in this enumeration because it will be used to specify the notification class value that is sent to the Notification Services when sending out a notification.

```
enum NotificationClass{
    Tile_RealTime=1,
    Tile_Priority=11,
    Tile_Regular = 21,
    Toast_RealTime=2,
    Toast_Priority=12,
    Toast_Regular=22,
    Raw_RealTime=3,
    Raw_Priority=13,
    Raw_Regular=23
}
```

The three priority levels are Real Time, Priority, and Regular, which correspond to the message being sent immediately, within 450 seconds and within 900 seconds. You have to remember that no matter what the priority level is called, the Notification Service is only a best-effort service. This means that not only is delivery not guaranteed, but there is also no service level agreement around the time it takes for a notification to reach its destination. If you want more precision around message delivery, then you should consider either rolling your own solution or using a third party, such as the Windows Azure AppFabric. The Real Time priority level should be used for notifications that you would like to get to the device as soon as possible.

Once you have determined what type of notification and at what priority level you are going to be sending, the actual process of sending a notification is the same for all notification types. The following code provides a wrapper function that accepts the notification message to be sent, the channel URI, and the type of notification:

```
private static void SendNotification(string notificationMessageText,
                                     string channelUri,
                                     NotificationClass notificationType){
    byte[] notificationMessage =
                    Encoding.Default.GetBytes(notificationMessageText);
    SendNotification(notificationMessage, channelUri, notificationType);
}

private static void SendNotification(byte[] notificationMessage,
                                     string channelUri,
                                     NotificationClass notificationType){
    var request = (HttpWebRequest)WebRequest.Create(channelUri);

    request.Method = "POST";

    //Indicate that you'll send toast notifications!
    request.Headers = new WebHeaderCollection();
    request.Headers.Add("X-NotificationClass", ((int)notificationType).ToString());
    var targetHeader = notificationType.TargetHeader();
    if (!string.IsNullOrEmpty(targetHeader)){
        request.ContentType = "text/xml";
        request.Headers.Add("X-WindowsPhone-Target", targetHeader);
    }

    // Sets the web request content length.
    request.ContentLength = notificationMessage.Length;
    using (Stream requestStream = request.GetRequestStream()){
        requestStream.Write(notificationMessage, 0, notificationMessage.Length);
    }

    // Sends the notification and gets the response.
    HttpWebResponse response = (HttpWebResponse)request.GetResponse();
    string notificationStatus = response.Headers["X-NotificationStatus"];
    string notificationChannelStatus = response.Headers["X-SubscriptionStatus"];
    string deviceConnectionStatus = response.Headers["X-DeviceConnectionStatus"];
}

private static string TargetHeader(this NotificationClass notificationType)
{
    switch (notificationType)
    {
        case NotificationClass.Tile_Priority:
        case NotificationClass.Tile_RealTime:
        case NotificationClass.Tile_Regular:
            return "token";
        case NotificationClass.Toast_Priority:
        case NotificationClass.Toast_RealTime:
        case NotificationClass.Toast_Regular:
            return "toast";
    }
    return null;
}
```

Code snippet from NotificationHelper.cs

You need to be aware that there is an implicit relationship between the structure of the notification message and the notification type. For each type of notification there is a specific message structure that you have to adhere to.

Tile Notifications

Earlier you saw how you can set up a schedule to update the tile background on the Start for a pinned application. This relies on the remote image being periodically updated. In a lot of cases, you will want the remote image to be updated based on some change in the data associated with the application. For example, it may be that a long-running operation in Windows Azure has completed. In these cases, a Tile Notification can be sent to the Windows Phone to update the tile for that application to reflect the new status.

The sender of a Tile Notification can control the Background Image, the Count, and the Title of the tile. As you saw earlier, these three are combined together to generate the Application Tile that appears on the Start. When an Application Tile changes, the tile may animate or the device may beep or vibrate, depending on how the Windows Phone is configured. Note that this setting is applied across all tiles on the Start.

Sending a Tile Notification to the Notification Service is just a matter of creating a message with the right structure. The following snippet illustrates the structure that needs to be sent to the Notification Service in order to send out a Tile Notification:

Available for download on Wrox.com

```
public static void SendTileNotification(string channelUri,
                                        string backgroundImageUri,
                                        int updateCount, string tileTitle){
    var messageTemplate = "<?xml version=\"1.0\" encoding=\"utf-8\"?>" +
                          "<wp:Notification xmlns:wp=\"WPNotification\">" +
                          "<wp:Tile>" +
                          "<wp:BackgroundImage>{0}</wp:BackgroundImage>" +
                          "<wp:Count>{1}</wp:Count>" +
                          "<wp:Title>{2}</wp:Title>" +
                          "</wp:Tile> " +
                          "</wp:Notification>";

    var message = string.Format(messageTemplate, backgroundImageUri,
                                updateCount, tileTitle);
    SendNotification(message, channelUri, NotificationClass.Tile_RealTime);
}
```

Code snippet from NotificationHelper.cs

The three parameters that are merged into the message template to generate the message to be sent to the Notification Service are the remote URI for the Application Tile image, the Count, and the Title to be added as an overlay over the tile.

Within the Windows Phone application there is one additional step that needs to be made when setting up the notification channel. After opening the channel, you need to `BindToShellTile` — in other words, connect the channel to the tile that appears on the Start screen. There are two alternatives to bind the channel. The first is to simply call `BindToShellTile` with no parameters.

This wires up the channel so that the Title and the Count are updated and applied to the Application Tile. You can still update the background image but it has to reference an image that has been deployed within the Windows Phone application (ie no images from remote URLs):

```
this.Channel.Open();
this.Channel.BindToShellTile();
```

The other alternative is to create a Uri collection that represents a list of domains that the background image can be sourced from. When a tile notification is received that has the BackgroundImage element set to a remote URL, the system checks to ensure the image is being loaded from a domain in this list.

```
this.Channel.Open();
var allowedDomains = new Collection<Uri> {
                                  new Uri("http://www.builttoroam.com") };
this.Channel.BindToShellTile(allowedDomains);
```

Code snippet from MainPage.xaml.cs

Once you have set up the binding between the notification channel and the application's tile, that's all you need to do for the tile to be updated by an incoming tile notification.

Toast Notifications

Sending a Toast Notification is similar to sending a Tile Notification, although the structure of the message differs slightly, as shown in the following code:

```
public static void SendToastNotification(string channelUri,
                                         string toastText1,
                                         string toastText2){
    var messageTemplate = "<?xml version=\"1.0\" encoding=\"utf-8\"?>" +
                          "<wp:Notification xmlns:wp=\"WPNotification\">" +
                          "<wp:Toast>" +
                          "<wp:Text1>{0}</wp:Text1>" +
                          "<wp:Text2>{1}</wp:Text2>" +
                          "</wp:Toast>" +
                          "</wp:Notification>";

    var message = string.Format(messageTemplate, toastText1, toastText2);
    SendNotification(message, channelUri, NotificationClass.Toast_RealTime);
}
```

Code snippet from NotificationHelper.cs

From the code, you can see that there are two holes in the message template that are replaced by strings to generate the actual message. As with Tile Notifications, there is an additional step required for a Windows Phone application to accept Toast Notifications. This time, instead of binding to the `BindToShellTile`, you need to call the `BindToShellToast` method. This can be used in conjunction with a call to `BindToShellTile` to allow an application to receive both Tile and Toast Notifications:

```
this.Channel.Open();
this.Channel.BindToShellToast();
```

When users see a Toast Notification, they can click on the notification to navigate directly to that application. As the toast will appear over any currently running application, it is possible for the users to rapidly jump between applications, using the Back button to step back to where they've come from.

If the application is already running when a Toast Notification is received, then the toast will not be displayed. Instead, the application can intercept the notification, allowing it to be processed and optionally displayed to the user.

Available for download on Wrox.com

```
void httpChannel_ShellToastNotificationReceived(object sender,
                                              NotificationEventArgs e){
    if (e.Collection != null){
        var messageBuilder = new System.Text.StringBuilder();

        foreach (string key in e.Collection.Keys){
            messageBuilder.AppendLine(key + " - " + e.Collection[key]);
        }

        this.Dispatcher.BeginInvoke(
             () => MessageBox.Show(messageBuilder.ToString()));
    }
}
```

Code snippet from MainPage.xaml.cs

Raw Notifications

Quite often when an application is running, it is useful to be able to send it a notification to tell it that it needs to perform some action. It may be to prompt the user to do an action, or it might be a silent notification prompting the application to synchronize some data from a server. This is the function of the third form of Push Notification — Raw Notifications. This form of notification is aptly named as it allows you to send a notification containing a byte array of data. This could be a set of numeric product identifiers, a JPEG photo, an encoded string, and the like. As you can see from the following code, there is no need to embed the data within a message template as you do with Tile or Toast Notifications:

Available for download on Wrox.com

```
public static void SendRawNotification(string channelUri, string data){
    SendNotification(data, channelUri, NotificationClass.Raw_RealTime);
}
```

Code snippet from NotificationHelper.cs

Raw Notifications are only useful if the application is still running when the notification is sent to the device. If the application has been terminated, the notification will be discarded when it reaches the device. There is no caching in the Notification Service, other than to account for lack of device connectivity. Once the Notification Service can communicate with the device, it will attempt to deliver all pending notifications.

If your application is running when a Raw Notification is received, it will automatically be routed through to the event handler for the `HttpNotificationReceived` event:

```
void httpChannel_HttpNotificationReceived(object sender, HttpNotificationEventArgs e){
    var strm = e.Notification.Body;
    var reader = new System.IO.StreamReader(strm);
    var str = reader.ReadToEnd();
    this.Dispatcher.BeginInvoke(() =>{
        MessageBox.Show(str);
        });
}
```

Code snippet from MainPage.xaml.cs

This event handler works on the assumption that the message being sent is a string. If you are working with some other data type, you will need to provide your own serialization and deserialization functionality for sending and receiving the data.

If the application is not running and you want to force the application to perform an update, you may want to consider using both Raw and Toast Notifications. You can attempt to communicate via a Raw Notification initially. If you fail to get a response back from your application, you can then send a Toast Notification. The Toast Notification will require action from the user when your application is not running, but at least it will ensure that your application is updated.

Examples

In the previous sections you saw wrapper methods for sending Tile, Toast, and Raw notifications from a WPF application. To put these in context, let's look at examples of each.

Tile Notification

Firstly, let's send a notification that updates the application's tile on the Start screen. The notification will update the background image, specify that there is a randomly assigned number of updates pending and will update the title of the tile to "Notif Updates".

Available for download on Wrox.com

```
private Random random = new Random((int)DateTime.Now.Ticks);
private void SendTileNotificationButton_Click(object sender, RoutedEventArgs e){
    var updates = random.Next(0, 100);
    NotificationHelper.SendTileNotification(this.ChannelUriLabel.Content as string,
      "http://www.builttoroam.com/books/devwp7/chapter7/notificationtile.png",
      updates, "Notif Updates");
}
```

Code snippet from MainWindow.xaml.cs

Figure 7-8 illustrates the default tile for the Notifications application on the left, and the updated tile after the tile notification has been received. In this case the Count property was set to 98, as indicated in the top right corner of the tile.

FIGURE 7-8

Toast Notification

Sending a toast notification is just a matter of specifying the two pieces of text that will be displayed in the toast notification.

Available for download on Wrox.com

```
private void SendToastNotificationButton_Click(object sender, RoutedEventArgs e){
    NotificationHelper.SendToastNotification(this.ChannelUriLabel.Content as string,
                                            "App Update",
                                "Data has changed - Click to open application");
}
```

Code snippet from MainWindow.xaml.cs

Figure 7-9 illustrates the toast notification appearing over the Start screen. Notice that the first piece of text, "App Update," is in bold at the start of the toast, followed by the second piece of text. Remember that if the application is in the foreground the toast will be captured by the application. You might initially think it a good idea to limit the text to only what will be displayed on the screen. However, it may be worth including additional information in a second text element that your application can decode if it is running.

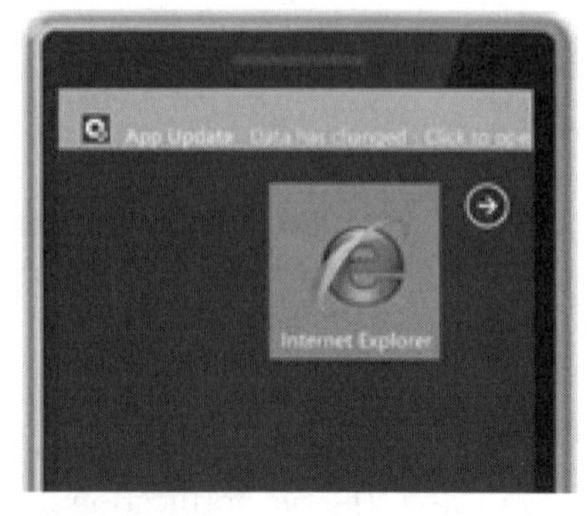

FIGURE 7-9

Raw Notification

Lastly, send a raw notification you need to pass in the piece of text you want sent.

Available for download on Wrox.com

```
private void SendRawNotificationButton_Click(object sender, RoutedEventArgs e)
{
    NotificationHelper.SendRawNotification(
                 this.ChannelUriLabel.Content as string, "Data Changed");
}
```

Code snippet from MainWindow.xaml.cs

If you look back at the SendNotification method there were two overloads. The first accepts a text string as the message to be sent. The string is converted into a byte array which is appended to the HTTP request to the Notification Service. To send binary data you can simply pass a byte array into the second overload of the SendNotification method. Of course the Windows Phone application logic would need to be adjusted to expect a byte array.

Long Running Web Service

If your application has to perform a long computation, do heavy data processing or wait for an event to happen on the server, the best practice for mobile applications is to hand off this work to the server. The server can then use Push Notification to notify the Windows Phone application when there is an update or when the work is complete. Take the following example of a long running WCF service.

```
public class LongRunningService : ILongRunningService{
    public void DoWork(){
        Thread.Sleep(30 * 1000);
    }
}
```

If this service was invoked from within a Windows Phone application, the connection would stay open for the 30 seconds that it takes this method to complete. This would not only drain battery life, but could result in lost updates should the user navigate away from the application. Instead, if this service was modified to initiate a background operation to do the work, it would return almost immediately. When the work has been completed, the background thread can send a notification back to the Windows Phone application.

Available for download on Wrox.com

```
public class LongRunningService : ILongRunningService{
    public void DoWork(string responseUri){
        var thread = new Thread((ParameterizedThreadStart)Work);
        thread.Start(responseUri);
    }

    private void Work(object responseUri){
        Thread.Sleep(30 * 1000);
        NotificationHelper.SendToastNotification(responseUri.ToString(),
                                                  "DoWork","Work completed");
    }
}
```

Code snippet from LongRunningService.cs

In this case the service is sending a toast notification which will either be captured by the application, or notify the user that the operation has completed.

Errors

As with any complex system there is always the potential for something to go wrong. The Push Notification service involves three components, being the Windows Phone application, the Notification Service (hosted by Microsoft), and the Notification Source (the source of the notifications). There is the potential for any of these components to generate errors, resulting in the failure to deliver or process notifications.

The `HttpNotificationChannel` exposes an `ErrorOccurred` event, which you should add an event handler to in order to process the different types of errors generated by the channel.

Available for download on Wrox.com

```
void httpChannel_ErrorOccurred(object sender, NotificationChannelErrorEventArgs e){
    switch (e.ErrorType){
        case ChannelErrorType.ChannelOpenFailed:
            break;
        case ChannelErrorType.MessageBadContent:
            break;
        case ChannelErrorType.NotificationRateTooHigh:
            break;
        case ChannelErrorType.PayloadFormatError:
            break;
        case ChannelErrorType.PowerLevelChanged:
            break;
    }
}
```

Code snippet from MainPage.xaml.cs

Table 7-2 provides an overview of the different types of errors defined by the `ChannelErrorType` enumeration.

TABLE 7-2: Channel Errors

ERROR	DESCRIPTION
ChannelOpenFailed	Indicates there were issues opening the channel. The channel was not established correctly so it should be reinitialized.
MessageBadContent	This type of error is raised when a tile notification is received that contains an image from a domain that isn't in the list of acceptable domains (specified in the `BindToShellTile` method).
NotificationRateTooHigh	If you attempt to send notifications in quick succession (for example if you use raw notifications to send a binary stream), you may receive this error indicating that the Notification Source is sending notifications too quickly. In this case you may need a mechanism for the Windows Phone application to be able to message the Notification Source to tell it to slow down the transmission rate.
PayloadFormatError	Indicates that the format of the notification message is invalid. In this case the current channel is disconnected and must be reopened.
PowerLevelChanged	When the battery level on the phone changes, this error type is raised to indicate to the Windows Phone application that it may not receive all types of messages. The ErrorAdditionalData property holds additional information about the new power level that can be used to work out which notification types will still be received by the application.

```
var powerLevel = (ChannelPowerLevel)e.ErrorAdditionalData;
switch (powerLevel){
    case ChannelPowerLevel.NormalPowerLevel:
        // All notifications sent to device
        break;
    case ChannelPowerLevel.LowPowerLevel:
        // Battery < 30% - only raw notification sent
        break;
    case ChannelPowerLevel.CriticalLowPowerLevel:
        // No notifications sent to device
        break;
}
```

SUMMARY

The Start area of a Windows Phone is the first experience the user has whenever he or she looks at their device. It's important for you to make optimum use of this space. You need to craft background tiles that not only have a meaningful Title and Count, but also convey meaning through dynamically changing the Background Image of the tile.

You can leverage the Push Notification system to update the Application Tile on the Start and to send Toast Notifications to the user when priority updates are available for an application. While your application is active, you can make use of Raw Notifications to perform immediate updates within your application.

Tasks

WHAT'S IN THIS CHAPTER

- How Windows Phone Tasks provide applications with the ability to integrate with services on the device
- The difference between a Launcher and Chooser
- How the Windows Phone execution model affects the way you work with Choosers
- Building and registering a photo Extras application

Continuing the theme of integrating into the Windows Phone experience, *Tasks* provide another mechanism through which your application can hook into information and functionality available within other applications on the phone. In this chapter, you will learn how to invoke launchers in order to commence an activity — for example, sending an SMS — and choosers to retrieve some data, such as taking a photo or accessing contacts. You will also learn how your application can register as an extension for certain activities on the device. For example, if your application can be used to modify images, you can register it as an image editor to appear within the built-in Pictures application.

WINDOWS PHONE TASKS

One of the things that make a Windows Phone application different from a regular desktop application is that it is running on a mobile device. In fact, a Windows Phone is not just any old mobile device — it is, of course, foremost a *phone* that can be used to make phone calls and send text messages. Without a way to tap into these unique device capabilities, you would be limited to building small screen versions of desktop applications.

Unlike its predecessor, Windows Mobile, which had a hotchpotch of disjointed interfaces for querying different types of data from the phone, Windows Phone introduces a generic concept called *Tasks*. When an application requires information from the phone, such as the phone number of a contact or a photo from the camera, it needs to launch an appropriate Task.

Windows Phone Tasks are all structured essentially the same way. However, they can be placed into two groups referred to as *launchers* and *choosers*. As you can imagine, a *launcher* is a Task that launches another application on the phone. It might be to send an e-mail, send a text message, or display a web page within the browser. Essentially, a launcher doesn't return any data to the application. In fact, when an application triggers a launcher, it should be aware that the user may not return to the application. A *chooser*, on the other hand, is a Task that the application can expect to return with a piece of information — for example, requesting that the user take a photo or select a phone number. Table 8-1 lists all the Windows Phone Tasks, including the type of any data returned.

TABLE 8-1: Windows Phone Tasks

CHOOSERS	DESCRIPTION	RETURN TYPE
`CameraCaptureTask`	Opens camera application to take a photo.	`PhotoResult`
`PhotoChooserTask`	Selects an image from your Picture Gallery.	`PhotoResult`
`EmailAddressChooserTask`	Selects an e-mail address from your Contacts List.	`EmailResult`
`PhoneNumberChooserTask`	Selects a phone number from your Contacts List.	`PhoneNumberResult`
`SaveEmailAddressTask`	Saves an e-mail address to an existing or new contact.	
`SavePhoneNumberTask`	Saves a phone number to an existing or new contact.	

LAUNCHERS	DESCRIPTION
`EmailComposeTask`	Composes a new e-mail.
`PhoneCallTask`	Initiates a phone call to a specified number.
`SmsComposeTask`	Composes a new text message.
`SearchTask`	Launches Bing Search with a specified search term.
`WebBrowserTask`	Launches Internet Explorer browsing to a specific URL.
`MarketplaceDetailTask`	Launches Marketplace with the details of a specific application.
`MarketplaceHubTask`	Launches Marketplace at one of the three hubs: Applications, Music or Podcasts.
`MarketplaceReviewTask`	Launches Marketplace to provide a review of the current application.
`MarketplaceSearchTask`	Launches Marketplace and performs a search for content.
`MediaPlayerLauncher`	Launches Media Player.

The general pattern for invoking a Task is to create an instance of the Task, set any necessary properties, and then call its `Show` method. For example, you can invoke the Task to request the user to select an e-mail address using the following two lines:

```
EmailAddressChooserTask addressTask = new EmailAddressChooserTask();
this.addressTask.Completed += addressTask_Completed;
addressTask.Show();

void addressTask_Completed(object sender, EmailResult e){...}
```

In the case of choosers, you will also need to attach an event handler to the `Completed` event. When the chooser application closes, the `Completed` event is invoked, and any return value can be accessed from the event arguments. As mentioned earlier, there is no information returned from a launcher so there is no method to indicate that the launcher has completed.

Where Did My Application Go?

Before going through each of the different Tasks it's important to understand how Windows Phone applications behave when they are placed into the background. This was covered in Chapter 6 in the context of the navigation system but is equally applicable when working with Tasks. If you recall, when your application goes into the background, the Deactivated event is raised and then the application is marked as "Eligible for Termination." At this point it is highly likely that your application will be terminated. This is true even if your application invokes a chooser task that is to return data.

The following code creates an instance of the EmailAddressChooserTask within the EmailAddressButton_Click method. When this method is invoked, the EmailAddressChooserTask will be create, the event handler wired up, and the appropriate chooser displayed. The last part of this process takes the focus away from the application, putting it into the background and making it "Eligible for Termination."

```
private void EmailAddressButton_Click(object sender, RoutedEventArgs e){
    EmailAddressChooserTask addressTask = new EmailAddressChooserTask();
    addressTask.Completed += addressTask_Completed;
    addressTask.Show();
}
```

So what happens if the application does get terminated when the chooser is displayed? More importantly, what happens when the user has selected the contact's e-mail address that they want returned to the application. As you saw in Chapter 6 the application is restarted and the page that the application was on is navigated to. This is where you will run into issues with defining the chooser task inside a method scope (as in the previous code snippet). Because the instance is only created, and the event handler wired up, within the scope of a method, there is no way for the system to know to invoke the `addressTask_Completed` method with the results of the chooser task.

The correct way to work with chooser tasks is to create the chooser as an instance level variable. In the following code the `EmailAddressChooserTask` is instantiated during the construction of the `MainPage` and the event handler for the `Completed` event is wired up at the end of the constructor.

Available for download on Wrox.com

```
public partial class MainPage : PhoneApplicationPage{
    EmailAddressChooserTask addressTask = new EmailAddressChooserTask();

    public MainPage(){
        InitializeComponent();

        this.addressTask.Completed += addressTask_Completed;
    }

    private void EmailAddressButton_Click(object sender, RoutedEventArgs e){
        addressTask.Show();
    }

    void addressTask_Completed(object sender, EmailResult e){...}
}
```

Code snippet from MainPage.xaml.cs

With this code, when the application is re-launched after the chooser has been completed, the EmailAddressChooserTask is created and the Completed event wired up at the same time as the MainPage. The process of wiring up the event handler for the Completed event does a check to see if there are any pending events to be raised. When the application returns from a chooser there is such an event, so the event handler is invoked.

In the next sections you will learn about the different Tasks that are available. Make sure you remember that any Task that returns a value, in other words a chooser, needs to be instantiated as an instance level variable.

Camera and Photos

The CameraCaptureTask allows your application to retrieve a photo from the camera, while the PhotoChooserTask can be used to select an image that resides on the device.

CameraCaptureTask

Taking an image with the camera is as simple as creating a new instance of the `CameraCaptureTask` and then calling `Show`. When the camera application closes, if an image has been taken, a `PhotoResult` instance will be returned that contains a reference to the image as a ready-to-read stream and the filename where it resides on the device:

Available for download on Wrox.com

```
CameraCaptureTask cameraTask = new CameraCaptureTask();

public MainPage(){
    InitializeComponent();
    this.cameraTask.Completed+=cameraTask_Completed;
}

private void CameraCaptureButton_Click(object sender, RoutedEventArgs e){
    cameraTask.Show();
}

Private void cameraTask_Completed(object sender, PhotoResult e){
```

```
        if (e.TaskResult == TaskResult.OK){
            CompleteCameraTask(e);
        }
    }
```

Code snippet from MainPage.xaml.cs

The Windows Phone emulator includes a mock camera application that consists of a white background and a black rectangle that roams around the screen. Figure 8-1 shows the camera application running. The black rectangle moves around the screen simulating a moving object, thus allowing you to simulate taking a photo.

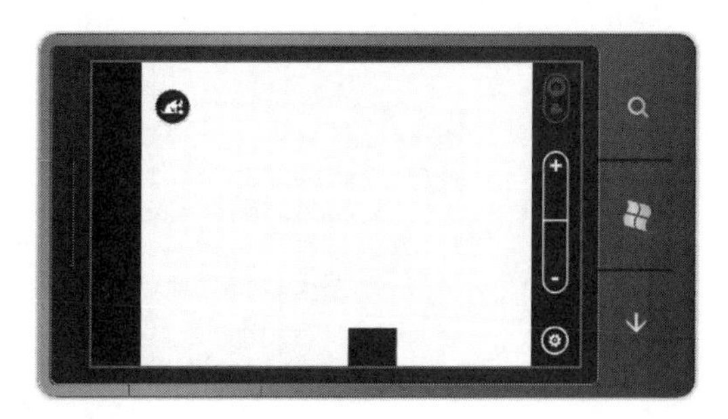

FIGURE 8-1

In order to take a picture with the camera application, you click the icon in the top-left corner of the image (it's in the top-right corner when running the application in portrait). When the photo has been taken, the camera application will close and your application will return to the foreground. The `Completed` event will then be raised, giving you an opportunity to handle the image that has been captured.

The `PhotoResult`, that is passed to the `CompleteCameraTask` method, has two properties that you can access. In the following code the `OriginalFileName` property is used to create a `BitmapImage` instance that can then be used as the Source of an `Image` control. This will simply display the image taken by the camera in the `Image` control.

```
private bool CompleteCameraTask(TaskEventArgs<PhotoResult> e){
    // Display the photo as taken by the camera
    this.CameraImage.Source = new BitmapImage(new Uri(e. OriginalFileName));
}
```

As an alternative, you can also access the image contents of the specified file by reading the stream provided by the `ChosenPhoto` property, this is useful if you want to manipulate the contents of the image. To demonstrate using this property, the following code sample creates a `WriteableBitmap` instance with the contents of the selected image:

Available for download on Wrox.com

```
private void CompleteCameraTask(PhotoResult e){
    // Display the photo as taken by the camera
    var bm = new BitmapImage();
    bm.SetSource(e.ChosenPhoto);
    this.CameraImage.Source = bm;

    var img = new BitmapImage();
    img.SetSource(e.ChosenPhoto);
    WriteableBitmap writeableBitmap = new WriteableBitmap(img);
    GenerateMirrorImage(writeableBitmap);
}
```

```
private void GenerateMirrorImage(WriteableBitmap writeableBitmap){

    writeableBitmap.Invalidate();

    // Code to reflect pixels
    int pixelPosition = 0;
    int reversePosition = writeableBitmap.PixelWidth * writeableBitmap.PixelHeight;
    int pixelValue;
    for (int i = 0; i < writeableBitmap.PixelHeight / 2; i++){
        reversePosition -= writeableBitmap.PixelWidth;
        for (int j = 0; j < writeableBitmap.PixelWidth; j++){
            pixelValue = writeableBitmap.Pixels[reversePosition];
            writeableBitmap.Pixels[reversePosition] =
                                  writeableBitmap.Pixels[pixelPosition];
            writeableBitmap.Pixels[pixelPosition] = pixelValue;

            pixelPosition++;
            reversePosition++;
        }
        reversePosition -= writeableBitmap.PixelWidth;
    }

    this.MirrorImage.Source = writeableBitmap;
}
```

Code snippet from MainPage.xaml.cs

A standard `BitmapImage` instance is read-only. It is not possible to alter the contents of the image once it is loaded. A `WriteableBitmap`, on the other hand, allows the color of individual pixels to be queried and replaced. The code sample above illustrates how you can walk through the pixels of the image and modify them one by one. In this case, the code applies a horizontal flip. This is illustrated in Figure 8-2.

FIGURE 8-2

There is much more that you can do with a `WriteableBitmap`, most of which revolves around pixel manipulation. WriteableBitmapEx is an Open Source library that provides extensions for drawing shapes, lines, curves, and much more. It's available via CodePlex at `http://writeablebitmapex.codeplex.com/`.

PhotoChooserTask

Similar to the `CameraCaptureTask`, the `PhotoChooserTask` returns a `PhotoResult` object in the `Completed` event that contains a reference to an image. However, instead of displaying the camera application to capture a new image, the `PhotoChooserTask` displays the Picture Picker application to select from the user's gallery. The following code displays the Picture Picker, but also specifies that the returned image should be cropped to 50 × 50 and that the user should have the option to elect to take a new photo with the camera:

Available for download on Wrox.com

```
PhotoChooserTask choosePhoto = new PhotoChooserTask();

public MainPage(){
    InitializeComponent();
    this. choosePhoto.Completed+= choosePhoto_Completed;
}

private void ChoosePhotoButton_Click(object sender, RoutedEventArgs e){
    choosePhoto.PixelHeight = 50;
    choosePhoto.PixelWidth = 50;
    choosePhoto.ShowCamera = true;
    choosePhoto.Show();
}

void choosePhoto_Completed(object sender, PhotoResult e){
    if (e.TaskResult == TaskResult.OK){
        CompleteCameraTask(e);
    }
}
```

Code snippet from MainPage.xaml.cs

Figure 8-3 illustrates the Picture Picker. If the `ShowCamera` property is set to `true`, the user has the option of launching the camera application to take a photo. This will be added to the images stored on the device and returned as the image for the `PhotoChooserTask`.

Since the `PhotoChooserTask` is similar to the `CameraCaptureTask` and returns a familiar `PhotoResult`, it is easy to reuse the existing `CompleteCameraTask` method to handle the response in the same way.

FIGURE 8-3

Phone and SMS

Being true to what it is at heart, Windows Phone provides good integration with the phone and instant messaging capabilities of the device. Through the `PhoneNumberChooserTask` and associated `SavePhoneNumberTask`, `PhoneCallTask`, and `SmsComposeTask`, you can request that the user select or save phone numbers in his or her contacts, initiate a phone call, or send text messages.

SavePhoneNumberTask

In order for your application to record a new phone number in the built-in Contacts application, you will need to use the `SavePhoneNumberTask`, as shown in the following code. Note that while you might think that this would be a launcher, it is in fact a chooser. It doesn't actually return any information to the application other than whether the number was saved or not. This is indicated by the TaskResult being set to OK or Cancel, in the case that the user has cancelled the save operation.

```
SavePhoneNumberTask saveNumber = new SavePhoneNumberTask();

public MainPage(){
    InitializeComponent();
    this.saveNumber.Completed += saveNumber_Completed;
}

private void SaveNumberButton_Click(object sender, RoutedEventArgs e){
    saveNumber.PhoneNumber = "+1 425 001 0001";
    saveNumber.Show();
}

void saveNumber_Completed(object sender, TaskEventArgs e){
    if (e.TaskResult == TaskResult.OK){
        MessageBox.Show("Phone number saved!");
    }
    else{
        MessageBox.Show("Phone number not saved");
    }
}
```

Code snippet from MainPage.xaml.cs

As you can see, the application doesn't get control to specify where the phone number is saved or which contact it will be associated with. When the `Show` method is invoked, the Contact Selector is displayed, as shown in the leftmost image of Figure 8-4. Compared to the Contact Selector shown when selecting a phone number, one difference in this Contact Selector is that there is an icon to "Add a Contact."

FIGURE 8-4

Regardless of whether the user selects an existing contact or creates a new one, the next screen allows the user to edit the number being saved and to specify which phone number category it should be saved against. The third (from left to right) image of Figure 8-4 illustrates the list of phone number properties available for a Windows Phone contact. Once the user has confirmed the new number to be saved, the full Contact Editor is displayed. From here the user can add a photo or add or edit additional phone and e-mail properties for the contact.

Finally, once the user hits the Save button on the Application Bar, the contact summary is displayed. The contact summary not only lists the contact information available for the contact, but it also contains the "What's New" feed, which will draw in updates from various social networks to keep you up-to-date with that contact. To return to your application the user needs to click the Back button. This isn't immediately obvious to the user, so it may be that they click the Start button and do other things on the phone before returning to your application. For these kinds of reasons, it's important to remember this and to persist any relevant information prior to invoking any of the Windows Phone Tasks.

PhoneNumberChooserTask

To allow users to select a phone number from their list of contacts on the phone, you just need to create an instance of the `PhoneNumberChooserTask` and invoke its `Show` method:

Available for download on Wrox.com

```
PhoneNumberChooserTask chooseNumber = new PhoneNumberChooserTask();

public MainPage(){
    InitializeComponent();
    this.chooseNumber.Completed += chooseNumber_Completed;
}
private void PhoneNumberButton_Click(object sender, RoutedEventArgs e){
    chooseNumber.Show();
}

void chooseNumber_Completed(object sender, PhoneNumberResult e){
    PhoneNumberText.Text = e.PhoneNumber;
}
```

Code snippet from MainPage.xaml.cs

The `PhoneNumberChooserTask` uses the Contact Selector application, which displays all the contacts on the phone, as shown in the images of Figure 8-5. When the selector application is first opened, it will display the list of contacts as a vertical list, as shown in the leftmost image of Figure 8-5.

FIGURE 8-5

As the number of contacts grows, users will be able to scroll the screen to flip through their contacts. The contacts are listed alphabetically, with the blue squares to the left of the list indicating the start of a new letter of the alphabet. In the case of a large number of contacts, users can jump to a particular letter in the alphabet by clicking on one of the blue squares. This will display the Alphabet Jump page, shown in the second-from-the-left image of Figure 8-5. The letters that have one or more contacts starting with that letter are indicated with a blue square. These can be clicked on to navigate directly to that letter in the Contacts List. An alternative way to locate contacts is via the Search box, shown in the third-from-the-left image of Figure 8-5. The Search box is displayed when the hardware Search button is pressed on the Windows Phone, illustrating the use of context-relevant searching.

When a contact is selected, if there is only one phone number for that contact, the Selector application closes immediately, returning the phone number wrapped in a `PhoneNumberResult`. However, if the contact has multiple phone numbers, the rightmost image of Figure 8-5 is displayed, allowing the user to select which phone number to return. In the `Completed` event the returned number is simply displayed on the screen.

One thing you'll notice is that the `PhoneNumberResult` only returns the phone number that the user selected. There is no other identifying information about the contact such as their name or a unique identifier. This is by design because it ensures the privacy of personal data on the device, while also avoiding the possibility of applications duplicating core Windows Phone functionality such as storing lists of contacts.

PhoneCallTask

Once a phone number has been obtained, it is as simple as creating an instance of the `PhoneCallTask` class and setting the `PhoneNumber` property to initiate an outgoing phone call:

Available for download on Wrox.com

```
private void CallNumberButton_Click(object sender, RoutedEventArgs e){
    PhoneCallTask callNumber = new PhoneCallTask();
    callNumber.DisplayName = "Fake Number";
    callNumber.PhoneNumber = "+1 425 001 0001";
    callNumber.Show();
}
```

Code snippet from MainPage.xaml.cs

It's recommended that you also set the `DisplayName` property, which is displayed through the duration of the call, as shown in Figure 8-6. This sequence of images shows the stages of a call invoked using the `PhoneCallTask`. As with most tasks that require access to device capabilities, users are prompted to confirm that they wish to dial the number. You should rely on this behavior within your application and not provide a second confirmation dialog. For example, if a user opens a client record within a Customer Relationship Management (CRM) application and taps the client's office phone number, the CRM application should not display a dialog confirming the intent to call the number. Instead, allow the `PhoneCallTask` to do this using the built-in prompt shown in the leftmost image of Figure 8-6.

FIGURE 8-6

The next three images (from left to right) of Figure 8-6 show the call in progress. In the top-left corner you can see the call duration, and there are three buttons to "End Call," display the keypad, and display other call actions, such as put the call on speaker, mute, or hold the call. The final image shows the call summary that is displayed when the call has ended. This is only displayed for a short time before disappearing, returning users to the applications they were working in prior to the phone call.

SmsComposeTask

Unlike previous versions of Windows Mobile wherein an application could send text messages without the user's permission, or even knowledge, the `SmsComposeTask` launches the messaging application with either (or both) the phone number or the message body specified:

```
private void SendSMSButton_Click(object sender, RoutedEventArgs e){
    SmsComposeTask sendSMS = new SmsComposeTask();
    sendSMS.To = "+1 425 001 0001";
    sendSMS.Body = "Hello from my Windows Phone";
    sendSMS.Show();
}
```

Code snippet from MainPage.xaml.cs

When the `Show` method is invoked, the messaging application is displayed with the text message pre-populated and ready to send. The leftmost image of Figure 8-7 shows the message waiting for the user to hit the Send icon in the Application Bar. If users wish to change the number that the message is being sent to, they can simply click on the phone number at the top of the screen. This will display the dropdown selection shown in the second (from left to right) image of Figure 8-7, where users can open the contact or remove the number from the message being sent. Alternatively, if users wish to add another recipient of the message, they can click the small plus icon at the end of the "To:" line. The third image shows that an additional contact, Sally, has been added as a recipient of this message.

FIGURE 8-7

Finally, when the user clicks the Send button, the message scrolls up the screen and changes color to indicate that it has been sent.

WHAT'S THE EMULATOR'S PHONE NUMBER?

In the preceding example, a text message was sent to two numbers. When you hit Send on the emulator, you may notice that a Windows Phone Toast Notification is displayed, similar to that in Figure 8-8.

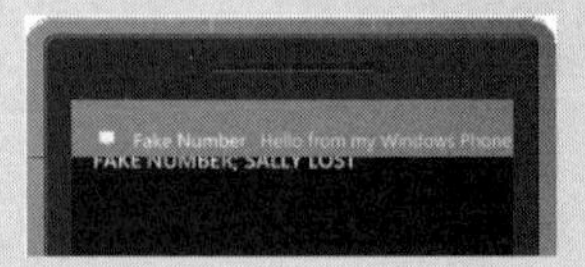

FIGURE 8-8

This toast is being displayed not to inform you that the message has been sent, but rather that a text message has been received on the emulator. If you've ever worked with the Windows Mobile emulators, you may have known that the emulator's phone number is +1 425-001-0001. This is a fake number located in Redmond (indicated by the area code of 425) in the United States (indicated by the +1). Although the Windows Phone emulator has the same phone number, it isn't particularly useful as there is no Windows Phone equivalent to the Windows Mobile SDK's Cellular emulator that allowed you to intercept outgoing, or generate fake incoming, text messages.

E-Mail

Over the past couple of years, it has almost become standard for a mobile phone to be able to send and receive not only text messages, but also e-mail. Windows Phone supports traditional e-mail services such as POP3 (Post Office Protocol), IMAP (Internet Message Access Protocol), and SMTP (Simple Mail Transfer Protocol), as well as synchronizing with Exchange Server. The e-mail Tasks allow your application to compose e-mails and interact with the user's contacts to save and retrieve e-mail addresses.

SaveEmailAddressTask

Your application can add an additional e-mail address to a contact within the Windows Phone Contact List via the `SaveEmailAddressTask`. Simply create an instance of the `SaveEmailAddressTask`, set the `Email` property, and invoke the `Show` method:

Available for download on Wrox.com

```
SaveEmailAddressTask saveEmailTask = new SaveEmailAddressTask();

public MainPage(){
    InitializeComponent();
    this.saveEmailTask.Completed += saveEmailTask_Completed;
}

private void SaveEmailButton_Click(object sender, RoutedEventArgs e){
    saveEmailTask.Email = "nick@builttoroam.com";
    saveEmailTask.Show();
}

void saveEmailTask_Completed(object sender, TaskEventArgs e){
    if (e.TaskResult == TaskResult.OK){
        MessageBox.Show("Email saved!");
    }
    else{
        MessageBox.Show("Email not saved");
    }
}
```

Code snippet from MainPage.xaml.cs

The work flow for saving an e-mail address is the same as for a phone number. Figure 8-9 illustrates this process, with the only difference being the selector for the type of e-mail address that is being saved.

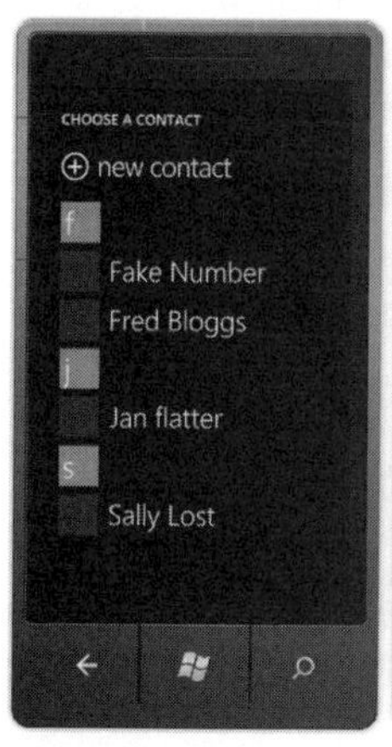

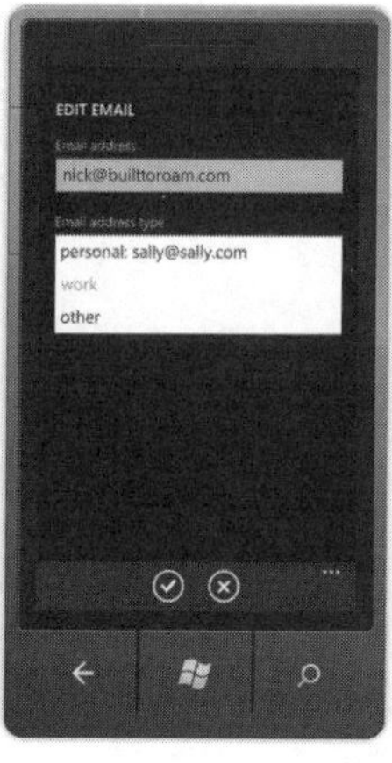

FIGURE 8-9

As you can see from the second (from left to right) image of Figure 8-9, the selector for the type of e-mail only contains three items, so it doesn't take up the full screen. You can also see that in this case there is already a personal e-mail address assigned to this contact and that the work e-mail is currently selected.

EmailAddressChooserTask

The `EmailAddressChooserTask` again makes use of the Contact Selector, first demonstrated with the `PhoneNumberChooserTask`, to return an e-mail address. As with the `PhoneNumberChooserTask`, you use the `EmailAddressChooserTask` by creating an instance and then invoking the `Show` method:

```
EmailAddressChooserTask addressTask = new EmailAddressChooserTask();

public MainPage(){
    InitializeComponent();
    this.addressTask.Completed += addressTask_Completed;
}

private void EmailAddressButton_Click(object sender, RoutedEventArgs e){
    addressTask.Show();
}

void addressTask_Completed(object sender, EmailResult e){
    if (e.TaskResult == TaskResult.OK){
        EmailAddressText.Text = e.Email;
    }
}
```

Code snippet from MainPage.xaml.cs

If the user selects a contact with only a single e-mail address, the Contact Selector application will immediately close, returning the selected e-mail address via the `Completed` event. If the selected contact has multiple e-mail addresses, users will be prompted to select which e-mail addresses they wish to use. The `EmailResult` object contains a single property, `Email`, which contains the e-mail address returned from the Contact Selector application. As with the `PhoneNumberChooserTask`, no identifying information about the selected contact is returned to your application. In this case the returned e-mail address is simply displayed on the screen in a TextBlock.

EmailComposeTask

Sending an e-mail from your application is possible using the `EmailComposeTask`. As with initiating a phone call or sending a text message, this action requires the user's permission. You can generate a pre-populated e-mail by setting the `To`, `Subject`, and `Body` fields of an instance of the `EmailComposeTask`:

```
private void ComposeEmailButton_Click(object sender, RoutedEventArgs e){
    EmailComposeTask composeTask = new EmailComposeTask();
    composeTask.Body = "This is the first email you'll ever want to send.....";
    composeTask.To = "nick@builttoroam.com";
```

```
        composeTask.Subject = "Welcome to WP7";
        composeTask.Show();
    }
```

Code snippet from MainPage.xaml.cs

Invoking `Show` will prompt users to select which account they wish to send the e-mail from, as shown in the leftmost image of Figure 8-10. After selecting the account to use, the user is given the opportunity to review the contents of the e-mail. At this point, they can change to whom the e-mail is addressed, the subject or body, and even add an attachment. Clicking the ellipsis on the Application Bar reveals the ability to adjust the priority of the message and to display the cc and bcc address fields. Once they have double-checked what is being sent, users can either click the Send button or cancel the message.

FIGURE 8-10

You will notice from the second (from left to right) image of Figure 8-10 that there is a cross icon in the Application Bar. This is a Cancel button and will exit out of the Compose E-Mail application, after asking users if they wish to save the message. The Cancel button is not required, since users could have just used the Back button. However, this is an example of where it makes sense to have a more intuitive interface to perform an action. Using the Back button has two issues. First, while the users can probably deduce that by going back they will be effectively canceling the message, it isn't immediately obvious. The second issue is that to cancel the action of sending a message, users would have to press the Back button twice, once to go back to the account selection page, then again to return to the previous application. These issues are mitigated by providing an explicit way to cancel the message. The Cancel button also prompts if you want the message to be saved in the Drafts folder so that it can be sent at a later stage.

In order for the user to send an e-mail they have to have an e-mail account configured on the device. If they don't have an account setup, launching this task will display an error advising the user to setup an account. As the EmailComposeTask is a launcher, there is no notification within the Windows Phone application that the task has failed.

One point to be aware of with the use of both the `SmsComposeTask` and `EmailComposeTask` Tasks is that they are both launchers. As such, there is no value returned when they close and the application returns to the foreground. This also means that there is no way for your application to detect whether the user has sent or canceled the operation. Nor can you detect if they have modified any of the parameters of the message being sent. If you require more control over these operations, you should investigate third-party messaging options that have a Web Service interface through which you can send messages directly from within your application.

Launchers

You have already seen how a few of the built-in Windows Phone launchers are used to initiate phone, SMS, or e-mail functions on the phone from within your application. There are seven other launchers that your application can use to initiate other activities on the phone.

SearchTask

The `SearchTask` provides you with a way to invoke a Bing Web Search for a particular search string that you supply via the `SearchQuery` property:

Available for download on Wrox.com

```
private void SearchButton_Click(object sender, RoutedEventArgs e){
    SearchTask search = new SearchTask();
    search.SearchQuery = "Microsoft Windows Phone";
    search.Show();
}
```

Code snippet from MainPage.xaml.cs

When you invoke the `Show` method on the `SearchTask` instance, users will be presented with a dialog to allow Search access to their current location, shown in the leftmost image of Figure 8-11. If the users don't permit access to their location, they will still see search results, but the results won't be filtered based on users' locations. In the middle image of Figure 8-11, you can see that there is a Callout icon on the right of the Search box at the top of the screen. Clicking on this will invoke the speech input that will permit users to enter a search query verbally. This can be particularly useful if users have to use their phone one-handed, for example, if they are carrying shopping bags.

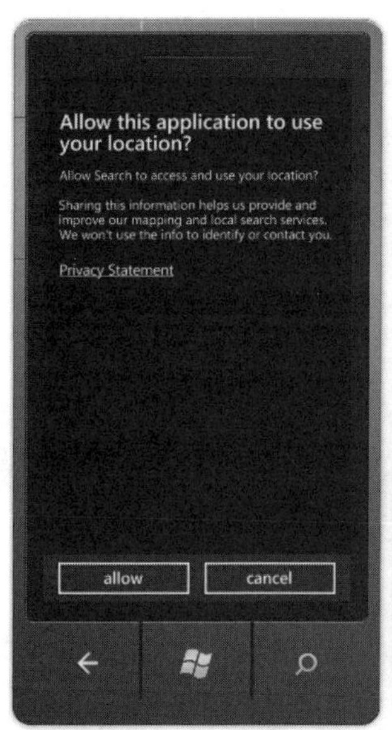

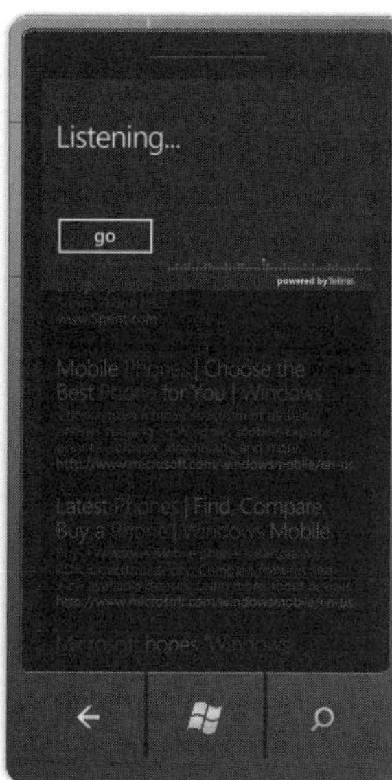

FIGURE 8-11

Use of the `SearchTask` is perhaps limited. At any stage, the user can invoke a Web search via the hardware Search button that all Windows Phones are required to have. As such, it is not necessary for you to provide a mechanism to open search, unless you are going to pre-populate the search with a particular search query.

WebBrowserTask

To open Internet Explorer, you can create an instance of the `WebBrowserTask` and invoke the `Show` method. The web page that will initially be displayed can be specified via the `URL` property:

```
private void WebBrowserButton_Click(object sender, RoutedEventArgs e){
    WebBrowserTask browser = new WebBrowserTask();
    browser.URL = "http://www.builttoroam.com";
    browser.Show();
}
```

Code snippet from MainPage.xaml.cs

Figure 8-12 illustrates a web page loaded in Internet Explorer while in Landscape orientation.

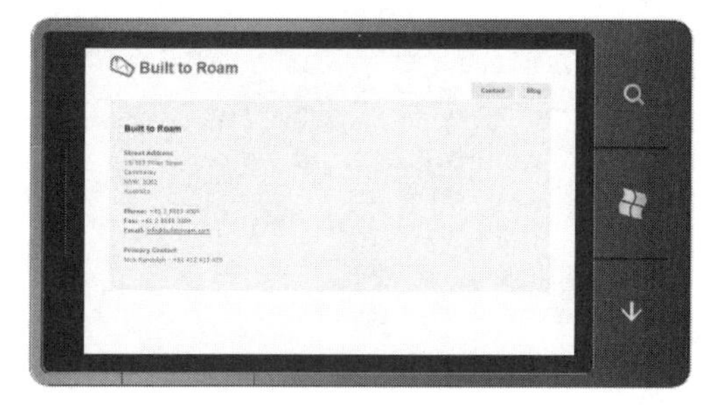

FIGURE 8-12

One of the limitations of the Windows Phone platform is that there is no way to invoke other applications that reside on the device. For example, you might want to open Word to read a report, or Excel for a spreadsheet. A work-around to this is to save the document to a Web repository and then to open Internet Explorer, pointing it at the URL for that document. The following snippet is similar to opening a normal website, except the URL is for a document that resides on the Web Server:

Available for download on Wrox.com

```
private void WebBrowserButton_Click(object sender, RoutedEventArgs e){
    WebBrowserTask browser = new WebBrowserTask();
    browser.URL = "http://www.builttoroam.com/books/devwp7/chapter8/test.docx";
    browser.Show();
}
```

Code snippet from MainPage.xaml.cs

When the `Show` method is invoked, Internet Explorer is still displayed. However, as you will see in the left image of Figure 8-13, instead of just displaying the document within the browser, it displays a prompt within the application indicating the file that will be viewed.

Tapping the icon in the middle of the screen will download the file to the device and open it with Word, as shown in the right image of Figure 8-13. Using Word, the user can edit and save the document to the device. This technique can be used for other Office applications wherein the file extension is associated with a particular application — for example, XLSX with Excel and PPTX for PowerPoint.

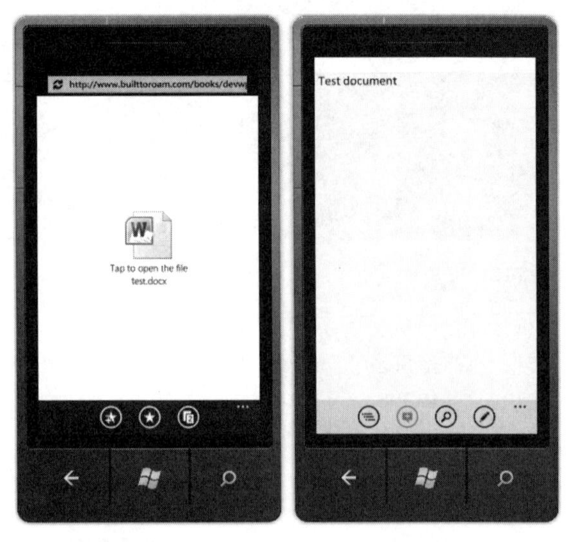

FIGURE 8-13

MediaPlayerLauncher

You have seen in Chapter 3 how you can use the `MediaElement` to display media within your application. An alternative is to use the `MediaPlayerLauncher` and allow the media to be played via the built-in Media Player on the device. The `Media` property is a URI and can be either Web-based media, as in the code snippet, or media residing on the device.

Available for download on Wrox.com

```
private void MediaPlayerButton_Click(object sender, RoutedEventArgs e){
    MediaPlayerLauncher mediaPlayer = new MediaPlayerLauncher();
    mediaPlayer.Controls = MediaPlaybackControls.Pause | MediaPlaybackControls.Stop;
    mediaPlayer.Media =
            new Uri("http://www.builttoroam.com/books/devwp7/chapter8/wildlife.wmv");
    mediaPlayer.Show();
}
```

Code snippet from MainPage.xaml.cs

The media player will be loaded in landscape and you can adjust which controls are displayed by adjusting the `Controls` property, as shown in Figure 8-14.

FIGURE 8-14

Marketplace

Your application is also able to invoke the Marketplace application on the device by using one of four launcher tasks: `MarketplaceHubTask`, `MarketplaceDetailTask`, `MarketplaceReviewTask`, and the `MarketplaceSearchTask`.

Available for download on Wrox.com

```
private void MarketplaceHubButton_Click(object sender, RoutedEventArgs e){
    MarketplaceHubTask hubTask = new MarketplaceHubTask();
    hubTask.ContentType = MarketplaceContentType.Applications;
    hubTask.Show();
}

private void MarketplaceButton_Click(object sender, RoutedEventArgs e){
    MarketplaceDetailTask detailTask = new MarketplaceDetailTask();
    detailTask.ContentIdentifier = "2f7bb8df-dc80-df11-a490-00237de2db9e";
    detailTask.ContentType = MarketplaceContentType.Applications;
    detailTask.Show();
}


private void MarketplaceReviewButton_Click(object sender, RoutedEventArgs e){
    MarketplaceReviewTask reviewTask = new MarketplaceReviewTask();
    reviewTask.Show();
}

private void MarketplaceSearchButton_Click(object sender, RoutedEventArgs e){
    MarketplaceSearchTask searchTask = new MarketplaceSearchTask();
    searchTask.ContentType = MarketplaceContentType.Applications;
    searchTask.SearchTerms = "Weather";
    searchTask.Show();
}
```

Code snippet from MainPage.xaml.cs

There are three different types of content that can be displayed within the Marketplace application on a Windows Phone device. The hub, detail, and search Tasks all have a `ContentType` property which can take on values `Application`, `Music`, and `Podcasts`.

The `MarketplaceHubTask` opens the Marketplace application at the main panoramic experience (also known as the Marketplace hub). The `ContentType` property will determine whether this displays applications, music, or podcasts.

If you want to up-sell other applications you have written then the `MarketplaceDetailTask` allows you to link to the information page in Marketplace for those applications. This page contains the logo, description, rating, and screenshots for the application, and, most importantly, allows the user to purchase the application.

You should encourage users to rate your application, as this will hopefully improve downloads and/or sales of your application. The MarketplaceReviewTask provides a way to open up the application review page for your application. There are no properties for this task as it can only be used to link to the review page for the current application.

Lastly, the MarketplaceSearchTask gives you a quick way to allow users to look up all applications, music or podcasts that match the specified search terms. If you have a large number of applications this is a convenient way to allow users to see what other applications you have for sale in the Marketplace.

EXTRAS

So far you've seen how you can seamlessly integrate the functionality of built-in Windows Phone applications into your own applications using Tasks, but not how your application can become integrated into others. There is no support for integrating your application into any of the Windows Phone hubs, nor is there support for third-party applications, written by people like yourself, to launch your application. However, Windows Phone does include a concept called *Extras*.

An *Extra* is an application that provides extra functionality to an existing Windows Phone application. For example, if you register your application as a PhotosExtrasApplication, it can be invoked to provide extra functionality within the Photos application.

The LaunchersAndChoosers application has been registered as a PhotosExtrasApplication. Figure 8-15 illustrates the work flow a user might go through when viewing and working with photos. After selecting an image from the Pictures hub, the user can tap the image to bring up a list of actions to perform. At the bottom of this list is the Extras item.

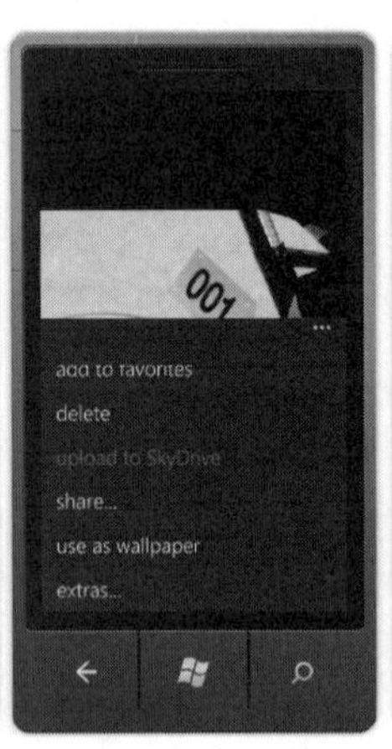

FIGURE 8-15

When the user selects Extras, a list of all applications that have been registered as PhotosExtrasApplications is displayed. Selecting an item launches that application, and the image that was being displayed is passed through to the application so that it can be manipulated. For example, an application might resize, crop, rotate, or apply special effects to the image in some way, before saving it back to the Pictures library.

To register your application as an Extra, you need to include an additional XML file within your application called *Extras.xml*. The format of this file simply includes a list of the particular type of Extras the application is being registered for. In the following snippet, the application is being registered as a PhotosExtrasApplication:

Available for download on Wrox.com

```
<?xml version="1.0" encoding="utf-8" ?>
<Extras>
  <PhotosExtrasApplication>
    <Enabled>true</Enabled>
    <StorageFolder>Photos</StorageFolder>
  </PhotosExtrasApplication>
</Extras>
```

Code snippet from Extras.xml

The `StorageFolder` element specifies the folder within IsolatedStorage (see Chapter 16) where the image being edited will be copied. After you include the Extras.xml file within your application, when you next run your application it will appear within the Extras list within the existing Photos application, as illustrated in the last image of Figure 8-15.

When the user clicks on the application in the Extras list, the application will be launched. Within the `OnNavigatedTo` method, which you need to override from the `PhoneApplicationPage` base class, you can extract the filename of the picture that was being viewed when the application was launched. Notice that the filename is combined with the path "Photos" which is the same as the `StorageFolder` element in the Extras.xml file that was packaged with the application.

```
protected override void OnNavigatedTo(NavigationEventArgs e){
    string filename = string.Empty;

    this.NavigationContext.QueryString.TryGetValue("file",out filename);

    if (!string.IsNullOrEmpty(filename)){
        filename = System.IO.Path.Combine("Photos", filename);

        //The following opens the file in the isolated storage folder.
        using (var isolatedStorageFileStream =
                        IsolatedStorageFile.GetUserStoreForApplication().
                            OpenFile(filename, FileMode.Open, FileAccess.Read)){
            WriteableBitmap picLibraryImage =
                        PictureDecoder.DecodeJpeg(isolatedStorageFileStream);
            GenerateMirrorImage(picLibraryImage);
        }
    }
}
```

This method loads the contents of the file and then calls the GenerateMirrorImage method that was created earlier to invert and display the image.

SUMMARY

In this chapter, you've learned about the Windows Phone Tasks that your application can call in order to initiate different activities on the device. Your application can also provide extra functionality to some of the built-in applications by registering using the Extras.xml file.

There is clearly a balance between exposing all the capabilities of the device and risking the privacy and security of the user's personal data.

Touch Input

WHAT'S IN THIS CHAPTER

- Understanding the user experience guidelines for touch, layout, and gestures
- Handling touch events
- Working with multi-touch

It seems that everywhere you look, people are talking about the advantages of multi-touch input. Every device from desktop computers running Windows 7, to Surface computing, through to Windows Phone is including support to detect one or more finger presses on the screen. Windows Mobile has long had support for single touch and gestures, but it has been held back through the use of resistive screens that were optimized to allow precise input with a stylus. Windows Phone moves away from requiring the use of a stylus, to supporting the ability to detect up to four simultaneous touch points.

Moving from a single mouse or stylus point to multiple touches enables a wide range of gestures and natural user interface (UI) designs to be implemented. The guidance from Microsoft around the Metro experience provides a great starting point on how to handle different touch inputs from a simple tap all the way through to flicks and other gestures. In this chapter, you will learn how to process both single- and multi-touch input events, and how you can couple this to your user interface to control navigation and increase interactivity.

USER EXPERIENCE

Since the release of the iPhone, it has become evident that users want to be able to use their fingers to do the walking. Rather than having to reach for a stylus to tap at a barely legible on-screen keyboard or lug around a bulky device supporting a hardware keyboard, users want the freedom to use their fingers to navigate, select, and enter data. If you look at the mobile landscape, there are devices that cater to all types of users — from the keyboard-driven Blackberry devices, which lack a touch screen; to dual-form Windows Mobile devices, which sport both a touch screen and a keyboard; all the way through to the iPhone, which only provides an on-screen keyboard. Although the iPhone has changed the market perception on touch screens, some view the lack of a keyboard as an unacceptable limitation. Windows Phone seeks to embrace the lessons learned through multiple iterations of the Windows Mobile platform, coupled with the touch revolution invoked by Apple with the release of the iPhone and subsequent iPad devices.

To understand the implications of building a touch-friendly platform, let's revisit the Windows Mobile–to–Windows Phone comparison. Figure 9-1 illustrates Windows Mobile 6 alongside a Windows Phone. Owing to differences in dimensions, it appears that the Windows Mobile device, on the left, has a larger screen. But this is not the case, and, in fact, the Windows Phone sports a much higher resolution. What's interesting to note is that while the Windows Mobile device appears larger, the screen is much less readable than that of the Windows Phone. The home screen of Windows Mobile is much more cluttered in an attempt to present more information to the user.

Beyond being harder to read, the Windows Mobile interface is also harder to navigate using a finger. Selecting any of the items on the home screen requires the user to jab at the screen using a fingertip. Compare this with the Start of Windows Phone, where each tile is immediately identifiable, even at a distance, and easy to tap on without having to worry about accidentally tapping an adjacent tile. Throughout the Windows Phone user experience there has been great care taken to the detail of how to make it touch-friendly. Figure 9-2 illustrates the Day view from the Calendar application for both Windows Mobile (left) and Windows Phone (right).

FIGURE 9-1

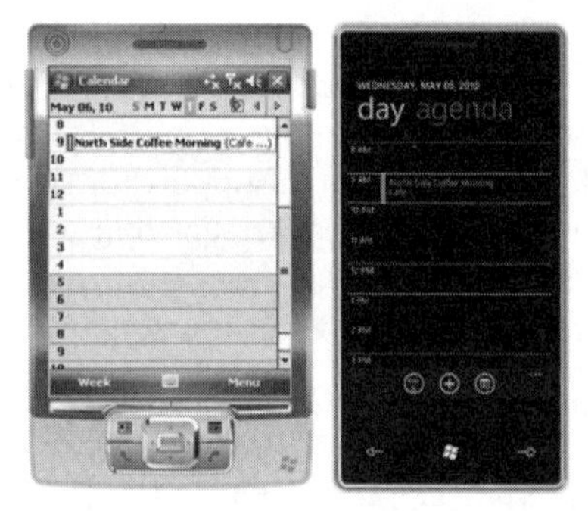

FIGURE 9-2

Again, you can see how the interface on Windows Mobile has been optimized for the presentation of a large amount of data and to be operated using a stylus. In contrast, the Windows Phone interface has been optimized for touch input, allowing actions to be performed with a single tap, at the expense of requiring additional scrolling because of the increased amount of screen space allocated to each element.

Guidelines

In order for your Windows Phone application to be successful, it will need to embrace touch as a primary form of user interaction. Although there will be Windows Phones that have hardware keyboards, the primary form of user interaction should be via touch input. Designing an interface that is optimized for touch can be quite challenging for developers who have spent years building applications for computers that have both a mouse and a keyboard. Recognizing that the large majority of developers haven't built a mobile application before, Microsoft has released a set of guidelines (the Windows Phone 7 UI Design and Interaction Guide, downloadable from `http://developer.windowsphone.com`) on how to tailor an application to make it easier to operate using one or multiple touch points.

Target Size

It might seem obvious, but a user's finger is significantly larger and less accurate than the tip of a stylus. This means that you need to provide a much larger target for them to hit when they touch the screen. You also need to consider the positioning of items to limit the chance that the user makes an error.

The guidance released by Microsoft suggests that the minimum dimension of a visual element that will respond to touch input should be 7 millimeters (mm, or 26 pixels). You can think of this as the control's *interaction area*, and it represents the visual part of the control that guides users to where they can tap the screen.

The important thing here is that the interaction area is visible to the user. In most cases, you will want to make the *touch target*, or hit zone, larger than the interaction area. The *interaction area* is the area of the screen that actually responds to touch input for the control. The guidance indicates that this should be no smaller than 9 mm, or 34 pixels. Figure 9-3 indicates the relationship between the interaction area (7 mm × 7 mm) and the touch target (9 mm × 9 mm).

In the lower portion of Figure 9-3, you can see that the separation between the interaction areas of two adjacent objects is 2 mm, or 8 pixels. The guidance from Microsoft states that this should be the minimum separation between two interaction areas to avoid users accidentally tapping the wrong item.

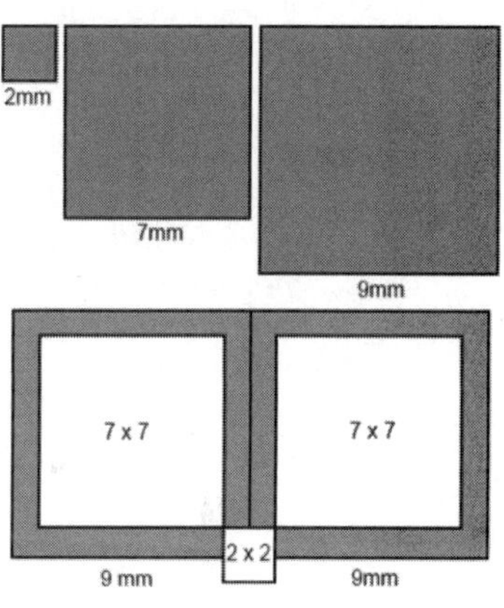

FIGURE 9-3

The default controls that ship with the Windows Phone Developer Tools have been styled in accordance to the guidance outlined above. It is recommended that you take advantage of these controls and extend them to ensure that your applications provide a similar touch experience to the standard applications.

Layout

When determining the size and location of visual controls, you should consider how frequently the control will be used. High-frequency controls should typically be larger and positioned toward the middle of the screen, where they are easier to hit.

You should also consider the order in which a user will use the controls and whether the control invokes a destructive action. Take, for example, the phone dialer shown in Figure 9-4. This is a case in which each key in the numeric pad is large to make it easy to hit. However, the spacing between the keys has been kept to an absolute minimum, which ensures that a sequence of numbers can quickly be entered without having to be too accurate in the location of your finger presses. In contrast, the "Delete Appointment" confirmation dialog uses large buttons with a clear separation between the Delete and Cancel operations. Deleting an appointment is a destructive operation so it is important to reduce the possibility of a misplaced tap on the Cancel button accidentally selecting the Delete button.

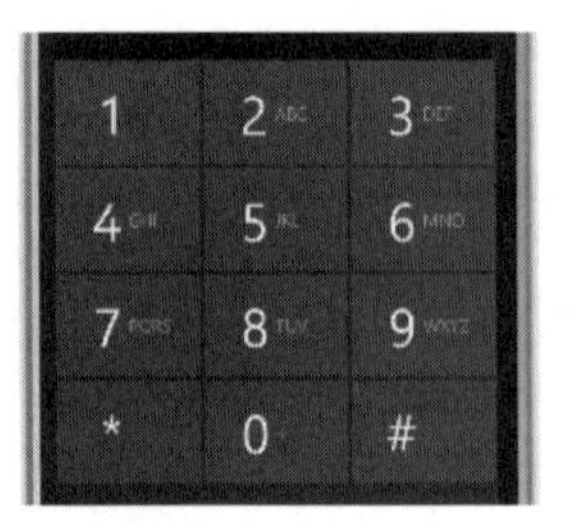

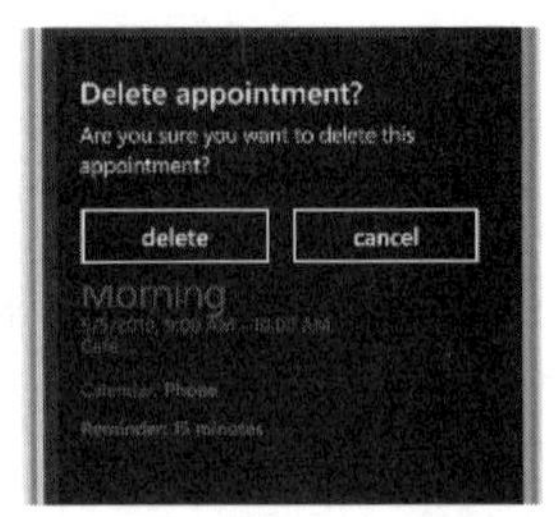

FIGURE 9-4

User actions can be classified as destructive *or* non-destructive *based on whether the action deletes or destroys data. Desktop applications generally perform destructive actions without a confirmation dialog as they typically also include the ability for a user to undo the last action. This effectively negates the need for a confirmation dialog. If users accidentally perform a destructive action, they can simply undo it. On the other hand, within a Windows Phone application, it is difficult to provide an undo capability in an intuitive way, so destructive actions should provide a confirmation dialog.*

Gestures

Building a user experience that is optimized for touch input isn't just about increasing the size of controls to make them more easily selectable with a single finger. A human hand is a more versatile input device than a traditional stylus or mouse. As well as detecting a simple touch-and-release of a single finger, it is also possible to detect more complex gestures, potentially even including more than one finger. Throughout the Metro user experience, a set of standard, or at least well understood, gestures are used. See Table 9-1.

There is no requirement that you use these gestures in the manner described. However, it is advisable not to alter the behavior that occurs when these gestures are detected, unless essential to your application. Windows Phone users who become familiar with the Metro user experience will connect with applications that behave in a similar manner. Using standard gestures in a non-standard way may cause uncertainty and frustration among your users.

TABLE 9-1: Standard Gestures

GESTURE	COMMENT
Single-Touch	
Tap	User touches the screen and then releases almost immediately. Most controls that respond to a Tap will expose this gesture via a `Click` event. When scrolling or executing some continuous action, the Tap gesture can be used to stop the current action.
Double-Tap	Two Tap gestures within a short period of time Most controls don't have a corresponding event for the Double-Tap gesture. The guidelines from Microsoft indicate that a Double-Tap should toggle between a zoomed-in and a zoomed-out state of the application. This is demonstrated in Internet Explorer, where double-clicking the screen auto-zooms the current page to make it more readable.
Pan	User touches the screen and moves his or her finger in one or more directions before releasing. The Pan gesture is implicitly supported within list controls, correlating to a scroll action. The Metro user experience uses the Pan gesture on most screens to either scroll or move content. For example, on the Start, you can move tiles by selecting a tile (Touch and Hold gesture) and then using a Pan gesture to move the tile. Within lists, the user pans vertically to scroll through items. Alternatively, within the hubs, the user pans horizontally to scroll between different states.
Flick	The Flick gesture corresponds to a quick movement in a single direction followed by the finger being lifted off the screen. This gesture can start either with the user touching the screen, or at the end of a Pan gesture. As with the Pan gesture, the Windows Phone list controls implicitly support the Flick gesture to scroll a list continuously in the direction of the flick. The Flick gesture should be used to scroll content in the direction of the flick. It is up to you whether the content continues to scroll until the user taps the screen, or whether the content has a natural inertia, in which case, it will progressively slow to a halt.

(continues)

TABLE 9-1 *(continued)*

GESTURE	COMMENT
Touch and Hold	The user touches the screen and holds his or her finger in the same location for a specified period. The guidance indicates that this should display a context menu or additional options for the selected item. There is no built-in support for the Touch and Hold gesture in the Windows Phone controls. The guidance for the Touch and Hold gesture is somewhat at odds with how this gesture is used within the Metro user interface. In some places, such as the Calendar application, this gesture invokes a context menu. However, on the Start, this gesture is used to select a tile so that it can be moved. One thing you have to be aware of is that the Touch and Hold gesture is somewhat of a hidden gesture and only the right-clickers of the world may find functionality that is only available via this gesture.
Multi-Touch	
Pinch and Stretch	User touches the screen in multiple places and moves the touch points either closer (pinch) or further away (stretch). The Windows Phone controls don't have built-in support for the Pinch and Stretch gesture. There are some applications within the Metro user experience in which the Pinch and Stretch gesture is used to continuously zoom in or out on the current content around the center of the touch points.

Luke Wroblewski, Chief Design Architect at Yahoo! Inc., has published the Touch Gesture Reference Guide *(by Craig Villamor, Dan Willis, and Luke Wroblewski, self- published, April 15, 2010;* `www.lukew.com/touch/TouchGestureGuide.pdf`*), which covers both the standard gestures given in this chapter as well as several other single and multi-touch gestures. More information, including a downloadable PDF, can be found at his website at* `www.lukew.com/touch`.

Unfortunately, the standard Windows Phone controls only provide support for a limited number of gestures such as a simple Tap or Double-Tap. However, the controls do expose basic touch events that you can process further in order to detect additional gestures.

TOUCH EVENTS

There is only limited support within the Windows Phone Developer Tools for the automatic detection of touch gestures. This is not as much of a hindrance as you might first imagine, as many gestures are easily detectable using a small amount of code. In this section, you'll see how to use the basic touch events to determine which gesture has been entered.

Single Touch

The simplest types of gestures to detect are those that involve a single touch on the screen by one finger.

Tap

Several of the Windows Phone controls already detect a single Tap gesture to perform a certain action, for example, to click a button or select an item in a list. The `Button` control exposes the Tap gesture via a `Click` event that you can add an event handler to.

XAML

```
<Button Content="Simple Gestures" Name="SimpleGestureButton"
        Click="SimpleGestureButton_Click" />
```

C#

```
private void SimpleGestureButton_Click(object sender, RoutedEventArgs e){
    MessageBox.Show("Button has been tapped!");
}
```

However, not all controls handle the Tap gesture. Take, for example, the `Border` control, which inherits from `Panel` and as such doesn't expose a `Click` event. If you inspect the inheritance list, you will notice that it inherits from `UIElement`. This means that it provides two events named `MouseLeftButtonDown` and `MouseLeftButtonUp`. These events are actually the equivalent to touch and release actions for a user's finger and are triggered at appropriate times when a user taps the control. One way to handle the tap for a control that doesn't expose a `Click` event would be to explicitly handle both these mouse events. A good way to implement this in a manner that makes it generic and reusable is to use a custom behavior, as discussed in Chapter 4.

In order to use or create behaviors in your Windows Phone application you need to add a reference to System.Windows.Interactivity.dll. Right-click on your project in Solution Explorer and select Add Reference. Select System.Windows .Interactivity and click OK. Within your code you may need to add a using statement for the System.Windows.Interactivity namespace.

The following code creates a behavior that wraps both mouse events, exposing a single `Tap` event that can be handled to perform an action when the Tap gesture is performed over a control:

Available for download on Wrox.com

```
public class TapAction:Behavior<UIElement>{
    public event EventHandler Tap;

    protected bool MouseDown { get; set; }

    protected override void OnAttached(){
        base.OnAttached();

        this.AssociatedObject.MouseLeftButtonDown += AO_MouseLeftButtonDown;
```

```
            this.AssociatedObject.MouseLeftButtonUp += AO_MouseLeftButtonUp;
        }

        protected override void OnDetaching(){
            this.AssociatedObject.MouseLeftButtonDown -= AO_MouseLeftButtonDown;
            this.AssociatedObject.MouseLeftButtonUp -= AO_MouseLeftButtonUp;

            base.OnDetaching();
        }

        void AO_MouseLeftButtonUp(object sender, MouseButtonEventArgs e){
            if (MouseDown){
                OnTap();
            }
            MouseDown = false;
        }

        void AO_MouseLeftButtonDown(object sender, MouseButtonEventArgs e){
            MouseDown = true;
        }

        protected virtual void OnTap(){
            if (Tap != null){
                Tap(this.AssociatedObject, EventArgs.Empty);
            }
        }
    }
```

Code snippet from TapAction.cs

To declaratively use this behavior within XAML, you simply add a `TapAction` instance to the `Interaction.Behaviors` attached property of the control you want to handle the Tap gesture for. In this case, the `TapAction` behavior has been added to a `Border` element, and an event handler, `TapGesture_Tap`, has been added for the `Tap` event.

Available for download on Wrox.com

XAML

```
<Border BorderBrush="#FFF11717" BorderThickness="2" Height="90" Margin="20,0,0,409"
        VerticalAlignment="Bottom" Background="#FFCC8787" HorizontalAlignment="Left"
        Width="160">
    <i:Interaction.Behaviors>
        <local:TapAction x:Name="TapGesture" Tap="TapGesture_Tap" />
    </i:Interaction.Behaviors>
</Border>
```

Code snippet from MainPage.xaml

C#

```
private void TapGesture_Tap(object sender, EventArgs e){
    MessageBox.Show("Border has been tapped!");
}
```

Code snippet from MainPage.xaml.cs

BEHAVIOR REFRESHER

Once you have created the TapAction behavior in code, in order to use it you need to switch over to Expression Blend. If your TapAction behavior doesn't appear in the Assets window under the Behaviors node you may need to build the project. Figure 9-5 illustrates how you can add the TapAction behavior to the Border element.

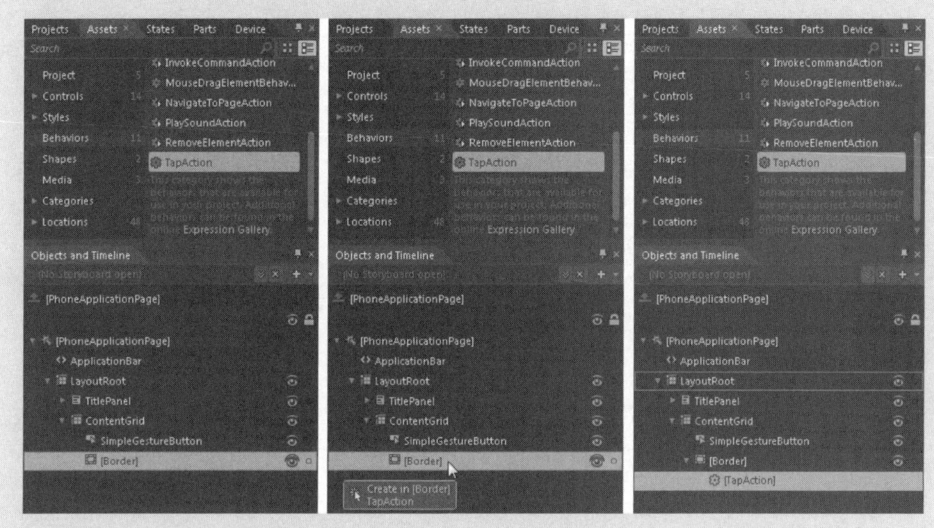

FIGURE 9-5

Simply select the TapAction and drag it onto the element you want to add it to, in this case the [Border] node in the Objects and Timeline window. Unlike the behaviors discussed in Chapter 4, the TapAction doesn't have any configurable properties. To attach an event handler to the Tap event you will need to edit the XAML for the TapAction element in either Visual Studio or Blend.

Double-Tap

Since the Double-Tap gesture is just two single Tap gestures within a short period, you can simply extend the `TapAction` behavior to raise a `DoubleTap` event if two taps are received within a specified time-out period. The following code creates a behavior called `DoubleTapAction`, which inherits from `TapAction` and has a property aptly named `DoubleTapTimeoutInMilliseconds`. This property has a 1-second (1,000 milliseconds) default value, so in most cases, it won't need to be modified but will be exposed via the Properties Tool window in either Visual Studio or Blend for the times when you may want to alter it.

Available for download on Wrox.com

```
public class DoubleTapAction:TapAction{
    public event EventHandler DoubleTap;

    public int DoubleTapTimeoutInMilliseconds{
        get { return (int)GetValue(DoubleTapTimeoutInMillisecondsProperty); }
        set { SetValue(DoubleTapTimeoutInMillisecondsProperty, value); }
    }

    public static readonly DependencyProperty
        DoubleTapTimeoutInMillisecondsProperty =
        DependencyProperty.Register("DoubleTapTimeoutInMilliseconds", typeof(int),
                                    typeof(DoubleTapAction),
                                    new PropertyMetadata(1000));

    protected DateTime? FirstTap { get; set; }

    protected override void OnTap(){
        base.OnTap();

        if (FirstTap.HasValue &&
    FirstTap.Value.AddMilliseconds(DoubleTapTimeoutInMilliseconds) > DateTime.Now){
            OnDoubleTap();
            FirstTap = null;
        }
        else{
            FirstTap = DateTime.Now;
        }
    }

    protected virtual void OnDoubleTap(){
        if (DoubleTap != null){
            DoubleTap(this.AssociatedObject, EventArgs.Empty);
        }
    }
}
```

Code snippet from DoubleTapAction.cs

In order to detect a Double Tap gesture, this behavior records the time that the first tap was made (by overriding the `OnTap` method of the underlying `TapAction` behavior). Then, when the second tap is made, it compares the time between taps to determine if the second tap was made within the specified time-out. If it is, the `DoubleTap` event is raised.

To demonstrate this, add the `TapAction` behavior to a `Border` element, and add an event handler, `DoubleTapGesture_Tap`, to the `DoubleTap` event.

Available for download on Wrox.com

XAML

```
<Border BorderBrush="#FFF11717" BorderThickness="2" Margin="0,118,21,0"
        Background="#FFCC8787" HorizontalAlignment="Right" Width="160" Height="90"
        VerticalAlignment="Top">
    <i:Interaction.Behaviors>
        <local: DoubleTapAction DoubleTap="DoubleTapGesture_DoubleTap" />
    </i:Interaction.Behaviors>
</Border>>
```

Code snippet from MainPage.xaml

C#

```
private void DoubleTapGesture_DoubleTap(object sender, EventArgs e){
    MessageBox.Show("Border has been double tapped!");
}
```

Code snippet from MainPage.xaml.cs

Pan

The Pan gesture is slightly different from either the Tap or Double Tap gestures in that it needs to track when the user moves a finger across the screen. If you've ever built a drag-and-drop interface for a desktop application, you will be familiar with tracking the mouse movement using the mouse `move` event. Detecting a swipe or Pan gesture on a Windows Phone device can similarly be implemented by tracking the `MouseMove` event.

In the same way as you handled the Tap and Double Tap gestures, you can create a `PanAction` behavior. The idea of this behavior is that it will automatically detect when the user performs a `Pan` action on a canvas and then pan all of its children controls in the direction indicated by the user. Alternatively, if the user selects a specific child, then only that control should be moved.

Figure 9-6 illustrates a `Canvas` with several `Border` controls that have been styled by rounding the corners and have been positioned randomly. The larger circle indicates where the user has touched the `Canvas`. If you recall, the `Canvas` is unlike other panels in that it defines the `Left` and `Top` positions for all its child control. The `Left` and `Top` positions are recorded for each child control using an attached property as discussed in Chapter 3. When the user pans the canvas to the left, as in the second (from left to right) image of Figure 9-6, the `Left` position of all children is adjusted to move them to the left. Note that when a child object hits the edge of the `Canvas`, it doesn't disappear off the `Canvas`; instead, it sticks to the wall of the `Canvas`. This is a feature of this implementation; you can decide whether this is how you want your behavior to work. The third (from left to right) image shows the user panning the `Canvas` up. Again, the controls are panned in the direction that the user pans the `Canvas`, and controls stick to the top of the `Canvas` rather than disappearing off the edge.

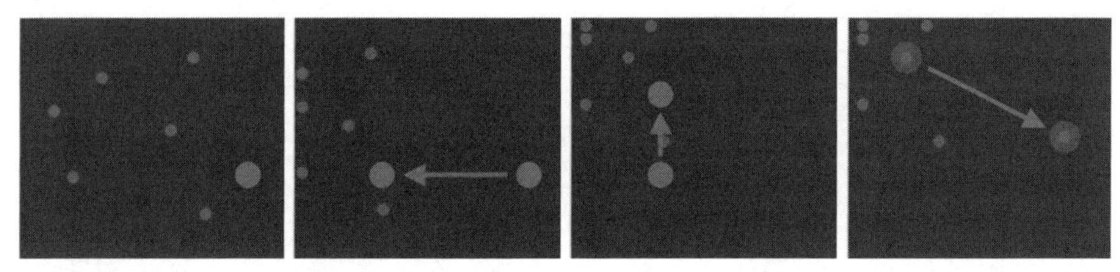

FIGURE 9-6

The final image of Figure 9-6 shows the user panning a single control. In this case, the user has touched the control he wishes to move, rather than touching an empty space on the canvas. Let's see how this is done in code. The first point to note is that the `PanAction` behavior is specific to the `Canvas` control (deriving from `Behavior<Canvas>`). You can choose to create a Pan gesture behavior that works with other control types, but in this case, you're going to be using the way that a `Canvas` positions its child controls to allow you to reposition child controls in response to the Pan gesture being detected. In addition to handling the `MouseLeftButtonDown` and `MouseLeftButtonUp` events, the `PanAction` behavior also attached event handles to the `MouseMove` event:

Available for download on Wrox.com

```
using System.Linq;
public class PanAction : Behavior<Canvas>{
    protected Point? MouseDown;
    protected UIElement SelectedItem { get; set; }

    protected override void OnAttached(){
        base.OnAttached();

        this.AssociatedObject.MouseLeftButtonDown += AO_MouseLeftButtonDown;
        this.AssociatedObject.MouseLeftButtonUp += AO_MouseLeftButtonUp;
        this.AssociatedObject.MouseMove += AO_MouseMove;
    }

    protected override void OnDetaching(){
        this.AssociatedObject.MouseLeftButtonDown -= AO_MouseLeftButtonDown;
        this.AssociatedObject.MouseLeftButtonUp -= AO_MouseLeftButtonUp;
        this.AssociatedObject.MouseMove -= AO_MouseMove;

        base.OnDetaching();
    }

    private void AO_MouseLeftButtonDown(object sender, MouseButtonEventArgs e){
        this.MouseDown = e.GetPosition(null);
        this.SelectedItem =  VisualTreeHelper.FindElementsInHostCoordinates(
                      e.GetPosition(null),
                      this.AssociatedObject).FirstOrDefault();
        if (this.SelectedItem == this.AssociatedObject){
            this.SelectedItem = null;
        }
    }

    protected virtual void AO_MouseMove(object sender, MouseEventArgs e){
        var pos = e.GetPosition(null);
        var xdiff = pos.X - MouseDown.Value.X;
        var ydiff = pos.Y - MouseDown.Value.Y;
        if (MouseDown.HasValue){
            MoveSelectedItems(xdiff, ydiff);
        }
        MouseDown = pos;
    }

    private void AO_MouseLeftButtonUp(object sender, MouseButtonEventArgs e){
            this.SelectedItem = null;
            this.MouseDown = null;
    }

    protected void MoveSelectedItems(double xdiff, double ydiff){
        if (this.SelectedItem != null){
            MoveItem(this.SelectedItem, xdiff, ydiff);
        }
```

```
        else{
            foreach (var child in this.AssociatedObject.Children){
                MoveItem(child, xdiff, ydiff);
            }
        }
    }

    private void MoveItem(UIElement item, double xdiff, double ydiff){
        var left = Canvas.GetLeft(item) + xdiff;
        left = Math.Min(Math.Max(0, left),this.AssociatedObject.ActualWidth);
        var top = Canvas.GetTop(item) + ydiff;
        top = Math.Min(Math.Max(0, top),this.AssociatedObject.ActualHeight);
        Canvas.SetLeft(item, left);
        Canvas.SetTop(item, top);
    }
}
```

Code snippet from PanAction.cs

In the mouse button down event handler, the behavior now tracks not only that the user is touching the screen, but also the position where they touched the `Canvas`. It also uses the `VisualTreeHelper` to identify the child control, if any, that is directly under the touch point.

VISUAL TREE

When you add controls to a page, you start by adding a control directly to the page, generating a single-level hierarchy. However, since the page can only take a single child, you will most likely use a `Grid` or `StackPanel`; subsequent controls are then added as nested controls, generating a multi-level hierarchy, with each control having a single parent but potentially multiple child controls.

Within XAML terminology, the controls you nest within one another are collectively termed the *Logical Tree*. A page has a stack panel, which has two buttons, for example.

However, if you consider a `Button` control, you can also apply a custom control template, or style, which alters the visual appearance of the button by introducing additional visual elements such as an internal border, a StackPanel or multiple TextBlocks, and so on. The tree of nested controls that affect the visual appearance of the application is known as the *Visual Tree*.

The `VisualTreeHelper` is a utility class that allows you to interrogate the Visual Tree. In this case, the `PanAction` behavior is using the `FindElementsInHostCoordinates` method to do a hit test to work out which elements include the point at which the user touched the canvas. There are also helper methods for accessing child controls and a control's parent.

When the user pans across the canvas, this raises the `MouseMove` event, which is used to calculate how far the user has moved. Note that this event is periodically fired until the user stops panning. Each time the event is raised, the position of the relevant controls is updated based on how far the user has panned incrementally.

To apply the PanAction all you need to do is add an instance of the behavior to the Canvas element. This will add the ability to pan all of the nested child elements. The layout of Figure 9-6 was created with the following code.

Available for download on Wrox.com

```
<Border HorizontalAlignment="Left" Height="368" Margin="19,231,0,0"
        VerticalAlignment="Top" Width="441" BorderBrush="White" BorderThickness="2"
        Background="#FF313131">
    <Canvas Background="Transparent" >
        <Canvas.Resources>
            <Style x:Key="RoundedBorder" TargetType="Border">
                <Setter Property="BorderThickness" Value="3"/>
                <Setter Property="CornerRadius" Value="20"/>
                <Setter Property="Background" Value="#FF287E3D"/>
                <Setter Property="BorderBrush" Value="#FF0F451C"/>
                <Setter Property="Width" Value="25"/>
                <Setter Property="Height" Value="25"/>
            </Style>
        </Canvas.Resources>
        <i:Interaction.Behaviors>
            <local:PanAction />
        </i:Interaction.Behaviors>
        <Border Canvas.Left="43" Canvas.Top="137"
                Style="{StaticResource RoundedBorder}" >
        </Border>
        <Border Canvas.Left="122" Canvas.Top="83"
                Style="{StaticResource RoundedBorder}" >
        </Border>
        <Border Canvas.Left="294" Canvas.Top="301"
                Style="{StaticResource RoundedBorder}" >
        </Border>
        <Border Canvas.Left="273" Canvas.Top="51"
                Style="{StaticResource RoundedBorder}" >
        </Border>
        <Border Canvas.Left="74" Canvas.Top="242"
                Style="{StaticResource RoundedBorder}" >
        </Border>
        <Border Canvas.Left="236" Canvas.Top="167"
                Style="{StaticResource RoundedBorder}" >
        </Border>
    </Canvas>
</Border>
```

Code snippet from MainPage.xaml.cs

In this code snippet you will notice that the Background of the Canvas is set to Transparent. By default the Canvas has no Background which means that it won't intercept any mouse (ie touch) events. Setting the Background to Transparent causes the Canvas to raise the mouse and manipulation events whilst still allowing the color of the parent control to be displayed.

Flick

Conceptually, there are two types of flicks, one that starts when the user touches the screen, and one that occurs at the end of a pan. However, when you translate this to code, you can think of both as occurring at the end of a pan. The first case is simply a pan of zero distance. For this reason, it seems logical to build the `FlickAction` behavior by extending the `PanAction` behavior.

In order to detect a `Flick` action, you need to know the speed that the user's finger is moving when it detaches from the screen. The basic mouse handling events you have used so far do not provide enough detail, as they are primarily designed for code-level compatibility with desktop-based Silverlight. Windows Phone exposes an additional set of three events that provide a much finer level of control for touch events — `ManipulationStarted`, `ManipulationDelta`, and `ManipulationCompleted`. `ManipulationStarted` is raised at the start of a manipulation event, usually triggered by the user touching the screen at one or more points. As the user changes the position of his or her fingers on the screen, the `ManipulationDelta` event is periodically raised. Finally, when one or more of the fingers is released, the `ManipulationCompleted` event is raised. For the `FlickAction` behavior, you are interested in the `ManipulationCompleted` event because it exposes a property called `FinalVelocities` that includes the velocity (in screen units per second) that the user's finger was moving at when it left the screen. This will be used to determine how far the child controls should be shifted as a result of the flick.

The calculations in the `FlickAction` *provide a very basic drift effect to controls as a result of the flick. If you want to use a flick to spin through elements in a list or trigger other animations, you may want to reference a more realistic physics engine.*

The `FlickAction` exposes a `FlickDurationInMilliseconds` property that determines the length of time that controls will continue to move after the user's finger has left the screen. This time period is split into 10 equal intervals with the controls moving less distance in each interval, thus giving the effect of them decelerating. The timer raises an event at the end of each time interval, allowing the position of the controls to be updated and the velocity to be updated. An alternative approach would be to use an animation storyboard to achieve a similar effect. This could leverage some of the built-in easing functions to add different animation effects.

Available for download on Wrox.com

```
public class FlickAction:PanAction{
    // The number of time increments over which deceleration will occur
    private const int Increments = 10;
    // The deceleration multiplier
    private const double Deceleration = 0.4;

    // A timer used to periodically update the position of controls
    //during deceleration
    DispatcherTimer timer = new DispatcherTimer();

    // The number of increments remaining
    public int Counter { get; set; }

    // The velocity at which the user released the pan (ie to generate a flick)
    private Point ReleaseVelocity { get; set; }

    // The duration of the flick deceleration
    public int FlickDurationInMilliseconds{
        get { return (int)GetValue(FlickDurationInMillisecondsProperty); }
        set { SetValue(FlickDurationInMillisecondsProperty, value); }
    }

    public static readonly DependencyProperty FlickDurationInMillisecondsProperty =
                DependencyProperty.Register("FlickDurationInMilliseconds",
                                            typeof(int), typeof(FlickAction),
                                            new PropertyMetadata(500));

    public FlickAction(){
        timer.Tick += new EventHandler(timer_Tick);
    }

    protected override void OnAttached(){
        base.OnAttached();
        this.AssociatedObject.ManipulationStarted += AO_ManipulationStarted;
        this.AssociatedObject.ManipulationCompleted += AO_ManipulationCompleted;
    }

    protected override void OnDetaching(){
        this.AssociatedObject.ManipulationStarted -= AO_ManipulationStarted;
        this.AssociatedObject.ManipulationCompleted -= AO_ManipulationCompleted;
        base.OnDetaching();
    }

    void timer_Tick(object sender, EventArgs e){
        Counter--;
        if (Counter == 0){
            timer.Stop();
        }

        MoveSelectedItems(ReleaseVelocity.X / 100, ReleaseVelocity.Y / 100);

        ReleaseVelocity = new Point(ReleaseVelocity.X * Deceleration,
                                    ReleaseVelocity.Y * Deceleration);
    }
```

```
    void AO_ManipulationStarted(object sender, ManipulationStartedEventArgs e) {
        timer.Stop();
    }

    void AO_ManipulationCompleted(object sender,
                              ManipulationCompletedEventArgs e) {
        ReleaseVelocity = e.FinalVelocities.LinearVelocity;
        timer.Interval = new TimeSpan(0,0,0,0,
                                  FlickDurationInMilliseconds / Increments);
        Counter = Increments;
        timer.Start();
    }
}
```

Code snippet from FlickAction.cs

As this behavior extends the `PanAction`, you should remove the `PanAction` behavior from the `Canvas` control before adding the `FlickAction` behavior. Once added, you should see that controls continue to move beyond the release point when you use a flick at the end of a pan movement.

Available for download on Wrox.com

```
<i:Interaction.Behaviors>
    <!--<local:PanAction />-->
    <local:FlickAction />
</i:Interaction.Behaviors>
```

Code snippet from MainPage.xaml.cs

Touch and Hold

Desktop developers often use the second, or right, mouse button to provide additional functionality such as context menus within their applications. In the absence of a mouse, contextual menus and actions are often exposed in a touch-based interface via a Touch and Hold gesture. The `TouchAndHoldAction` behavior can be applied to any `UIElement` and simply uses a timer to determine whether the user has been holding the control for the specified time-out. There is an additional tolerance property that can be tweaked to control how sensitive the behavior is to the user moving a finger on the screen during the Touch and Hold gesture.

Available for download on Wrox.com

```
public class TouchAndHoldAction : DoubleTapAction {
    public event EventHandler TouchAndHold;

    private Point? TapLocation;
    private DispatcherTimer timer = new DispatcherTimer();

    public int HoldTimeoutInMilliseconds{
        get { return (int)GetValue(HoldTimeoutInMillisecondsProperty); }
        set { SetValue(HoldTimeoutInMillisecondsProperty, value); }
    }

    public static readonly DependencyProperty HoldTimeoutInMillisecondsProperty =
                       DependencyProperty.Register("HoldTimeoutInMilliseconds",
```

```
                                                    typeof(int), typeof(TouchAndHoldAction),
                                                    new PropertyMetadata(2000));

        public int Tolerance{
            get { return (int)GetValue(ToleranceProperty); }
            set { SetValue(ToleranceProperty, value); }
        }

        public static readonly DependencyProperty ToleranceProperty =
                        DependencyProperty.Register("Tolerance", typeof(int),
                                                    typeof(TouchAndHoldAction),
                                                    new PropertyMetadata(2));

        protected override void OnAttached(){
            base.OnAttached();

            this.AssociatedObject.MouseLeftButtonDown += AO_MouseLeftButtonDown;
            this.AssociatedObject.MouseMove += AO_MouseMove;
            this.timer.Tick += timer_Tick;
        }

        protected override void OnDetaching(){
            this.AssociatedObject.MouseLeftButtonDown -= AO_MouseLeftButtonDown;
            this.AssociatedObject.MouseMove -= AO_MouseMove;
            this.timer.Tick -= timer_Tick;

            base.OnDetaching();
        }

        void AO_MouseLeftButtonDown(object sender, MouseButtonEventArgs e){
            var pos = e.GetPosition(null);
            OnTouchAndHoldStarted(pos);
        }

        void AO_MouseMove(object sender, MouseEventArgs e){
            if (TapLocation.HasValue){
                var pos = e.GetPosition(null);
                if (Math.Abs(TapLocation.Value.X - pos.X) > Tolerance ||
                    Math.Abs(TapLocation.Value.Y - pos.Y) > Tolerance){
                     OnTouchAndHoldStarted(pos);
                }
            }
        }

        void timer_Tick(object sender, EventArgs e){
            OnTouchAndHoldCompleted();
            timer.Stop();
            TapLocation = null;
        }

        protected override void OnTap(){
            timer.Stop();
            base.OnTap();
        }
```

```
        protected virtual void OnTouchAndHoldStarted(Point pt){
             TapLocation = pt;

             timer.Stop();
             timer.Interval = new TimeSpan(0, 0, 0, 0, HoldTimeoutInMilliseconds);
             timer.Start();
        }
        protected virtual void OnTouchAndHoldCompleted(){
            MouseDown = false;
            if (TouchAndHold != null){
                TouchAndHold(this.AssociatedObject, EventArgs.Empty);
            }
        }
    }
```

Code snippet from TouchAndHoldAction.cs

Applying the `TouchAndHoldAction` behavior is again just a matter of dragging the behavior onto a control within Blend. However, wiring up the event handler for the `TouchAndHold` event needs to be done within Visual Studio because of a limitation of the Blend Designer. Here you've attached this behavior to each of the borders that is in the canvas used in the previous section.

```
<Border Canvas.Left="122" Canvas.Top="83" Style="{StaticResource RoundedBorder}" >
    <i:Interaction.Behaviors>
        <local:TouchAndHoldAction TouchAndHold="TouchAndHoldAction_TouchAndHold" />
    </i:Interaction.Behaviors>
</Border>
<Border Canvas.Left="294" Canvas.Top="301" Style="{StaticResource RoundedBorder}" >
    <i:Interaction.Behaviors>
        <local:TouchAndHoldAction TouchAndHold="TouchAndHoldAction_TouchAndHold" />
    </i:Interaction.Behaviors>
</Border>
<Border Canvas.Left="273" Canvas.Top="51" Style="{StaticResource RoundedBorder}" >
    <i:Interaction.Behaviors>
        <local:TouchAndHoldAction TouchAndHold="TouchAndHoldAction_TouchAndHold" />
    </i:Interaction.Behaviors>
</Border>
<Border Canvas.Left="74" Canvas.Top="242" Style="{StaticResource RoundedBorder}" >
    <i:Interaction.Behaviors>
        <local:TouchAndHoldAction TouchAndHold="TouchAndHoldAction_TouchAndHold" />
    </i:Interaction.Behaviors>
</Border>
<Border Canvas.Left="236" Canvas.Top="167" Style="{StaticResource RoundedBorder}" >
    <i:Interaction.Behaviors>
        <local:TouchAndHoldAction TouchAndHold="TouchAndHoldAction_TouchAndHold" />
    </i:Interaction.Behaviors>
</Border>
```

Code snippet from MainPage.xaml

In this case each of the `TouchAndHold` behaviors is connected to the same event handler for the `TouchAndHold` event. When the user invokes the `TouchAndHold` event, the border and background colors are inverted.

Available for download on Wrox.com

```
private void TouchAndHoldAction_TouchAndHold(object sender, EventArgs e){
    var element = sender as Border;
    if (element != null){
        element.Background = InvertColor(element.Background);
        element.BorderBrush = InvertColor(element.BorderBrush);
    }
}

private Brush InvertColor(Brush input){
    var color = (input as SolidColorBrush).Color;
    var brush = new SolidColorBrush(Color.FromArgb(color.A,
                                                (byte)(255 - color.R),
                                                (byte)(255 - color.G),
                                                (byte)(255 - color.R)));
    return brush;
}
```

Code snippet from MainPage.xaml.cs

Since the Touch and Hold gesture can take a while to perform, it may be worth providing visual feedback to the user when he or she touches the screen to indicate that this gesture is a possibility. For example, Windows Mobile progressively displays a series of circles that surround the touch point. The behavior created previously could be expanded to support this functionality by starting and stopping the animation within the `OnTouchAndHoldStarted` and `OnTouchAndHoldCompleted` methods, respectively.

Multi-Touch

In the last couple of years, multi-touch devices have become a hot topic. Multi-touch devices come in a range of shapes and sizes from large stationary devices, such as Microsoft Surface, to multi-touch desktop and laptops, to mobile devices. Windows Phone is no exception with support for up to four concurrent points of touch.

One of the things to be aware of when building a multi-touch application for Windows Phone is that you are building an application to be used on a mobile phone. Although mobile devices are becoming more sophisticated and powerful, you still can't escape the size constraint. Building a mobile multi-touch interface that requires more than two touch points can make the application hard to use without really benefitting the user experience. That said, there are many cases in which using multiple touch points does makes a lot of sense, such as using a two-finger Pinch or Stretch gesture to zoom in or out of an image.

Simulating Multi-Touch

Before you get in and look at how you can create a behavior that allows a user to make use of multi-touch gestures, you'll want a way to develop for multiple points of touch. The easiest way, of course, is to use a real device, but in the absence of a device, you can use the Windows Phone emulator to simulate the multi-touch experience. If you have a Windows 7 device that supports multi-touch, you don't need to do anything extra to configure the emulator as it will automatically support multiple points of contact. Simply use the touch screen and place two fingers on the emulator at once.

For a lot of you, having a multi-touch Windows 7 device isn't a viable option. Luckily, there is a way to simulate having a multi-touch device through the use of multiple USB mice. The Multi-Touch Vista project (`http://multitouchvista.codeplex.com`) was initially started to provide multi-touch support to Vista. With the advent of Windows 7, which includes built-in support for multi-touch, this project simply provides the conduit between multiple USB mice and the native Windows 7 multi-touch support. This means that any application, such as Paint and the Windows Phone emulator that supports multi-touch under Windows 7, will work after running Multi-Touch Vista. These instructions are for installing Multi-Touch Vista under Windows 7:

1. The first step is to go to the Multi-Touch Vista CodePlex project (`http://multitouchvista.codeplex.com`) and download the latest release.
2. Unblock (right-click the file ➪ Properties ➪ Unblock) and then Extract the contents of the downloaded zip file.
3. Open a command prompt using "Run As Administrator" (this is important — otherwise, the driver will fail to load).
4. Navigate to the Driver\x32 or Driver\x64 folder, depending on your system architecture.
5. Run "Install drive.cmd."
6. Confirm that it is OK to install even though Windows can't verify the publisher. (See Figure 9-7.) You should only do this if you are confident you understand the potential effects this may have on your computer.
7. When the previous step completes, you should see a new line appear in the Command prompt, allowing you to enter more commands. At this point, you can close the Command Prompt window.
8. Open the Device Manager (Start ➪ Right-click Computer ➪ Properties ➪ Device Manager).
9. Locate the Universal Software HID device that should appear under the Human Interface Devices node. Right-click and select Disable (as shown in Figure 9-8), and then confirm that you want to disable this device.
10. Right-click on the same device and select Enable. This may seem a little silly, but in some cases, the device doesn't start properly and this process of disabling and then enabling it again ensures that it is operating correctly.

FIGURE 9-7

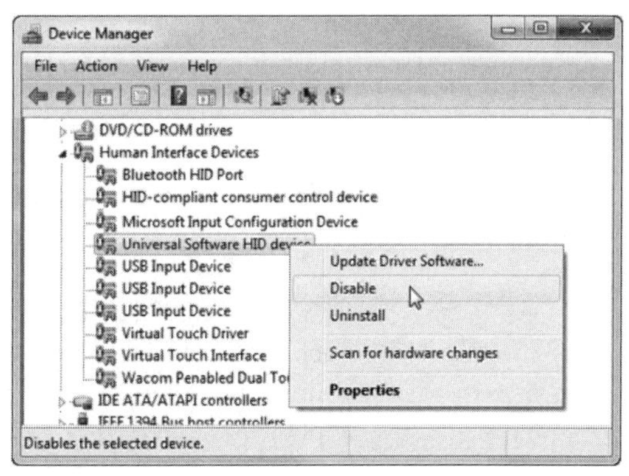

FIGURE 9-8

11. Open Pen and Touch settings (Start ➪ Control Panel ➪ Pen and Touch), locate the Touch tab, and check the "Show the touch pointer when I'm interacting with items on the screen" checkbox, as shown in Figure 9-9.

12. Open Windows Explorer and navigate to the folder that you extracted Multi-Touch Vista to. Double-click Multitouch.Service.Console.exe to start the multi-touch service (Figure 9-10).

    At this point, you should see one or more red dots appear on the screen. There is one for each mouse that you have connected to your computer. They won't become active until you run the multi-touch driver (next step).

13. Double-click Multitouch.Driver.Console.exe to start the multi-touch driver (Figure 9-11).

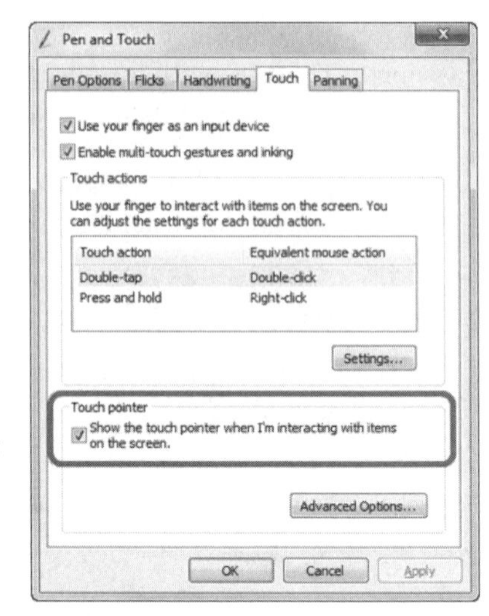

FIGURE 9-9

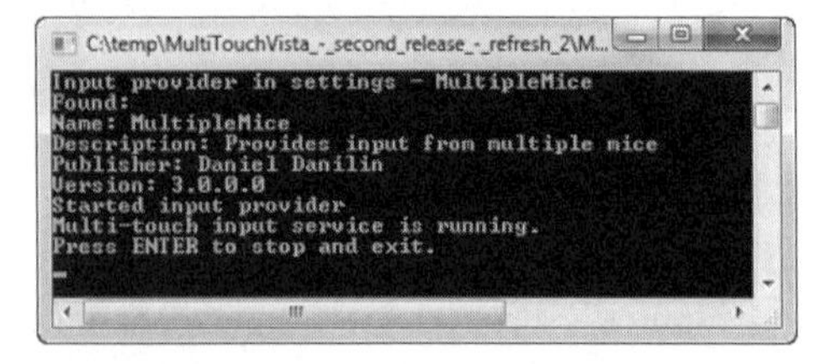

FIGURE 9-10

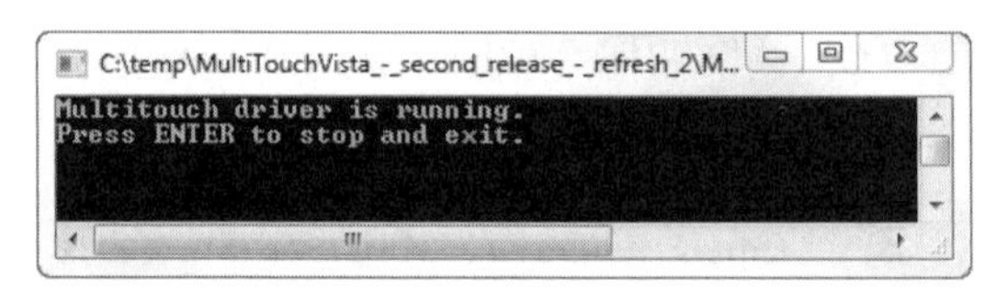

FIGURE 9-11

14. Double-click Multitouch.Configuration.WPF.exe to start the configuration tool for Mulit-Touch Vista. (See Figure 9-12.) This tool allows you to configure the inputs, specifically the ability to disable the native mouse input. It's recommended that you check this option as you will be using the red dots as your mouse cursors, rather than the native arrow cursor.

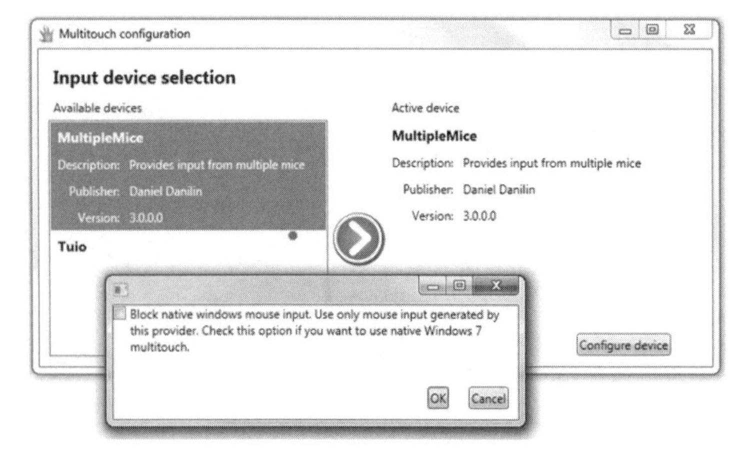

FIGURE 9-12

Be aware that when you disable the native windows mouse input, the next time you run the multi-touch service it will disable the native mouse input, and the red dots won't operate until you run the multi-touch driver. Unfortunately, since you have to start the service before the driver, there will be a short time when you are without a mouse and will have to operate the computer using your keyboard.

15. Open Paint and confirm that you are able to draw with multiple mouse cursors, as shown in Figure 9-13.

You are now set up and ready to simulate multi-touch with the Windows Phone emulator. You will find that within the Windows Phone emulator there are several applications such as Internet Explorer and the Picture Viewer that support multi-touch and allow you to zoom in and out on content.

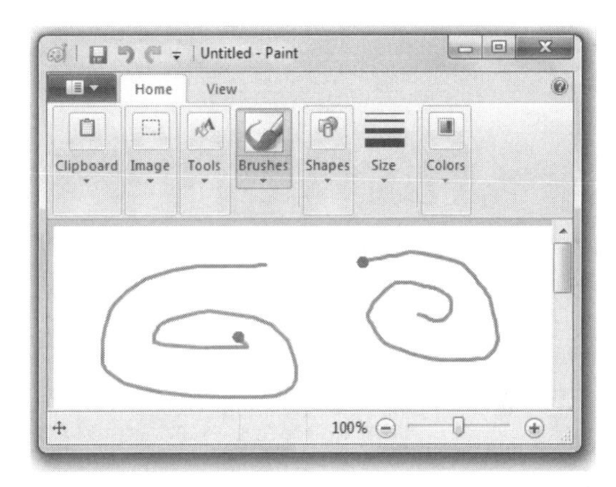

FIGURE 9-13

Pinch and Stretch

The last gesture you'll cover in this chapter is Pinch and its inverse, Stretch. These gestures are usually reserved for zooming in or out on content, and in this case, you'll see them applied to the Canvas example from earlier. Pinching will collapse the `Border` controls to the center of the `Canvas`, and conversely, Stretching will explode the controls toward the outskirts of the `Canvas`. This is illustrated in Figure 9-14 with the two circles in the center being stretched apart. The remaining circles explode toward the edge of the `Canvas`.

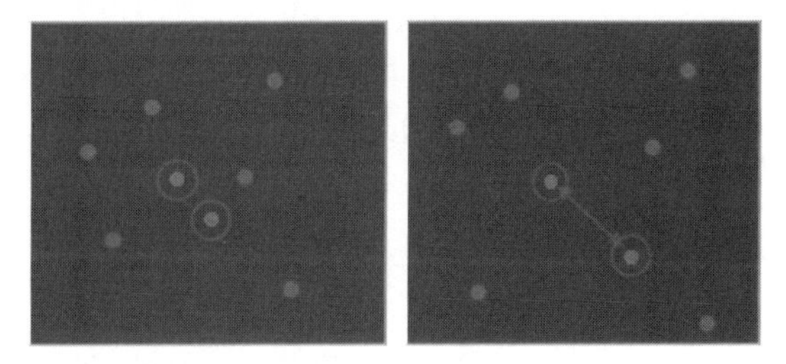

FIGURE 9-14

As you want to preserve both the Pan and Flick gestures, the new `ZoomAction` behavior will inherit from the `FlickAction` behavior. When multiple points of touch are active, the `ManipulationDelta` event is raised whenever one or more of the points move. The data supplied as part of this event gives an indication as to whether the touch points are moving toward each other (*pinch*) or away (*stretch*). This is reflected in the `DeltaManipulation.Scale` property and has values for the scale factor to be applied in both the X and Y directions. Where each child control is located determines whether this scale factor is applied by multiplying or dividing.

Available for download on Wrox.com

```
public class ZoomAction : FlickAction{
    private bool DisablePan;

    protected override void OnAttached(){
        base.OnAttached();
```

```
        this.AssociatedObject.ManipulationDelta += AO_ManipulationDelta;
        this.AssociatedObject.ManipulationCompleted += AO_ManipulationCompleted;
    }

    protected override void OnDetaching(){
        this.AssociatedObject.ManipulationDelta -= AO_ManipulationDelta;
        this.AssociatedObject.ManipulationCompleted -= AO_ManipulationCompleted;

        base.OnDetaching();
    }

    protected override void AO_MouseMove(object sender, MouseEventArgs e){
        if (DisablePan){
            return;
        }
        base.AO_MouseMove(sender, e);
    }

    void AO_ManipulationCompleted(object sender, ManipulationCompletedEventArgs e){
        DisablePan = false;
    }

    void AO_ManipulationDelta(object sender, ManipulationDeltaEventArgs e){
        if (e.CumulativeManipulation.Scale.X > 0 ||
            e.CumulativeManipulation.Scale.Y > 0){
            DisablePan = true;

            var scaleX = e.DeltaManipulation.Scale.X;
            var scaleY = e.DeltaManipulation.Scale.Y;
            foreach (var child in this.AssociatedObject.Children){
                if (scaleX != 0){
                    var left = Canvas.GetLeft(child);
                    if (left > this.AssociatedObject.ActualWidth/2){
                        left *= scaleX;
                    }
                    else{
                        left /= scaleX;
                    }
                    left = Math.Min(Math.Max(0, left),
                                    this.AssociatedObject.ActualWidth);
                    Canvas.SetLeft(child, left);
                }
                if (scaleY != 0){
                    var top = Canvas.GetTop(child);
                    if (top > this.AssociatedObject.ActualHeight / 2){
                        top *= scaleY;
                    }
                    else{
                        top /= scaleY;
                    }
                    top = Math.Min(
```

```
                            Math.Max(0, top), this.AssociatedObject.ActualHeight);
                    Canvas.SetTop(child, top);
                }
            }
        }
    }
}
```

Code snippet from ZoomAction.cs

When multiple touch points are active, you no longer want the Flick or Pan gestures to be operational because they interfere with the Pinch and Stretch gestures. In this code snippet, you will notice that the `DisablePan` flag is set whenever the cumulative scale factor in either direction is not zero.

The `ZoomAction` behavior should be added to the Canvas in the same way as you did for the `PanAction` and `FlickAction`. If you already have one of either of these behaviors attached to the Canvas, they should be removed as they will conflict with the `ZoomAction`.

```
<i:Interaction.Behaviors>
    <!--<local:PanAction />-->
    <!--<local:FlickAction />-->
    <local:ZoomAction />
</i:Interaction.Behaviors>
```

Code snippet from MainPage.xaml.cs

Touch Frames

The manipulation events you've seen so far are great for working with a large range of gestures. However, sometimes you need to get even more granular and work at the individual touch-point level. This is why there is a `Touch` class that exposes a static `FrameReported` event, which is raised whenever there is a change to the number or position of touch points.

As the `FrameReported` event is a static event, it is not associated with any particular control. This means that you can either use the raw position of the touch points relative to the Windows Phone screen, as shown in the following code snippet, or you can retrieve the points relative to a `UIElement` on the current page:

```
public MainPage(){
    InitializeComponent();

    Touch.FrameReported += new TouchFrameEventHandler(Touch_FrameReported);
}

void Touch_FrameReported(object sender, TouchFrameEventArgs e){
    var points = e.GetTouchPoints(null);
    foreach (var point in points){
        var x = point.Position.X;
```

```
            var y = point.Position.Y;

            // Do action with touch information
        }
    }
```

Code snippet from MainPage.xaml.cs

By handling the `FrameReported` event, you can do your own processing to determine when users touch or release from the screen, when they move their fingers across the screen, and how fast their fingers are traveling. This is particularly useful when you are building games and you want hi-fidelity control over the processing of touch input.

SUMMARY

In this chapter, you've learned about the different standard touch gestures and how you can use them within a Windows Phone application. You should consider whether your application calls for single- or multiple-touch gestures and how you can implement gestures through reusable behaviors that can easily be applied to controls.

10

Shake, Rattle, and Vibrate

WHAT'S IN THIS CHAPTER

- Understanding how to interact with the accelerometer
- How to simulate the accelerometer in the Windows Phone emulator
- How to use a Wii remote to test applications that use the accelerometer
- How to vibrate the Windows Phone

I wouldn't recommend that you dunk your Windows Phone in a martini, but your application can definitely let you know if it's being shaken or stirred. All Windows Phone devices will have an integrated accelerometer that you can use to track the motion of the device.

In this chapter you will see how to register for events that fire whenever the direction and physical orientation of the device change. You will also learn how you can cause the device to vibrate in order to provide tactile feedback to the user.

ACCELEROMETER

Applications designed for Windows Phone can accept input from the user in various ways. Traditional desktop applications may accept input from the keyboard, mouse, and, more recently, a touch screen. On a mobile device, the user is no longer desk-bound, and thus there are other input devices that can be used to interact with the application's user interface (UI). The *accelerometer* is one such input sensor — it detects the acceleration of the device as the user moves or rotates it.

The Windows Phone accelerometer reports a three-dimensional (3D) vector, in the form of individual *x*, *y*, and *z* values. These values indicate the direction and magnitude of the force currently applied to the device.

If the Windows Phone device is sitting in an upright or vertical position on a flat surface such as a desk, it will feel a force that results in an acceleration of 1*g* (1*g* is the acceleration that results from a mechanical force being applied to a stationary object on the Earth's surface due to the effects of gravity, approximately equal to 9.8ms^2) pushing up on it. This is the mechanical force required to counterbalance the effects of gravity. The Windows Phone accelerometer will report this as a vector of (0, –1, 0).

As you move your Windows Phone device into a number of orientations and positions, you will see the accelerometer report a set of vectors similar to those shown in Table 10-1. This is a useful table to have handy because it provides a quick reference to see how the accelerometer reports acceleration, depending on the device orientation.

TABLE 10-1: Accelerometer Vectors

SIDE	ACCELERATION VECTOR (*X, Y, Z*)
Vertical/Upright	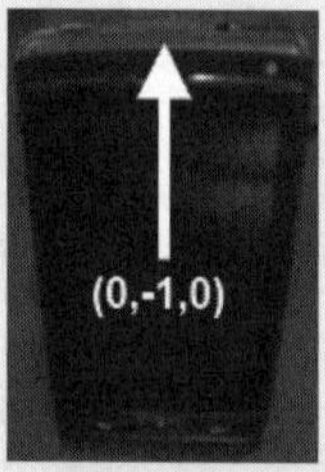 (0,-1,0)
Right Side Face Up	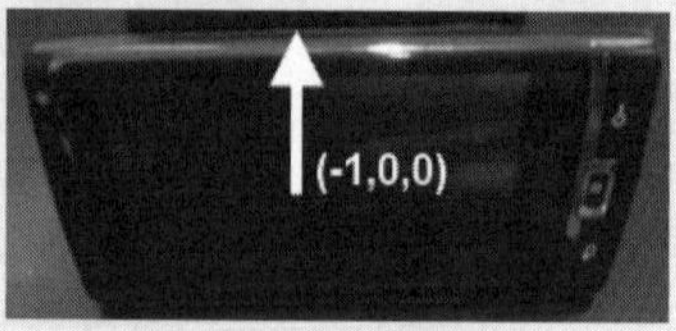 (-1,0,0)
Upside Down	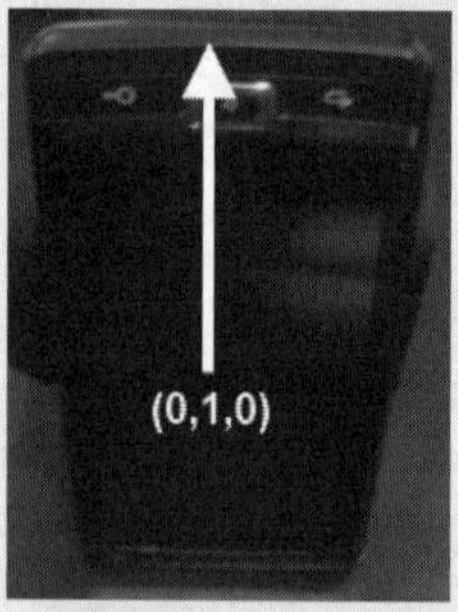 (0,1,0)
Left Side Face Up	 (1,0,0)

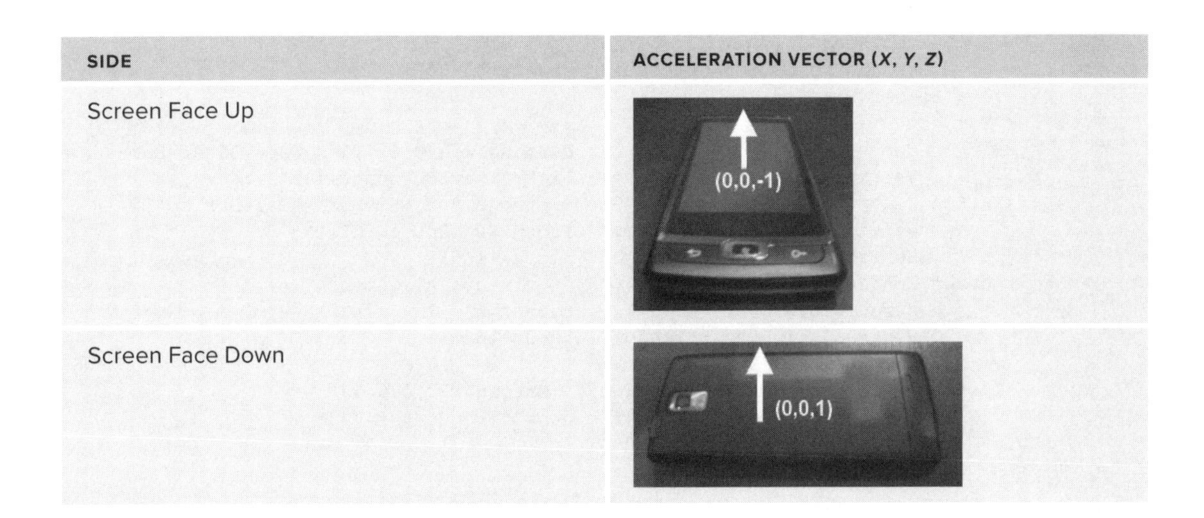

SIDE	ACCELERATION VECTOR (*X, Y, Z*)
Screen Face Up	(0,0,-1)
Screen Face Down	(0,0,1)

Any orientation between these states will yield acceleration vectors ranging from –1 to 1 for the three dimensions. You'll see later how you can use the acceleration vector to give you an approximation of the orientation of the device.

One thing to note about the vectors returned by the accelerometer is that they do not use the same coordinate space used by Silverlight. When you place a Windows Phone with its right side facing up, you may expect the acceleration to be 1g in the x *direction; instead, it is –1g. Similarly, with the screen facing up, you would expect an acceleration of 1g in the* z *direction; it's actually –1g.*

Applications will often make use of the accelerometer to adjust the layout of the current page. If your application supports multiple orientations, you need to be careful that the user doesn't become confused when these two features are intermixed. For example, if the user tilts the device on its side, this will cause a change in page orientation, but will also result in the accelerometer indicating a different force vector. If you change the position of items on the screen, it will be very difficult for the user to keep track of what's going on. For this reason, it is recommended that when you use the accelerometer to adjust layout, you should restrict the `SupportedOrientations` property of the page to either `Portrait` or `Landscape` but not both.

To integrate the accelerometer into your application, you need to create an instance of the `Accelerometer` class and invoke the `Start` method. To demonstrate this, create a new project using the Windows Phone Application project template, and replace the `ContentGrid` control within `MainPage` with the following XAML:

Available for download on Wrox.com

```
<Grid x:Name="ContentGrid" Grid.Row="1">
    <Border BorderBrush="Silver" BorderThickness="1" Margin="41,1,39,0"
            Name="border1" Background="#FFA70000" RenderTransformOrigin="0.5,0.5"
            Height="400" VerticalAlignment="Top">
```

```
            <Border.Projection>
                <PlaneProjection CenterOfRotationZ="0.5"/>
            </Border.Projection>
        </Border>
        <Button Content="Start" Height="70" Margin="163,0,157,66" Name="StartButton"
                VerticalAlignment="Bottom" Click="StartButton_Click" />
        <Button Content="Stop" Height="70" Margin="0,0,6,66" Name="StopButton"
                VerticalAlignment="Bottom" Click="StopButton_Click"
                HorizontalAlignment="Right" Width="160" />
        <Button Content="Create" Height="70" HorizontalAlignment="Left"
                Margin="6,0,0,66" Name="CreateButton"
                VerticalAlignment="Bottom" Width="160"
                Click="CreateButton_Click" />
        <TextBlock HorizontalAlignment="Left" Margin="21,0,0,17"
                Name="SensorText" Text="" Width="439" Height="43"
                VerticalAlignment="Bottom" />
    </Grid>
```

Code snippet from MainPage.xaml

The first child element is the `Border` control, which will eventually be reoriented based on the most recent accelerometer reading. For this reason, it has an associated `PlaneProjection` that will be used to rotate the `Border`. The remaining buttons create, start, and stop the accelerometer. There is also a `TextBlock` that will be used to display the most recent accelerometer data.

The `Accelerometer` class is located in the `Microsoft.Devices.Sensors` assembly. You will need to add a reference to this library, as well as adding appropriate using statements to your code. Then you can create an instance of the `Accelerometer` class as follows:

```
Accelerometer sensor;

private void CreateButton_Click(object sender, RoutedEventArgs e){
    sensor = new Accelerometer();
    sensor.ReadingChanged += sensor_ReadingChanged;
}
```

Code snippet from MainPage.xaml.cs

As with most hardware sensors, the accelerometer is interrupt-based, which translates very nicely into the event model provided by .NET. In this case, the `Accelerometer` class exposes a `ReadingChanged` event that is raised every time the acceleration of the device changes. The above code attaches the `sensor_ReadingChanged` method as an event handler for the `ReadingChanged` event. This can be implemented as follows, which simply updates a `TextBlock` called `SensorText` with the current acceleration vector:

```
void sensor_ReadingChanged(object sender, AccelerometerReadingEventArgs e){
    this.Dispatcher.BeginInvoke(
        () =>{
```

```
                this.SensorText.Text =
                    string.Format("[{0:0.00},{1:0.00},{2:0.00}] at {3}", e.X, e.Y, e.Z,
                                e.Timestamp.TimeOfDay.ToString());
            });
    }
```

Code snippet from MainPage.xaml.cs

You will notice that the method accepts an `AccelerometerReadingEventArgs` argument. This argument consists of a Timestamp that indicates when a reading was made, and a tuple of X, Y and Z reading values.

The `AccelerometerSensor` class exposes methods for starting (`Start`) and stopping (`Stop`) the sensor, as well as its current state (`SensorState`). The possible states a sensor can find itself in are listed in the following enumeration, which is defined in the Microsoft.Devices.Sensors namespace.:

```
public enum SensorState{
    NotSupported,
    Ready,
    Initializing,
    NoData,
    NoPermissions,
    Disabled
}
```

If you run your Windows Phone application on the emulator and create an instance of the `Accelerometer` class, you will see that the `SensorState` property is set to `Ready`, despite there not being an accelerometer in the emulator itself (The emulator will periodically raise the `ReadingChanged` event, always with an accelerator vector of (0,0,-1), which is the equivalent of the device sitting face up on a table). For cases where the accelerometer isn't supported you will need to adapt your application to handle this scenario. In the following code that starts the accelerometer, a check is done to ensure that the application is being run on a platform that supports an accelerometer:

Available for download on Wrox.com

```
private void StartButton_Click(object sender, RoutedEventArgs e){
    if (sensor.State == SensorState.NotSupported){
        this.SensorText.Text = "Accelerometer not supported on this platform";
        return;
    }

    try{
        sensor.Start();
    }
    catch (AccelerometerFailedException ex){
        this.SensorText.Text = ex.Message.ToString();
    }
}
```

Code snippet from MainPage.xaml.cs

Note that it is possible for the sensor to fail to start, throwing an `AccelerometerStartFailedException`, so you should always wrap the `Start` method in a `Try-Catch` block. Similarly, when stopping the accelerometer, you should both check to ensure that the platform supports the accelerometer sensor and trap any exceptions raised when stopping the sensor.

Available for download on Wrox.com

```
private void StopButton_Click(object sender, RoutedEventArgs e){

    if (sensor.State == SensorState.NotSupported){
        this.SensorText.Text = "Accelerometer not supported on this platform";
        return;
    }

    try{
        sensor.Stop();
    }
    catch (AccelerometerFailedException ex){
        this.SensorText.Text = ex.Message.ToString();
    }
}
```

Code snippet from MainPage.xaml.cs

When you run this you can start the accelerometer by clicking the Create button, followed by the Start button. Figure 10-1 illustrates this example running in the emulator, showing an acceleration vector of (0,0,-1).

FIGURE 10-1

Working with the Emulator

The Windows Phone emulator doesn't really emulate the behavior of a real accelerometer. If you're using the accelerometer in your application and don't have access to a real device, you will need to mock the functionality of an accelerometer during development. There are numerous ways to do this; in this section, you will see how to implement a mock accelerometer using random data or the accelerometer data taken from a Wii controller.

In order to work with a mock accelerometer, you need to refactor your code so that it can easily switch between the mock and real accelerometers without requiring constant code changes. Ideally, you shouldn't have to modify your code at all while switching between running in the emulator and running on a real device. On the former the mock accelerometer should be loaded, while on the real device the real accelerometer should be loaded. This can be achieved by detecting the type of device the application is running on and creating the correct type of accelerometer driver.

Start by creating a new interface, `IAccelerometer`, which has methods `Start` and `Stop`, and an event, `ReadingChanged`. You'll also need to create an `AccelerometerEventArgs` class that inherits from `EventArgs` and has a property, `Reading`, of type `Vector3`. `Vector3` is found in the

`Microsoft.XNA.Framework` assembly, which you should add a reference to and add as a `using` statement. Unfortunately it's not possible to reuse the existing AccelerometerReadingEventArgs that belongs in the Microsoft.Devices.Sensors namespace as all the property setters are private.

Available for download on Wrox.com

```
using Microsoft.Xna.Framework;

public interface IAccelerometer {
    void Start();
    void Stop();
    event EventHandler<AccelerometerEventArgs> ReadingChanged;
}

public class AccelerometerEventArgs:EventArgs{
    public DateTimeOffset Timestamp { get; set; }
    public Vector3 Reading { get; set; }
}
```

Code snippet from IAccelerometer.cs

The `IAccelerometer` interface defines how your application will interact with an arbitrary accelerometer. The next step is to create a class that will wrap the real accelerometer, using the adapter pattern, in order to provide an implementation of the `IAccelerometer` interface:

Available for download on Wrox.com

```
public class RealAccelerometer: IAccelerometer{
    protected Accelerometer Sensor { get; set; }
    public event EventHandler<AccelerometerEventArgs> ReadingChanged;

    public RealAccelerometer(){
        this.Sensor = new Accelerometer();
        this.Sensor.ReadingChanged += Sensor_ReadingChanged;
    }

    void Sensor_ReadingChanged(object sender, AccelerometerReadingEventArgs e){
        if (ReadingChanged != null){
            ReadingChanged(this, new AccelerometerEventArgs() {
                                        Reading = new Vector3((float)e.X,
                                                              (float)e.Y,
                                                              (float)e.Z),
                                        Timestamp=e.Timestamp});
        }
    }

    public void Start(){
        this.Sensor.Start();
    }

    public void Stop(){
        this.Sensor.Stop();
    }
}
```

Code snippet from RealAccelerometer.cs

Instead of directly creating an instance of the `RealAccelerometer` class, you can use a factory method. This method will eventually contain the logic to determine which accelerometer, the real or the mock, to load:

Available for download on Wrox.com

```
public class Utilities{
    public static IAccelerometer LoadAccelerometer(){
        return new RealAccelerometer();
    }
}
```

Code snippet from Utilities.cs

Lastly, you need to update the existing setup and reading event handler methods to use the new `IAccelerator` functionality instead of referring directly to the `Accelerometer` class:

Available for download on Wrox.com

```
private IAccelerometer sensor;

private void CreateButton_Click(object sender, RoutedEventArgs e){
    sensor = Utilities.LoadAccelerometer();
    sensor.ReadingChanged += sensor_ReadingChanged;
}

void sensor_ReadingChanged(object sender, AccelerometerEventArgs e){
    this.Dispatcher.BeginInvoke(
                () =>{
                this.SensorText.Text =
                   string.Format("[{0:0.00},{1:0.00},{2:0.00}] at {3}",
                                 e.Reading.X, e.Reading.Y, e.Reading.Z,
                                 e.Timestamp.TimeOfDay.ToString());

                });
}

private void StartButton_Click(object sender, RoutedEventArgs e){
    try{
         sensor.Start();
    }
    catch (AccelerometerFailedException ex){
         this.SensorText.Text = ex.Message.ToString();
    }
}

private void StopButton_Click(object sender, RoutedEventArgs e){
    try{
         sensor.Stop();
    }
    catch (AccelerometerStopFailedException ex){
         this.SensorText.Text = ex.Message.ToString();
    }
}
```

Code snippet from MainPage.xaml.cs

With these changes in place, you should currently see no change in behavior when the application is run. However, by inserting the `IAccelerator` interface into the application, you are now ready to create alternative mock accelerometer services that can be easily exchanged for the `RealAccelerometer` class created above.

Random Accelerometer Data

To create a mock accelerometer that returns random data, it's just a matter of providing an alternative implementation of the `IAccelerometer` interface. The interface requires an event called `ReadingChanged` that will be raised whenever there is a new reading. It also requires both a `Start` and a `Stop` method to be implemented. The implementation, `FakeAccelerometer`, uses a simple flag to track whether it is in a running state or not. If it is not running and the `Start` method is called, a new thread is spawned that invokes the `Run` method:

```
public class FakeAccelerometer:IAccelerometer{
    public event EventHandler<AccelerometerEventArgs> ReadingChanged;

    private bool IsRunning { get; set; }
    private Random random = new Random();

    public void Start(){
        if (IsRunning) return;

        IsRunning = true;
        var runThread = new Thread(Run);
        runThread.Start();
    }

    public void Stop(){
        if (!IsRunning) return;
        IsRunning = false;
    }

    public void Run(){
        double deltaX = 0.0, deltaY = 0.8, deltaZ = -0.6;

        while (IsRunning){
            Vector3 reading = new Vector3((float)Math.Sin(deltaX),
                                          (float)Math.Cos(deltaY * 1.1),
                                          (float)Math.Sin(deltaZ * .7));
            reading.Normalize();

            if (ReadingChanged != null){
                var readingArgs = new AccelerometerEventArgs()
                                           { Reading = reading,
                                             Timestamp=DateTimeOffset.Now };
                ReadingChanged(this, readingArgs);
            }

            Thread.Sleep(random.Next(100));
```

```
                deltaX += 0.001;
                deltaY -= 0.003;
                deltaZ += 0.002;
            }
        }
    }
```

Code snippet from FakeAccelerometer.cs

The Run method consists of a single while loop that continues until the IsRunning flag is set to false (i.e., when Stop is called). Each iteration of the loop increments the three direction variables, deltaX, deltaY, and deltaZ; raises the ReadingChanged event; and sleeps for a random amount of time.

Now that you have an alternative accelerometer class, you need to alter the LoadAccelerometer method to load an instance of FakeAccelerometer if the application is being run in the emulator. As an alternative to creating an instance of RealAccelerometer and testing to see if it is supported on the current platform, the LoadAccelerometer method queries the DeviceType to determine whether the current device is an emulator. Since all Windows Phones must have an accelerometer, this is an acceptable way to determine whether the RealAccelerometer class has suitable hardware available.

Available for download on Wrox.com

```
public class Utilities{
    public static IAccelerometer LoadAccelerometer(){
        if (Microsoft.Devices.Environment.DeviceType ==
                                 Microsoft.Devices.DeviceType.Device){
            return new RealAccelerometer();
        }

        return new FakeAccelerometer();
    }
}
```

Code snippet from Utilities.cs

The FakeAccelerometer isn't a particularly good simulation of an accelerometer because it assumes a continual change of direction and does not introduce any spikes or jitter in the data as you might expect to see on a real device. In the next section, you'll see how you can use a Wii controller as another way to simulate accelerometer information within the emulator.

Wii Simulation

The purpose behind using a Wii controller to simulate accelerometer information is so that you can test your application with more realistic data. Although the FakeAccelerometer can be used to test your application behavior under basic conditions, it's hard to model real device movement or precisely control when the simulated movements occur. The controllers for the Nintendo Wii are easily connectable to a desktop PC and have a built-in accelerometer that can be used to generate a realistic data stream. As such, they make an ideal mock accelerometer device that can be used to test your application within the Windows Phone 7 emulator.

Connecting Your Wiimote

To start with, you will need a Wii controller (also referred to as a *Wiimote*, for "Wii Remote"). These can be purchased independently of the Nintendo Wii console for which they were originally designed. You will also need a computer that is Bluetooth-capable — you can purchase a USB-based Bluetooth adapter if your computer isn't already Bluetooth-capable. The other component you will need is *WiimoteLib*, which is a managed library for receiving input from the Wiimote. It's an Open Source project, hosted on CodePlex (`http://wiimotelib.codeplex.com`). Download the latest version of this project, unblock the downloaded file, and extract the contents to your computer.

In Windows Vista and Windows 7, when you download a file from the Internet it may be labeled "Blocked." This places restrictions on the files to indicate that they are from an untrusted source. To unblock a file, right-click on the file in Windows Explorer and select Properties. If the file has been Blocked, there will be an Unblock button located toward the bottom of the Properties window. Clicking this button will Unblock the file. You should only do this on files that are from a location that you trust.

Before you get to writing code, you need to set up the Wiimote by establishing a Bluetooth connection between your computer and the controller. On your computer load up your Bluetooth settings (this will vary depending on the specific Bluetooth stack used by your computer) and launch the wizard to create a new connection. Step through the wizard until you get to the point where it begins searching for Bluetooth devices as shown in the example in Figure 10-2.

At this point, pick up your Wiimote and simultaneously press buttons 1 and 2, as indicated in Figure 10-3.

FIGURE 10-2

FIGURE 10-3

You should see the four lights at the base of the device start to flash repetitively. The new Bluetooth connection wizard on your computer should detect the Wiimote and list it as an available device to connect to (Figure 10-4). The name of the remote should be similar to *Nintendo RVL-CNT-01*.

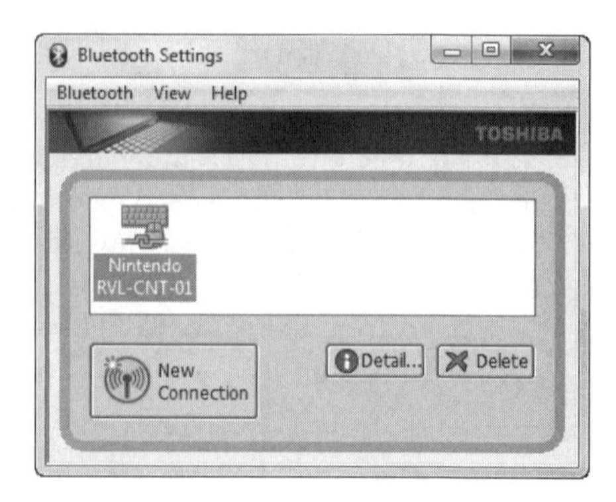

FIGURE 10-4

Complete the new Bluetooth connection wizard. The device doesn't need a pin to connect to, so if prompted, just leave the pin field empty. Once complete, the Bluetooth configuration should list the Wiimote as a paired device. Check to ensure that it appears in the list and there is an active connection. If it is listed, but not currently connected, make sure that the blue lights are flashing on the device, and attempt to connect to the device.

At this point, the Wiimote should be paired with your computer and have an active connection. To verify that this is the case and that data can be received from the device, run the WiimoteTest.exe located at the root of the WiimoteLib directory. Figure 10-5 illustrates the test program running.

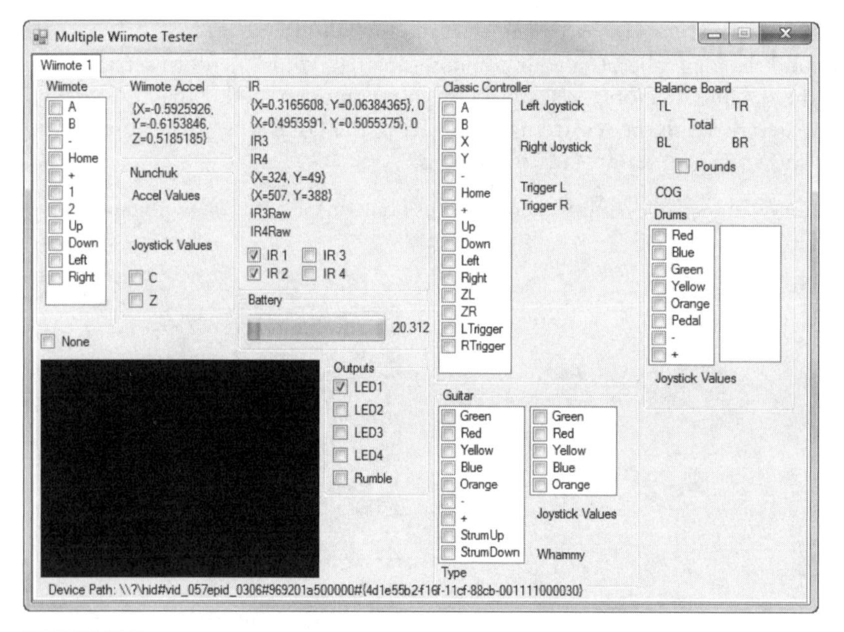

FIGURE 10-5

The information you are interested in is the "Wiimote Accel"[erometer], located in the top-left corner of Figure 10-5. This gives accelerometer information in the *x*, *y*, and *z* directions. You can close the test application once you have verified that these values change as you move your Wiimote.

Publishing the Accelerometer Values

The next step in getting the Wiimote accelerometer data into the Windows Phone emulator is to create a simple Windows Communication Foundation (WCF) Service. As the WCF Service will be registering a URI (Uniform Resource Identifier) at which the service can be addressed, you will need to be running Visual Studio with administrator privileges. Start by creating a new project based on the Console Application template, calling it *WiiDataService*. To this project add a new item, *WiimoteSensor*, based on the WCF Service template. Add a reference to WiimoteLib.dll, located in the folder where you expanded the WiimoteLib download. Now, change the `IWiimoteSensor` interface to the following:

Available for download on Wrox.com

```
[ServiceContract]
public interface IWiimoteSensor{
    [OperationContract]
    SensorData CurrentState();
}

public class SensorData{
    public float X { get; set; }
    public float Y { get; set; }
    public float Z { get; set; }
}
```

Code snippet from IWiimoteSensor.cs

This code includes a class called `SensorData` that is used to represent the tuple of accelerometer information. Before you implement this interface, you need to create a wrapper class that will connect to and retrieve data from the Wiimote. Add a new class called `WiiWrapper` to the project:

Available for download on Wrox.com

```
internal class WiiWrapper{
    public static readonly WiiWrapper Instance = new WiiWrapper();
    static WiiWrapper(){}

    public Wiimote Wiimote = new Wiimote();
    public SensorData state= new SensorData();

    private WiiWrapper(){}

    void TryConnect() {
        try{
            Wiimote.Connect();
            Wiimote.WiimoteChanged += Wiimote_WiimoteChanged;
            Wiimote.SetReportType(InputReport.IRAccel, true);
            Wiimote.SetLEDs(true, false, false, false);
            Console.WriteLine("Wiimote connected and ready");
        }
        catch (Exception e){
            Console.WriteLine(e.Message);
        }
    }
```

```
        public void Initialize(){
            Wiimote = new Wiimote();
            TryConnect();
        }

        public void Stop(){
            this.Wiimote.Disconnect();
        }

        void Wiimote_WiimoteChanged(object sender, WiimoteChangedEventArgs e){
            var state = e.WiimoteState.AccelState;
            this.state = new SensorData { X=state.Values.X,
                                          Y=state.Values.Y,
                                          Z=state.Values.Z };
        }
    }
```

Code snippet from WiiWrapper.cs

`WiiWrapper` is a singleton class that establishes a connection to the Wiimote in the `TryConnect` method. An event handler is registered for the `WiimoteChanged` event that simply updates the current state information based on the current accelerometer reading. Now you can go back and complete the implementation of `IWiimoteSensor` in the `WiimoteSensor` class. This is relatively straightforward as it simply returns the current state from the `WiiWrapper` instance:

Available for download on Wrox.com

```
    public class WiimoteSensor : IWiimoteSensor{
        public SensorData CurrentState(){
            var state = WiiWrapper.Instance.state;
            return new SensorData() { X = state.X, Y = state.Y, Z = -state.Z };
        }
    }
```

Code snippet from WiimoteSensor.cs

You will notice that the Wiimote accelerometer in the *z* direction is reversed before being returned to the Windows Phone simulator. This is to correct for a difference in coordinate systems used by the Wiimote and Windows Phone accelerometers, respectively. Compared to Windows Phone, the Wiimote has the *z* axis upside down.

To complete the WiimoteSensor WCF Service, you need to update Program.cs to initialize the `WiimoteWrapper` class in order to establish a connection to the Wiimote, and then host the WiimoteSensor Service:

Available for download on Wrox.com

```
    class Program{
        static void Main(string[] args){
            // Create the ServiceHost.
            using (ServiceHost host = new ServiceHost(typeof(WiimoteSensor))){
                WiiWrapper.Instance.Initialize();

                host.Open();

                Console.WriteLine("The service is ready");
```

```
                Console.WriteLine("Press <Enter> to stop the service.");
                Console.ReadLine();

                // Close the ServiceHost.
                host.Close();
            }
        }
    }
```

Code snippet from Program.cs

The WiiDataService project should also include an app.config file that declares the configuration for the WiimoteSensor Service. The contents of this file should look similar to the following:

```
<?xml version="1.0" encoding="utf-8" ?>
<configuration>
    <system.serviceModel>
        <behaviors>
            <serviceBehaviors>
                <behavior name="WiiDataService.WiimoteSensorBehavior">
                    <serviceMetadata httpGetEnabled="true" />
                    <serviceDebug includeExceptionDetailInFaults="false" />
                </behavior>
            </serviceBehaviors>
        </behaviors>
        <services>
            <service behaviorConfiguration="WiiDataService.WiimoteSensorBehavior"
                name="WiiDataService.WiimoteSensor">
                <endpoint address="" binding="basicHttpBinding"
                          bindingConfiguration=""
                    contract="WiiDataService.IWiimoteSensor">
                </endpoint>
                <endpoint address="mex" binding="mexHttpBinding"
                          contract="IMetadataExchange" />
                <host>
                    <baseAddresses>
                        <add baseAddress="http://localhost:8080/Wiimote" />
                    </baseAddresses>
                </host>
            </service>
        </services>
    </system.serviceModel>
</configuration>
```

Code snippet from app.config

If you compare this configuration with the one that is created in the app.config file when you created the WCF Service, you'll notice that it uses the basicHttpBinding instead of the wsHttpBinding. This is because Windows Phone only supports the basicHttpBinding for WCF Services. The most important part of this configuration is the `baseAddress` setting which determines the URI at which the service can be accessed while it is running. Make sure that the Bluetooth connection to your Wiimote is active and run the WiiDataService project. You should see a command prompt appear, stating that the "Wiimote is connected and ready," as illustrated in Figure 10-6.

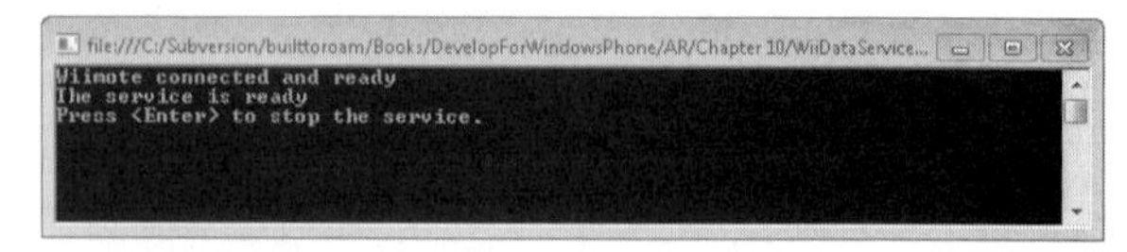

FIGURE 10-6

There are two potential issues you may come across when running the WiiDataService:

- *If you get an* `AddressAccessDeniedException`, *make sure you are running Visual Studio as an administrator.*
- *If the console displays "Error reading data from Wiimote . . . is it connected?" you need to make sure that your Wiimote is paired and the connection hasn't been accidentally terminated. In some cases, you may have to re-pair the Wiimote with your computer.*

Consuming the Accelerometer Values

To be able to connect to the WiiDataService from your Windows Phone application, you need to have the WiiDataService running (either start a second instance of Visual Studio or run the service from outside Visual Studio), then select "Add Service Reference." Enter the address of the service, shown in Figure 10-7, and click Go. Set the Namespace to **WiimoteData** and click OK to add the service reference to your project.

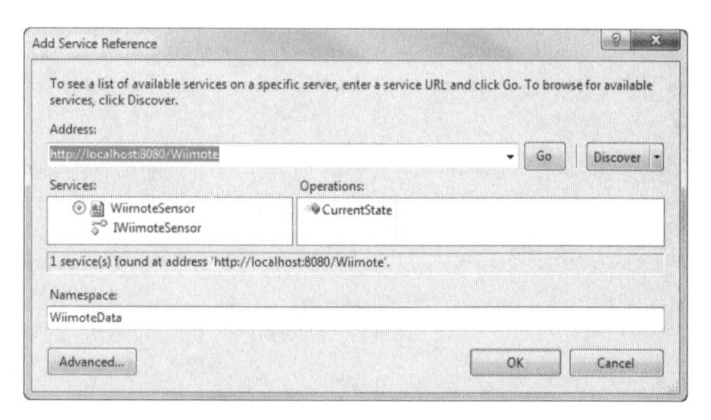

FIGURE 10-7

Now, all that is left is to wrap access to the WiimoteSensor WCF Service within a class that implements the `IAccelerometer` interface:

Available for download on Wrox.com

```
public class WiiAccelerometer: IAccelerometer{
    public event EventHandler<AccelerometerEventArgs> ReadingChanged;

    private Vector3 reading;
    private AutoResetEvent waitForCurrentState = new AutoResetEvent(false);

    private bool IsRunning { get; set; }
    private  WiimoteData.WiimoteSensorClient WiiSensor { get; set; }
```

```
    public void Start(){
        if (IsRunning) return;

        IsRunning = true;
        WiiSensor = new WiimoteData.WiimoteSensorClient();
        WiiSensor.CurrentStateCompleted += WiiData_CurrentStateCompleted;
        var runThread = new Thread(Run);
        runThread.Start();
    }

    public void Stop(){
        if (!IsRunning) return;

        IsRunning = false;
    }

    public void Run(){
        while (IsRunning){
            WiiSensor.CurrentStateAsync();
            waitForCurrentState.WaitOne();

            if (ReadingChanged != null) {
                 var readingArgs = new AccelerometerEventArgs()
                                     { Reading = reading,
                                       Timestamp = DateTimeOffset.Now};
                 ReadingChanged(this, readingArgs);
            }
        }
    }

    private void WiiData_CurrentStateCompleted(object sender,
                                WiimoteData.CurrentStateCompletedEventArgs e){
        if (e.Error==null && !e.Cancelled){
            reading = new Vector3(e.Result.X, e.Result.Y, e.Result.Z);
        }
        waitForCurrentState.Set();
    }
}
```

Code snippet from WiiAccelerometer.cs

The `WiiAccelerometer` class continually polls the `CurrentState` method of the WiiSensor WCF Service to obtain the latest accelerometer data.

Now you can update the `LoadAccelerometer` method to create an instance of the `WiiAccelerometer` class instead of the `FakeAccelerometer` class when running on the emulator:

Available for download on Wrox.com

```
public class Utilities{
    public static IAccelerometer LoadAccelerometer(){
        if (Microsoft.Devices.Environment.DeviceType ==
                              Microsoft.Devices.DeviceType.Device){
```

```
                return new RealAccelerometer();
            }

            return new WiiAccelerometer();
        }
    }
```

Code snippet from Utilities.cs

Running the application within the emulator, you should now notice that the `TextBlock` updates based upon movement of the Wiimote.

Reactive Extensions for .NET

Although the `ReadingChanged` event you exposed on the `IAccelerometer` interface provides you with a way to obtain data from an accelerometer device, the approach is not without its downsides. For example, by exposing an event, you cannot easily use a LINQ query to filter or summarize changes in the accelerometer information. To resolve this situation, the Windows Phone SDK includes the *Reactive Extensions for .NET (Rx)* framework.

The concept behind Rx is to model a series of events firing within an application as a *stream of data* (also referred to as an *observable sequence*). An application can then subscribe to an observable sequence and receive notifications whenever new items are added. Using Rx also allows the stream of events to be easily filtered using a LINQ expression.

To use Rx within an application, you must first add references to the `Microsoft.Phone.Reactive` and `System.Observable` assemblies. Once this is done, you are ready to create an Observable collection. Since the accelerometer data source is an event, you can create a suitable Observable collection by calling the static `FromEvent` method on the `Observable` class. You can also take the opportunity to use a simple LINQ statement to report only the actual accelerometer data, rather than the entire `AccelerometerEventArgs` provided by the `ReadingChanged` event.

Once you have a suitable Observable collection, you can *subscribe* ("listen for changes") to it by calling the `Subscribe` method. You can think of `Subscribe` as a way to wire up an event handler that will be invoked whenever new data is added to the sequence.

Available for download on Wrox.com

```
private void StartButton_Click(object sender, RoutedEventArgs e){
    var accelerometerEvents = Observable.FromEvent<AccelerometerEventArgs>(
                ev => sensor.ReadingChanged += ev,
                ev => sensor.ReadingChanged -= ev);

    var readings = from args in accelerometerEvents
                   select args.EventArgs.Reading;

    readings.Subscribe(ObservableReadingChanged);
    try{
        // Start the accelerometer
        sensor.Start();
    }
```

```
        catch (AccelerometerStartFailedException ex){
            this.SensorText.Text = ex.Message.ToString();
        }
    }

    void ObservableReadingChanged(Vector3 data)
    {
        this.Dispatcher.BeginInvoke(() => {
            (this.border1.Projection as PlaneProjection).RotationX = 90 * data.X;
            (this.border1.Projection as PlaneProjection).RotationY = -90 * data.Y;
            (this.border1.Projection as PlaneProjection).RotationZ = 90 * data.Z;
        });
    }
```

Code snippet from MainPage.xaml.cs

Whenever the accelerometer detects changes, it will raise `ReadingChanged` events that Rx will use to populate the accelerometerEvents Observable collection. Each time a reading is added to the sequence, the `ObservableReadingChanged` method is invoked. In this case, this method adjusts the rotation of the `PlaneProjection` for the `Border`, as demonstrated by Figure 10-8.

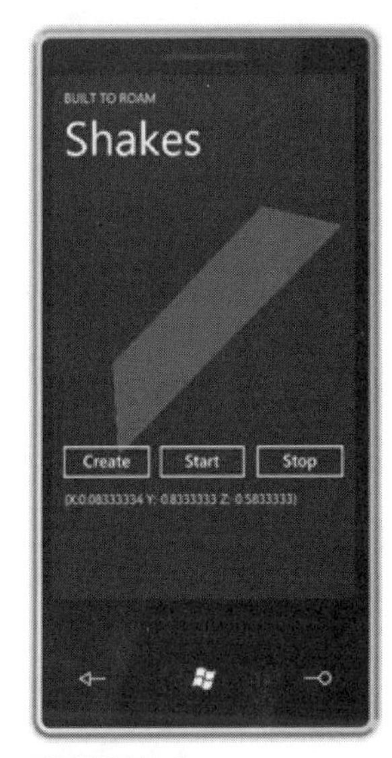

FIGURE 10-8

Initially, this code appears to be overkill in order to subscribe to a simple event, but it forms the basis to do more complex filtering of the sequence. You can do this by making use of more complex LINQ expressions to filter the raw accelerometer data.

Available for download on Wrox.com

```
var readings = from args in accelerometerEvents
               where args.EventArgs.Reading.X>0.2 ||
                     args.EventArgs.Reading.Y>0.2 ||
                     args.EventArgs.Reading.Z>0.2
               select args.EventArgs.Reading;
```

Code snippet from MainPage.xaml.cs

This LINQ expression filters the readings so that events in which the acceleration in any direction is less than 0.2 are ignored. This filter is applied to the observable sequence so that only elements that meet this condition will invoke the `ObservableReadingChanged` method.

VIBRATION

Most applications provide visual and occasionally audio feedback to the user. Another way that is particularly useful on a mobile device is using a short vibration. This can be used to attract attention to the device (e.g., most mobile devices have a vibrate setting for the ring volume) or to simulate some other event (e.g., getting hit in an action game on the device).

Vibrating the device is done using the `VibrateController` class that is located in the `Microsoft.Devices` namespace. There are only two methods on the `VibrateController` class, `Start` and `Stop`. The `Start` method takes a `TimeSpan`, which tells the device how long to vibrate for. If you want the device to stop vibrating before this time span is complete (e.g., if the user presses a Dismiss button), you can call the `Stop` method to cancel the vibration.

```
private void VibrateStartButton_Click(object sender, RoutedEventArgs e){
    // Create a TimeSpan of 3 seconds
    var duration = new TimeSpan(0,0,3);
    VibrateController.Default.Start(duration);
}

private void VibrateStopButton_Click(object sender, RoutedEventArgs e){
    VibrateController.Default.Stop();
}
```

This code will execute on the emulator even though there is no ability for the emulator to vibrate.

SUMMARY

In this chapter, you have seen how you can interact with the accelerometer on a Windows Phone. Although there is no support in the Windows Phone emulator, you can provide a mock accelerometer that will allow you to test your application. You also learned how you can trigger a Windows Phone to vibrate for a specified duration, which can be useful for providing tactile feedback in your application.

11

Who Said That?

WHAT'S IN THIS CHAPTER

- Playing media using the MediaElement
- Playing and combining SoundEffects
- Recording audio from the microphone
- Integrating into the media hub
- Tuning the FM radio

Windows Phone is an amazing playback device for both audio and video. Your application can make use of the rich media playback capabilities offered by Silverlight as well as the XNA audio framework for playing polyphonic sound throughout your application.

In this chapter, you will learn how to integrate audio playback into your application and how you can access the microphone audio stream to do your own audio processing. You'll also see how you can leverage cloud-based services to extend the Windows Phone audio capabilities to do text-to-speech and language translation.

MEDIA PLAYBACK

There are two ways to play media within a Windows Phone application. If you want to do simple video or audio playback, you can add a `MediaElement` to your `PhoneApplicationPage` and use the `Source` property to specify the media to play. Alternatively, you can load an audio file into memory using the `SoundEffect` class and then play the sound on demand.

MediaElement

Let's start with a simple example of using the `MediaElement` to play back an audio file. Create a new project within Visual Studio using the Windows Phone Application project template and add an audio file (e.g., an mp3 or wav file), making sure that the `Build Action` property is set to `Content` and the `Copy to Output Directory` property is set to `Copy Always`. Drag a `MediaElement` instance from the Toolbox into the ContentGrid area of the MainPage. As you're going to be playing audio, you don't need to worry about the size or location of the `MediaElement`. However, to ensure that it doesn't appear visually, you should set its `Visibility` property to `Collapsed`. Lastly, set the `Source` property to be the name of the audio file you added to the project and the `AutoPlay` property to `True`. When the application runs, the audio file, Kalimba.mp3 in the example below, will automatically be played when the page is loaded:

Available for download on Wrox.com

```
<Grid x:Name="ContentGrid" Grid.Row="1">
    <MediaElement x:Name="AudioPlayer" HorizontalAlignment="Left"
                  VerticalAlignment="Top" Source="Kalimba.mp3"
                  AutoPlay="True" Visibility="Collapsed" />
</Grid>
```

Code snippet from MainPage.xaml

The `Source` property is actually a Uri and can be set to either a local file (deployed within your application xap file) or on a remote server. For example, you could change the `MedieElement` as follows:

```
    <MediaElement x:Name="AudioPlayer" HorizontalAlignment="Left"
                  VerticalAlignment="Top" AutoPlay="True" Visibility="Collapsed"
          Source="http://www.builttoroam.com/books/devwp7/chapter11/Kalimba.mp3" />
```

One thing to be aware of when loading media from a remote server is that there will be a significant delay while the media is loaded and possible gaps in the audio if the network connection drops out or slows down during playback. To avoid such issues, you may want to consider downloading remote media onto the device before playing it. The following example uses the `WebClient` class to download the entire remote file Kalimba.mp3 into isolated storage (see Chapter 16 for more on working with isolated storage) before beginning playback. After downloading the file use the `SetSource` method to pass a stream to the `MediaElement`:

Available for download on Wrox.com

```
private void DownloadAndPlayButton_Click(object sender, RoutedEventArgs e){
    var downloadClient = new WebClient();
    downloadClient.OpenReadCompleted += downloadClient_OpenReadCompleted;
    downloadClient.OpenReadAsync(
      new Uri("http://www.builttoroam.com/books/devwp7/chapter11/Kalimba.mp3"));
}

void downloadClient_OpenReadCompleted(object sender,
                                      OpenReadCompletedEventArgs e){
    using (var strm = e.Result)
    using (var isoFile = new IsolatedStorageFileStream("Kalimba.mp3",
             FileMode.Create, IsolatedStorageFile.GetUserStoreForApplication())){
        int read;
        var buffer = new byte[10000];
        while ((read = strm.Read(buffer, 0, buffer.Length)) > 0){
```

```
                isoFile.Write(buffer, 0, read);
            }
            isoFile.Seek(0, SeekOrigin.Begin);
            this.AudioPlayer.SetSource(isoFile);
        }
    }
```

Code snippet from MainPage.xaml.cs

To display video using the `MediaElement`, you simply need to set the `Visibility` property to `Visible` and set the `Source` property to be a video file. Add a second `MediaElement` to the page, and set the `Source` property to the name of a video file that you have added to the project. Again, make sure that the file's `Build Action` is set to `Content` and that the `Copy to Output Directory` property is set to `Copy Always`.

Available for download on Wrox.com

```
<MediaElement x:Name="VideoPlayer" AutoPlay="True"
              Margin="188,140,70,0" Height="144" VerticalAlignment="Top"
              Source="Wildlife.wmv" />
```

Code snippet from MainPage.xaml

Before you run this example, you need to explicitly set the `AutoPlay` attribute to `False` for the existing `MediaElement` named `AudioPlayer`. While you can have multiple `MediaElement` instances on a page, only one can be playing at any given time. The default value for `AutoPlay` is `True`, so simply removing this attribute from the XAML isn't enough.

The `MediaElement` has several methods and properties (shown in Table 11-1) that can be used to control the playback of media.

TABLE 11-1: MediaElement Properties and Methods

PROPERTY	DESCRIPTION
`AutoPlay`	Indicates whether media should be played as soon as it has been loaded (`true`), or not (`false`).
`BufferingProgress` (read only)	The percentage (on a scale of 0 to 1) of the `BufferingTime` that has been buffered
`BufferingTime`	The amount of time that should be buffered before playback commences. This is a `TimeSpan` property, so if you want to specify this in XAML, the value needs to be specified in the format `[days.]hours:minutes:seconds[fractionalSeconds]`, where the [ ] indicate optional components (rather than literals).

continues

TABLE 11-1 *(continued)*

PROPERTY	DESCRIPTION
`CacheMode`	Specifies whether the caching of rendered content should be performed. The only supported value for Windows Phone is `BitmapCache`, wherein elements are cached as bitmaps. This significantly improves performance as the caching is done by the hardware graphics processing unit. Silverlight for the Desktop requires GPU acceleration to be enabled in the HTML hosting the Silverlight plug-in. In contrast, hardware acceleration is automatically enabled for all Windows Phone applications.
`CanPause` (read only)	Indicates whether the media playback can be paused and resumed.
`CanSeek` (read only)	Indicates whether the media playback position can be modified via the `Position` property.
`CurrentState` (read only)	`MedieElementState` enumeration consisting of: ➤ `Closed` — No media has been specified, or the Uri specified via the `Source` property has failed to load. ➤ `Opening` — The `Source` property has been set, and the Uri is being opened in preparation for playback. Any actions taken in this state, such as `Play` or `Stop`, will be buffered until `Opening` has completed. ➤ `Buffering` — Media is being buffered in preparation for playback either at the beginning or during playback. While in this state, the `Position` property will not change, no audio will be played, and the last visible video frame will continue to be displayed. ➤ `Playing` — Media has been loaded, and either the `Play` method has been invoked or the `MediaElement` is set to `AutoPlay`. The `Position` property advances during this state. ➤ `Paused` — Media playback is suspended (`Play` will resume playback from the current position) with no audio and the last frame of video being displayed until playback resumes. ➤ `Stopped` — Media has been loaded, but playback has either stopped or hasn't commenced. If the Source is a video, then the first frame will be displayed. There are also states for the Individualizing and Acquiring License that relate to how DRM (Digital Rights Management)-protected content is loaded and validated.

PROPERTY	DESCRIPTION
`DownloadProgress` (read only)	The percentage (on scale of 0 to 1) of media that has been downloaded
`DownloadProgressOffset` (read only)	The percentage of video that was skipped before playback began. For example, if the `Position` property is moved to 30%, downloading will commence from that point. When `DownloadProgress` returns 40%, this means that 10% of the file (i.e., the difference between `DownloadProgress` and `DownloadProgressOffset`) has been downloaded.
`IsMuted`	Audio can be disabled by setting `IsMuted` to `true`.
`NaturalDuration` (read only)	The `NaturalDuration` returns a `Duration` object that wraps a `TimeSpan` indicating the length of the media. If no media has been loaded, the `Duration` is set to the Automatic static value.
`NaturalVideoHeight` and `NaturalVideoWidth` (read only)	The native height and width of the video content. This is useful if you want to adjust the `MediaElement` for optimum video output without scaling or letterboxing.
`Position`	Returns a `TimeSpan` indicating the position through the media. If `CanSeek` is `true`, then this property can be used to adjust the current position in the playback.
`Source`	The Uri of the media to be played
`Volume`	The volume (on a scale of 0 to 1, with default 0.5) of the media being played. As each media will have a different baseline, this property can be difficult to get right. In addition, the user can control the overall device volume using the hardware volume keys.
METHOD	**DESCRIPTION**
`Pause`	If `CanPause` returns `true`, this method will suspend playback. The last video frame will continue to be displayed, but no audio will sound.
`Play`	Plays the current media either from the beginning, if the media was stopped, or from the point where the media had been paused.
`SetSource`	Can be used to set the media source to a stream. This can be useful if you have stored media in `IsolatedStorage`, which is covered in more depth in Chapter 16.
`Stop`	Stops the current media from playing. If the media is video, then the first frame will be displayed.

MEDIAELEMENT FOR WINDOWS PHONE

There are several properties that exist on the `MediaElement` control that are only present for compatibility with the Desktop version of Silverlight:

- `AudioStreamCount` — This will only ever return a value of 1, even if the media contains multiple audio streams.
- `AudioStreamIndex` — This will always be set to 0, selecting the first audio stream, in line with the `AudioStreamCount` value of 1.
- `Balance` — This will always return a value of 1, which would usually correspond to all the sound going to the right speaker. Windows Phone only supports a single speaker, so this is directing all the audio to that one speaker.
- `CanPause` — This property will always return `false` when loading media from a stream, but also returns `false` for some types of media. For example, `CanPause` will return `true` when a wmv file is being played but `false` when an mp3 file is playing. If you are going to give the user the ability to pause the playback, you should consider using this value to enable/disable that control.
- `RenderedFramesPerSecond` — This property exists but is not supported on Windows Phone.

Media Controls

Let's make use of a couple of these properties to add some controls that will control the behavior of the `VideoPlayer MediaElement` that was added to the page. Start off by adding two buttons and setting their `Content` properties to label the controls "Play" and "Stop," respectively:

Available for download on Wrox.com

```
<Button x:Name="VideoPlayButton" Content="Play" Height="70"
        HorizontalAlignment="Left" Margin="0,140,0,0" Width="168"
        VerticalAlignment="Top" Click="VideoPlayButton_Click" />
<Button x:Name="VideoStopButton" Content="Stop" Height="70"
        HorizontalAlignment="Left" Margin="0,212,0,0" Width="168"
        VerticalAlignment="Top" Click="VideoStopButton_Click" />
```

Code snippet from MainPage.xaml

Create the corresponding event handlers to start and stop the media playback on the `VideoPlayer`. At this point, you can remove the `AutoPlay` attribute on the `VideoPlayer` element since you now have on-screen controls that can be used instead:

Available for download on Wrox.com

```
private void VideoPlayButton_Click(object sender, RoutedEventArgs e){
    this.VideoPlayer.Play();
}
```

```
private void VideoStopButton_Click(object sender, RoutedEventArgs e){
    this.VideoPlayer.Stop();
}
```

Code snippet from MainPage.xaml.cs

One of the immediate issues with this basic interface is that both buttons are always enabled regardless of the playback status. What should happen is that when there is no media playing, the Play button should be enabled, and when media is playing, the Stop button should be enabled. Rather than wiring up event handlers to the `MediaElement` and then manually adjusting the IsEnabled state of the buttons, you can use data-binding to effectively connect the two buttons to the `CurrentState` property of the `MediaElement`. You would have seen a little bit of this in Chapter 3, and you'll learn much more about data-binding in Chapter 15. Here you're going to focus on using a subset of data-binding that will allow you to connect the property of one control to a property of another control.

There are a couple of small hurdles that you have to overcome before you can bind the `IsEnabled` property of the buttons to the `CurrentState` property of the `MediaElement`. Firstly, the `CurrentState` property is of type `MediaElementState`, which is an enumeration, whereas the `IsEnabled` property is a `Boolean`. Therefore, you need to convert a `CurrentState` value into a `Boolean` value indicating whether the button should be enabled. Silverlight supports the conversion of a value into another data type via the `IValueConverter` interface. Since the two buttons, Play and Stop, should be enabled for different `CurrentState` values, you can create a converter that accepts a parameter indicating which state should result in `True` being returned:

Available for download on Wrox.com

```
public class AudioStateConverter:IValueConverter{
    public object Convert(object value, Type targetType,
                          object parameter, CultureInfo culture){
        MediaElementState testState;
        if (parameter is MediaElementState){
            testState = (MediaElementState)parameter;
        }
        else{
            testState = (MediaElementState)Enum.ToObject(
                           typeof(MediaElementState), parameter);
        }
        var state = (MediaElementState)value;
        var matched = state == testState;
        return matched;
    }

    public object ConvertBack(object value, Type targetType,
                              object parameter, CultureInfo culture){
        throw new NotSupportedException();
    }
}
```

Code snippet from AutoStateConverter.cs

The `AudioStateConverter` implements `IValueConverter`, which requires both the `Convert` and `ConvertBack` methods to be implemented. In a lot of cases, you will only want to provide a one-way conversion, so it is common for the other method to throw a `NotSupportedException`, or `NotImplementedException`.

The `Convert` method of the `AudioStateConverter` class tests the value and parameter arguments to determine if they are both of type `MediaElementState` and equal in value. The result of this test is the `Boolean` value, which will be returned as the converted value.

Unfortunately the `MediaElement` doesn't actually support data binding directly to the `CurrentState` property. Luckily it does have a `CurrentStateChanged` event which can be used to track when the `CurrentState` changes. Rather than simply adding an event handler and then updating the states of the Play and Stop buttons, you're going to create a proxy class that will track the `CurrentState` property of the `MediaElement` in a way that it can be used for data binding. This might seem like overkill when you're only updating one or two elements but if you have more elements that require updating according to different state values then it will pay to be able to do this using data binding.

The `MediaElementBinder` has two dependency properties: The first is used to specify which `MediaElement` it is tracking the `CurrentState` property for; the second is of course the current `CurrentState` property for the `MediaElement`. When the `MediaElement` property is set, the `MediaElementBinder` attaches an event handler to the `CurrentStateChanged` event. This is used to update the `CurrentState` dependency property, which in turn will update any element that is data binding to this property.

```
public class MediaElementBinder : FrameworkElement{
    public MediaElement MediaElement{
        get { return (MediaElement)GetValue(MediaElementProperty); }
        set { SetValue(MediaElementProperty, value); }
    }

    public static readonly DependencyProperty MediaElementProperty =
        DependencyProperty.Register(
            "MediaElement",
            typeof(MediaElement),
            typeof(MediaElementBinder),
            new PropertyMetadata(OnMediaElementChanged));

    private static void OnMediaElementChanged(DependencyObject o,
                                    DependencyPropertyChangedEventArgs e){
        var mediator = (MediaElementBinder)o;
        var mediaElement = (MediaElement)(e.NewValue);
        if (null != mediaElement){
            mediaElement.CurrentStateChanged += mediator.MediaPlayer_CurrentStateChanged;
        }
    }

    public MediaElementState CurrentState{
        get { return (MediaElementState)GetValue(CurrentStateProperty); }
        set { SetValue(CurrentStateProperty, value); }
    }
```

```
    public static readonly DependencyProperty CurrentStateProperty =
        DependencyProperty.Register(
            "CurrentState",
            typeof(MediaElementState),
            typeof(MediaElementBinder),
            new PropertyMetadata(null));

    private void MediaPlayer_CurrentStateChanged(object sender,
                                                 RoutedEventArgs e) {
        this.CurrentState = (sender as MediaElement).CurrentState;
    }
}
```

Code snippet from MediaElementBinder.cs

After adding both the `AudioStateConverter` and `MediaElementBinder` classes to your Windows Phone project, switch over to Expression Blend as this will make wiring up the data-binding much easier. Start by making sure that the project has been rebuilt, to ensure that the new classes will be located. Then create an instance of the `MediaElementBinder` by dragging it onto the ContentGrid node of the Objects and Timeline window, after the VideoPlayer element. With the new element selected, locate the MediaElement property in the Miscellaneous of the Properties window (left image of Figure 11-1).

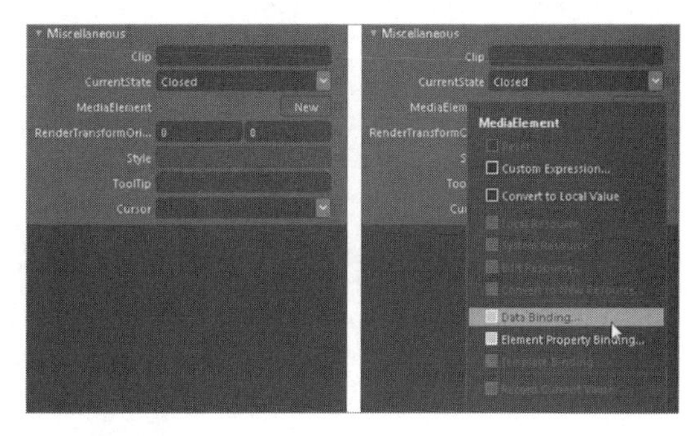

FIGURE 11-1

Click the small square on the far right side of the MediaElement row. This will display the data-binding shortcut menu. In the Create Data Binding dialog that is displayed, select the Element Property tab and select the VideoPlayer element and click OK. This has set the `MediaElement` property of the `MediaElementBinder` to be the VideoPlayer element.

The next thing to do is to data bind the `IsEnabled` property of the `VideoPlayButton` element to the `CurrentState` property of the `MediaElementBinder`. Select the `VideoPlayButton` element and find the `IsEnabled` property in the Properties window. You may need to expand the Common Properties area as the `IsEnabled` property is hidden by default. Again, click the small square to the right of the property which will display the "Create Data Binding" dialog as before (Figure 11-2).

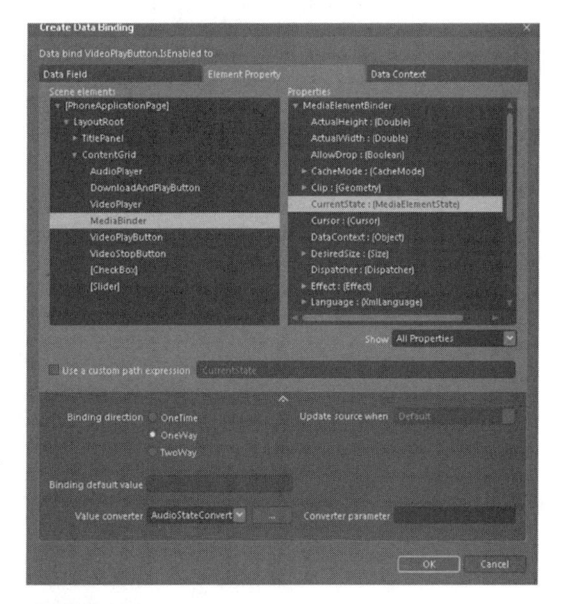

FIGURE 11-2

In Chapter 15 you'll learn more about using the Data Field and Data Context tabs of this window. For the time being, you should select the "MediaBinder" element in the left pane, followed by the "CurrentState" property in the right pane. You will probably have to change the Show dropdown to list "All Properties," rather than just those that match the type of the `IsEnabled` property (i.e., the properties that return a Boolean value). Also, expand the lower region of this window and click the ellipses button next to "Value converter." This will open the "Add Value Converter" dialog of Figure 11-3.

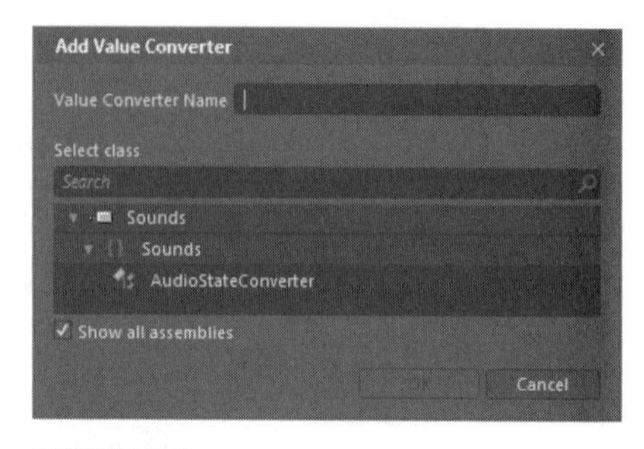

FIGURE 11-3

Select the "AudioStateConverter" class you created earlier, and click OK twice to confirm the new data-binding. The one thing you didn't do was to specify the parameter value to be passed into the converter. Unfortunately, this is one of the few things that you have to write manually in XAML as the Blend user interface (UI) only allows the converter parameter to be specified as a simple string, whereas the `AudioStateConverter` class requires a `MediaElementState` enumeration value.

Switch to the XAML view and locate the VideoPlayButton. In order to specify an enumeration value as a converter parameter, you first need to create an instance of the enumeration. This is done by creating a static resource within the `ContentGrid`. This XAML snippet assumes that the audio prefix is associated with the `System.Windows.Media` namespace:

Available for download on Wrox.com

```
<Grid x:Name="ContentGrid" Grid.Row="1">
    <Grid.Resources>
        <audio:MediaElementState
                x:Key="playing_state">Playing</audio:MediaElementState>
        <audio:MediaElementState
                x:Key="stopped_state">Stopped</audio:MediaElementState>
    </Grid.Resources>
    ...
</Grid>
```

Code snippet from MainPage.xaml

Here you will have defined two instances of the `MediaElementState` enumeration with the values `Playing` and `Stopped`. These can then be referenced elsewhere in the XAML via their Key values.

In order to reference the MediaElementState enumeration in XAML you have to import the System.Windows.Media namespace. Add the following xmlns attribute to the PhoneApplicationPhone node of your page:

```
xmlns:audio="clr-namespace:System.Windows.Media;
assembly=System.Windows"
```

The XAML for the `VideoPlayer`, `MediaElementBinder` and the two buttons, `VideoPlayButton` and `VideoStopButton`, can then be updated as follows:

```
<MediaElement x:Name="VideoPlayer" AutoPlay="True"
              Margin="188,140,70,0" Height="144"
              VerticalAlignment="Top" Source="Wildlife.wmv" >
</MediaElement>

<local:MediaElementBinder x:Name="MediaBinder"
                          MediaElement="{Binding ElementName=VideoPlayer}"/>

<Button x:Name="VideoPlayButton" Content="Play" Height="70"
        Margin="0,140,0,0" Width="168" VerticalAlignment="Top"
        HorizontalAlignment="Left" Click="VideoPlayButton_Click"
        IsEnabled="{Binding CurrentState,
                    ConverterParameter={StaticResource stopped_state},
                    Converter={StaticResource AudioStateConverter},
                    ElementName=MediaBinder}" />
<Button x:Name="VideoStopButton" Content="Stop" Height="70"
        Margin="0,212,0,0" VerticalAlignment="Top" Width="168"
        HorizontalAlignment="Left" Click="VideoStopButton_Click"
        IsEnabled="{Binding CurrentState,
                    ConverterParameter={StaticResource playing_state},
                    Converter={StaticResource AudioStateConverter},
                    ElementName=MediaBinder}" />
```

Code snippet from MainPage.xaml

When you run this example, you will see that initially the Play button is disabled (note that `AutoPlay` on the `VideoPlayer` is set to True). After clicking the Stop button to stop the media playback, the Play button is enabled and the Stop button is disabled.

Another nice feature to add is the ability for the user to adjust the volume or mute the audio altogether. For this you'll use a `Slider` and `CheckBox` control, respectively, and demonstrate data-binding again to hook these up without any source code.

In Blend, add these two controls from the Assets window, and adjust the `Orientation` property of the `Slider` to `Vertical` and the `Content` of the `CheckBox` to `Mute`. Wiring up the Mute CheckBox is relatively straightforward as you can data-bind the `IsChecked` property of the CheckBox directly to the `IsMuted` property of the `VideoPlayer MediaElement`. One thing to note is that you want to make sure the binding direction is set to "TwoWay," as shown in Figure 11-4. This ensures that changes in either the CheckBox's `IsChecked` property or the MediaElement's `IsMuted` property will cause the other control to update to match. In contrast, OneWay data-binding would only ensure that a change in the `IsChecked` property updates the `IsMuted` property. It would not cause the CheckBox's `IsChecked` property to change value if the `IsMuted` property changed in code.

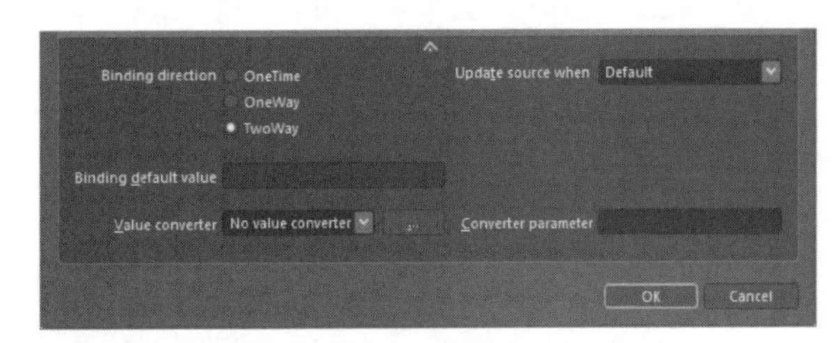

FIGURE 11-4

You can wire up the volume slider in a similar way by connecting the `Value` property of the `Slider` to the `Volume` property of the `VideoPlayer MediaElement`. Since the `Volume` property is on a scale of 0 to 1, you will also need to change the `Maximum` property of the `Slider` to 1.

If you connect these properties to each other directly, you may notice that the volume slider doesn't work particularly well. As you change the slider position, instead of the volume increasing and decreasing smoothly, it will appear as if the volume changes on an exponential scale, making it hard to find a suitable volume level at which to listen. This is due to the way the human ear perceives differences in volume. A solution to this problem is to apply a logarithmic scale to the volume slider. Again, this can be done by creating a custom value converter and injecting it into the data-binding process to convert our raw volume slider value into the actual value applied to the `MediaElement`:

Available for download on Wrox.com

```
public class VolumeConverter:IValueConverter{
    private static double MaximumConvertedValue = Math.Log10((1 * 10) + 1);

    public object Convert(object value, Type targetType,
                          object parameter, CultureInfo culture) {
        double returnValue = 0.0;
        if (value is double){
            var doubleValue = (double)value;
            returnValue =
               (Math.Pow(10, MaximumConvertedValue * doubleValue) - 1) / 10.0;
        }
        return returnValue;
    }

    public object ConvertBack(object value, Type targetType,
                              object parameter, CultureInfo culture) {
        double returnValue = 0.0;
        if (value is double){
            var doubleValue = (double)value;
            returnValue =
               Math.Log10((doubleValue * 10) + 1) / MaximumConvertedValue;
        }
        return returnValue;
    }
}
```

Code snippet from VolumeConverter.cs

This is an example of how a logarithmic scale can be applied. Note that you need to implement both the `Convert` and the `ConvertBack` methods in order for the two-way binding of the volume slider to work properly. Figure 11-5 illustrates how the logarithmic scale used by the `VolumeConverter` differs from a raw data-binding.

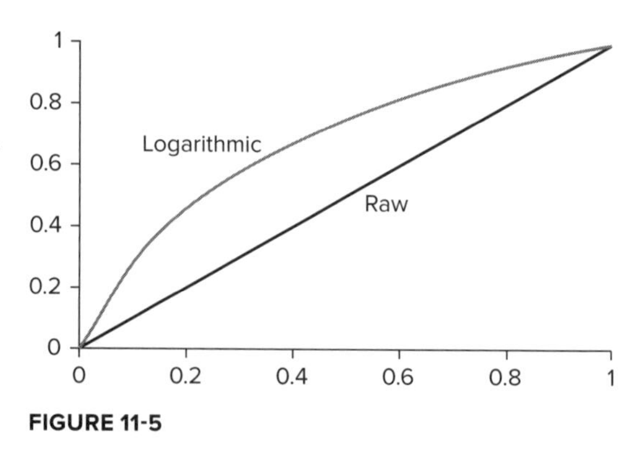

FIGURE 11-5

Figure 11-5 can be interpreted by thinking of the x value (horizontal axis) as the volume slider value and the y value (vertical axis) being the corresponding value that is passed to the

VideoPlayer MediaElement. When the VolumeConverter is used, the volume Slider should give you a much smoother adjustment of the apparent volume. The final XAML for both the Muted CheckBox and the Volume Slider should be similar to the following:

Available for download on Wrox.com

```
<Grid.Resources>
    <local:VolumeConverter x:Key="VolumeConverter"/>
</Grid.Resources>
<CheckBox Content="Mute" HorizontalAlignment="Right"
          Margin="0,74,12,0" VerticalAlignment="Top"
          IsChecked="{Binding IsMuted, ElementName=VideoPlayer, Mode=TwoWay}"/>
<Slider Margin="416,140,12,0" VerticalAlignment="Top" Height="144"
        Orientation="Vertical" Maximum="1"
        Value="{Binding Volume,
                        Converter={StaticResource VolumeConverter},
                        ElementName=VideoPlayer, Mode=TwoWay}"/>
```

Code snippet from MainPage.xaml

The final element to add is a progress bar that indicates the playback progress. Unfortunately, there isn't a single property that indicates the percentage of playback completion. There is the Position property, which returns a TimeSpan indicating the current time since the beginning of the media, and there is a NaturalDuration property, which provides a TimeSpan indicating the total length of the media. In order to wire up a progress bar, you need to use both of these properties to set the Maximum and Value properties of the progress bar.

It is not recommended to data bind directly to the Position property as this can cause performance problems with the media playback. Each time the Position updates, data binding would attempt to update other elements. Depending on the number and complexity of the data binding operations this can adversely affect the smoothness of the media playback. You'll extend the MediaElementBinder to poll the MediaElement control for an updated Position value when it is playing.

Available for download on Wrox.com

```
public class MediaElementBinder : FrameworkElement{
    private DispatcherTimer timer = new DispatcherTimer();

    public MediaElementBinder(){
        // Timer interval 200 milliseconds
        timer.Interval = new TimeSpan(0, 0, 0, 0, 200);
        timer.Tick += timer_Tick;
    }

    void timer_Tick(object sender, EventArgs e){
        this.Position = this.MediaElement.Position;
    }

    public TimeSpan Position{
        get { return (TimeSpan)GetValue(PositionProperty); }
        set { SetValue(PositionProperty, value); }
    }

    public static readonly DependencyProperty PositionProperty =
        DependencyProperty.Register("Position", typeof(TimeSpan),
```

```
                                          typeof(MediaElementBinder),
                                          new PropertyMetadata(null));

        public Duration Duration{
            get { return (Duration)GetValue(DurationProperty); }
            set { SetValue(DurationProperty, value); }
        }

        public static readonly DependencyProperty DurationProperty =
            DependencyProperty.Register("Duration", typeof(Duration),
                                          typeof(MediaElementBinder),
                                          new PropertyMetadata(null));

        private void MediaPlayer_MediaOpened(object sender, RoutedEventArgs e){
            var element =sender as MediaElement;
            this.Duration = element.NaturalDuration;
        }


        private void MediaPlayer_CurrentStateChanged(object sender,
                                                     RoutedEventArgs e){
            this.CurrentState = (sender as MediaElement).CurrentState;

            if (this.CurrentState == MediaElementState.Playing){
                timer.Start();
            }
            else{
                timer.Stop();
            }
        }
        // Existing properties omitted for brevity
        ...
    }
```

Code snippet from MediaElementBinder.cs

You'll also need to add an additional value converter that is capable of converting a `TimeSpan` or `Duration` object into an integer value (essentially returning the corresponding `Ticks` value):

Available for download on Wrox.com

```
    public class TimeSpanConverter : IValueConverter{
        public object Convert(object value, Type targetType, object parameter,
                              CultureInfo culture) {
            long retValue = 0;
            if (value is TimeSpan){
                var durValue = (TimeSpan)value;
                retValue = durValue.Ticks;
            }
            else if (value is Duration){
                var durValue = (Duration)value ;
                if (durValue.HasTimeSpan && durValue != Duration.Automatic
                    && durValue != Duration.Forever){
                    retValue = durValue.TimeSpan.Ticks;
                }
```

```
        }
        return retValue;
    }

    public object ConvertBack(object value, Type targetType,
                              object parameter, CultureInfo culture){
        throw new NotImplementedException();
    }
}
```

Code snippet from TimeSpanConverter.cs

Now you can simply add a `ProgressBar` to the page and data-bind the `Maximum` property to the `Duration` property of the `MediaElementBinder` and the `Value` property to the `Position` property of the `MediaElementBinder`, using a `TimeSpanConverter` in both cases to deal with the required data-type conversions. This should yield XAML similar to the following:

```
<ProgressBar x:Name="MediaProgress" Margin="8,286,8,301"
             Value="{Binding Position,
                     Converter={StaticResource TimeSpanConverter},
                     ElementName=MediaBinder}"
             Maximum="{Binding Duration,
                     Converter={StaticResource TimeSpanConverter},
                     ElementName=MediaBinder}"/>
```

Code snippet from MainPage.xaml

Transforms

Before you move on to learning how you can use the XNA `SoundEffects` class to play back audio, let's take a look at how you can apply transforms to the `MediaElement` to adjust the way your video appears. At this stage, switch to Expression Blend as this will make working with transforms much easier. First, add a `Border` by right-clicking on the `VideoPlayer` `MediaElement` and selecting Group Into ➪ Border. With the border selected, pick a solid color for the `BorderBrush` and set the `BorderThickness` to 2 on each side. Adding a border will make it easier to see the transforms that are applied to the `MediaElement`.

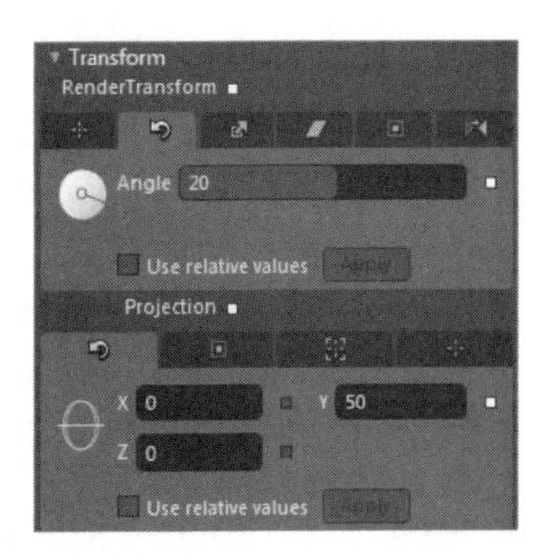

FIGURE 11-6

You'll start by adding a rotation transform to rotate the `MediaElement`. Instead of applying this transform to the `MediaElement` directly, the transform will actually be applied to the `Border` that wraps the `MediaElement`. This ensures that the border follows the `MediaElement` as it rotates. The two transforms you're going to apply are a `RenderTransform` and a `Projection`. Figure 11-6 illustrates a rotation of 20 degrees for the `RenderTransform` and a `Projection` rotation about the *y*-axis of 50 degrees.

This will generate the following XAML, which illustrates the `MediaElement` nested within a `Border` that has both a `RenderTransform` and a `Projection` applied to it:

Available for download on Wrox.com

```
<Border Height="144" Margin="188,140,70,0" VerticalAlignment="Top"
        BorderThickness="2" BorderBrush="#FFFF2B2B"
        RenderTransformOrigin="0.5,0.5" CacheMode="BitmapCache">
    <Border.RenderTransform>
        <CompositeTransform Rotation="20"/>
    </Border.RenderTransform>
    <Border.Projection>
        <PlaneProjection RotationY="50"/>
    </Border.Projection>

    <MediaElement x:Name="VideoPlayer" AutoPlay="True" />
</Border>
```

Code snippet from MainPage.xaml

When you start applying transforms to a `MediaElement`, *you must set the* `CacheMode` *property on the element being transformed (in this case, the* `Border`*). There is only one value supported in a Windows Phone application, which is* `BitmapCache`. *If your* `MediaElement` *doesn't display any video after you have applied a transform, double-check that this property has been correctly set.*

When you run this, you will see that the video is still displayed. However, it will be twisted according to the transforms that have been applied to it, as shown in Figure 11-7.

FIGURE 11-7

Clipping

You can apply a clip region to the `MediaElement` so that it only displays a portion of the video. Windows Phone only supports clipping using a `Rectangle` and doesn't support rounded corners (i.e., the `RadiusX` and `RadiusY` properties of the `Rectangle` have to be set to `0`). It does, however, support applying transforms, such as rotation, to the clip region.

Start by adding a rectangle anywhere on the page. It doesn't matter where or how big the rectangle is. Right-click the rectangle and select Path ➪ Make Clipping Path. This will display a dialog requesting that you select an element to apply the clipping to. After selecting the `VideoPlayer MediaElement`, you will notice that the rectangle disappears from the page. Unfortunately, there is no designer support for arranging the clip region after it has been created, nor does Blend do a good job of converting the current layout of the rectangle into the clip region. As such, in order to tweak the clipping region, you are forced to edit XAML manually. The complete XAML for the `Border` and `MediaElement` is as follows:

Available for download on Wrox.com

```
<Border Height="144" Margin="188,140,70,0" VerticalAlignment="Top"
        BorderThickness="2" BorderBrush="#FFFF2B2B"
        RenderTransformOrigin="0.5,0.5" CacheMode="BitmapCache">
    <Border.RenderTransform>
        <CompositeTransform Rotation="20"/>
    </Border.RenderTransform>
    <Border.Projection>
        <PlaneProjection RotationY="50"/>
    </Border.Projection>

    <MediaElement x:Name="VideoPlayer" AutoPlay="True">
        <MediaElement.Clip>
            <RectangleGeometry RadiusY="0" RadiusX="0" Rect="30,30,100,85">
                <RectangleGeometry.Transform>
                    <CompositeTransform Rotation="-5" CenterX="130" CenterY="75" />
                </RectangleGeometry.Transform>
            </RectangleGeometry>
        </MediaElement.Clip>
    </MediaElement>
</Border>
```

Code snippet from MainPage.xaml

Running this example will give you a rather unusual `MediaElement` showing the clipped video inside the `Border`, as shown in Figure 11-8.

FIGURE 11-8

SoundEffects with XNA

An alternative to using the `MediaElement` is to use the `SoundEffect` class to load and play back sounds. The `SoundEffect` class, which is part of the XNA `Audio` namespace, has the appearance of being quite simple to use, and for playing WAV-based audio files it couldn't really get any easier. However, there is much more to working with XNA Audio than first meets the eye.

Let's start with a simple example of playing a WAV file. The simplest approach is to include the wav file within your application's XAP package (in this case "blockrock.wav"). As with content added to be used with the `MediaElement`, the `Build Action` property needs to be set to `Content` and the `Copy to Output Directory` needs to be set to `Copy Always`.

Available for download on Wrox.com

```
private void XNALocalButton_Click(object sender, RoutedEventArgs e){
    var sound = SoundEffect.FromStream(TitleContainer.OpenStream("blockrock.wav"));
    sound.Name = "Loaded from Local Content";
    sound.Play();
}
```

Code snippet from XNAAudioPage.xaml.cs

XNA AUDIO

In order to work with the XNA audio classes, such as SoundEffect, you need to add a reference to Microsoft.Xna.Framework.dll. You will also need to add a using statement to any file where you wish to access these classes.

Before you can play any audio using the SoundEffect class or record sound with the Microphone class you will also need to periodically call FrameworkDispatcher.Update. To do this, add a new class called XNADispatcher to your Windows Phone application and add the following code to periodically call the Update method whenever the DispatcherTimer raises the Tick event.

Available for download on Wrox.com

```
public class XNADispatcher : IApplicationService{
   private DispatcherTimer frameworkDispatcherTimer =
              new DispatcherTimer();

   public int DispatchIntervalInMilliseconds { get; set; }

   void IApplicationService.StartService(ApplicationServiceContext
                                          context){
       this.frameworkDispatcherTimer.Tick
             += new EventHandler(frameworkDispatcherTimer_Tick);
       this.frameworkDispatcherTimer.Interval =
            new TimeSpan(0, 0, 0, 0, DispatchIntervalInMilliseconds);
       this.frameworkDispatcherTimer.Start();
   }

   void IApplicationService.StopService(){
       this.frameworkDispatcherTimer.Stop();
   }

   void frameworkDispatcherTimer_Tick(object sender, EventArgs e){
       FrameworkDispatcher.Update();
   }
}
```

Code snippet from XNADispatcher.cs

You only need to create a single instance of the XNADispatcher class when your application launches, so it can be added to the ApplicationLifecycleObjects in the App.xaml file.

Available for download on Wrox.com

```
<Application.ApplicationLifetimeObjects>
   <local:XNADispatcher DispatchIntervalInMilliseconds="50"/>
   ...
</Application.ApplicationLifetimeObjects>
```

Code snippet from App.xaml

In this case the timer interval is set to 50 milliseconds, which should work for most scenarios.

The static `OpenStream` method on XNA's `TitleContainer` class can be used to access the contents of any file, with the `Build Action` property set to `Content`, as a stream. The static `FromStream` method on the `SoundEffect` class can then be used to load the wav file contained within the stream.

An alternative way to obtain a `SoundEffect` is to fetch the audio file from a remote server. Loading content from a remote server is done in two parts. The first is to use a WebClient to open a connection to the remote media. Then, once a connection has been opened, the returned stream can be used to create the `SoundEffect` instance.

Available for download on Wrox.com

```
private void XNARemoteButton_Click(object sender, RoutedEventArgs e){
    var client = new WebClient();
    client.OpenReadCompleted += ((s, args) =>
            {
                var sound = SoundEffect.FromStream(args.Result);
                sound.Name = "Loaded from Remote Content";
                sound.Play();
            });
    client.OpenReadAsync(
    new Uri("http://www.builttoroam.com/books/devwp7/chapter11/blockrock.wav"));
}
```

Code snippet from XNAAudioPage.xaml.cs

As mentioned earlier, one of the dangers of loading content directly from a remote store is that it makes the application conditional on there being a network connection. It also makes the application bandwidth-hungry as each time the `SoundEffect` is required it will be re-downloaded. The final technique demonstrated here resolves these problems by storing the downloaded audio file within Isolated Storage so that it can be reused each time the application loads the sound effect, rather than downloading it again from the remote server:

Available for download on Wrox.com

```
private void XNAIsolatedStorageButton_Click(object sender, RoutedEventArgs e){
    if (!IsolatedStorageFile.GetUserStoreForApplication()
                            .FileExists("blockrock.wav")){
        var client = new WebClient();
        client.OpenReadCompleted += ((s, args) =>
                {
                    using (var iss = new IsolatedStorageFileStream(
                                "blockrock.wav", FileMode.OpenOrCreate,
                                IsolatedStorageFile.GetUserStoreForApplication())){
                        byte[] buffer = new byte[args.Result.Length];
                        args.Result.Read(buffer, 0, buffer.Length);
                        iss.Write(buffer, 0, buffer.Length);
                    }

                    PlayFromIsolatedStorage("blockrock.wav");
                });
        client.OpenReadAsync(
    new Uri("http://www.builttoroam.com/books/devwp7/chapter11/blockrock.wav"));
    }
```

```
        else{
            PlayFromIsolatedStorage("blockrock.wav");
        }
    }

    private void PlayFromIsolatedStorage(string fileName){
        using (var iss = new IsolatedStorageFileStream(
                                fileName, FileMode.OpenOrCreate,
                                IsolatedStorageFile.GetUserStoreForApplication())){
            iss.Seek(0, SeekOrigin.Begin);
            var sound = SoundEffect.FromStream(iss);
            sound.Name = "Loaded from Isolated Storage";
            sound.Play();
        }
    }
```

Code snippet from XNAAudioPage.xaml.cs

Within some sound effect scenarios, you may find another overload of the SoundEffect Play method to be useful. This overload allows you to specify the volume (scale of 0 to 1), pitch (scale of –1 to 1 representing a transpose of one octave down and up, respectively), and pan (scale of –1 to 1 representing left–right balance) of the sound effect.

Volume, Balance, and Looping

Playing audio using the SoundEffect class is very primitive and doesn't give you any ability to control the playback. Once started, the audio can't be stopped, paused, or adjusted. As an alternative, the SoundEffect class also exposes a CreateInstance method that is used to create one or more SoundEffectInstance objects. The audio media is loaded into memory once, when the SoundEffect is created, and then referenced by each created SoundEffectInstance. From a single SoundEffect you can create one or more SoundEffectInstances and then control them independently.

A SoundEffectInstance exposes methods to Play, Pause, and Stop playback. There is also a corresponding State property that indicates which of the three states — Stopped, Paused, and Playing — the instance is in. Unlike the SoundEffect class, where the Volume, Pitch, and Pan were specified via arguments to the Play method, a SoundEffectInstance exposes these three aspects as properties. There is one more property that is of interest, which is the IsLooped property. Setting this to true will result in the media being repeated, or looped, continuously until the Stop method is invoked.

For the purpose of demonstrating the SoundEffectInstance class in action, you're going to use a new PhoneApplicationPage. Update the ContentGrid with the following XAML:

```
<Grid Grid.Row="1" x:Name="ContentGrid">
    <Grid.Resources>
        <DataTemplate x:Key="ItemWithNameTemplate">
            <TextBlock TextWrapping="Wrap" Text="{Binding Name}"/>
        </DataTemplate>
    </Grid.Resources>
```

```
    <Button Content="Load" Height="67" HorizontalAlignment="Left" Margin="-3,0,0,0"
            x:Name="LoadButton" VerticalAlignment="Top" Width="166"
            Click="LoadButton_Click" />
    <ListBox HorizontalAlignment="Left" Margin="1,88,0,398" x:Name="SoundsList"
             Width="474" ItemTemplate="{StaticResource ItemWithNameTemplate}" />
    <ListBox Height="122" HorizontalAlignment="Left" Margin="3,311,0,0"
             x:Name="SoundInstancesList" VerticalAlignment="Top" Width="474"
             ItemTemplate="{StaticResource ItemWithNameTemplate}"
             SelectionChanged="SoundInstancesList_SelectionChanged" />
    <Button Content="Create Instance" Height="72" HorizontalAlignment="Left"
            Margin="3,233,0,0" Name="CreateSoundInstanceButton"
            VerticalAlignment="Top" Width="243"
            Click="CreateSoundInstanceButton_Click" />
    <TextBox Height="72" HorizontalAlignment="Left" Margin="252,233,0,0"
             Name="SoundInstanceNameText" VerticalAlignment="Top" Width="222" />

    <Grid Height="162" HorizontalAlignment="Left" Margin="1,439,0,0"
          Name="SoundEffectInstanceGrid" VerticalAlignment="Top" Width="479"
          Grid.Row="1">
        <Button Content="Play" Height="70" HorizontalAlignment="Left"
                Name="PlayButton" VerticalAlignment="Top" Width="160"
                Margin="0,6,0,0" Click="PlayButton_Click" />
        <Button Content="Stop" Height="70" HorizontalAlignment="Left"
                Name="StopButton" VerticalAlignment="Top" Width="160"
                Margin="153,6,0,0" Click="StopButton_Click" />
        <Slider HorizontalAlignment="Left" Margin="2,75,0,0"
                VerticalAlignment="Top" Width="460" Maximum="1" Minimum="-1"
                LargeChange="0.01" Value="{Binding Instance.Pan, Mode=TwoWay}"
                Name="PanSlider" />
        <CheckBox Content="Loop" HorizontalAlignment="Left"
                  Margin="313,13,0,0" Name="checkBox1" VerticalAlignment="Top"
                  IsChecked="{Binding Instance.IsLooped, Mode=TwoWay}"/>
    </Grid>
</Grid>
```

Code snippet from SoundEffectsPage.xaml

The Grid.Resource section defines an `ItemTemplate` that will be used by the two `ListBoxes` that will be used. The template defines a single `TextBlock`, which is bound to the `Name` property of the items in the list. Next is the Load button, which is used to load the initial set of `SoundEffect` objects:

Available for download on Wrox.com

```
private void LoadButton_Click(object sender, RoutedEventArgs e){
    var sound = SoundEffect.FromStream(TitleContainer.OpenStream("blockrock.wav"));
    sound.Name = "Block Rock";
    this.SoundsList.Items.Add(sound);

    sound = SoundEffect.FromStream(TitleContainer.OpenStream("puncher.wav"));
    sound.Name = "Puncher";
    this.SoundsList.Items.Add(sound);
```

```
        sound = SoundEffect.FromStream(TitleContainer.OpenStream("fastbreak.wav"));
        sound.Name = "Fast Break";
        this.SoundsList.Items.Add(sound);
    }
```

Code snippet from SoundEffectsPage.xaml.cs

The `SoundsList` `ListBox` will display the three instances of the `SoundEffect` class. When the "Create Instance" button is pressed, the currently selected `SoundEffect` will be used to create a new `SoundEffectInstance`. This will then be added to the `SoundInstancesList`. As there is no `Name` property within the `SoundEffectInstance` class to display in the listbox, it is wrapped using the `InstanceWrapper` class, which exposes a `Name` property:

Available for download on Wrox.com

```
    private void CreateSoundInstanceButton_Click(object sender, RoutedEventArgs e){
        var sound = this.SoundsList.SelectedItem as SoundEffect;
        if (sound == null || string.IsNullOrEmpty(SoundInstanceNameText.Text)) return;
        var instance = new InstanceWrapper(){
                            Instance = sound.CreateInstance(),
                            Name = SoundInstanceNameText.Text
                           };
        this.SoundInstancesList.Items.Add(instance);
    }

    public class InstanceWrapper{
        public SoundEffectInstance Instance { get; set; }
        public string Name { get; set; }
    }
```

Code snippet from SoundEffectsPage.xaml.cs

When an item is selected in the `SoundInstancesList`, that item becomes the `DataContext` for the `SoundEffectInstanceGrid`:

Available for download on Wrox.com

```
    private void SoundInstancesList_SelectionChanged(object sender,
                                                      SelectionChangedEventArgs e){
        this.SoundEffectInstanceGrid.DataContext = this.SoundInstancesList.SelectedItem;
    }
```

Code snippet from SoundEffectsPage.xaml.cs

Without getting into the intricacies of data-binding, this allows all the controls nested within the `Grid` to reference the item using the Binding syntax seen earlier. For example, the `Value` property on the `Slider` is data-bound to the `Instance.Pan` property on the `InstanceWrapper` using the following syntax: `Value="{Binding Instance.Pan, Mode=TwoWay}"`. Through data-binding, the `Slider` is wired up to the `Pan` (i.e., the left–right balance), and the `CheckBox` is wired to the `IsLooped` property. These properties can be adjusted on each `SoundEffectInstance` independently.

Of course, you will want to be able to Play and Stop the `SoundEffectInstance` you have selected. The interesting thing about this is that you can create multiple `SoundEffectInstances` based on one or more `SoundEffects` and have them all play concurrently. While they are playing, you can dynamically adjust the `Pan` (and if you were to add appropriate controls, the `Volume` and `Pitch`) of each effect independently:

```
private void PlayButton_Click(object sender, RoutedEventArgs e){
    var instance = this.SoundEffectInstanceGrid.DataContext as InstanceWrapper;
    if (instance == null) return;
    instance.Instance.Play();
}

private void StopButton_Click(object sender, RoutedEventArgs e){
    var instance = this.SoundEffectInstanceGrid.DataContext as InstanceWrapper;
    if (instance == null) return;
    instance.Instance.Stop();
}
```

Code snippet from SoundEffectsPage.xaml.cs

Figure 11-9 illustrate the sound effects page. The Load button causes the three audio files, Block Rock, Puncher and Fast Break, to be loaded. You can then create one or more instances of these sounds by selecting the audio file, giving it a name and clicking the Create Instance button. The instances can then be played and stopped using the Play and Stop buttons.

FIGURE 11-9

3D Sound

There is one method on the `SoundEffectInstance` that hasn't been covered. This is the `Apply3D` method, which can be used to dynamically adjust the sound Volume, Pan, and Pitch based on the apparent position, in three dimensions, of a listener from where the sound is being emitted. If this doesn't make sense, then imagine that you are at a music festival with multiple concerts going on at the same time in different tents. As you walk around, you can hear the concerts coming from different directions and, depending on how far you are from each tent, at different volumes. You can think of each tent as being a sound emitter, and yourself as a listener. These are reflected in the XNA Audio classes, `AudioEmitter` and `AudioListener`.

You're going to use the example of a three-concert festival to demonstrate how you can use the `Apply3D` method to provide realistic audio effects within your application. Although this is clearly designed for XNA games, there is no reason why you can't make use of this to enhance your Silverlight application.

Again, you're going to start with a new PhoneApplicationPage. This time you need to add a Canvas onto which you add four circular Ellipses, each of a different color. You should end up with XAML similar to the following:

```
<Grid Grid.Row="1" x:Name="ContentGrid">
    <Canvas Height="350" HorizontalAlignment="Left" Margin="26,46,0,0"
            Name="LocationCanvas" VerticalAlignment="Top" Width="429"
            Background="#72FF2E2E"
            MouseLeftButtonDown="LocationCanvas_MouseLeftButtonDown"
            MouseLeftButtonUp="LocationCanvas_MouseLeftButtonUp"
            MouseMove="LocationCanvas_MouseMove">
        <Ellipse Canvas.Left="36" Canvas.Top="26" Height="49" Name="SoundSource1"
                 Stroke="Black" StrokeThickness="1" Width="50" Fill="Blue" />
        <Ellipse Canvas.Left="154" Canvas.Top="263" Height="49" Name="SoundSource2"
                 Stroke="Black" StrokeThickness="1" Width="50" Fill="DarkViolet" />
        <Ellipse Canvas.Left="303" Canvas.Top="37" Height="49" Name="SoundSource3"
                 Stroke="Black" StrokeThickness="1" Width="50" Fill="Green" />
        <Ellipse Canvas.Left="154" Canvas.Top="107" Height="49" Name="Listener"
                 Stroke="Black" StrokeThickness="1" Width="50" Fill="Yellow" />
    </Canvas>
</Grid>
```

Code snippet from 3DAudioPage.xaml

Label the first three ellipses *SoundSource1* through *SoundSource3*. Each ellipse represents an AudioEmitter allocated to a particular SoundEffectInstance. The last ellipse is the *Listener* and represents an AudioListener object. The AudioEmitters and AudioListener are created along with the page. Then, when the page has completed loading, the SoundEffectInstances will be created and correlated to the position of each Ellipse:

```
AudioEmitter emitter1 = new AudioEmitter();
AudioEmitter emitter2 = new AudioEmitter();
AudioEmitter emitter3 = new AudioEmitter();
AudioListener listener = new AudioListener();

private void PhoneApplicationPage_Loaded(object sender, RoutedEventArgs e){
    UpdateSounds();


    var sound = SoundEffect.FromStream(TitleContainer.OpenStream("blockrock.wav"));
    var instance = sound.CreateInstance();
    instance.IsLooped = true;
    instance.Apply3D(listener, emitter1);
    instance.Play();
    this.SoundSource1.Tag = instance;

    sound = SoundEffect.FromStream(TitleContainer.OpenStream("puncher.wav"));
    instance = sound.CreateInstance();
    instance.IsLooped = true;
    instance.Apply3D(listener, emitter2);
```

```
        instance.Play();
        this.SoundSource2.Tag = instance;

        sound = SoundEffect.FromStream(TitleContainer.OpenStream("fastbreak.wav"));
        instance = sound.CreateInstance();
        instance.IsLooped = true;
        instance.Apply3D(listener, emitter3);
        instance.Play();

        this.SoundSource3.Tag = instance;
    }

    private void UpdateSounds(){
        emitter1.Position = new Vector3(
          (float)((LocationCanvas.ActualWidth -
                  Canvas.GetLeft(SoundSource1)) / LocationCanvas.ActualWidth),
          (float)(Canvas.GetTop(SoundSource1) / LocationCanvas.ActualHeight),
          0.0f);

        emitter2.Position = new Vector3(
          (float)((LocationCanvas.ActualWidth -
                  Canvas.GetLeft(SoundSource2)) / LocationCanvas.ActualWidth),
          (float)(Canvas.GetTop(SoundSource2) / LocationCanvas.ActualHeight),
          0.0f);
        emitter3.Position = new Vector3(
          (float)((LocationCanvas.ActualWidth -
                  Canvas.GetLeft(SoundSource3)) / LocationCanvas.ActualWidth),
          (float)(Canvas.GetTop(SoundSource3) / LocationCanvas.ActualHeight),
          0.0f);
        listener.Position = new Vector3(
          (float)((LocationCanvas.ActualWidth -
                  Canvas.GetLeft(Listener)) / LocationCanvas.ActualWidth),
          (float)(Canvas.GetTop(Listener) / LocationCanvas.ActualHeight),
          0.0f);

        if (SoundSource1.Tag != null){
            (SoundSource1.Tag as SoundEffectInstance).Apply3D(listener, emitter1);
        }
        if (SoundSource2.Tag != null){
            (SoundSource2.Tag as SoundEffectInstance).Apply3D(listener, emitter2);
        }
        if (SoundSource3.Tag != null){
            (SoundSource3.Tag as SoundEffectInstance).Apply3D(listener, emitter3);
        }
    }
```

Code snippet from 3DAudioPage.xaml.cs

The `UpdateSounds` method is used to set the position of the `AudioEmitters` and the `AudioListener` based on the position of the `Ellipses`. The updated positions are used by the `Apply3D` method to determine the acoustic attributes of the `SoundEffectInstance`. Of course, this requires there to be a way for the user to reposition the `Ellipses` and hence adjust the location of the `AudioEmitters` and `AudioListener`:

Available for download on Wrox.com

```
private UIElement SelectedElement;
private System.Windows.Point MousePosition;
private void LocationCanvas_MouseLeftButtonDown(object sender,
                                                MouseButtonEventArgs e){
    var elements = VisualTreeHelper.FindElementsInHostCoordinates
                                    (e.GetPosition(null), this.LocationCanvas);
    SelectedElement = (from element in elements
                       where element is Ellipse
                       select element).FirstOrDefault();
    MousePosition = e.GetPosition(null);
}

private void LocationCanvas_MouseLeftButtonUp(object sender,
                                              MouseButtonEventArgs e){
    SelectedElement = null;
}

private void LocationCanvas_MouseMove(object sender, MouseEventArgs e){
    if (SelectedElement == null) return;
    var newPosition = e.GetPosition(null);
    Canvas.SetLeft(SelectedElement,
            Canvas.GetLeft(SelectedElement) + (newPosition.X - MousePosition.X));
    Canvas.SetTop(SelectedElement,
            Canvas.GetTop(SelectedElement) + (newPosition.Y - MousePosition.Y));
    MousePosition = newPosition;
    UpdateSounds();
}
```

Code snippet from 3DAudioPage.xaml.cs

When the user touches the screen, the `MouseLeftButtonDown` event handler will get called. This records the mouse position, and the `SelectedElement` variable remembers which `Ellipse` the user has selected. The `MouseLeftButtonUp` event handler does the reverse by setting the `SelectedElement` to `null`. Finally, the `MouseMove` event handler is used to reposition the selected `Ellipse` and update the position of the associated `AudioEmiter` object. What you should end up with is something that looks similar to Figure 11-10.

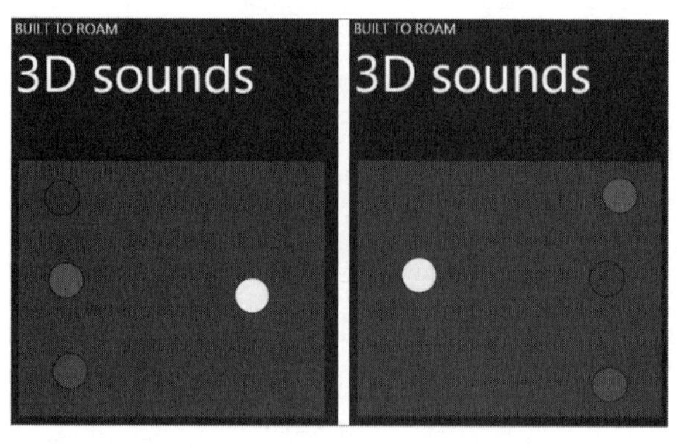

FIGURE 11-10

In the left image, Figure 11-10 shows the listener on the right side, while the right image shows the listener on the left side. To someone listening to the audio output, the audio should appear to come from the left in the first image and from the right in the second image.

3D audio is used to give the perception of sound emitting from different locations relative to the listener. As this is contingent on the listener being able to isolate left and right speakers the effect of 3D audio is dependent upon having two speakers that are physically separate. It is up to the hardware manufacturer as to whether a Windows Phone has a single speaker or multiple speakers on the device. On devices that have a single speaker it may not be possible to detect audio positioning as defined using 3D audio. Instead, by plugging a headset in you should be able to detect the location of sounds being emitted.

Microsoft Translator

Before moving on to discussing the audio recording capabilities of Windows Phone, it's worth pointing out how easily you can use the cloud-based Microsoft Translator Service to translate text into another language or to perform text-to-speech. This example will show how you can retrieve a list of languages supported by Microsoft Translator for the Speech API. The selected language will then be used to convert some text into a wav file using the Speech API. Let's start by building the user interface, which will consist of a TextBox, for the text to be translated; a ListBox, to select the language to use; and a Button to invoke the translation and subsequent conversion to Speech:

```
<Grid Grid.Row="1" x:Name="ContentGrid">
    <Grid.Resources>
        <DataTemplate x:Key="LanguageTemplate">
            <TextBlock Foreground="White" Margin="0,0,1,0" Text="{Binding Name}"
                       TextWrapping="Wrap" />
        </DataTemplate>
    </Grid.Resources>
    <Button Content="Speak" Height="66" HorizontalAlignment="Right"
            Margin="0,0,0,197" Name="Speak" VerticalAlignment="Bottom"
            Width="474" Click="Speak_Click" />
    <TextBox Margin="0,0,0,427" Name="TextToSpeachText" Text="This is a very
long sentence with lots of text, a full stop or two and a whole bunch of
nothing that will go on and on and on and on and on. Someone please stop
this audio loop before I get confused." TextWrapping="Wrap" />
    <ListBox HorizontalAlignment="Left"
             ItemTemplate="{StaticResource LanguageTemplate}" Margin="20,209,0,275"
             Name="ListLanguages" Width="441">
    </ListBox>
</Grid>
```

Code snippet from TranslateAndRecordPage.xaml

To make use of the Microsoft Translator Service, you need to add a service reference to your Windows Phone project. Right-click the "Windows Phone Project" in solution explorer, and select "Add Service Reference."

Figure 11-11 illustrates the "Add Service Reference" dialog into which you should enter **http://api.microsofttranslator.com/V2/Soap.svc** in the `Address` field and click the Go button. When the service is located, select the LanguageService node, provide a meaningful Namespace, in this case **BingTranslator**, and click OK. This will add a reference to the Microsoft Translator Service to your application. Services are covered in more detail in Chapter 14.

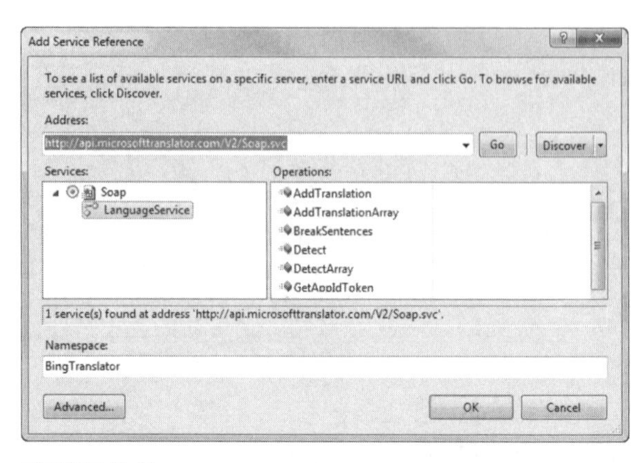

FIGURE 11-11

Before you can use Microsoft Translator in your application, you will need to go to `http://msdn.microsoft.com/en-us/library/ff512386(v=MSDN.10).aspx` *and follow the instructions to obtain a valid Bing API AppID. This is used whenever you access one of the Microsoft Translator Services. In this chapter, placeholders such as* `<AppID>` *have been used to signify where you have to enter your AppID.*

When the page is first loaded, the `languages` `ListBox` has to be populated with the languages that Microsoft Translator supports for the `Speak` method. These are accessible via the `GetLanguagesForSpeak` method call, as shown in the following code:

Available for download on Wrox.com

```
private void PhoneApplicationPage_Loaded(object sender, RoutedEventArgs e){
    FrameworkDispatcher.Update();

    var translator = new BingTranslator.LanguageServiceClient();
    translator.GetLanguagesForSpeakCompleted
                += translator_GetLanguagesForSpeakCompleted;
    translator.GetLanguagesForSpeakAsync("<AppID>", translator);
}
```

Code snippet from TranslateAndRecordPage.xaml.cs

The `GetLanguagesForSpeak` method only returns the language codes, for example, *en* for English and *fr* for French, which isn't particularly good for displaying in a `ListBox`. Luckily, there is a second method, `GetLanguageNames`, which can be called in order to get the friendly names for the languages. In this case, they're being requested in English by using the en language code:

Available for download on Wrox.com

```
void translator_GetLanguagesForSpeakCompleted(object sender,
                     BingTranslator.GetLanguagesForSpeakCompletedEventArgs e){
    var translator = e.UserState as BingTranslator.LanguageServiceClient;
    translator.GetLanguageNamesCompleted += translator_GetLanguageNamesCompleted;
    translator.GetLanguageNamesAsync("<AppID>", "en", e.Result, e.Result);
}
```

Code snippet from TranslateAndRecordPage.xaml.cs

To populate the `Languages ListBox`, you can use some LINQ magic to create instances of the `Language` class made up of the language code and friendly name. This array of languages is then set as the `ItemsSource` for the `ListBox`:

Available for download on Wrox.com

```
void translator_GetLanguageNamesCompleted(object sender,
                          BingTranslator.GetLanguageNamesCompletedEventArgs e){
    var codes = e.UserState as ObservableCollection<string>;
    var names = e.Result;

    var languages = (from code in codes
                          let cindex = codes.IndexOf(code)
                          from name in names
                          let nindex = names.IndexOf(name)
                          where cindex == nindex
                          select new TranslatorLanguage () { Name = name,
                                                        Code = code}).ToArray();
    this.Dispatcher.BeginInvoke(() => {
              this.ListLanguages.ItemsSource = languages;
          });
}

public class TranslatorLanguage{
    public string Name { get; set; }
    public string Code { get; set; }
}
```

Code snippet from TranslateAndRecordPage.xaml.cs

Now that you have the `Language ListBox` populated, all you need to do is to call the `Speak` method on the Microsoft Translator Service in order to convert the entered text into a wav file that contains speech in the chosen language:

Available for download on Wrox.com

```
private void Speak_Click(object sender, RoutedEventArgs e){
    var languageCode = "en";
    var language = this.ListLanguages.SelectedItem as TranslatorLanguage;
    if (language != null){
        languageCode = language.Code;
    }

    var translator = new BingTranslator.LanguageServiceClient();
    translator.SpeakCompleted += translator_SpeakCompleted;
```

```
        translator.SpeakAsync("<AppID>", this.TextToSpeachText.Text,
                              languageCode, "audio/wav");
    }
```

Code snippet from TranslateAndRecordPage.xaml.cs

The Speak method returns the URL to the wav audio file representing the translated, spoken text. This can then be played back via the use of either of the MediaElement or SoundEffect classes. Of course, you could always decide to download the audio to Isolated Storage if you expect it to be reused in the future.

Available for download on Wrox.com

```
    void translator_SpeakCompleted(object sender,
                                   BingTranslator.SpeakCompletedEventArgs e){
        var client = new WebClient();
        client.OpenReadCompleted += ((s, args) =>
                {
                    SoundEffect se = SoundEffect.FromStream(args.Result);
                    se.Play();
                });
        client.OpenReadAsync(new Uri(e.Result));
    }
```

Code snippet from TranslateAndRecordPage.xaml.cs

Figure 11-12 illustrates the simple translation page in action. When the page is loaded the language list is downloaded. The user just has to select a language and tap the Speak button to hear the text in the chosen language.

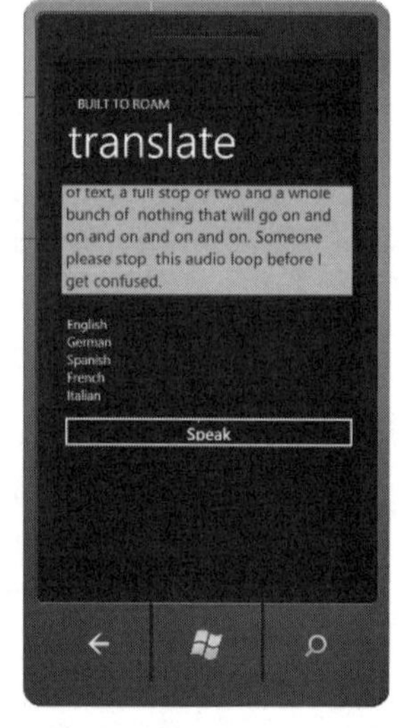

FIGURE 11-12

AUDIO RECORDING

In addition to being able to play back audio and video in a Windows Phone application, you can also record audio using the built-in microphone. The Microphone class is also part of the XNA Audio Framework and is accessible as a singleton via the Microphone.Default property. In this example, you'll build a simple sound recorder that is capable of playing back and saving audio:

Available for download on Wrox.com

```
    <Grid Grid.Row="1" x:Name="ContentGrid">
        ...
        <Button Content="Start" Height="70" HorizontalAlignment="Left"
                Name="StartButton" VerticalAlignment="Top" Width="160"
                Margin="3,477,0,0" Click="StartButton_Click" />
        <Button Content="Stop" Height="70" HorizontalAlignment="Left"
                Name="StopButton" VerticalAlignment="Top" Width="160"
                Margin="152,477,0,0" Click="StopButton_Click" />
```

```
    <Button Content="Play" Height="70" HorizontalAlignment="Left"
            Name="PlayRecordingButton" VerticalAlignment="Top" Width="155"
            Margin="306,477,0,0" Click="PlayRecordingButton_Click" />
</Grid>
```

Code snippet from TranslateAndRecordPage.xaml

To work with the microphone, you simply need to create an event handler for the `Microphone`'s `BufferReady` event, which will be called each time an audio buffer is ready for you to process. The buffer is nothing more than an array of bytes. While you can do whatever you want with this buffer, the simplest approach is simply to write this into a memory stream. This memory stream can then be accessed when the recording ends to either play back the recorded samples or to save them to Isolated Storage or a remote server.

To commence recording, the `BufferDuration` property needs to be set. The `Microphone` class uses this property to determine how big the buffers passed via the `BufferReady` event should be. With the `BufferDuration` set, calling the `Start` method will invoke the `BufferReady` event handler periodically as audio is captured. Within your `BufferReady` event handler, you can call the `GetData` method to retrieve the buffered data:

Available for download on Wrox.com

```
Microphone microphone = Microphone.Default;
byte[] microphoneBuffer;
MemoryStream audioStream;
byte[] recording;

private void StartButton_Click(object sender, RoutedEventArgs e){
    audioStream = new MemoryStream();
    microphone.BufferDuration = TimeSpan.FromMilliseconds(1000);
    microphoneBuffer =
  new byte[microphone.GetSampleSizeInBytes(microphone.BufferDuration)];
    microphone.BufferReady += microphoneBuffer_BufferReady;
    microphone.Start();
}

void microphoneBuffer_BufferReady(object sender, EventArgs e){
    microphone.GetData(microphoneBuffer);
    audioStream.Write(microphoneBuffer, 0, microphoneBuffer.Length);
}

private void StopButton_Click(object sender, RoutedEventArgs e){
    microphone.Stop();
    microphone.BufferReady -= microphoneBuffer_BufferReady;
    audioStream.Flush();
    recording= audioStream.ToArray();
    audioStream.Dispose();
}
```

Code snippet from TranslateAndRecordPage.xaml.cs

When stopping the recording, it is important both to invoke the `Stop` method on the `Microphone` as well as to remove the `BufferReady` event handler. This will ensure that the microphone can be reused the next time the Start button is pressed.

Playback

Once the recorded audio samples have been stored, playback is relatively straightforward as it's simply a matter of loading the recorded samples into a `SoundEffect` instance and invoking the `Play` method:

Available for download on Wrox.com

```
private void PlayRecordingButton_Click(object sender, RoutedEventArgs e){
    var effect = new SoundEffect(recording, microphone.SampleRate,
                                        AudioChannels.Mono);
    effect.Play();
}
```

Code snippet from TranslateAndRecordPage.xaml.cs

The `SoundEffect` constructor takes three parameters that include the samples, as a byte array; the sample rate; and whether the samples represent Mono or Stereo data. Looking at the `Microphone` class, you'll see that it exposes a `SampleRate` property, but it doesn't specify if the recording is of Mono- or Stereo-based audio. However, if you take a look in the debugger and expand out the private format property of the `Microphone` instance, as shown in Figure 11-13, you can confirm that the Microphone provides only a single channel (i.e., Mono).

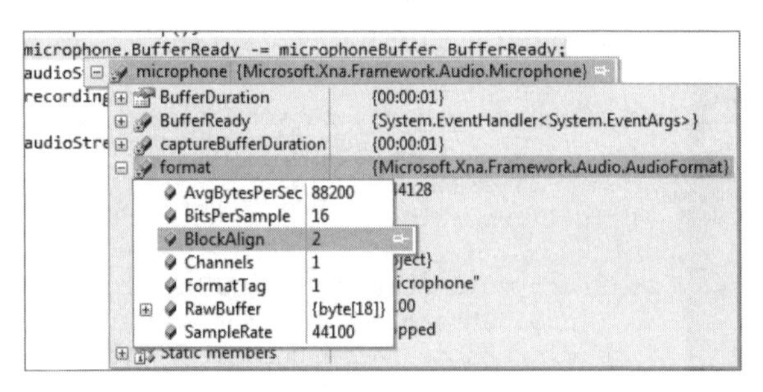

FIGURE 11-13

Saving

When you record audio from the microphone, what you get is an array of raw sample data. This is useful if you simply want to play it back using the `SoundEffect` class. However, if you want to use it within another media application or share it with a desktop PC, you may need to save it in a more standardized file format so that questions such as the recording's sample rate or number of channels can easily be answered by the recipient's device. Since the WAV file format consists of a small header before the raw audio samples, this is perhaps the easiest file format to work with within your Windows Phone 7 application. To create a WAV file, you can include the following `SoundUtility` class that exposes a `ConvertRawSamplesToWav` method. This method accepts an array of raw sample data and returns an array of bytes representing a wav file, complete with header information:

Available for download on Wrox.com

```
public static class SoundUtilities{
    public class WaveFormat{
        public int Encoding { get; set; }
        public int AverageBytesPerSecond { get; set; }
```

```
    public int BlockAlign { get; set; }
    public int Channels { get; set; }
    public int SamplesPerSecond { get; set; }
}

public static byte[] ConvertRawSamplesToWav(byte[] samples, WaveFormat format){
    var ascii = Encoding.UTF8;
    var byteArrayLength = 36 + samples.Length;
    var byteArray = new byte[byteArrayLength];
    var index = 0;

    // Specify that this is a RIFF file and the length of the file in bytes
    index += byteArray.CopyInto(index, ascii.GetBytes("RIFF"));
    index += byteArray.CopyInto(index, byteArrayLength.AsFixedByteArray(4));

    // Specify that this is a WAVE and start the format header
    index += byteArray.CopyInto(index, ascii.GetBytes("WAVE"));
    index += byteArray.CopyInto(index, ascii.GetBytes("fmt "));

    // Format header is fixed size of 16
    index += byteArray.CopyInto(index, (16).AsFixedByteArray(4));

    // Encoding: "1" for PCM
    index += byteArray.CopyInto(index, (format.Encoding).AsFixedByteArray(2));

    // Number of Channel
    index += byteArray.CopyInto(index, (format.Channels).AsFixedByteArray(2));

    // Samples per second
    index += byteArray.CopyInto(index,
                        (format.SamplesPerSecond).AsFixedByteArray(4));

    // Average bytes per second
    index += byteArray.CopyInto(index,
                        (format.AverageBytesPerSecond).AsFixedByteArray(4));

    // Block Align
    index += byteArray.CopyInto(index, (format.BlockAlign).AsFixedByteArray(4));

    // Bits per sample
    index += byteArray.CopyInto(index,
     ((8*format.AverageBytesPerSecond)/
       format.SamplesPerSecond).AsFixedByteArray(2));

    // The Samples themselves
    index += byteArray.CopyInto(index, ascii.GetBytes("data"));
    index += byteArray.CopyInto(index, samples.Length.AsFixedByteArray(2));
    index += byteArray.CopyInto(index, samples);

    return byteArray;
}

public static int CopyInto(this byte[] byteArray, int offset,byte[] bytes ){
    bytes.CopyTo(byteArray, offset);
    return byteArray.Length;
}
```

```
        public static byte[] AsFixedByteArray(this int number, int fixedByteArraySize){
            int remainder, result;
            var returnarray = new byte[fixedByteArraySize];

            result = DivRem(number, 256, out remainder);

            if (result >= 1){
                returnarray[0] = Convert.ToByte(remainder);
                var tmpArray = result.AsFixedByteArray(fixedByteArraySize - 1);
                tmpArray.CopyTo(returnarray, 1);
            }
            else{
                returnarray[0] = Convert.ToByte(number);
            }
            return returnarray;
        }

        public static int DivRem(int a, int b, out int result){
            result = a % b;
            return (a / b);
        }
    }
```

Code snippet from SoundUtilities.cs

Once the recording is stopped, this method can be used to convert the recording into a wav file before saving it into Isolated Storage:

```
    using (var iss = new IsolatedStorageFileStream("recording.wav",
                                                   FileMode.OpenOrCreate,
                                           IsolatedStorageFile.GetUserStoreForApplication())){
        var bytesToWrite = SoundUtilities.ConvertRawSamplesToWav(recording,
                                                     new SoundUtilities.WaveFormat(){
                        AverageBytesPerSecond = microphone.SampleRate * 2,
                        BlockAlign = 2,
                        Channels = 1,
                        Encoding = 1,
                        SamplesPerSecond = microphone.SampleRate
                    });
        iss.Write(bytesToWrite, 0, bytesToWrite.Length);
    }
```

Code snippet from TranslateAndRecordPage.xaml.cs

MUSIC AND VIDEO HUB

Part of the Windows Phone user experience is the ability to manage and play media on the Music and Video hub. Your application can access the MediaLibrary to list and play music. Let's start by adding a simple list to display all the songs that are in the media library. Each item in the list will consist of the song name and the artist.

```
<Grid x:Name="ContentGrid" Grid.Row="1">
    <ListBox x:Name="MediaList" SelectionChanged="MediaList_SelectionChanged">
        <ListBox.Resources>
            <DataTemplate x:Key="MediaItemTemplate">
                <StackPanel Orientation="Horizontal">
                    <TextBlock Margin="0,0,1,7" TextWrapping="Wrap"
                               Text="{Binding Name}"
                               Style="{StaticResource PhoneTextNormalStyle}"/>
                    <TextBlock Margin="0,0,1,7" TextWrapping="Wrap" Text="["
                               Style="{StaticResource PhoneTextNormalStyle}"/>
                    <TextBlock Margin="0,0,1,7" TextWrapping="Wrap"
                               Text="{Binding Artist.Name}"
                               Style="{StaticResource PhoneTextNormalStyle}"/>
                    <TextBlock Margin="0,0,1,7" TextWrapping="Wrap" Text="]"
                               Style="{StaticResource PhoneTextNormalStyle}"/>
                </StackPanel>
            </DataTemplate>
        </ListBox.Resources>
        <ListBox.ItemTemplate>
            <StaticResource ResourceKey="MediaItemTemplate"/>
        </ListBox.ItemTemplate>
    </ListBox>
</Grid>
```

Code snippet from MediaLibraryPage.xaml

Wiring up the list of songs to the ListBox is just a matter of setting the ItemsSource to the Songs property of a new instance of the MediaLibray.

```
void MediaLibraryPage_Loaded(object sender, RoutedEventArgs e){
    var library = new MediaLibrary();
    this.MediaList.ItemsSource = library.Songs;
}

private void MediaList_SelectionChanged(object sender, SelectionChangedEventArgs e){
    if (e.AddedItems.Count == 0) return;
    var song = e.AddedItems[0] as Song;
    if (song == null) return;

    MediaPlayer.Play(song);
}
```

Code snippet from MediaLibraryPage.xaml.cs

This code also illustrates how you can play one of the songs when the user selects it in the list. This uses the static methods on the MediaPlayer class to control what is currently playing. Songs that are played via the MediaPlayer appear in the Music and Video hub in the media history list.

FM TUNER

Every Windows Phone will be equipped with an FM radio tuner built-in. You can turn on and tune the radio from within your application by setting a couple of properties on the FMRadio singleton object.

```
FMRadio.Instance.CurrentRegion = RadioRegion.UnitedStates;
FMRadio.Instance.PowerMode = RadioPowerMode.On;
FMRadio.Instance.Frequency = 96.7;
```

You can query the strength of the current frequency by accessing the SignalStrength property (ranges from 0 to 1). In order to data bind to the properties of the radio you can again create a proxy class that will fetch the current radio properties and expose them in a way that they can be data bound to.

Available for download on Wrox.com

```
public class RadioBinder : FrameworkElement{
    private DispatcherTimer timer = new DispatcherTimer();
    public RadioBinder(){
        if (!System.ComponentModel.DesignerProperties.IsInDesignTool){
            timer.Interval = new TimeSpan(0, 0, 1);
            timer.Tick += new EventHandler(timer_Tick);
            timer.Start();

            FMRadio.Instance.CurrentRegion = RadioRegion.UnitedStates;
            FMRadio.Instance.PowerMode = RadioPowerMode.On;
        }
    }

    void timer_Tick(object sender, EventArgs e){
        if (FMRadio.Instance.PowerMode == RadioPowerMode.On){
            SignalStrength = FMRadio.Instance.SignalStrength;
            Frequency = FMRadio.Instance.Frequency;
        }
        else{
            SignalStrength = 0.0;
        }
    }

    public double Frequency{
        get { return (double)GetValue(FrequencyProperty); }
        set { SetValue(FrequencyProperty, value); }
    }

    public static readonly DependencyProperty FrequencyProperty =
        DependencyProperty.Register("Frequency", typeof(double),
                                typeof(RadioBinder),
                                new PropertyMetadata(95.0,ChangedFrequency));

    private static void ChangedFrequency(DependencyObject d,
                                          DependencyPropertyChangedEventArgs e){
        if (!System.ComponentModel.DesignerProperties.IsInDesignTool){
            try{
                FMRadio.Instance.Frequency = (double)e.NewValue;
            }
```

```
                catch (Exception ex){
                    (d as RadioBinder).Error = ex.Message + "-" + e.NewValue.ToString();
                }
            }
        }

        public double SignalStrength{
            get { return (double)GetValue(SignalStrengthProperty); }
            set { SetValue(SignalStrengthProperty, value); }
        }

        public static readonly DependencyProperty SignalStrengthProperty =
                DependencyProperty.Register("SignalStrength", typeof(double),
                                            typeof(RadioBinder),
                                            new PropertyMetadata(0.0));
    }
```

Code snippet from RadioBinder.cs

To allow someone to control the frequency of the radio you can use a slider and data bind it to the Frequncy property on the RadioBinder. When the user changes the slider the Frequency property will be updated, which in turn will set the Frequency on the FMRadio. The following layout includes an instance of the RadioBinder, a Slider (for controlling the frequency), a rotated ProgressBar (to indicate the current signal strength), and a TextBlock (to show the current frequency in text). The Slider has a two-way binding to the Frequency property on the RadioBinder, whereas the other controls only have one-way bindings.

```
<Grid x:Name="ContentGrid" Grid.Row="1">
    <local:RadioBinder x:Name="Radio"/>
    <Slider  Value="{Binding Frequency, Mode=TwoWay, ElementName=Radio}"
             Height="104" HorizontalAlignment="Left" Margin="0,58,0,0"
             VerticalAlignment="Top" Width="330" Minimum="90" Maximum="110" />
    <ProgressBar Value="{Binding SignalStrength, ElementName=Radio}" Height="83"
                 HorizontalAlignment="Left" Margin="336.5,57.5,0,0"
                 Name="SignalStrength"
                 VerticalAlignment="Top" Width="135" Maximum="10" SmallChange="0.01"
                 Minimum="0" RenderTransformOrigin="0.5,0.5" UseLayoutRounding="False">
        <ProgressBar.RenderTransform>
            <CompositeTransform Rotation="-90"/>
        </ProgressBar.RenderTransform>
    </ProgressBar>
    <TextBlock Height="40" HorizontalAlignment="Left" Margin="18,133,0,0"
               Text="{Binding Frequency, ElementName=Radio}"
               VerticalAlignment="Top" Width="83" />
    <Button Content="Start Tune" Height="72" HorizontalAlignment="Left"
            Margin="0,318,0,0" Name="TuneStartButton" VerticalAlignment="Top"
            Width="210"
            Click="TuneStartButton_Click" />
    <Button Content="Stop Tune" Height="72" HorizontalAlignment="Left"
            Margin="270,318,0,0" Name="StopTuneButton" VerticalAlignment="Top"
            Width="210" Click="StopTuneButton_Click" />
</Grid>
```

Code snippet from RadioPage.xaml

This layout also includes two buttons to start and stop tuning. There isn't built-in support within the FMRadio class for tuning, but you can do this by incrementing the Frequency.

Available for download on Wrox.com

```
DispatcherTimer timer;
private void TuneStartButton_Click(object sender, RoutedEventArgs e){
    if (timer == null){
        timer = new DispatcherTimer();
        timer.Interval = new TimeSpan(0, 0, 2);
        timer.Tick += new EventHandler(timer_Tick);
    }
    timer.Start();
}

void timer_Tick(object sender, EventArgs e){
    Radio.Frequency += 0.1;
    if (Radio.Frequency == 105){
        Radio.Frequency = 89;
    }
}

private void StopTuneButton_Click(object sender, RoutedEventArgs e){
    timer.Stop();
}
```

Code snippet from RadioPage.xaml.cs

There are a couple of points that you should be aware of with regard to working with the radio. Firstly, if the application doesn't have the correct capabilities assigned to it (details in Chapter 20), attempting to access the radio with throw an exception. The second point is that you can't debug an application that uses the radio on a real device while Zune is running (or in fact any of the media library services).

SUMMARY

Windows Phone applications can make use of both audio and video playback. You've seen how you can manipulate the `MediaElement` and how you can overlay multiple sounds using the `SoundEffect` and `SoundEffectInstance` classes. The built-in microphone also adds an extra dimension to applications, allowing them to capture audio that can be stored and played back on the device. Alternatively, the recorded audio could be sent to a remote server for conversion to text.

12

Where Am I? Finding Your Way

WHAT'S IN THIS CHAPTER

- Determining the current location
- Simulating location in the Windows Phone emulator
- Working with the Bing Maps control

Over the last couple of years more and more devices have a built-in GPS device. Unfortunately, this typically doesn't work well when you are located in an office or inside shopping. In these scenarios the GPS either takes a long period to resolve a signal or is blocked by the surrounding buildings. Luckily, in these cases, triangulation, based on available Wi-Fi networks or cellphone towers, will give a good approximation to your location.

In this chapter, you will learn how to connect to and work with the Windows Phone Location Service, which hides the complexities of communicating with GPS devices and deciding when to perform Wi-Fi or cellular tower–based triangulation behind an easy-to-use interface. This can be used not only to retrieve the user's current location, but also to detect when they are on the move. You will also see how easily you can integrate a map into your application in order to display geographically based data.

GEO-LOCATION

In order to protect the user's privacy, Windows Phone takes a consent-based approach to location. This means that in order for an application to access location information, the user will have to actively agree to the application's request. The first time the application attempts to access location information, it must display a prompt requesting that the user consent to the use of location information. For example Figure 12-1 illustrates the consent dialog for Bing search (which is displayed the first time you press the Search button).

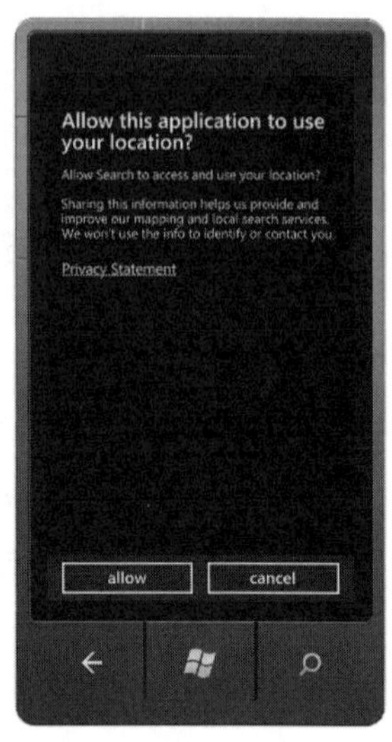

FIGURE 12-1

The Windows Phone Location Service provides access to raw information about the location of the device. The quality of a location update can vary depending on whether it is provided by the built-in GPS or is a result of cellphone tower or Wi-Fi triangulation. By combining information from multiple sources, Windows Phone is able to optimize the location information supplied to your application, improving accuracy when requested, or reducing power consumption when lower-accuracy location information is adequate for an application's needs. You should be aware that the accuracy of the location information available can be affected by several factors. These include the number of cell towers that are within range, whether there are known Wi-Fi networks in range, and whether the GPS signal is being interfered with by buildings.

GeoCoordinateWatcher

Accessing the Location Service on Windows Phone is via an instance of the `GeoCoordinateWatcher` class. This class is located in the System.Device.dll, which you will need to reference in your Windows Phone application. The `GeoCoordinateWatcher` provides a wrapper around the Location Service, watching for any changes in state of the service (as listed by the `GeoPositionStatus` enumeration) or changes in the location of the device. It also exposes methods to Start and Stop the Location Service. Initially, the Location Service starts in the NoData state. Once `Start` is called, the service enters the Initializing state during which it attempts to access the current location information from the device hardware. If this succeeds, the state changes to Ready, which means that location information is available to be read. Alternatively, if the user has disabled the location service on the device, the state will change to Disabled:

Available for download on Wrox.com

```
GeoCoordinateWatcher geowatcher;
private void PhoneApplicationPage_Loaded(object sender, RoutedEventArgs e){
    geowatcher = new GeoCoordinateWatcher();
    geowatcher.StatusChanged += geowatcher_StatusChanged;
}

private void StartButton_Click(object sender, RoutedEventArgs e){
    geowatcher.Start();
}

private void StopButton_Click(object sender, RoutedEventArgs e){
    geowatcher.Stop();
}

private void geowatcher_StatusChanged(object sender,
                                    GeoPositionStatusChangedEventArgs e){
    this.Dispatcher.BeginInvoke(() => {
```

```
                this.StatusText.Text = e.Status.ToString();
            });
}
```

Code snippet from MainPage.xaml.cs

If the Location Service enters the Disabled state, you can check the `Permission` property of the `GeoCoordinateWatcher` instance. This will return Granted, Denied, or Unknown (the values of the `GeoPositionPermission` enumeration), depending on whether the user has given permission for the application to access his or her position. You can use this value to adjust the way your application behaves.

Location (Latitude, Longitude, and Altitude)

Once the Location Service is in the Ready state, the most recent location update can be accessed via the `GeoCoordinateWatcher`'s `Position` property. Alternatively, an event handler can be attached to the `PositionChanged` event, which will be invoked whenever the device detects a change in position. The current location is represented via latitude, longitude, and altitude values.

In the following code snippet, an event handler is attached to the `PositionChanged` event. Since the `GeoCoordinateWatcher` operates on a background thread, it is important to remember that if you want to update any visual elements from within the `PositionChanged` event handler, you will need to cross back to the UI thread through the use of the dispatcher:

Available for download on Wrox.com

```
GeoCoordinateWatcher geowatcher;
private void PhoneApplicationPage_Loaded(object sender, RoutedEventArgs e){
    geowatcher = new GeoCoordinateWatcher();
    geowatcher.PositionChanged += geowatcher_PositionChanged;
}

private void geowatcher_PositionChanged(object sender,
                                    GeoPositionChangedEventArgs<GeoCoordinate> e){
    var locationText = PositionString(e.Position.Location);

    this.Dispatcher.BeginInvoke(() => this.GeoLocationText.Text = locationText);
}

private string PositionString(GeoCoordinate position){
    return "Latitude: " + position.Latitude.ToString() + Environment.NewLine +
           "Longitude: " + position.Longitude.ToString() + Environment.NewLine +
           "Altitude: " + position.Altitude.ToString();
}
```

Code snippet from MainPage.xaml.cs

Handling the `PositionChanged` event can be useful if your application requires constant updates of the device's location. In fact, this is the most efficient technique, as the `GeoCoordinateWatcher` will only interrupt you when a change has been detected. On the other hand, if you only require a single location update, it can be overkill. In this scenario, the `GeoCoordinateWatcher` class provides the `Position` property, which can be queried as follows:

Available for download on Wrox.com

```
private void ReadLocationButton_Click(object sender, RoutedEventArgs e){
    MessageBox.Show(PositionString(geowatcher.Position.Location));
}
```

Code snippet from MainPage.xaml.cs

The `Position` property returns a `GeoPosition<GeoCoordinate>` object, which, in turn, has both a `Location` property (seen in this code snippet) and a `Timestamp` that indicates the device time when the location information was recorded. This is particularly useful as this information may not change for an extended period if the GPS signal is lost or the cell network becomes unreachable.

Heading (Course and Speed)

In addition to tracking the geo-position of the device, the location service also provides the current course and speed the device is moving at:

Available for download on Wrox.com

```
private string PositionString(GeoCoordinate position){
    return "Latitude: " + position.Latitude.ToString() + Environment.NewLine +
           "Longitude: " + position.Longitude.ToString() + Environment.NewLine +
           "Altitude: " + position.Altitude.ToString() + Environment.NewLine +
           "Course: " + position.Course.ToString() + Environment.NewLine +
           "Speed: " + position.Speed.ToString();
}
```

Code snippet from MainPage.xaml.cs

Immediate Location

In some cases, you will want to ensure that the current device location is available before the application proceeds. The `GeoCoordinateWatcher` instance has a `TryStart` method, which you can think of as a synchronous way to start the Location Service. It takes a `TimeSpan` argument, which determines how long the `TryStart` method should wait for the Location Service to start. If you want to wait indefinitely, which is not recommended, you can set this argument to `TimeSpan.MaxValue`.

Available for download on Wrox.com

```
private void FindMeNowButton_Click(object sender, RoutedEventArgs e){
    var t = new Thread((ThreadStart)(() =>
            {
                geowatcher.TryStart(true, TimeSpan.FromSeconds(30));
                string locationText;
                if (geowatcher.Status ==
                    System.Device.Location.GeoPositionStatus.Ready){
                    var pos = geowatcher.Position;
                    locationText = PositionString(pos.Location);
                }
                else{
                    locationText = "Unable to retrieve location";
```

```
                }

                this.Dispatcher.BeginInvoke(() =>
                                this.GeoLocationText.Text = locationText);
            }));
        t.Start();
    }
```

Code snippet from MainPage.xaml.cs

In this example, a new thread is created to start up the `GeoCoordinateWatcher` and then report the current location of the device. As a general rule of thumb, you should never block the UI thread because this is guaranteed to lead to a poor experience for your users, since animations and other effects temporarily freeze. You should always look for ways to perform long-running actions on a background thread. This is particularly important if they are part of a complex or time-consuming startup routine.

MovementThreshold and DesiredAccuracy

The Windows Phone Location Service draws on data from the GPS, Wi-Fi, and cellular networks. With each of these sources, there is a trade-off between accuracy, the time taken to resolve the location, and power consumption. Depending on your application, you can decide whether to accept lower accuracy (`GeoPositionAccuracy.Default`) location updates. In most cases, this will result in the location of the device being made available much quicker than if you'd have specified high accuracy (`GeoPositionAccuracy.High`). The default `DesiredAccuracy` for the `GeoCoordinateWatcher` is low. In order to increase the accuracy, you should specify `GeoPositionAccuracy.High` as a parameter when creating the `GeoCoordinateWatcher` instance.

Once the initial location of the device has been returned, you may decide to improve the accuracy of the reading to progressively return a more accurate value. To do this, you need to make sure you stop and dispose of the existing `GeoCoordinateWatcher` instance before creating a new instance specifying a higher accuracy.

The other property that you may want to adjust is the `MovementThreshold`. This is the minimum distance that the device has to move before the next `PositionChanged` event will fire. Once the initial `PositionChanged` event has been raised to indicate that the position of the device has been acquired, the event will not be raised again until the device has traveled a minimum of the distance specified by the `MovementThreshold` property in meters.

If the `MovementThreshold` property is not specified, it will retain its default value of 0, which means that any position change will cause the `PositionChanged` event to be raised. Owing to inaccuracies in the different techniques and hardware used to capture location information, this can result in the `PositionChanged` event being raised because of noise in the location information, rather than an actual change in device location. It is therefore recommended that you set this property to the minimal change in location that your application is prepared to deal with. If your application is using the returned location information to detect which city the device is currently located in, it is hardly important for you to be notified when the device has moved 4 meters to the east, for example.

Accuracy

The GeoCoordinate value that is returned by the location service also indicates the accuracy of the data in both the horizontal (i.e., latitude and longitude) and vertical (i.e., altitude) directions. The following code updates the PositionString method to report both the HorizontalAccuracy and VerticalAccuracy values.

Available for download on Wrox.com

```
private string PositionString(GeoCoordinate position){
    return "Latitude: " + position.Latitude.ToString() + Environment.NewLine +
           "Longitude: " + position.Longitude.ToString() + Environment.NewLine +
           "Lat-Long Accuracy: " + position.HorizontalAccuracy.ToString() +
                                   Environment.NewLine +
           "Altitude: " + position.Altitude.ToString() + Environment.NewLine +
           "Altitude Accuracy: " + position.VerticalAccuracy.ToString() +
                                   Environment.NewLine +
           "Course: " + position.Course.ToString() + Environment.NewLine +
           "Speed: " + position.Speed.ToString();
}
```

Code snippet from MainPage.xaml.cs

IGeoPositionWatcher

As with the accelerometer, the Windows Phone emulator doesn't simulate the Location Service. Although you can instantiate the class, it will always indicate the inability to obtain location updates. However, the `System.Device.Location` assembly does provide an interface called `IGeoPositionWatcher<GeoCoordinate>` that is designed to serve a similar purpose to the `IAccelerometer` interface you developed in Chapter 9. Using the `IGeoPositionWatcher<GeoCoordinate>` interface will allow you to provide an alternative geo-location watcher implementation that can be used within the emulator to test your application. This section covers two such implementations.

Mock: Windows 7 Sensor API

Windows 7 included a standard set of interfaces for interacting with sensors. There are still very few laptop, or desktop, computers that include a GPS or alternative hardware for determining the location of the device. Rafael Rivera and Long Zheng came up with Geosense for Windows library (`www.geosenseforwindows.com/`), which makes use of a combination of Wi-Fi, cell tower triangulation, and reverse IP lookup. Once installed, Geosense for Windows will determine the location of the computer and report it to Windows 7 so that it can be queried by any other application via the standard sensor interface.

In order to use this feature to supply location information to the Windows Phone emulator, you need to create a simple console application that will expose a WCF Service. This service will connect to the Windows 7 sensor interface to retrieve the current location information and report it to the Windows Phone application when it calls the WCF Service to retrieve the latest location information. This is very similar in design to what you did with the Wiimote accelerometer data within Chapter 9.

Start by creating a new application using the Console Application template. Next, add a new WCF Service item called *MockLocationService* to the project. Download the Windows API Code Pack for Microsoft .NET Framework (`http://code.msdn.microsoft.com/WindowsAPICodePack`), and add a reference to both Microsoft.WindowsAPICodePack.dll and Microsoft.WindowsAPICodePack.Sensors.dll. The following code illustrates how you can host the WCF Service within the console application and how you can initialize the Windows 7 sensor API to start receiving location information:

```
class Program{
    public static LocationGpsSensor Sensor;
    public static MockGeoLocation CurrentLocation;

    static void Main(string[] args){
        if(!Initialize()) return;

        // Create the ServiceHost.
        using (ServiceHost host = new ServiceHost(typeof(MockLocationService))){
            host.Open();

            Console.WriteLine("The service is ready");
            Console.WriteLine("Press <Enter> to stop the service.");
            Console.ReadLine();

            // Close the ServiceHost.
            host.Close();
        }
    }

    public static bool Initialize(){
        try{
            var sensorsByTypeId = SensorManager.GetSensorsByTypeId<LocationGpsSensor>();
            if (sensorsByTypeId != null){
                var sensor = sensorsByTypeId[0] as LocationGpsSensor;
                if (sensor.State == SensorState.AccessDenied){
                    SensorList<Sensor> sensors = new SensorList<Sensor>();
                    sensors.Add(sensorsByTypeId[0]);
                    SensorManager.RequestPermission(IntPtr.Zero, false, sensors);
                    sensor.UpdateData();
                }
                else{
                    sensor.UpdateData();
                }
                CurrentLocation = new MockGeoLocation(){
                                          Address1 = sensor.Civic_Address1,
                                          Address2 = sensor.Civic_Address2,
                                          City = sensor.Civic_City,
                                          PostCode = sensor.Civic_PostCode,
                                          State = sensor.Civic_State,
                                          Region = sensor.Civic_Region,
                                          Latitude = sensor.Geo_Latitude,
                                          Longitude = sensor.Geo_Longitude,
                                          Altitude = sensor.Geo_Altitude,
```

```
                                        Speed = 0.0,
                                        Course = 0.0
                                    };
                Sensor = sensor;
            }

            return true;
        }
        catch (Exception){
            return false;
        }
    }
}
```

Code snippet from Program.cs

What's not shown in this code snippet is the `LocationGpsSensor` type, which you supply when calling `GetSensorsByTypeId` and the MockGeoLocation class, which is the data structure that is going to be returned by the service.

```
[SensorDescription("{ED4CA589-327A-4FF9-A560-91DA4B48275E}")]
public class LocationGpsSensor : Sensor{
    public string Civic_Address1 { get; private set; }
    public string Civic_Address2 { get; private set; }
    public string Civic_City { get; private set; }
    public string Civic_PostCode { get; private set; }
    public string Civic_Region { get; private set; }
    public string Civic_State { get; private set; }

    public double Geo_Latitude { get; private set; }
    public double Geo_Longitude { get; private set; }
    public double Geo_Altitude { get; private set; }

    public LocationGpsSensor(){
        base.DataReportChanged += this.GeosenseSensor_DataReportChanged;
    }

    private void GeosenseSensor_DataReportChanged(Sensor sender, EventArgs e){
        var locationData = base.DataReport.Values[
                  SensorPropertyKeys.SENSOR_DATA_TYPE_LATITUDE_DEGREES.FormatId];
        this.Geo_Latitude = (double)locationData[0];
        this.Geo_Longitude = (double)locationData[1];
        this.Geo_Altitude = (double)locationData[2];
        this.Civic_Address1 = (string)locationData[4];
        this.Civic_Address2 = (string)locationData[5];
        this.Civic_City = (string)locationData[6];
        this.Civic_Region = (string)locationData[7];
        this.Civic_State = (string)locationData[8];
        this.Civic_PostCode = (string)locationData[9];
    }
}
```

Code snippet from LocationGpsSensor.cs

```
public class MockGeoLocation
{
    public string Address1 { get;  set; }
    public string Address2 { get;  set; }
    public string City { get;  set; }
    public string PostCode { get;  set; }
    public string Region { get;  set; }
    public string State { get;  set; }

    public double Latitude { get;  set; }
    public double Longitude { get;  set; }
    public double Altitude { get; set; }

    public double  Course { get; set; }
    public double  Speed { get; set; }
}
```

Code snippet from MockGeoLocation.cs

You also need to implement the MockLocationService WCF Service. As you can see, this is relatively simple as it just returns the static `CurrentLocation` property of the program that's hosting the service:

```
[ServiceContract]
public interface IMockLocationService{
    [OperationContract]
    MockGeoLocation CurrentLocation();
}
```

Code snippet from IMockLocationService.cs

```
public class MockLocationService : IMockLocationService{
    public MockGeoLocation CurrentLocation(){
        return Program.CurrentLocation;
    }
}
```

Code snippet from MockLocationService.cs

After building and running this service (you will need to be running Visual Studio as Administrator for the service to start successfully), you can reference this service from your Windows Phone application by right-clicking on the project in the Solution Explorer window and selecting "Add Service Reference." Once added, you will need to implement the `IGeoPositionWatcher<GeoCoordinate>` interface. The Win7GeoPositionWatcher implementation periodically calls the WCF Service to retrieve the latest location information:

```
public class Win7GeoPositionWatcher : IGeoPositionWatcher<GeoCoordinate>{
   public event EventHandler<GeoPositionChangedEventArgs<GeoCoordinate>>
                                PositionChanged;
   public event EventHandler<GeoPositionStatusChangedEventArgs> StatusChanged;

   private Windows7LocationAPI.MockLocationServiceClient mockClient;

   private AutoResetEvent TryStartWait = new AutoResetEvent(false);
   private Timer timer;

   public Win7GeoPositionWatcher(){
       mockClient = new Windows7LocationAPI.MockLocationServiceClient();
       mockClient.CurrentLocationCompleted += mockClient_CurrentLocationCompleted;
       timer = new Timer(TimerCallback,null,Timeout.Infinite,Timeout.Infinite);
       Status = GeoPositionStatus.NoData;
   }

   public GeoPositionPermission Permission{
       get { return GeoPositionPermission.Granted; }
   }

   public GeoPosition<GeoCoordinate> Position { get; private set; }

   public GeoPositionStatus Status { get; private set; }

   private void TimerCallback(object state){
       timer.Change(Timeout.Infinite, Timeout.Infinite);
       mockClient.CurrentLocationAsync();
   }

   public bool TryStart(bool suppressPermissionPrompt, TimeSpan timeout){
       this.Start(suppressPermissionPrompt);
       return TryStartWait.WaitOne(timeout);
   }

   public void Start(bool suppressPermissionPrompt){
       Start();
   }

   public void Start(){
       if (Status == GeoPositionStatus.Disabled){
           Status = GeoPositionStatus.Initializing;
           RaiseStatusChanged();
       }

       mockClient.CurrentLocationAsync();
   }

   public void Stop(){
       this.Status = GeoPositionStatus.NoData;
       RaiseStatusChanged();
       this.Position = null;
   }

   private void mockClient_CurrentLocationCompleted(object sender,
                 Windows7LocationAPI.CurrentLocationCompletedEventArgs e){
```

```
            if (e.Error == null){
                if (Status!= GeoPositionStatus.Ready){
                    Status = GeoPositionStatus.Ready;
                    RaiseStatusChanged();
                }
                var currentLocation = e.Result;
                this.Position = new GeoPosition<GeoCoordinate>(DateTimeOffset.Now,
                                           new GeoCoordinate(currentLocation.Latitude,
                                                             currentLocation.Longitude,
                                                             currentLocation.Altitude));
                RaisePositionChanged();
                timer.Change(0, 100);
            }
            else{
                Status = GeoPositionStatus.Disabled;
                RaiseStatusChanged();
            }
            TryStartWait.Set();
        }

        private void RaiseStatusChanged(){
            if (StatusChanged != null){
                StatusChanged(this, new GeoPositionStatusChangedEventArgs(this.Status));
            }
        }

        private void RaisePositionChanged(){
            if (PositionChanged != null){
                PositionChanged(this,
                   new GeoPositionChangedEventArgs<GeoCoordinate>(this.Position));
            }
        }
    }
```

Code snippet from Win7GeoPositionWatcher.cs

You will notice in this implementation that both the `StatusChanged` and `PositionChanged` events are raised at appropriate times. Within a mock GeoPositionWatcher implementation, it is important to simulate the way a real Windows Phone behaves as much as possible so that you can accurately test the behavior of your application. To make use of the `Win7GeoPositionWatcher` class, all you need to do is determine whether the application is running on a real device or the simulator, switching between an instance of the `GeoCoordinateWatcher` and `Win7GeoPositionWatcher` classes as required:

Available for download on Wrox.com

```
public static IGeoPositionWatcher<GeoCoordinate> PickDeviceGeoPositionWatcher(){
    if (Microsoft.Devices.Environment.DeviceType == DeviceType.Device){
        return new GeoCoordinateWatcher();
    }
    else{
        return new Win7GeoPositionWatcher();
    }
}
```

Code snippet from Utilities.cs

You will also need to modify your Windows Phone application to update your code to use `IGeoPositionWatcher<GeoCoordinate>` instead of `GeoCoordinateWatcher`.

```
IGeoPositionWatcher<GeoCoordinate> geowatcher;

private void PhoneApplicationPage_Loaded(object sender, RoutedEventArgs e){
    geowatcher = Utilities.PickDeviceGeoPositionWatcher();
    geowatcher.StatusChanged += geowatcher_StatusChanged;
    geowatcher.PositionChanged += geowatcher_PositionChanged;
}
```

Code snippet from MainPage.xaml.cs

While Win7GeoPositionWatcher provides a realistic approach to simulating the Windows Phone application's location, it will typically report a single location. This is, of course, useless unless you happen to be working on the bus or train, in which case, the location reported by the host Windows 7 computer may update from time to time. To improve the Win7GeoPositionWatcher, you can extend the MockLocationService WCF Service to allow the course and speed to be manually adjusted at runtime. The new speed and course can then be used to adjust the location recorded by the service over time in order to simulate the device moving. Start by extending the WCF Service to include an `UpdateCourseAndSpeed` method:

```
[ServiceContract]
public interface IMockLocationService{
    [OperationContract]
    MockGeoLocation CurrentLocation();

    [OperationContract]
    void UpdateCourseAndSpeed(double course, double speed);
}
```

Code snippet from IMockLocationService.cs

```
public class MockLocationService : IMockLocationService{
    public MockGeoLocation CurrentLocation(){
        return Program.CurrentLocation;
    }

    public void UpdateCourseAndSpeed(double course, double speed){
        Program.CurrentLocation.Course = course;
        Program.CurrentLocation.Speed = speed;
    }
}
```

Code snippet from MockLocationService.cs

You also need to update the host console application to periodically update the location based on the course and speed specified. In the following code, a `Timer` raises an event after `TimeoutInMilliseconds`. If the device is moving (i.e., `Speed` is not zero), a new location is calculated. The formula for calculating the new location uses the dimensions of the Earth to determine the new latitude and longitude based on how far the device traveled between timer events (this and other geo-location formulas can be found at `www.movable-type.co.uk/scripts/latlong.html`):

```
private const int TimeoutInMilliseconds = 100;
private const double EarthRadiusInMeters = 6378.1 * 1000.0;

private static System.Timers.Timer timer;

static Program(){
    CurrentLocation = new MockGeoLocation();

    timer = new System.Timers.Timer(TimeoutInMilliseconds);
    timer.Elapsed += new System.Timers.ElapsedEventHandler(timer_Elapsed);
    timer.Enabled = true;
}

static void timer_Elapsed(object sender, System.Timers.ElapsedEventArgs e){
    if(Sensor == null) return;
    timer.Enabled = false;
    if (CurrentLocation.Speed != 0){
        var lat1 = CurrentLocation.Latitude*(Math.PI/360.0);
        var lon1 = CurrentLocation.Longitude * (Math.PI / 360.0);
        var d = CurrentLocation.Speed * ((double)TimeoutInMilliseconds) / 1000.0;
        var R = EarthRadiusInMeters;
        var brng = CurrentLocation.Course;

        var lat2 = Math.Asin(Math.Sin(lat1) * Math.Cos(d / R) +
                             Math.Cos(lat1) * Math.Sin(d / R) * Math.Cos(brng));
        var lon2 = lon1+Math.Atan2(Math.Sin(brng) *
                                   Math.Sin(d / R) * Math.Cos(lat1),
                                   Math.Cos(d / R) -
                                   Math.Sin(lat1) * Math.Sin(lat2));

        CurrentLocation.Latitude = lat2 * (360.0/Math.PI);
        CurrentLocation.Longitude = lon2 * (360.0/Math.PI);
    }
    else{
        var sensor = Sensor;
        CurrentLocation = new MockGeoLocation(){
              Address1 = sensor.Civic_Address1,
              Address2 = sensor.Civic_Address2,
              City = sensor.Civic_City,
              PostCode = sensor.Civic_PostCode,
              State = sensor.Civic_State,
              Region = sensor.Civic_Region,
              Latitude = sensor.Geo_Latitude,
              Longitude = sensor.Geo_Longitude,
              Altitude = sensor.Geo_Altitude,
              Speed = 0.0,
              Course = 0.0
```

```
            };
        }
        timer.Enabled = true;
    }
```

Code snippet from Program.cs

The next step is to create a simple application, based on the WPF Application template. This will be used to adjust the `Course` and `Speed` values used by the WCF Service. Within the application, add a reference to the MockLocationService WCF Service and then add two sliders and a button. The first slider will determine the `Course`, so the `Maximum` property should be `360`. The second slider will determine the `Speed` that the device is traveling in meters per second, so a `Maximum` of `100` is suitable. In the event handler for the button, call the `UpdateCourseAndSpeed` method on an instance of the MockLocationServiceClient proxy, passing in the current values of the two sliders:

```
namespace GeoChanger{
    public partial class MainWindow : Window{
        public MainWindow(){
            InitializeComponent();
        }

        private void ChangeButton_Click(object sender, RoutedEventArgs e){
            Windows7LocationAPI.MockLocationServiceClient client =
                        new Windows7LocationAPI.MockLocationServiceClient();
            client.UpdateCourseAndSpeed(this.CourseSlider.Value,
                                        this.SpeedSlider.Value);
        }
    }
}
```

Code snippet from MainWindow.xaml.cs

Initially the location returned by the MockLocationService will be the actual location of the computer, as returned by the Windows 7 API. However, once, the course and speed have been set using the WPF application (shown in Figure 12-2). The position will be continually be updated with the course and speed set by the WPF application.

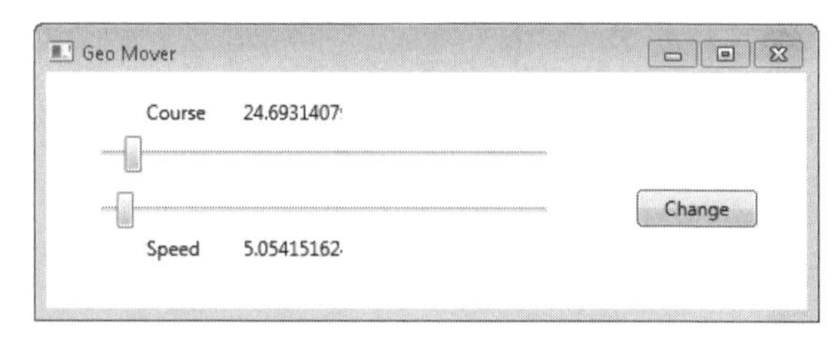

FIGURE 12-2

Mock: Time–Location

An alternative way to simulate the Windows Phone Location Service is to implement the `IGeoPositionWatcher<GeoCoordinate>` interface using a predefined set of locations and timings. The constructor for the `TimerGeoPositionWatcher` class accepts several `GeoEvent` objects that specify a location, along with a `Timespan` that indicates the amount of time to remain at that location. After `Start` is called, the location of the first `GeoEvent` in the list becomes the current location reported by the `TimerGeoPositionWatcher`. A timer is then used to raise an event after the corresponding time-out, which causes the current location to be updated to the location of the next `GeoEvent`. Once the end of the list has been reached, the current location returns to the beginning of the list:

```
public class GeoEvent{
    public double Longitude { get; set; }
    public double Latitude { get; set; }
    public TimeSpan Timeout { get; set; }
}

public class TimerGeoPositionWatcher : IGeoPositionWatcher<GeoCoordinate>{
   public event EventHandler<GeoPositionChangedEventArgs<GeoCoordinate>>
                                PositionChanged;
   public event EventHandler<GeoPositionStatusChangedEventArgs> StatusChanged;

    private List<GeoEvent> events;
    private int currentEventId;

    private AutoResetEvent TryStartWait = new AutoResetEvent(false);
    private Timer timer;

    public TimerGeoPositionWatcher(IEnumerable<GeoEvent> events){
        this.events = new List<GeoEvent>(events);
        this.currentEventId = -1;
        this.timer = new Timer(TimerCallback, null, Timeout.Infinite, Timeout.Infinite);

        Status = GeoPositionStatus.Disabled;
        RaiseStatusChanged();
    }

    public GeoPositionPermission Permission{
            get { return GeoPositionPermission.Granted; }
    }

    public GeoPosition<GeoCoordinate> Position { get; private set; }

    public GeoPositionStatus Status { get; private set; }

    public bool TryStart(bool suppressPermissionPrompt, TimeSpan timeout){
        Start();
        return TryStartWait.WaitOne(timeout);
    }

    public void Start(bool suppressPermissionPrompt){
        Start();
    }

    public void Start(){
        if (Status == GeoPositionStatus.Disabled){
            Status = GeoPositionStatus.Initializing;
```

```
            RaiseStatusChanged();
        }

        NextGeoEvent();
        timer.Change(Current.Timeout, Current.Timeout);
    }

    public void Stop(){
        timer.Change(Timeout.Infinite, Timeout.Infinite);
        if (Status != GeoPositionStatus.Disabled){
            Status = GeoPositionStatus.Disabled;
            RaiseStatusChanged();
        }
    }

    private void TimerCallback(object state){
        if (Status == GeoPositionStatus.Initializing){
            Status = GeoPositionStatus.NoData;
            RaiseStatusChanged();
        }

        NextGeoEvent();
        timer.Change(Current.Timeout, Current.Timeout);
    }

    private GeoEvent Current{
        get{ return events[currentEventId % events.Count];}
    }

    private void NextGeoEvent(){
        // Move to the next GeoEvent
        currentEventId++;

        this.Position = new GeoPosition<GeoCoordinate>(DateTimeOffset.Now,
                             new GeoCoordinate(Current.Latitude,
                                               Current.Longitude, 0.0));
        if (Status != GeoPositionStatus.Ready) {
            Status = GeoPositionStatus.Ready;
            RaiseStatusChanged();
        }

        RaisePositionChanged();
    }

    private void RaiseStatusChanged(){
        if (StatusChanged != null){
            StatusChanged(this, new GeoPositionStatusChangedEventArgs(this.Status));
        }
    }

    private void RaisePositionChanged(){
        if (PositionChanged != null){
```

```
                PositionChanged(this,
                        new GeoPositionChangedEventArgs<GeoCoordinate>(this.Position));
            }
        }
    }
```

Code snippet from TimerGeoPositionWatcher.cs

As with the `Win7GeoPositionWatcher` class, to use the `TimerGeoPositionWatcher` class, you simply return an instance of it in place of the standard `GeoCoordinateWatcher`:

```
private static IGeoPositionWatcher<GeoCoordinate> PickDeviceGeoPositionWatcher(){
    if (Microsoft.Devices.Environment.DeviceType == DeviceType.Device){
        return new GeoCoordinateWatcher();
    }
    else{
        GeoEvent[] events = new GeoEvent[] {
new  GeoEvent { Latitude=-37.998152, Longitude=145.013596,
                Timeout=new TimeSpan(0,0,3) },
new  GeoEvent { Latitude=-37.998352, Longitude=145.023596,
                Timeout=new TimeSpan(0,0,4) },
new  GeoEvent { Latitude=-37.998552, Longitude=145.033596,
                Timeout=new TimeSpan(0,0,3) },
new  GeoEvent { Latitude=-37.998752, Longitude=145.043596,
                Timeout=new TimeSpan(0,0,7) },
new  GeoEvent { Latitude=-37.998952, Longitude=145.053596,
                Timeout=new TimeSpan(0,0,3) },
new  GeoEvent { Latitude=-37.999152, Longitude=145.063596,
                Timeout=new TimeSpan(0,0,7) },
new  GeoEvent { Latitude=-37.999352, Longitude=145.073596,
                Timeout=new TimeSpan(0,0,6) }
};
        return new TimerGeoPositionWatcher(events);
    }
}
```

Code snippet from Utilities.cs

BING MAPS

Often when you are working with location information, you will want to be able to display a map showing the current location of the user and other nearby points of interest. The Map control, originally designed for use within Silverlight applications on the Web, has been updated to work within a Windows Phone application to not only display maps but also to display pushpins and routing information. To get started all you need to do is to add a reference to the Microsoft.Phone.Controls.Maps.dll.

Map Design

The easiest way to work with the Map control is to use Blend to add a Map to your `PhoneApplicationPage`. Go to the Assets window in Blend, you will see that there are several new items such as Map, MapLayer and MapPolyline, as shown in Figure 12-3.

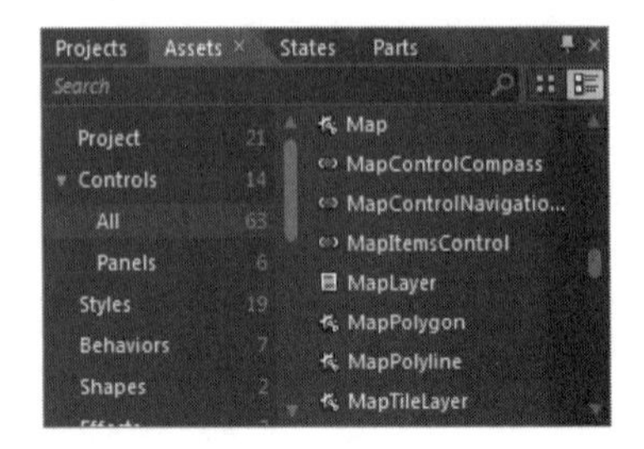

FIGURE 12-3

From the Assets window, drag the main `Map` control onto your page. You should see a rectangle appear where you dropped the control on the page. If you are connected to the Internet, you should gradually see a map of the world being displayed. This is not just a placeholder — it's the actual map that will be displayed when you run the application. If you want to change the center point and zoom of the map, you can do this by adjusting the `Center` and `ZoomLevel` properties of the control, as illustrated in Figure 12-4.

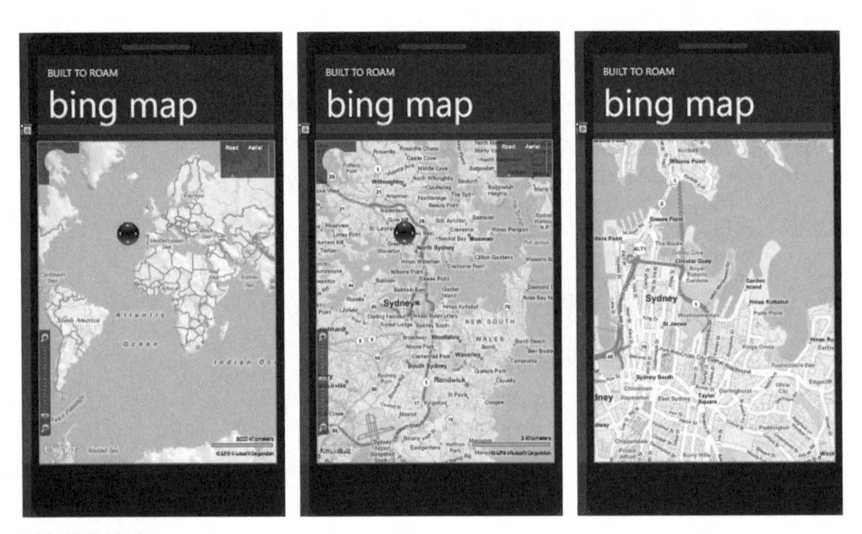

FIGURE 12-4

One of the downsides of reusing a control that was originally designed for desktop web applications that are primarily mouse-driven is that the default controls for adjusting the zoom and position of the map are just too small to be usable. The rightmost image of Figure 12-3 shows how you can remove these by setting the `ZoomBarVisibility` and `ScaleVisibility` properties to `Collapsed`.

Having hidden the built-in controls, you may want to provide your own buttons for controlling the map. If this is the case, you could consider using an Application Bar with an `Opacity` of less than 1. This would result in the buttons being overlaid on top of the map. Since the map can easily be scrolled by the user, the fact that the Application Bar partially obscures a small amount of the map is permissible.

Available for download on Wrox.com

XAML

```
<phone:PhoneApplicationPage.ApplicationBar>
    <shell:ApplicationBar Opacity="0.5">
        <shell:ApplicationBarIconButton IconUri="/zoomin.png"
                                        Text="Zoom In"
                                        Click="ZoomInButton_Click" />
        <shell:ApplicationBarIconButton IconUri="/zoomout.png"
                                        Text="Zoom Out"
                                        Click="ZoomOutButton_Click" />
    </shell:ApplicationBar>
</phone:PhoneApplicationPage.ApplicationBar>
```

Code snippet from MainPage.xaml

C#

```
private void ZoomInButton_Click(object sender, EventArgs e){
    BingMap.ZoomLevel++;
}

private void ZoomOutButton_Click(object sender, EventArgs e){
    BingMap.ZoomLevel--;
}
```

Code snippet from MainPage.xaml.cs

Map Credentials

In order to use Bing Maps within your application, you will require an application key, which can be obtained be signing up for an account at www.bingmapsportal.com. The following XAML snippet shows how you can assign your application key to the Bing Maps control you have added to your page. This is required; otherwise, an ugly notice is displayed across the middle of the map indicating that no valid credentials have been supplied.

Available for download on Wrox.com

```
<Grid x:Name="ContentPanel" Grid.Row="1" Margin="12,0,12,0">
    <Grid.Resources>
        <Microsoft_Phone_Controls_Maps:ApplicationIdCredentialsProvider
            ApplicationId="<AppID>" x:Key="MapCredentials" />
    </Grid.Resources>
    <Microsoft_Phone_Controls_Maps:Map x:Name="BingMap" ZoomLevel="14"
                      Center="-33.866567,151.219254,0"
                      ScaleVisibility="Collapsed"
                      CredentialsProvider="{StaticResource MapCredentials}" />

</Grid>
```

Code snippet from MapPage.xaml

Points of Interest and Lines

Adding points of interest to a map can be done in several ways. The easiest is to make use of the Pushpin control that comes with the SDK. From the Assets window in Blend, you can drag a pushpin onto the map and position it where you want it. However, you need to be aware that this is simply controlling the position of the pushpin relative to the visible portion of the Map control (i.e., by setting the Margin property), rather than relative to the Earth (i.e., a latitude and longitude). In most cases, you will want to position the pushpin by supplying a latitude and longitude. This will ensure that the pushpin moves with the map as the user scrolls, or zooms in and out. To do this, make sure that you reset the Margin property on the Pushpin and then specify the Location property as the Latitude and Longitude where the pushpin should be anchored:

```
<Microsoft_Phone_Controls_Maps:Map x:Name="BingMap" ZoomLevel="14"
            Center="-33.866567,151.219254,0"
            ScaleVisibility="Collapsed"
            CredentialsProvider="{StaticResource MapCredentials}">
    <Microsoft_Phone_Controls_Maps:Pushpin Location="-33.866567,151.219254"
                                           Content="5 />
</Microsoft_Phone_Controls_Maps:Map>
```

The Pushpin is somewhat limited in that it can only really be used to display two or three characters. Its primary use is to indicate the number of items of interest there are at a given point on the map, or to uniquely label each location in a set of results. For example, in the previous code snippet, the Content property was set to 5, indicating that there are five items of interest.

Before you go on to look at adding custom shapes and lines to the map, let's take a look at how you would add multiple pushpins to the map. Rather than having to write code that explicitly creates each pushpin and adds it to the map, you can use data-binding to automatically add and remove pushpins. In order to use data-binding, you need a view model to bind to. This will contain the current state of the map (i.e., center and zoom level) and the details of each pin that is to be placed on the map (i.e., position and content). Each pin will be represented by an instance of the following PinData class:

Available for download on Wrox.com

```
public class PinData{
    public GeoCoordinate PinLocation { get; set; }
    public string[] Data{get;set;}

    public string PinContent {
        get{
            if (Data == null) return "0";
            return Data.Length.ToString();
        }
    }
}
```

Code snippet from PinData.cs

The MapData class holds the current state of the map and pins. Note that although it is not required, all of the properties have an initial state. This provides a much better designer experience when working in Blend.

```
public class MapData:INotifyPropertyChanged{
    public event PropertyChangedEventHandler  PropertyChanged;

    private void RaisedPropertyChanged(string propertyName){
        if(PropertyChanged!=null){
            PropertyChanged(this,new PropertyChangedEventArgs(propertyName));
        }
    }

    private GeoCoordinate mapCenter = new GeoCoordinate(-33.866567, 151.219254);
    public GeoCoordinate MapCenter
        get{ return this.mapCenter; }
        set{
            if (this.mapCenter == value) return;
            this.mapCenter = value;
            this.RaisedPropertyChanged("MapCenter");
        }
    }

    private double zoom = 14.0;
    public double Zoom{
        get{ return this.zoom; }
        set{
            if (this.zoom == value) return;
            this.zoom = value;
            this.RaisedPropertyChanged("Zoom");
        }
    }

    private ObservableCollection<PinData> pins = new ObservableCollection<PinData>() {
            new PinData(){PinLocation= new GeoCoordinate(-33.866567, 151.219254),
                          Data=new string[]{"Mary", "Bob","Joe","Frank",
                                            "Beth","Nick","Jeff","Alex"} },
            new PinData(){PinLocation= new GeoCoordinate(-33.876567, 151.219254) ,
                          Data=new string[]{"Frank", "Beth"} }};
    public ObservableCollection<PinData> Pins{
        get{ return pins; }
    }

    private PinData selectedPin;
    public PinData SelectedPin{
        get{ return this.selectedPin; }
        set{
            if (this.selectedPin == value) return;
            this.selectedPin = value;
            this.RaisedPropertyChanged("SelectedPin");
        }
    }
}
```

Code snippet from MapData.cs

An instance of the `MapData` class will be the `DataContext` for the `PhoneApplicationPage` that the `Map` is placed on. In Blend, select the PhoneApplicationPage node in the "Objects and Timeline" window. From the Properties window, find the DataContext property and click the New button. Find "MapData" in the "Select Object" window (shown in Figure 12-5) and click OK.

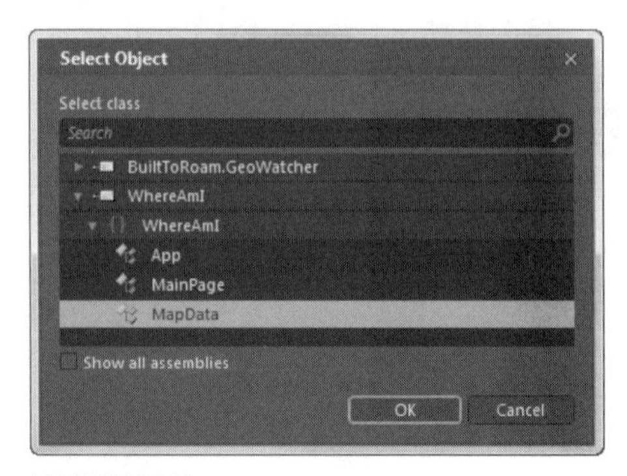

FIGURE 12-5

This process has created a new instance of the MapData class and configured it as the DataContext for the page. If you look at the XAML for the page, you will notice that the DataContext property has a nested MapData element:

Available for download on Wrox.com

```
<navigation:PhoneApplicationPage.DataContext>
    <local:MapData/>
</navigation:PhoneApplicationPage.DataContext>
```

Code snippet from MapPage.xaml

When an instance of this page is created, an instance of the MapData class is created and assigned to the DataContext property of the page. This propagates down to all elements of the page, meaning that the data is accessible from the Map control and its elements. Select the Map control and locate the Center and ZoomLevel properties. To bind these properties to the data held by the MapData object, click on the colored square to the right of the Property Value textbox and select "Data Binding" (left image of Figure 12-5). In the "Create Data Binding" window (right image of Figure 12-6), select the appropriate property from the Fields list. Make sure that the Binding direction is set to TwoWay. If the Binding direction property isn't visible, you will need to click the down-arrow to expand out the additional data-binding options area.

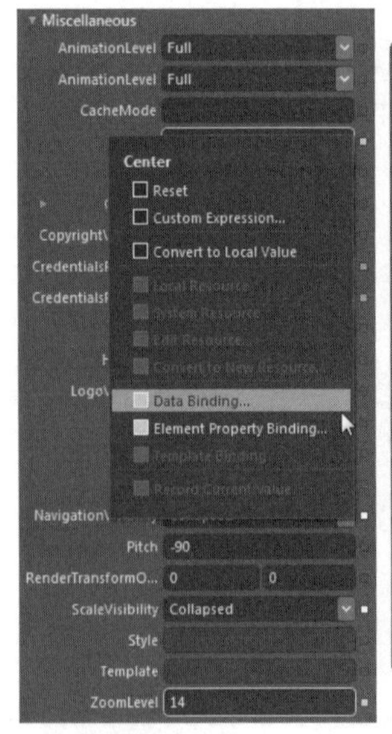

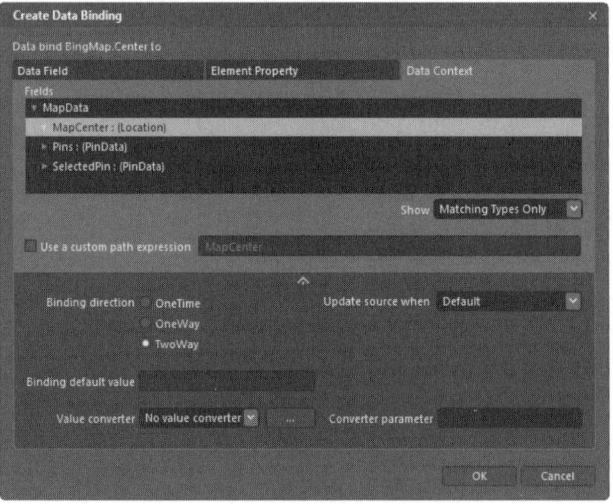

FIGURE 12-6

The next thing to do is to add a `Pushpin` for each of the locations specified by the `Pins` property of the `MapData` class. Rather than looping through these elements in code, you can use the `MapItemsControl`. This is similar to a `ListBox` control in that it expects a list of items. The difference is that the `MapItemsControl` has been designed to work with items that are to be positioned relative to the map they are displayed upon. Start by adding a `MapItemsControl` to the `Map` control. Locate the `ItemsSource` property and set up data binding to the `Pins` property of the `MapData` (follow the same process as used for the `Center` and `ZoomLevel` properties).

Initially, you won't see anything appear on the map. This is because you haven't specified what each of the pins will look like. To do this, right-click on "MapItemsControl" in the "Objects and Timeline" window, and select Edit Additional Templates ➪ Edit Generated Items (ItemTemplate) ➪ Create Empty. Delete the `Grid` that is added to the template by default, and add a `Pushpin`. Select the `Content` property for the `Pushpin` and set the data binding to the `PinContent` property (note that this is a property of the `PinData` class, which makes sense as you are modifying how each item in the list will appear). To set the position of the Pushpin you need to set the `MapLayer.Position` property in XAML. This is an attached property which is why it doesn't appear in the Blend designer. Your XAML should look similar to the following where the `MapLayer.Position` property is data bound to the `PinLocation` property.

Available for download on Wrox.com

```
<DataTemplate x:Key="PinTemplate">
    <Microsoft_Phone_Controls_Maps:Pushpin Content="{Binding PinContent}"
                          FontSize="14.667" Location="{Binding PinLocation}" />
</DataTemplate>
```

When you run the application, you will see the pushpins appear in the correct location on the map, as shown in Figure 12-7. If you scroll or zoom the map, you will see that the locations of the pins move with the map.

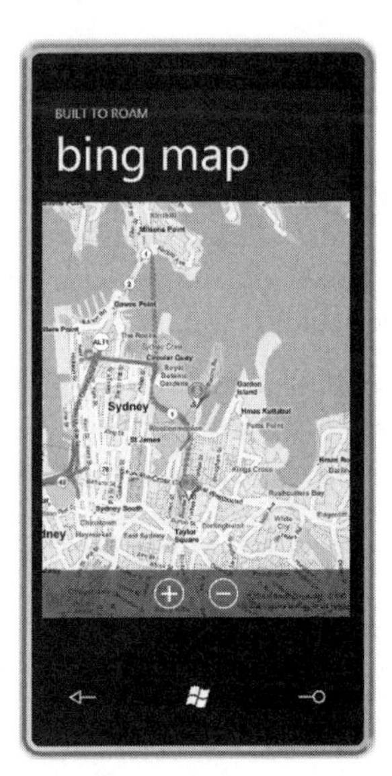

FIGURE 12-7

In this scenario, it may seem that the only benefit from using data-binding is that you don't need to iterate manually through the pins and create individual pushpins to place on the map. However, as you will learn in Chapter 15, data-binding gives you the ability to add, remove, and even modify the pins and have them automatically update on the map. With the `Center` *and* `ZoomLevel` *data-bound to the* `MapData` *instance, these properties can be queried and updated without your code having to interact with the map itself. This separation of concern is extremely important for the long-term architecture of your application.*

To change the look of the pins on the map, instead of using the `Pushpin` control, you can actually use any Windows Phone control. Simply edit the `ItemsTemplate` and replace the `Pushpin` control with the control that you want to be displayed. For example, if you wanted to just display a `TextBlock` with a red border, you might set the `ItemsTemplate` to the following:

```
<DataTemplate x:Key="PinTemplate">
    <Microsoft_Phone_Controls_Maps:Pushpin FontSize="14.667"
                        Location="{Binding PinLocation}" >
        <Border Width="30" Height="30" Background="#B3B1B1B1" CornerRadius="5"
                BorderThickness="3,3,3,0" BorderBrush="#FFFF0A0A">
            <TextBlock TextWrapping="Wrap" Text="{Binding PinContent }"/>
        </Border>
    </Microsoft_Phone_Controls_Maps:Pushpin>
</DataTemplate>
```

Events

The `Map` control raises several events that you can attach handlers to in order to respond to what the user is doing. Of particular note are the events `ViewChangeStart`, `ViewChangeOnFrame`, and `ViewChangeEnd`, which correspond to the Map view (including center and zoom level) starting to change, progressive events while the Map view changes, and the end of changes to the Map view. Take, for example, the user scrolling the map: The `ViewChangeStart` would be raised when the scroll starts; the `ViewChangeOnFrame` would be raised, possibly multiple times, during the scrolling; and, finally, the `ViewChangeEnd` event would be raised once scrolling had ended.

In addition to events raised by the `Map` control itself, you can also add handlers to events raised by the pushpins. The previous example used the `MapItemsControl` to automatically create `Pins` based on the provided `ItemsTemplate`. This template can be modified to attach an event handler for the `MouseLeftButtonDown` event to the `Pushpin` control:

Available for download on Wrox.com

```
<DataTemplate x:Key="PinTemplate">
    <Microsoft_Phone_Controls_Maps:Pushpin Content="{Binding PinContent}"
                    FontSize="14.667" Location="{Binding PinLocation}"
                    MouseLeftButtonDown="Pushpin_MouseLeftButtonDown" />
</DataTemplate>
```

Code snippet from MapPage.xaml

In the event handler you can then access the pin that was selected and update the `SelectedPin` property on the `MapData` object:

```
private void Pushpin_MouseLeftButtonDown(object sender, MouseButtonEventArgs e){
    (this.DataContext as MapData).SelectedPin =
                   (sender as Pushpin).DataContext as PinData;
}
```

When a pin is selected, it would be nice to be able to see the details of the pin. This can be done using a `Popup` control. The following snippet uses a `ListBox` to display the `Data` property of the selected pin inside the `Popup` control:

```
<Popup x:Name="PinDetailsPopup" HorizontalAlignment="Center" VerticalAlignment="Center"
       Width="300" Height="300" >
    <Border Height="300" Width="300" CornerRadius="10" BorderBrush="Black"
            BorderThickness="2" Background="#CAE5E5E5" >
        <Grid>
        <ListBox ItemsSource="{Binding SelectedPin.Data}" Background="Transparent">
               <ListBox.Resources>
        <DataTemplate x:Key="PinDetails">
    <TextBlock TextWrapping="Wrap" Text="{Binding Mode=OneWay}"/>
                   </DataTemplate>
    </ListBox.Resources>
    <ListBox.ItemTemplate>
        <StaticResource ResourceKey="PinDetails"/>
               </ListBox.ItemTemplate>
                <ListBox.ItemContainerStyle>
                    <Style TargetType="ListBoxItem">
                        <Setter Property="Background" Value="Transparent"/>
                        <Setter Property="Foreground" Value="Black"/>
                    </Style>
                </ListBox.ItemContainerStyle>
    </ListBox>
        </Grid>
    </Border>
</Popup>
```

Code snippet from MapPage.xaml

To display the `Popup` control, you simply need to set the `IsOpen` property to `true`. As the `ListBox` within the `Popup` control is data-bound to the `Data` property of the `SelectedPin`, the contents will automatically be updated when the user selects a different pin. You should also handle the Back button to hide the `Popup` control:

```
private void Pushpin_MouseLeftButtonDown(object sender, MouseButtonEventArgs e){
    (this.DataContext as MapData).SelectedPin =
                   (sender as Pushpin).DataContext as PinData;
    this.PinDetailsPopup.IsOpen = true;
}


protected override void OnBackKeyPress(System.ComponentModel.CancelEventArgs e){
```

```
        if (this.PinDetailsPopup.IsOpen){
            this.PinDetailsPopup.IsOpen = false;
            e.Cancel = true;
        }
        base.OnBackKeyPress(e);
    }
```

Code snippet from MapPage.xaml.cs

Bing Maps Web Services

In addition to the Bing Maps Silverlight control, there are other services provided by Bing Maps that you can connect to in order to do geocoding, routing, and searching-related tasks. More information on any of these services can be found via the Bing Maps Web Services SDK (`http://msdn.microsoft.com/en-us/library/cc980922.aspx`).

Route Service

One of the most useful non-visual Bing Maps services is the Route Service, which can be used to get a route between a start point and an end point. You can even include waypoints through which the route must pass. To get started, add a service reference to your application for the `http://dev.virtualearth.net/webservices/v1/routeservice/routeservice.svc` service endpoint. This will create the necessary proxy classes for you to call the Route Service.

Calculate Route

To calculate a route between two points, you need to create an instance of the `RouteServiceClient` proxy class and call the `CalculateRouteAsync` method, passing in the desired start and end geo-location points. As there are a few properties you need to set on the request object, it can be convenient to encapsulate this into a helper class, as shown in the following `RouteHelper` class:

```
public static class RouteHelper{
    private static RouteService.RouteServiceClient RouteClient;

    static RouteHelper(){
        RouteClient = new RouteService.RouteServiceClient();
        RouteClient.CalculateRouteCompleted += client_RouteCompleted;
    }

    public static void CalculateRoute(Location startLocation, Location endLocation,
                        Credentials serviceCredentials,
                        Action<ObservableCollection<Location>> routePathPointsCallback){
        var locations = new Waypoint[]{
                new Waypoint(){Description="Start",Location=startLocation},
                new Waypoint(){Description="End",Location=endLocation}};

        RouteRequest request = new RouteRequest();
        request.Waypoints = new ObservableCollection<Waypoint>();
```

```
            foreach (var location in locations){
                request.Waypoints.Add(location);
            }

            // Don't raise exceptions.
            request.ExecutionOptions = new ExecutionOptions();
            request.ExecutionOptions.SuppressFaults = true;

            // Only accept results with high confidence.
            request.Options = new RouteOptions();
            request.Options.RoutePathType = RoutePathType.Points;

            request.Credentials = serviceCredentials;

            // Make asynchronous call to fetch the data ... pass state object.
            RouteClient.CalculateRouteAsync(request, routePathPointsCallback);
        }

        private static void client_RouteCompleted(object sender,
                                          CalculateRouteCompletedEventArgs e){
            if (e.Result.ResponseSummary.StatusCode == ResponseStatusCode.Success &&
                e.Result.Result.Legs.Count > 0){
                var callback = e.UserState as Action<ObservableCollection<Location>>;
                callback(e.Result.Result.RoutePath.Points);
            }
        }
    }
```

Code snippet from RouteHelper.cs

To perform a routing calculation based on the locations of the first two pushpins in the `MapData` object, you can use the following code:

```
private void RouteButton_Click(object sender, EventArgs e){
    var data = this.DataContext as MapData;

    this.BingMap.CredentialsProvider.GetCredentials((creds)=>{
        RouteHelper.CalculateRoute(new Location() {
                                        Latitude = data.Pins[0].PinLocation.Latitude,
                                        Longitude = data.Pins[0].PinLocation.Longitude },
                                    new Location() {
                                        Latitude = data.Pins[1].PinLocation.Latitude,
                                        Longitude = data.Pins[1].PinLocation.Longitude },
                                    new Credentials() {
                                        ApplicationId = creds.ApplicationId },
                                    RoutingCallback);
        });
    }
```

```
private void RoutingCallback(ObservableCollection<Location> points){
    this.Dispatcher.BeginInvoke(() =>{
        var data = this.DataContext as MapData;
        data.RoutePoints.Clear();
        foreach (var point in points){
            data.RoutePoints.Add(new GeoCoordinate(point.Latitude, point.Longitude));
        }
    });
}
```

Code snippet from MapPage.xaml.cs

In the callback from the Route Service, the returned points are added to the `RoutePoints` collection, which is a new property of type `LocationCollection` that has been added to the `MapData` object:

```
public class MapData : INotifyPropertyChanged{
    ...

    private LocationCollection routePoints = new LocationCollection ();
    public LocationCollection RoutePoints{
        get{ return routePoints; }
    }
}
```

Code snippet from MapData.cs

Displaying the Route

The last thing to do with the route is to plot it on the map. To do this, add a `MapPolyline` to the `Map`. For the route to appear, you need to set the `Stroke` to a solid color and then data-bind the `Locations` property to the `RoutePoints` property of the `MapData`, as follows:

```
<Microsoft_Phone_Controls_Maps:Map x:Name="BingMap"
                      ZoomLevel="{Binding Zoom, Mode=TwoWay}"
                      Center="{Binding MapCenter, Mode=TwoWay}"
                      NavigationVisibility="Collapsed" ScaleVisibility="Collapsed"
                      CredentialsProvider="{StaticResource MapCredentials}">
    <Microsoft_Phone_Controls_Maps:MapPolyline
                      Stroke="#FF0000FF"
                      StrokeThickness="5"
                      Locations="{Binding RoutePoints}"/>
</Microsoft_Phone_Controls_Maps:Map>
```

Code snippet from MapPage.xaml

Now you are ready to run your application and tap the Route button to display the route between the two pins on the map, as shown in Figure 12-8.

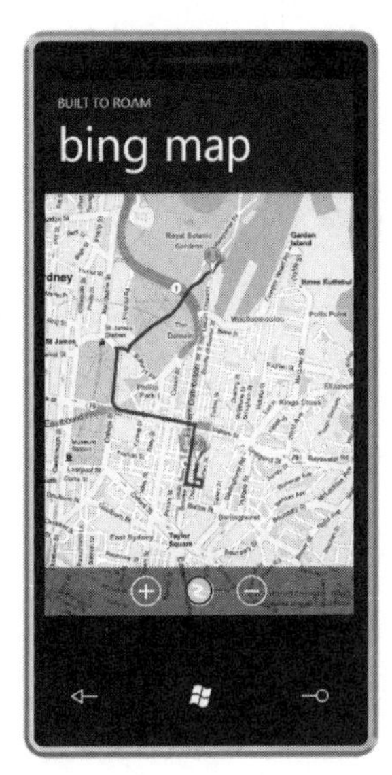

FIGURE 12-8

SUMMARY

Location and mapping can be used within your application to determine where the user is and where they might be going. Your application can use this data to provide context-relevant information to the user such as nearby shops and facilities, or display reminders for when the user has tasks to do in the nearby area.

In this chapter, you have seen how you can leverage the unified Windows Phone Location Services to get the geo-location of the device. The rich mapping capabilities of Bing Maps can be used to display the current location of the user or other points of interest within an application.

13

Connectivity and the Web

WHAT'S IN THIS CHAPTER

- Determining whether there is a network connection available
- Displaying content with the WebBrowser control
- Authenticating using Live ID

Although one of the Red Threads discussed in Chapter 3 was that users should feel connected, it is very common for a phone to find itself disconnected from the Internet. Whether the disconnection is for a short or an extended period, whether it was because the user stepped into an elevator or is roaming overseas, the effect is that your application will no longer be connected to services that are not running on the device itself.

This chapter will show how you can detect when the Windows Phone is disconnected and how you can integrate that knowledge into the behavior of your application. You will also see how easy it is to integrate Web content within your application using the `WebBrowser` control and how this can then be used to leverage third-party authentication services such as Windows Live ID.

CONNECTED STATUS

A common mistake made by developers new to building mobile applications is to make assumptions about the connectivity of the device. While it is not uncommon for mobile phones to have a data plan, the cost of such plans can still be prohibitive, resulting in users disabling their data connection or using it sparingly. In addition, there are still a large number of places where connectivity is a challenge. These may be elevators, tunnels, planes, or just remote areas that only have voice coverage at best. Rather than making the assumption that the phone — and thus your application — will always have connectivity, you should design your application to be network-aware and capable of operating without a constant connection.

Handling network availability isn't just about disabling functionality if the network is unavailable. It's about building your application to be less reliant on the network in the first place. The best mobile applications use a design principle that is often referred to as *Occasionally Connected*. The premise is that wherever possible, the application should perform operations such as reading and writing data locally and then in the background synchronize these changes with a server, if and when, available. There are many reasons why this principle leads to a better mobile application design. First, whenever the user is viewing or editing data, the application is reading/writing to the local data store. This makes each operation significantly quicker as there is very little latency in contrast to interacting with a remote server. Writing to the local data store has the added benefit that if the user exits the application, any information he or she has entered won't be lost. If the application had to wait for the server to be available in order to save the information, then there would be a risk that data would be lost if the application exited prematurely.

The Occasionally Connected design principle isn't without its challenges. Specifically, the added complexity of synchronizing data changes with the server. This can be difficult as you need to keep track of what changes have been made on the client, as well as any changes that have occurred on the server between synchronizations. These two distinct sets of changes need to be merged. Synchronization will be covered in more detail in Chapter 16.

Network Availability

In order to build an application that is network-aware, or Occasionally Connected, you need to be able to tell whether there is a network connection available to the Windows Phone device. This can be achieved via the static `GetIsNetworkAvailable` method found on the `NetworkInterface` class:

```
bool networkIsAvailable = NetworkInterface.GetIsNetworkAvailable();
```

Although you could set up your own thread and poll the `GetIsNetworkAvailable` method to determine when changes occur, it is more efficient to allow the operating system to perform this task for you. The `NetworkChange` class exposes a static event, `NetworkAddressChanged`, which is raised whenever the network address of the device changes:

```
NetworkChange.NetworkAddressChanged += NetworkChange_NetworkAddressChanged;

private void NetworkChange_NetworkAddressChanged(object sender, EventArgs e){
    bool networkIsAvailable = NetworkInterface.GetIsNetworkAvailable();
}
```

One thing to be aware of is that this event doesn't tell you anything about the current network connection, or what the change in network address was. In fact, the only information you can query is whether there is a network available via the `GetIsNetworkAvailable` method.

Service Reachability

The `GetIsNetworkAvailable` method is an oversimplification of the array of networking options available to Windows Phone. For example, it doesn't differentiate between the availability of a cellular or Wi-Fi network, which may be of importance for speed or cost-based decisions within your application. It is also important to note that, depending on where the service your application communicates with is located, having a network available may not necessarily indicate that your service is reachable. There are two approaches to detecting service reachability, but they both essentially rely on attempting to establish a connection to the server that your service is hosted on.

The first approach is to simply accept that service calls may fail when the service isn't reachable. Catching the resultant exception and determining the type of exception thrown can provide information as to whether the service is reachable. You should always handle service exceptions even if you think the service is currently available. Connectivity status can change at a moment's notice, and there is always a risk that the call may time-out or not reach its destination, so you should always plan to handle the worst-case scenarios.

The second, and more proactive, approach is to attempt to connect to the server that hosts the service whenever network availability changes. Rather than having to call a specific method on the server, it's simpler and quicker to have a static file located on the server that can be downloaded by the client to test reachability of the server:

```
private void IsServiceReachable(){
    var client = new WebClient();
    client.OpenReadCompleted += WebClient_OpenReadCompleted;
    client.OpenReadAsync(new Uri("http://www.builttoroam.com/ping.gif"));
}

private void WebClient_OpenReadCompleted(object sender, OpenReadCompletedEventArgs e){
    if (e.Error != null){
        // Service not reachable
        return;
    }

    var strm = e.Result;
    var buffer = new byte[1000];
    var cnt = 0;
    while (strm.Position < strm.Length){
        cnt = strm.Read(buffer, 0, buffer.Length);
        if (cnt == 0) break;
    }
    // Service is reachable
}
```

In this case, the file in question, ping.gif, is a 49-byte, 1 × 1 pixel GIF image that can be downloaded very quickly in order to verify that the server is reachable.

Some mobile developers advocate the actual calling of a simple zero-parameter service that simply returns a response of `true`*. This supposedly validates that the service is both reachable and that it is operating. However, all this does is introduce a further latency in testing for service reachability, while not really testing that the services you are going to call are available. The test service may be operational, because it doesn't do anything, but other services that rely on backend systems may not work as expected. In this case a call to these services is going to fail regardless of whether you've called the test service or not. The upshot is that you might as well just test for service reachability.*

Emulator Testing

Unfortunately, the Windows Phone emulator doesn't provide a mechanism to simulate different connectivity states. `GetIsNetworkAvailable` always returns a value of `true`, indicating that a network is available, regardless of the connectivity of the host operating system or settings such as Flight mode. As you have learned in some of the previous chapters, an approach to dealing with this limitation is to create an interface that allows for different implementations depending on whether the application is running on the device or the browser. The interface will cover both the `GetIsNetworkAvailable` method and the `NetworkAddressChanged` event:

Available for download on Wrox.com

```
public interface INetworkInterface{
    event NetworkAddressChangedEventHandler NetworkAddressChanged;

    bool GetIsNetworkAvailable();
}
```

Code snippet from NetworkInfo.cs

This interface can't be applied directly to the `NetworkChanged` or `NetworkInterface` classes, so you'll need to create a class that implements the interface and wraps the standard device functionality:

Available for download on Wrox.com

```
public class DeviceNetworkInterface : INetworkInterface{
    public event NetworkAddressChangedEventHandler NetworkAddressChanged;

    public DeviceNetworkInterface() {
        NetworkChange.NetworkAddressChanged += NetworkChange_NetworkAddressChanged;
    }

    private void NetworkChange_NetworkAddressChanged(object sender, EventArgs e) {
        if (NetworkAddressChanged != null) {
            NetworkAddressChanged(sender, e);
        }
    }

    public bool GetIsNetworkAvailable() {
        return NetworkInterface.GetIsNetworkAvailable();
    }
}
```

Code snippet from DeviceNetworkInterface.cs

When testing an application within the emulator, you will want to simulate changes in the connectivity status. You can do this by creating another implementation of the `INetworkInterface` that uses a timer to step through different connectivity states:

Available for download on Wrox.com

```
public class EmulatorNetworkInterface : INetworkInterface {
    public class ConnectionTimes {
        public bool Connected { get; set; }
        public TimeSpan ConnectionDuration { get; set; }
```

```
        }

        public event NetworkAddressChangedEventHandler NetworkAddressChanged;

        private ConnectionTimes[] Times { get; set; }
        private int currentTime = -1;
        private Timer timer;

        public EmulatorNetworkInterface(ConnectionTimes[] connectionTimes) {
            Times = connectionTimes;
            timer = new Timer(ChangeNetworkStatus, null,
                              Timeout.Infinite, Timeout.Infinite);
            MoveNext();
        }

        private void MoveNext() {
            timer.Change(Timeout.Infinite, Timeout.Infinite);
            currentTime = (currentTime+1) % Times.Length;
            var connection = Times[currentTime];

            if (NetworkAddressChanged != null) {
                NetworkAddressChanged(null, EventArgs.Empty);
            }

            timer.Change(connection.ConnectionDuration, connection.ConnectionDuration);
        }

        private void ChangeNetworkStatus(object state) {
            MoveNext();
        }

        public bool GetIsNetworkAvailable() {
            return Times[currentTime].Connected;
        }
    }
```

Code snippet from EmulatorNetworkInterface.cs

You'll also need some code to decide which implementation to load depending on whether the application is running in the emulator or on a Windows Phone:

Available for download on Wrox.com

```
    public class NetworkInfo {
        public readonly static INetworkInterface Instance;

        static NetworkInfo() {
            if (!DesignerProperties.IsInDesignTool) {
                Instance = PickRuntimeNetworkInterface();
            }
            else {
                Instance = CreateEmulatorInterface();
            }
        }

        private static INetworkInterface PickRuntimeNetworkInterface() {
```

```
        if (Microsoft.Devices.Environment.DeviceType == DeviceType.Device) {
            return new DeviceNetworkInterface();
        }
        else {
            return CreateEmulatorInterface();
        }
    }

    private static INetworkInterface CreateEmulatorInterface() {
        return new EmulatorNetworkInterface(
        new ConnectionTimes[] {
            new ConnectionTimes(){Connected=true,
                                  ConnectionDuration=new TimeSpan(0,0,1)},
            new ConnectionTimes(){Connected=true,
                                  ConnectionDuration=new TimeSpan(0,0,5)},
            new ConnectionTimes(){Connected=false,
                                  ConnectionDuration=new TimeSpan(0,0,5)},
            new ConnectionTimes(){Connected=true,
                                  ConnectionDuration=new TimeSpan(0,0,2)},
            new ConnectionTimes(){Connected=false,
                                  ConnectionDuration=new TimeSpan(0,0,5)},
            new ConnectionTimes(){Connected=true,
                                  ConnectionDuration=new TimeSpan(0,0,2)},
            new ConnectionTimes(){Connected=false,
                                  ConnectionDuration=new TimeSpan(0,0,8)},
            new ConnectionTimes(){Connected=false,
                                  ConnectionDuration=new TimeSpan(0,0,5)},
            new ConnectionTimes(){Connected=true,
                                  ConnectionDuration=new TimeSpan(0,0,3)}
        });
    }
}
```

Code snippet from NetworkInfo.cs

DESIGNER SUPPORT

In this code snippet you will notice that there are two conditional statements that determine which implementation should be used. The first checks whether the code is being executed during design time. The `IsInDesignTool` property returns `true` when the code is being executed within a designer such as Visual Studio or Expression Blend. This ensures that the `PickRuntimeNetworkInterface` method is only invoked at runtime. This is important as the `System.Environment.DeviceType` property can't be resolved at design time by the JIT compiler, resulting in an error. You can still use the designer, but there will be an error reported, and some design features will not work as expected. Moving the `DeviceType` conditional statement into a separate method means that the method will never be JIT compiled during design time.

Connectivity

Combining network availability, service reachability, and support for emulator testing, you can create a `Connectivity` class that can be used within your application to report on the current connectivity status of the Windows Phone. The `Connectivity` class attaches an event handler to the implementation of the `INetworkInterface` exposed by the `NetworkInfo` class. When this event is raised, it starts a test to see whether the specified `ServiceTestUrl` is reachable. A background timer is used to cancel the download after the specified `ServiceTimeout`. This ensures that service availability is reported within an acceptable period after the `NetworkAddressChanged` event is raised.

```
[Flags()]
public enum ConnectionStatus {
    Unknown = 0,
    NetworkAvailable = 1,
    ServiceReachable = 2,
    Disconnected = 4
}

public class ConnectionEventArgs : EventArgs {
    public ConnectionStatus Status { get; set; }
}

public static class Connectivity {
    public static event EventHandler<ConnectionEventArgs> ConnectivityChanged;

    public static Uri ServiceTestUrl { get; set; }
    public static TimeSpan ServiceTimeout { get; set; }

    private static ConnectionStatus status;
    private static int checkingConnectivity = 0;
    private static WebClient client;
    private static Timer serviceTestTimer;
    private static object serviceLock = new object();

    public static ConnectionStatus Status {
        get {
            return status;
        }
        private set {
            if (status == value) {
                return;
            }
            status = value;
            if (ConnectivityChanged != null) {
                ConnectivityChanged(null,
                                    new ConnectionEventArgs() { Status = status });
            }
        }
    }

    static Connectivity() {
        Status = ConnectionStatus.Unknown;
```

```
        NetworkInfo.Instance.NetworkAddressChanged += NetworkAddressChanged;
        serviceTestTimer = new Timer(ServiceTimeOutCallback, null,
                                     Timeout.Infinite, Timeout.Infinite);
    }

    private static void NetworkAddressChanged(object sender, EventArgs e) {
        TestConnectivity();
    }

    public static void TestConnectivity() {
        if (!DesignerProperties.IsInDesignTool){
            var t = new Thread(UpdateConnectivity);
            t.Start();
        }
    }

    private static void UpdateConnectivity() {
        if (Interlocked.CompareExchange(ref checkingConnectivity, 1, 0) == 1) {
            return;
        }

        var testingService = false;
        try {
            var connected = NetworkInfo.Instance.GetIsNetworkAvailable();
            if (connected) {
                if (ServiceTestUrl != null) {
                    lock (serviceLock) {
                        client = new WebClient();
                        client.OpenReadCompleted += WebClient_OpenReadCompleted;
                        serviceTestTimer.Change(ServiceTimeout, ServiceTimeout);
                        testingService = true;
                        client.OpenReadAsync(ServiceTestUrl);
                    }
                }
                else {
                    Status = ConnectionStatus.NetworkAvailable;
                }
            }
            else {
                Status = ConnectionStatus.Disconnected;
            }
        }
        finally {
            if (!testingService) {
                Interlocked.Decrement(ref checkingConnectivity);
            }
        }
    }

    private static void ServiceTimeOutCallback(object state) {
        lock (serviceLock) {
            serviceTestTimer.Change(Timeout.Infinite, Timeout.Infinite);
            client.CancelAsync();
        }
```

```
        }

        static void WebClient_OpenReadCompleted(object sender,
                                                OpenReadCompletedEventArgs e){
            try {
                serviceTestTimer.Change(Timeout.Infinite, Timeout.Infinite);

                if (e.Error != null) {
                    Status = ConnectionStatus.NetworkAvailable;
                    return;
                }

                var strm = e.Result;
                var buffer = new byte[1000];
                var cnt = 0;
                while (strm.Position < strm.Length) {
                    cnt = strm.Read(buffer, 0, buffer.Length);
                    if (cnt == 0) break;
                }

                Status = ConnectionStatus.ServiceReachable |
                         ConnectionStatus.NetworkAvailable;
            }
            catch (Exception) {
                Status = ConnectionStatus.NetworkAvailable;
            }
            finally {
                Interlocked.Decrement(ref checkingConnectivity);
            }
        }
    }
```

Code snippet from Connectivity.cs

When the `Status` property of the `Connectivity` class changes, a `ConnectivityChanged` event is also raised, which includes the current status of the connection as a property of the `ConnectionChangedEventArgs` argument. The Connectivity class will detect any changes in the network address information on the device and will automatically test the availability of a connection and the reachability of the service. However, in order to initiate the Status of the Connectivity class you need to ensure that the TestConnectivity method is invoked. You can do this by creating a class that implements the IApplicationService interface. This interface consists of two methods, one that will be called when the application starts up (in this case to call TestConnectivity the first time) and one that will be called when the application stops (in this case this method doesn't do anything).

Available for download on Wrox.com

```
public class ConnectivityService : IApplicationService
{
    public Uri ServiceTestUrl { get; set; }

    public TimeSpan ServiceTimeout { get; set; }

    public void StartService(ApplicationServiceContext context)
```

```
        {
            Connectivity.ServiceTestUrl = this.ServiceTestUrl;
            Connectivity.ServiceTimeout = this.ServiceTimeout;
            Connectivity.TestConnectivity();
        }

        public void StopService()
        {
        }
    }
```

Code snippet from ConnectivityService.cs

In order to register the ConnectivityService so that the appropriate methods get invoked during application startup and shutdown, you just need to add an instance to the ApplicationLifetimeObjects collection in the App.xaml. This also defines the ServiceTestUrl and ServiceTimeout values that will be passed through to the Connectivity class for service reachability testing.

```
<Application x:Class="GetConnected.App" ... >
    <Application.Resources>
    </Application.Resources>

    <Application.ApplicationLifetimeObjects>
       <network:ConnectivityService ServiceTestUrl="http://www.builttoroam.com/ping.gif"
                                    ServiceTimeout="00:00:15"/>
       <shell:PhoneApplicationService
            Launching="Application_Launching" Closing="Application_Closing"
            Activated="Application_Activated" Deactivated="Application_Deactivated"/>
    </Application.ApplicationLifetimeObjects>
</Application>
```

Code snippet from App.xaml

Here the ServiceTestUrl is ping.gif which is a small 1 x 1 pixel image that can be downloaded to test connectivity to the server. In your application this should be changed to a URL that is on the same server as the services your application connects to.

Data Binding

The `Connectivity` class is a useful wrapper and can be used anywhere in your code to determine the current state of network and service reachability. However, it is limited in that there are no dependency properties on which you can bind the user interface to. For this purpose you might want to consider exposing a wrapper class called `BindableConnectivity`:

```
public class BindableConnectivity : DependencyObject {
    public ConnectionStatus Status {
        get { return (ConnectionStatus)GetValue(StatusProperty); }
        set { SetValue(StatusProperty, value); }
    }

    public static readonly DependencyProperty StatusProperty =
```

```
        DependencyProperty.Register("Status", typeof(ConnectionStatus),
                                    typeof(BindableConnectivity),
                                    new PropertyMetadata(ConnectionStatus.Unknown));
    public bool NetworkAvailable {
        get { return (bool)GetValue(NetworkAvailableProperty); }
        set { SetValue(NetworkAvailableProperty, value); }
    }

    public static readonly DependencyProperty NetworkAvailableProperty =
        DependencyProperty.Register("NetworkAvailable", typeof(bool),
                                    typeof(BindableConnectivity),
                                    new PropertyMetadata(false));
    public bool ServiceReachable {
        get { return (bool)GetValue(ServiceReachableProperty); }
        set { SetValue(ServiceReachableProperty, value); }
    }

    public static readonly DependencyProperty ServiceReachableProperty =
        DependencyProperty.Register("ServiceReachable", typeof(bool),
                                    typeof(BindableConnectivity),
                                    new PropertyMetadata(false));
    public bool Disconnected {
        get { return (bool)GetValue(DisconnectedProperty); }
        set { SetValue(DisconnectedProperty, value); }
    }

    public static readonly DependencyProperty DisconnectedProperty =
        DependencyProperty.Register("Disconnected", typeof(bool),
                                    typeof(BindableConnectivity),
                                    new PropertyMetadata(false));

    public BindableConnectivity() {
        Connectivity.ConnectivityChanged += Connectivity_ConnectivityChanged;
        UpdateStatus(Connectivity.Status);
    }

    private void Connectivity_ConnectivityChanged(object sender,
                                                  ConnectionEventArgs e) {
        UpdateStatus(e.Status);
    }

    private void UpdateStatus(ConnectionStatus status) {
        this.Dispatcher.BeginInvoke(() =>{
           this.Status = status;
           this.NetworkAvailable =
                (this.Status & ConnectionStatus.NetworkAvailable) > 0;
           this.ServiceReachable =
                (this.Status & ConnectionStatus.ServiceReachable) > 0;
           this.Disconnected =
                (this.Status & ConnectionStatus.Disconnected) > 0;
        });
    }
}
```

Code snippet from BindableConnectivity.cs

To make use of the `BindableConnectivity` class, you can create an instance as a resource within App.xaml or within the XAML for a page. In this case, you're creating a single instance that can be referenced from anywhere within the application.

```
<Application x:Class="GettingConnected.App" ...
             xmlns:network="clr-namespace:GettingConnected">

    <Application.Resources>
        <network:BindableConnectivity x:Key="Connectivity" />
        ...
```

Code snippet from App.xaml

The `BindableConnectivity` instance can then be used within your application via the use of data-binding.

```
<Grid x:Name="ContentGrid" Grid.Row="1">
    <CheckBox Name="NetworkAvailableCheckBox" Content="Network Available"
              Margin="10,10,0,0" VerticalAlignment="Top" IsEnabled="False"
              DataContext="{StaticResource Connectivity}"
              IsChecked="{Binding NetworkAvailable}" />
    <CheckBox Name="ServiceReachableCheckBox" Content="Service Reachable"
              Margin="10,70,0,0" VerticalAlignment="Top" IsEnabled="False"
              DataContext="{StaticResource Connectivity}"
              IsChecked="{Binding ServiceReachable}" />
</Grid>
```

Code snippet from MainPage.xaml

In this case, two checkboxes are being used to indicate whether a network is available and whether the service is reachable, as shown in Figure 13-1.

FIGURE 13-1

WEBBROWSER CONTROL

There are several ways that you can present information to users that has been obtained from services on the Web. Most of the time it is preferable to use controls designed for Windows Phone to present this information as it can offer the most dynamic and interactive experience for the user. However, sometimes you may need to present information that is only available in the form of Web content. This might be a snippet of HTML or a remote website that you need the user to go to in order to authenticate or complete a transaction. For these scenarios, the `WebBrowser` control gives you the ability to load Web content directly within your application. You can think of the `WebBrowser` control as having an instance of Internet Explorer running within the confines of your application. Although it doesn't have any of the menus or the address bar that Internet Explorer does, it does behave and display content in the same way. For example, once you have loaded content, users can zoom in and out by double-tapping the screen; they can scroll by panning around the screen; and they can navigate to other pages by clicking on any hyperlinks.

To get started using the `WebBrowser` control, drag an instance from the ToolBox in Visual Studio or from the Assets window in Blend. The following XAML illustrates an instance of the `WebBrowser` control — a `TextBox` for the user to enter an address and a `Button` to navigate to the entered address. When the user clicks the `Button`, the `Navigate` method of the `WebBrowser` control is invoked to load the website specified in the `TextBox`:

Available for download on Wrox.com

XAML

```
<phone:WebBrowser Margin="0,135,0,210" Name="Browser" />
<TextBox Height="32" Margin="0,444,160,100" Name="NavigateText"
         Text="http://www.builttoroam.com" />
<Button Content="Go" Height="70" Margin="320,340,0,0" Name="NavigateButton" Width="160"
        Click="NavigateButton_Click" DataContext="{StaticResource Connectivity}"
        IsEnabled="{Binding NetworkAvailable}" />
```

Code snippet from MainPage.xaml

C#

```
private void NavigateButton_Click(object sender, RoutedEventArgs e){
    this.Browser.Navigate(new Uri(this.NavigateText.Text));
}
```

Code snippet from MainPage.xaml.cs

In this case, the `IsEnabled` property of the `Button` is data-bound to the `NetworkAvailable` property of the `BindableConnectivity` object you created earlier. As the `Button` will be navigating the `WebBrowser` to a web address, this should only be allowed if there is a network currently available. As the actual address the user is going to enter isn't known until they press the button, it is not possible to determine whether the server is going to be reachable in advance, which is why you've opted to use the `NetworkAvailable` property of the `BindableConnectivity` object instead of the `ServiceReachable` property.

As an alternative to navigating to an actual web address, you can also load your own HTML content into the Web browser via the `NavigateToString` method. In this case, you're going to load a very simple HTML document:

Available for download on Wrox.com

XAML

```
<Button x:Name="NavigateToHtmlButton" Content="Navigate to HTML" Margin="0,0,0,40"
        VerticalAlignment="Bottom" Click="NavigateToHtmlButton_Click"/>
```

Code snippet from MainPage.xaml

C#

```
private void NavigateToHtmlButton_Click(object sender, RoutedEventArgs e) {
    this.Browser.NavigateToString(
                @"<HTML>
                    <BODY>
                        Hello World!
                    </BODY>
                </HTML>");
}
```

Code snippet from MainPage.xaml.cs

Figure 13-2 illustrates this code in action showing the website `www.builttoroam.com` loaded in the leftmost image. The middle image shows the HTML snippet loaded via the `NavigateToString` method. The default view makes the text very difficult to read. While the user can zoom in by double-tapping the control, shown in the rightmost image, this isn't a particularly good user experience. You should consider styling the HTML snippet to ensure that it is readable without the user having to zoom in.

FIGURE 13-2

In addition to being able to set the web page or HTML content that is displayed within the `WebBrowser` control, you can also interact with the loaded content. The `WebBrowser` has an `IsScriptEnabled` property that controls whether the host application can invoke a JavaScript function within the loaded content, and whether the web page can trigger an event within the host application. These mechanisms can be used to pass data into and out of the `WebBrowser` control to allow the host application to interact with the Web-based content.

The first thing to do in order to start communicating between your Windows Phone application and the content within the embedded `WebBrowser` control is to enable scripting by setting the `IsScriptEnabled` property to `true`. You can also take this opportunity to wire up an event handler for the `ScriptNotify` event.

XAML

```
<phone:WebBrowser Margin="0,135,0,0" Name="Browser" Height="174"
                    VerticalAlignment="Top"
                    IsScriptEnabled="True" ScriptNotify="Browser_ScriptNotify" />
<TextBlock x:Name="ScriptPageOutputText" Height="43" Margin="6,0,10,0"
           TextWrapping="Wrap" VerticalAlignment="Bottom"/>
```

Code snippet from MainPage.xaml

C#

```
private void Browser_ScriptNotify(object sender, NotifyEventArgs e){
    this.ScriptPageOutputText.Text = e.Value;
}
```

Code snippet from MainPage.xaml.cs

The `ScriptNotify` event is triggered whenever the `window.external.notify` method is called from JavaScript executing within the `WebBrowser` control. In this example the event handler assigns the value passed into the `notify` method as the `Text` on a `TextBlock` element so that it can be displayed on screen:

HTML/JavaScript

```
<!DOCTYPE html PUBLIC "-//W3C//DTD XHTML 1.0 Transitional//EN"
   "http://www.w3.org/TR/xhtml1/DTD/xhtml1-transitional.dtd">
<html xmlns="http://www.w3.org/1999/xhtml">
<head>
    <title>Simple Script Page</title>
    <script type="text/javascript">
        function SilverlightOutput() {
            try {
                window.external.notify("Test content from the webpage");
                messageSent.innerHTML = "Message Sent";
            }
            catch (ex) {
                messageSent.innerHTML =
                "No Silverlight application to communicate with";
            }
        }
    </script>
</head>
```

```
<body>
  <input type="button" value="Send Text to Silverlight"
                       onclick="SilverlightOutput();" />
  <div id="messageSent"></div>
</body>
</html>
```

Code snippet from SimpleScript.htm

This is a very primitive message-passing mechanism as the `notify` method takes a single string argument. Whatever data you want to pass from within the `WebBrowser` control to the host Windows Phone application has to first be serialized to a string so that it can be passed as an argument to the `notify` method. In the event handler within the Windows Phone application, you would then need to deserialize the data.

One thing you may notice is that the JavaScript source code makes use of a try-catch block around the call to the `notify` method. The `notify` method only exists when the content is running within the `WebBrowser` control within a Silverlight or Windows Phone application. If the same web page is going to be viewable in a full browser, you should ensure that you handle any errors that may occur by calling `notify`. The use of a try-catch block is one way to detect if the call to `window.external.notify` fails. You could also detect the existence of the `notify` method and selectively call it if it is present.

Going the other direction, from the Windows Phone application into the `WebBrowser` control, you have a little more flexibility in that you can specify the name of the JavaScript function you want to invoke. You can also specify an array of strings to be passed in as arguments. For example the following code invokes the `SilverlightInput` JavaScript function, passing in three parameters:

XAML

```
<Button x:Name="SendToScriptPageButton" Content="Send Text" HorizontalAlignment="Right"
        Height="57" Margin="0,0,0,134" VerticalAlignment="Bottom" Width="221"
        Click="SendToScriptPageButton_Click"/>
```

Code snippet from MainPage.xaml

C#

```
private void SendToScriptPageButton_Click(object sender,RoutedEventArgs e){
    this.Browser.InvokeScript("SilverlightInput",
                              "Built to Roam", "Nick Randolph", "Copyright 2010");
}
```

Code snippet from MainPage.xaml.cs

In the HTML document there needs to be a corresponding JavaScript method, `SilverlightInput`. Again, this information is simply displayed on the screen via the `div` element within the `WebBrowser` content called `silverlightContent`.

Available for download on Wrox.com

HTML/JavaScript

```
<!DOCTYPE html PUBLIC "-//W3C//DTD XHTML 1.0 Transitional//EN"
          "http://www.w3.org/TR/xhtml1/DTD/xhtml1-transitional.dtd">
<html xmlns="http://www.w3.org/1999/xhtml">
<head>
    <title>Simple Script Page</title>
    <script type="text/javascript">
        function SilverlightInput(company, name, copyrightNotice) {
            silverlightContent.innerHTML = "<div>" +
                                                "<p>" + company +"</p>" +
                                                "<p>" + name + "</p>" +
                                                "<p>" + copyrightNotice + "</p>"
                                            "</div>";
            return true;
        }

        function SilverlightOutput() {
            try {
                window.external.notify("Test content from the webpage");
                messageSent.innerHTML = "Message Sent";
            }
            catch (ex) {
                messageSent.innerHTML =
                "No Silverlight application to communicate with";
            }
        }
    </script>
</head>
<body>
    <input type="button" value="Send Text to Silverlight"
           onclick="SilverlightOutput();" />
    <div id="silverlightContent"></div>
    <div id="messageSent"></div>
</body>
</html>
```

Code snippet from SimpleScript.htm

If you want to send more complex information from your Windows Phone application into the `WebBrowser` control, you can do so by serializing an object graph into a JSON-formatted string. The following code uses an instance of the `DataContractJsonSerializer` class (you will need to add a reference to System.Servicemodel.Web.dll to your project) to convert a `CompanyInfo` object into a JSON string:

Available for download on Wrox.com

XAML

```
<Button x:Name="SendJsonToScriptPageButton" Content="Send Json"
        HorizontalAlignment="Right" Height="57" Margin="0,0,0,62"
        VerticalAlignment="Bottom" Width="221"
        Click="SendJsonToScriptPageButton_Click"/>
```

Code snippet from MainPage.xaml

C#

```
private void SendJsonToScriptPageButton_Click(object sender, RoutedEventArgs e) {
    var company = new CompanyInfo(){
                Company = "Built to Roam",
                Name = "Nick Randolph",
                CopyrightNotice = "Copyright 2010"
            };

    var serializer = new DataContractJsonSerializer(typeof(CompanyInfo));
    var strm = new MemoryStream();
    serializer.WriteObject(strm, company);
    strm.Flush();
    strm.Seek(0, SeekOrigin.Begin);
    var reader = new StreamReader(strm);
    var serializedCompany = reader.ReadToEnd();
    this.Browser.InvokeScript("ComplexSilverlightInput", serializedCompany);
}
public class CompanyInfo{
    public string Company { get; set; }
    public string Name { get; set; }
    public string CopyrightNotice { get; set; }
}
```

Code snippet from MainPage.xaml.cs

This code invokes a JavaScript function called `ComplexSilverlightInput`, which accepts a single string argument. To convert this string back into an object graph, you can use the built-in JavaScript function `eval`. This will convert the JSON string in an object similar in shape to the C# `CompanyInfo` class. Note how the JavaScript object has the same properties — `Company`, `Name`, and `CopyrightNotice` — as the original C# class had:

Available for download on Wrox.com

```
function ComplexSilverlightInput(companyString) {
    var companyInfo  = eval('(' + companyString + ')');
    silverlightContent.innerHTML = "<div>" +
                                        "<p>" + companyInfo.Company + "</p>" +
                                        "<p>" + companyInfo.Name + "</p>" +
                                        "<p>" + companyInfo.CopyrightNotice + "</p>"
                                    "</div>";

    return true;
}
```

Code snippet from SimpleScript.htm

MULTISCALEIMAGE

SeaDragon (`www.seadragon.com`), also referred to as *Deep Zoom*, is an interesting piece of technology to emerge out of Microsoft Live Labs that allows for viewing of gigapixel images over a network connection. In essence, the technology works in a similar way to Bing Maps in that the image is available at a range of zoom levels, and at each level the image is segmented to allow for progressive downloading. The net effect is that the image can be viewed, zoomed, and scrolled without having the entire image downloaded to the device. Windows Phone applications can leverage this technology via the `MultiScaleImage` control. Figure 13-3 illustrates an image being displayed using the `MultiScaleImage` control. When the user taps the image, the control zooms in on that point. As you can see in the fourth image, there is legible text that wasn't even visible in the first image. This image, one of the samples available at `www.seadragon.com`, is 83 megapixels in size, making it prohibitive to download and display using a conventional Image control.

FIGURE 13-3

You can use the `MultiScaleImage` control in your application in a similar way to a simple Image control. Simply drag an instance from the Assets window in Blend onto the page. You can set the image to be loaded by specifying the `Source` property in the Properties window. In this case, the source is set to `http://static.seadragon.com/content/misc/milwaukee.dzi` (note that the file extension is not that of a standard image file):

```
<Grid Grid.Row="1" x:Name="ContentGrid">
    <MultiScaleImage x:Name="ScalingImage">
        <MultiScaleImage.Source>
            <DeepZoomImageTileSource
             UriSource="http://static.seadragon.com/content/misc/milwaukee.dzi" />
        </MultiScaleImage.Source>
    </MultiScaleImage>
</Grid>
```

Code snippet from ScalingImagePage.xaml

There's no built-in support within the `MultiScaleImage` control to handle interaction from the user. However, the usual mouse and manipulation events are available, so it's an easy process to trap and build appropriate behaviors into your application. For example, the following code zooms in when the user taps a point on the image and zooms back out when the user clicks the Back button:

Available for download on Wrox.com

```
private const double ZoomMultiplier = 1.4;
private double zoom = 1;
private Point lastZoom;

private void Zoom(double newzoom, Point p) {
    lastZoom = ScalingImage.ElementToLogicalPoint(
              new Point(0.5*ScalingImage.ActualWidth,0.5*ScalingImage.ActualHeight));
    ScalingImage.ZoomAboutLogicalPoint(newzoom / zoom, p.X, p.Y);
    zoom = newzoom;
}

private void ScalingImage_MouseLeftButtonDown(object sender, MouseButtonEventArgs e) {
    var point = e.GetPosition(ScalingImage);
    Zoom(zoom * ZoomMultiplier, ScalingImage.ElementToLogicalPoint(point));
}


protected override void OnBackKeyPress(System.ComponentModel.CancelEventArgs e) {
    var newZoom = zoom / ZoomMultiplier;
    if (newZoom >=1 ) {
        Zoom(newZoom, lastZoom);
        e.Cancel = true;
    }
    base.OnBackKeyPress(e);
}
```

Code snippet from ScalingImagePage.xaml.cs

This is a legitimate use of the Back button as it returns the user to the previous state he or she was in. Once the image has returned to its initial zoom level, the Back button will take the user back to the previous page. Although this is in theory acceptable, the reality might be that users don't understand that they need to go zoom out of the image before being able to go back to the previous page. As with all applications, you should conduct appropriate end-user testing to ensure that your application design is intuitive to your end users.

AUTHENTICATION

One of the primary tenants of building for Windows Phone is to think about how your application can connect into existing aspects of the user's life. Most users will have existing accounts using one or more online services such as Live ID, Facebook, or even Twitter. Rather than requiring the user to select and remember yet another set of credentials for your application, if you need to authenticate a user, it makes sense to leverage existing credentials that the user may already have. In this section, you'll learn how you can use Windows Live ID to authenticate a user and access information about the user and their contacts.

Windows Live ID

You might be wondering why you've selected Live ID as the credential source to cover in-depth, since Live ID is only one of many online credential providers. The reason for this is that for someone to use a Windows Phone device, they must enter a Live ID and password in order to authenticate the device. This means that you can be guaranteed that every Windows Phone owner has at least one Live ID, making it a good default choice for authenticating users within your application.

The process of working with Live ID has been well documented for web developers. However, since there is no Windows Phone–specific SDK available, you have to make use of the `WebBrowser` control to direct the user to the `login.live.com` website in order to log into your application.

The way that Live ID usually works is that when you want a user to log into a website that uses Live ID, they are redirected to `login.live.com`. The user signs in using their credentials, the credentials are verified, and the user is then directed back to the "Return URL" of the website that uses Live ID. Here the user token is decoded and used to uniquely identify the user. The user token is unique to both the Live ID that was used to sign in and the website that they signed into. If the same Live ID is used at a different website, that site will get a different user token, but each time the same Live ID logs into any given site, that site will always receive the same user token. This process is outlined in Figure 13-4.

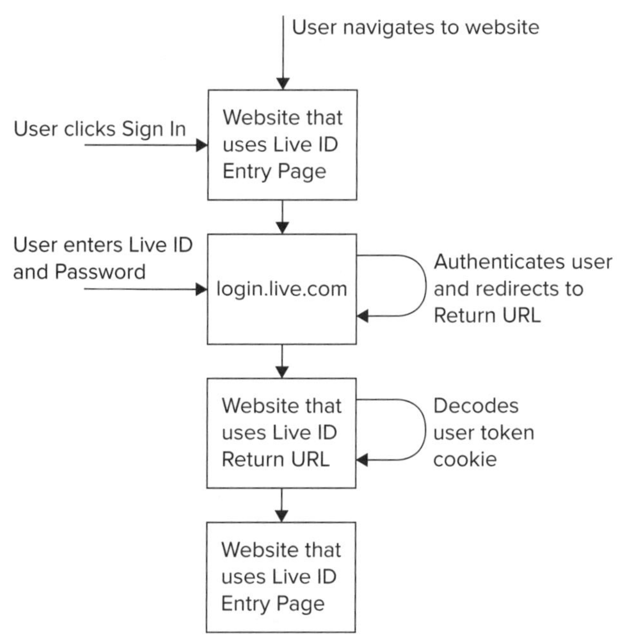

FIGURE 13-4

Clearly this process won't work without modification within your Windows Phone application. The user's Live ID token needs to somehow find its way back to the Windows Phone application. The trick is to use a WebBrowser control with IsScriptEnabled set to `true`, coupled with the `window.external.Notify` callback function discussed earlier.

To use Live ID within your Windows Phone application, when the user clicks the "Sign In" button, the application needs to display a `WebBrowser` control and navigate to `login.live.com`. The user signs in via the `WebBrowser` control and is redirected to a web page hosted on a site that you operate. This page decodes the user token cookie and then calls the `window.external.Notify` function, passing in the token as a parameter. The Windows Phone application receives the token by listening for the `ScriptNotify` event. Once the event has fired the `WebBrowser` can be hidden as it is no longer required. The token can then be used to uniquely identify the user within the application. Figure 13-5 illustrates the flow between the Windows Phone application and the content within the `WebBrowser` control.

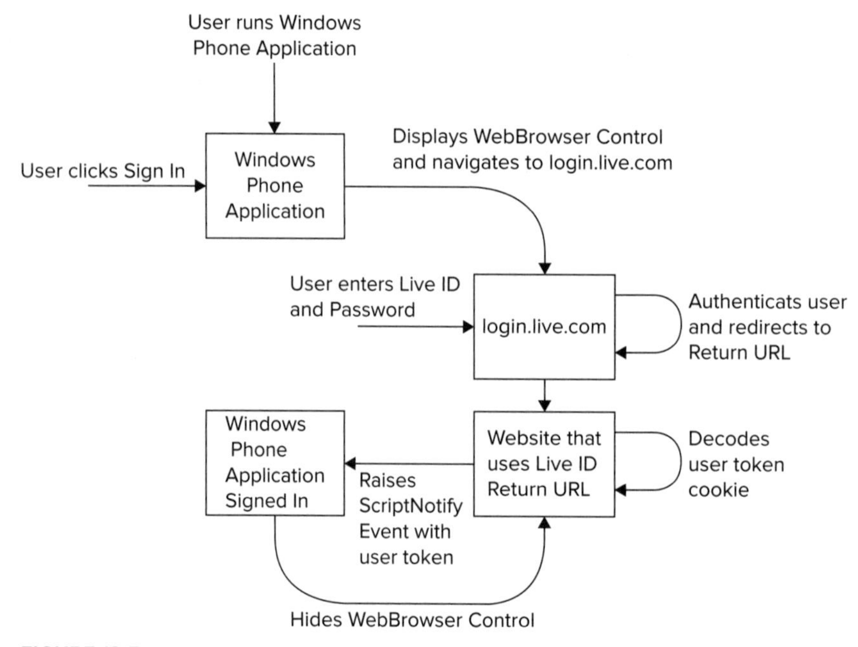

FIGURE 13-5

Windows Live ID supports two mechanisms for authentication depending on the level that your application needs to integrate with Windows Live. So far you have seen the process for using Web Authentication to validate that a user has a Live ID and to ensure their uniqueness within your application. Delegated Authentication extends Web Authentication to allow your application to query other personal information about the user. The basic concept of Delegated Authentication is that the user has to give your application permission to access their information. Hence, their is delegating authority of their information to your application for a limited time period.

You'll start by looking at how to implement Web Authentication as a simple mechanism for authenticating users and then extend that to show how you can use Delegated Authentication to request access to more information associated with the user's Live ID account.

Web Authentication

There are two parts to implementing Web Authentication for Windows Phone. First, you need to implement a website that is capable of capturing the Live ID information. This will contain the page, known as the *Return URL*, that will receive and decode the user token after the user has signed in. Once the token has been decoded, the page will redirect to another page that will invoke the JavaScript `Notify` function in order to pass the decoded user token back to the host Windows Phone application.

Start by creating a new ASP.NET web application in Visual Studio. You will need to change the application to run using IIS, rather than the default Visual Studio Development Server, because you need to specify a host name other than localhost for the ReturnURL. The Visual Studio Development Server will only accept incoming requests with localhost or the machine name as a host name. As you will see later on, when you set up the application ID for your Windows Phone application, you need to specify a valid domain name.

The use of IIS is actually only required for Delegated Authentication. If you only want to use Web Authentication to validate users, then you can continue to use localhost.

Figure 13-6 shows the Web tab of the project Properties window where you can set the application to use IIS. In this case, the application has been configured to run in a virtual directory called *AuthenticationWeb*.

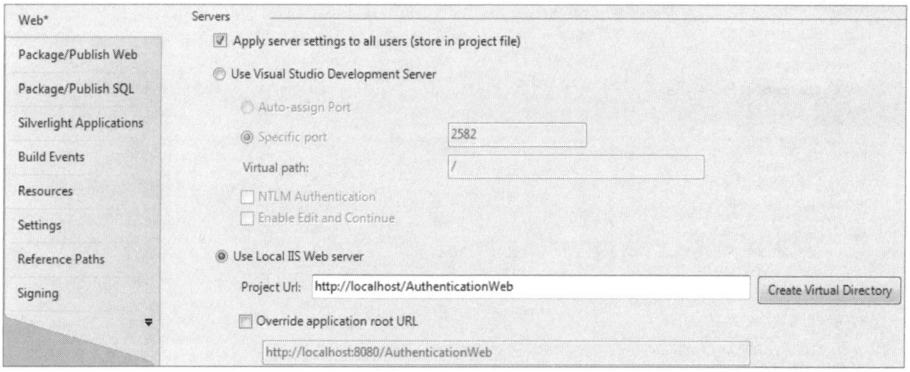

Web*
Package/Publish Web
Package/Publish SQL
Silverlight Applications
Build Events
Resources
Settings
Reference Paths
Signing
Servers
Apply server settings to all users (store in project file)
Use Visual Studio Development Server
Auto-assign Port
Specific port 2582
Virtual path: /
NTLM Authentication
Enable Edit and Continue
Use Local IIS Web server
Project Url: http://localhost/AuthenticationWeb
Create Virtual Directory
Override application root URL
http://localhost:8080/AuthenticationWeb

FIGURE 13-6

The next step is to select a fake domain name to use during development and testing. In this case, you're going to use `wptestsite.com`, which is completely fictitious and only going to be used during development. You need to set up your computer to recognize that requests to this domain should be routed to IIS running on your computer. There are a couple of ways to do this, but the easiest is to modify the Hosts file on your computer. To do this, open a command window as Administrator. Navigate to the directory c:\windows\system32\drivers\etc and type **"edit hosts"**. This will open Edit, allowing you to add an entry for the domain you have selected. In Figure 13-7, `wptestsite.com` has been configured to redirect to the local computer at `192.168.1.109` (change this value to be the address of your computer). Save the change to the Hosts file and exit the Edit application.

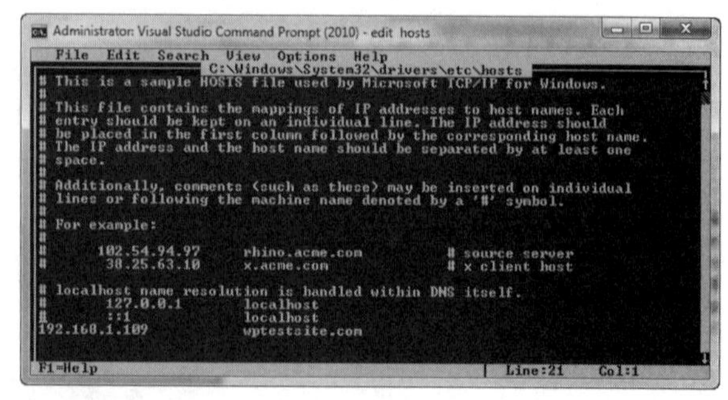

FIGURE 13-7

Before you go any further, you will need to register for an Application ID by going to the Live Services Developer Portal (`http://go.microsoft.com/fwlink/?LinkID=144070`). You will be prompted to sign in with your Windows Live ID, and then you will enter the Azure Services developer portal. From here you can configure your accounts for Windows Azure, SQL Azure, AppFabric (formerly .NET Services), and Live Services. In this instance, you want to create a new Live Services account, so make sure the "Live Services" tab is selected, then click the "New Service" link. Select "Live Services: Existing APIs," and after accepting the Terms of Use, you will see a form similar to Figure 13-8.

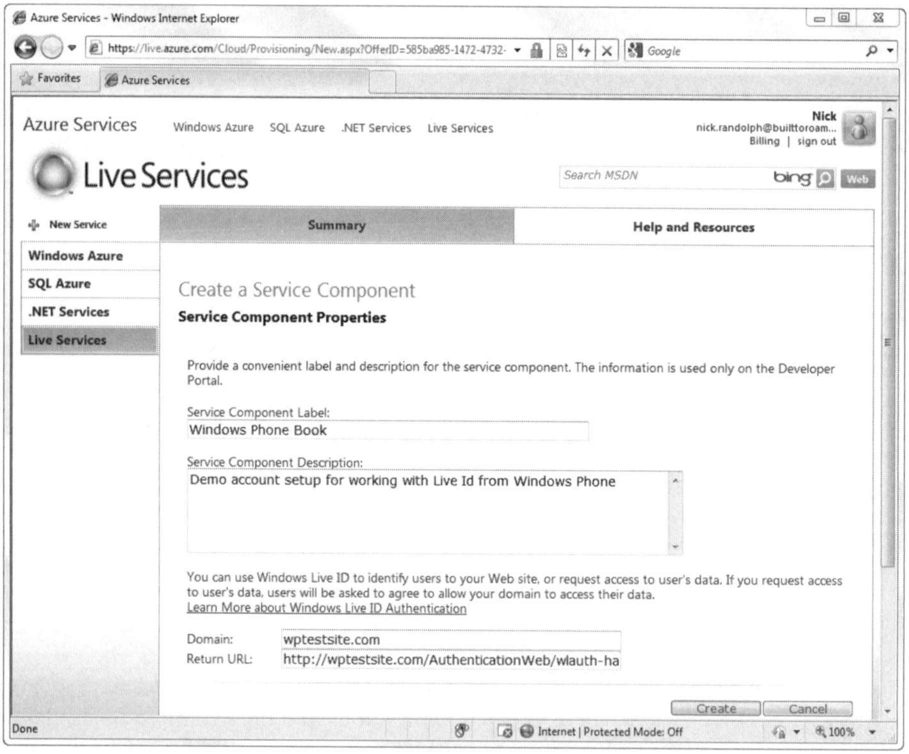

FIGURE 13-8

The Label and Description are simply there for you to identify this service within the developer portal. What's important is the name of the Domain and Return URL that you use. The Domain should be the same domain that you selected earlier when setting up the Hosts file, and the Return URL should specify the fully qualified URL of the page in your ASP.NET web application that will handle decoding the user token returned after the user has signed in at `login.live.com`. In this case, the Return URL is `http://wptestsite.com/AuthenticationWeb/wlauth-handler.aspx`, which is made up of the domain, the virtual directory you set up in IIS, and the name of the page. You'll create the wlauth-handler.aspx page a little later on.

Once you have created the Live Service, you will be presented with a summary of the new service that lists the Application ID and the Secret Key, as shown in Figure 13-9.

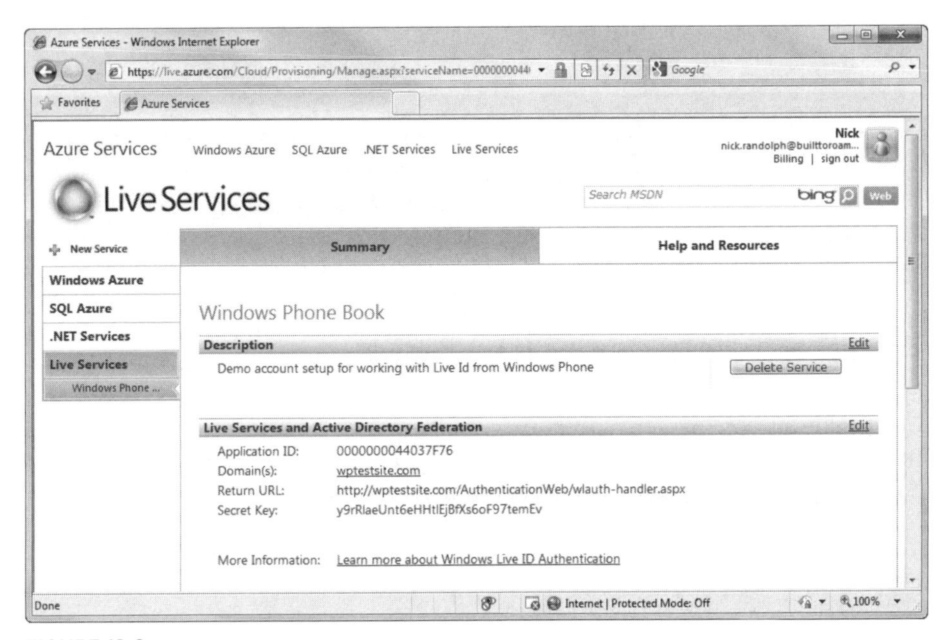

FIGURE 13-9

These values need to be recorded in the appSettings section of the web.config file for your ASP.NET web application as follows:

Available for download on Wrox.com

```
<appSettings>
    <add key="wll_appid" value="0000000044037F76"/>
    <add key="wll_secret" value="y9rRlaeUnt6eHHtlEjBfXs6oF97temEv"/>
    <add key="wll_securityalgorithm" value="wsignin1.0"/>
    <add key="wll_policyurl"
         value="http://wptestsite.com/AuthenticationWeb/policy.html"/>
    <add key="wll_returnurl"
         value="http://wptestsite.com/AuthenticationWeb/wlauth-handler.aspx"/>
</appSettings>
```

Code snippet from web.config

Note that there is an additional setting that points to a policy.html file that is required for Delegated Authentication.

The next step is to create the wlauth-handler.aspx page that will handle the response from `login.live.com`. There is quite a bit of code to do this, but luckily there are quick start samples available that can be reused within your application. There are samples for both Web and Delegated Authentication, but since the latter is a superset of the former, you only need to download and install the C# Delegated Authentication sample from `http://msdn.microsoft.com/en-us/library/cc287665(v=MSDN.10).aspx`. Once installed, copy the following files into your ASP.NET web application:

```
C:\Program Files\Windows Live ID\DelAuth\Sample2\wlauth-handler.aspx
C:\Program Files\Windows Live ID\DelAuth\Sample2\wlauth-handler.aspx.cs
C:\Program Files\Windows Live ID\DelAuth\Sample2\Policy.html
C:\Program Files\Windows Live ID\DelAuth\App_Code\WindowsLiveLogin.cs
```

Make sure that before you publish your application, you update the Policy.html file to include information about how the customer's information is used by your application.

Once the wlauth-handler.aspx page has extracted the user token, it will automatically redirect to default.aspx. If you want it to direct to a different page, you need to change the `LoginPage` constant at the top of the wlauth-handler.aspx.cs file. In the default.aspx page the user token needs to be extracted out of a cookie (used to pass the token between the wlauth-handler.aspx and default.aspx) and decoded into a `WindowsLiveLogin.User` object. The `ID` property corresponds to the user token that can be used to uniquely identify the user within the application:

```
using System;
using System.Web;
using WindowsLive;

public partial class DefaultPage : System.Web.UI.Page
{
    // user token key
    const string LoginCookie = "webauthtoken";

    // Initialize the WindowsLiveLogin module.
    static WindowsLiveLogin wll = new WindowsLiveLogin(true);

    protected string ReturnToken ="";

    protected void Page_Load(object sender, EventArgs e)
    {
        HttpRequest req = HttpContext.Current.Request;
        HttpCookie loginCookie = req.Cookies[LoginCookie];
        if (loginCookie != null)
        {
```

```
                string token = loginCookie.Value;

                if (!string.IsNullOrEmpty(token))
                {
                    WindowsLiveLogin.User user = wll.ProcessToken(token);

                    if (user != null)
                    {
                        ReturnToken = "UserToken " + user.Id;
                    }

                }
            }
        }
    }
```

Code snippet from Default.aspx.cs

The corresponding markup for the default.aspx page makes a call to `window.external.Notify`, passing in the `ReturnToken` generated in the `Page_Load` method:

```
<%@ Page Language="C#" AutoEventWireup="true" Inherits="DefaultPage"
    Codebehind="default.aspx.cs" %>

<!DOCTYPE HTML PUBLIC "-//W3C//DTD HTML 4.01 Transitional//EN">
<html>
<head>
   <meta http-equiv="Pragma" content="no-cache" />
   <meta http-equiv="Expires" content="-1" />
   <title>Windows Live ID</title>
</head>
<body>
    <script type="text/javascript">
        try{
            window.external.Notify("<%=ReturnToken%>");
        }
        catch(ex){
            alert('Host Silverlight application not available');
        }
    </script>

<a href="http://login.live.com/wlogin.srf?appid=0000000044037F76&alg=wsignin1.0">
   Login</a></br>
<a href="http://login.live.com/logout.srf?appid=0000000044037F76">Logout</a></br>
</body>
</html>
```

Code snippet from Default.aspx

At the end of this page there are two hyperlinks that allow you to test the functionality of the page (once the `appid` parameter has been replaced with the value you generated previously). Click the Login link and you should be directed to `login.live.com` and prompted to sign in. Once you do, you should be returned to the same page. Click Logout and you should see the page flicker, indicating multiple redirects — the first to `login.live.com` (which counter intuitively logs you out), then to wlauth-handler.aspx, and then finally back to default.aspx.

The appid used in the previous example must match the Application ID that was generated when you registered for a Live Services account. This is used in a number of places throughout this sample so make sure it is correct.

Most of the hard work is done; all you need to do now is add "Sign In" and "Sign Out" buttons and a `WebBrowser` control to your Windows Phone application. The following snippet includes a `TextBlock` so that you can see the user token that is returned from the `WebBrowser`. Note that the `WebBrowser` has `IsScriptEnabled` set to `true` to allow calls to be passed between the Windows Phone application and the content within the `WebBrowser` control:

```
<Grid x:Name="LayoutRoot" Background="{StaticResource PhoneBackgroundBrush}">
    <Button Content="Sign In" Height="70" HorizontalAlignment="Left"
            Name="SignInButton"
            VerticalAlignment="Bottom" Width="150" Click="SignInButton_Click" />
    <Button Content="Sign Out" Height="70" Margin="0" Name="SignOutButton"
            VerticalAlignment="Bottom" Width="150" Click="SignOutButton_Click"
            HorizontalAlignment="Right" />
    <phone:WebBrowser x:Name="AuthenticationBrowser"  IsScriptEnabled="True"
                      Grid.RowSpan="2" Margin="0,0,0,74" Visibility="Collapsed"
                      ScriptNotify="AuthenticationBrowser_ScriptNotify" />
    <TextBlock Name="UserIdText" Margin="0,0,0,75" VerticalAlignment="Bottom"
          TextWrapping="Wrap" Height="150"
          Style="{StaticResource PhoneTextNormalStyle}" />
</Grid>
```

Code snippet from AuthenticationPage.xaml

When the user clicks the "Sign In" button, the `WebBrowser` should be displayed and the `Navigate` method called to direct the user to `login.live.com`. You will notice that the URL specified is the same that was included earlier in the test hyperlinks within the ASP.NET website. The "Sign Out" button also launches the `WebBrowser`, this time navigating to the URL to log out of Live ID:

```
private void SignInButton_Click(object sender, RoutedEventArgs e){
    AuthenticationBrowser.Visibility = Visibility.Visible;
    AuthenticationBrowser.Navigate(
  new Uri("http://login.live.com/wlogin.srf?appid=0000000044037F76&alg=wsignin1.0"));
}

private void SignOutButton_Click(object sender, RoutedEventArgs e){
    UserIdText.Text = null;
    AuthenticationBrowser.Visibility = Visibility.Visible;
```

```
        AuthenticationBrowser.Navigate(
            new Uri("http://login.live.com/logout.srf?appid=0000000044037F76"));
    }
```

Code snippet from AuthenticationPage.xaml.cs

When the default.aspx page calls the `window.external.Notify` function, this raises the `ScriptNotify` event on the `WebBrowser` control. In this case, the returned value is just being used to update the `Text` property on the `TextBlock`. Since the `WebBrowser` is no longer required, it is hidden:

Available for download on Wrox.com

```
    private void AuthenticationBrowser_ScriptNotify(object sender, NotifyEventArgs e){
        UserIdText.Text = e.Value;

        AuthenticationBrowser.Visibility = Visibility.Collapsed;
    }
```

Code snippet from AuthenticationPage.xaml.cs

Figure 13-10 illustrates the process of a user signing in. The user taps the “Sign In” button, which displays the `WebBrowser` with the Live ID login page. Once he has entered their credentials and tapped the “Sign In” button, the `WebBrowser` disappears, leaving the user token in the `TextBlock`.

FIGURE 13-10

Delegated Authentication

The use of Web Authentication gives you the ability to verify the uniqueness of the user, but it gives you very little information about the actual user themselves. With Delegated Authentication you can request that the user give the application access to their personal information for a limited time period. In order to do this, you start off with the user signing in using Web Authentication. Once he has signed in, you need to prompt them for permission to access their information. This is done by again navigating the user to another external web page, referred to as the `ConsentUrl`. When the user approves access to their information, a consent or delegation token is returned via the same process as the user token for Web Authentication.

The first step is to update the server code to return the `ConsentUrl` with the initial user token after the user has signed in using Web Authentication. This will be picked up by the Windows Phone application and used to redirect the user to the page requesting their permission. The server code also has to handle extracting the delegation token. In the following code, the consent token is extracted from the application state and used to generate the `DelegationToken` and a `LocationId` (the `LocationId` is a reference to the information set you are being granted access to). These are both appended to the `ReturnToken` string, which is passed through to the `window.external.Notify` method:

Available for download on Wrox.com

```
using System;
using System.Web;
using WindowsLive;

public partial class DefaultPage : System.Web.UI.Page
{
    // user token key
    const string LoginCookie = "webauthtoken";
    // delegation token key
    const string AuthCookie = "delauthtoken";

    const string Offers = "Contacts.View";

    // Initialize the WindowsLiveLogin module.
    static WindowsLiveLogin wll = new WindowsLiveLogin(true);

    protected string ReturnToken ="";
    protected string ConsentUrl = "";


    protected void Page_Load(object sender, EventArgs e){
        ConsentUrl = wll.GetConsentUrl(Offers,
                                  "~/Content/DisplayAcctInfo/DisplayAcctInfo.aspx");

        HttpRequest req = HttpContext.Current.Request;
        HttpApplicationState app = HttpContext.Current.Application;
        HttpCookie loginCookie = req.Cookies[LoginCookie];
        if (loginCookie != null){
            string token = loginCookie.Value;
            if (!string.IsNullOrEmpty(token))
            {
                WindowsLiveLogin.User user = wll.ProcessToken(token);
                if (user != null) {
                    ReturnToken = "UserToken " + user.Id +
                                  ";ConsentUrl " + ConsentUrl;

                    if (user != null){
                        string cts = (string)app[user.Id];
```

```
                                WindowsLiveLogin.ConsentToken ct = wll.ProcessConsentToken(cts);

                                if (ct != null){
                                    if (!ct.IsValid()){
                                        if (ct.Refresh() && ct.IsValid()){
                                            app[user.Id] = ct.Token;
                                        }
                                    }

                                    if (ct.IsValid()){
                                        var Token = ct;
                                        if (!string.IsNullOrEmpty(ReturnToken))
                                            ReturnToken += ";";
                                        ReturnToken += "ConsentToken " +
                                                       Token.DelegationToken +
                                                        ";LocationId " + Token.LocationID;
                                    }
                                }
                            }
                        }
                    }
                }
            }
        }
```

Code snippet from Default.aspx.cs

There are also minor additions to the markup for the default.aspx page. The first is the addition of the `GoToConsentUrl` function, which will force the Web browser to navigate to the consent URL. The second is another hyperlink that allows you to manually test the consent process. Make sure that you log in first, then click the Consent link to take you through to the permissions page, where you can grant the application access to your personal information:

```
<%@ Page Language="C#" AutoEventWireup="true" Inherits="DefaultPage"
    Codebehind="default.aspx.cs" %>

<!DOCTYPE HTML PUBLIC "-//W3C//DTD HTML 4.01 Transitional//EN">
<html>
<head>
   <meta http-equiv="Pragma" content="no-cache" />
   <meta http-equiv="Expires" content="-1" />
   <title>Windows Live ID</title>
    <script type="text/javascript">
       function GoToConsentUrl(url) {
           window.location = url;
       }
    </script>
</head>
<body>
    <script type="text/javascript">
        try{
            window.external.Notify("<%=ReturnToken%>");
```

```
        }
        catch(ex){
            alert('Host Silverlight application not available');
        }
    </script>

<a href="http://login.live.com/wlogin.srf?appid=0000000044037F76&alg=wsignin1.0">
   Login</a></br>
<a href="http://login.live.com/logout.srf?appid=0000000044037F76">Logout</a></br>
<a href="<%=ConsentUrl%>">Consent</a></br>
</body>
</html>
```

Code snippet from Default.aspx

Once you have manually tested the delegated authentication process, the Windows Phone application can be updated to redirect the user to the consent URL once he or she has signed in. Before you can do that, you need to add some code to the `ScriptNotify` event handler to parse the additional information coming out of the `WebBrowser` control. In this case, it's paired data separated by semicolons, making it easy to split:

Available for download on Wrox.com

```
private const string UserTokenKey = "UserToken";
private const string ConsentUrlKey = "ConsentUrl";
private const string ConsentTokenKey = "ConsentToken";
private const string LocationIdKey = "LocationId";

private string UserToken = "";
private string ConsentToken = "";
private string LocationId = "";
private string ConsentUrl = "";
private void AuthenticationBrowser_ScriptNotify(object sender, NotifyEventArgs e){
    UserIdText.Text = e.Value;

    if (!string.IsNullOrEmpty(e.Value)){
        var values = e.Value.Split(';');
        foreach (var val in values){
            var bits = val.Split(' ');
            if (bits.Length == 2){
                switch (bits[0]){
                    case UserTokenKey:
                           UserToken = bits[1];
                           break;
                    case ConsentUrlKey:
                           ConsentUrl = bits[1];
                           break;
                    case ConsentTokenKey:
                           ConsentToken = bits[1];
                           break;
                    case LocationIdKey:
                           LocationId = bits[1];
                           break;
                }
            }
```

```
            }
        }

        AuthenticationBrowser.Visibility = System.Windows.Visibility.Collapsed;
    }
```

Code snippet from AuthenticationPage.xaml.cs

You should avoid attempting to invoke the `script` function while the page is still loading. The `WebBrowser` control raises a `LoadCompleted` event when the content has finished loading. In this case, you want to ensure that the `UserToken` has been set (in other words, the user has signed in) and that the `ConsentToken` hasn't already been set:

Available for download on Wrox.com

```
private bool ConsentRequestInProgress=false;
private void AuthenticationBrowser_LoadCompleted(object sender, NavigationEventArgs e){
    if (!string.IsNullOrEmpty(UserToken) &&
         string.IsNullOrEmpty(ConsentToken) &&
         !ConsentRequestInProgress){
        ConsentRequestInProgress = true;
        AuthenticationBrowser.Visibility = Visibility.Visible;
        AuthenticationBrowser.InvokeScript("GoToConsentUrl", ConsentUrl);
    }
}
```

Code snippet from AuthenticationPage.xaml.cs

Figure 13-11 illustrates the consent page that the user is presented with, requesting access to their information. The right image illustrates the application once consent has been granted. You can see that there is additional text in the `TextBlock` representing the delegation token.

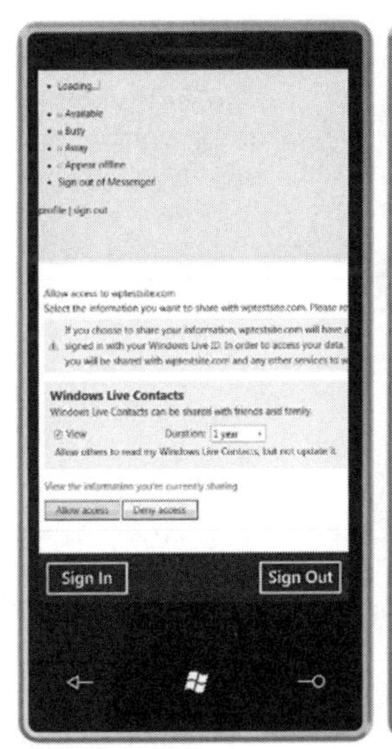

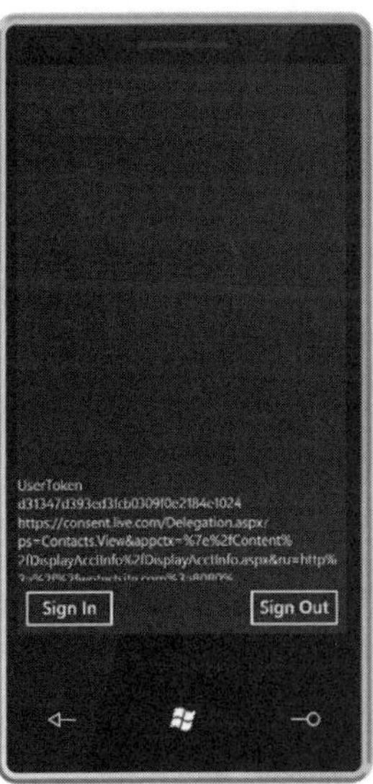

FIGURE 13-11

Personal and Contact Information

Now that you have a delegation token, what can you do with it? In actual fact, there is little that you can do directly from the device because of restrictions on how the Live ID services can be invoked. However, if you make use of a backend WCF Service, you can access a lot of information about the user and their Live ID contacts. Create a new WCF Service within your ASP.NET web application called *WindowsLiveIdService.svc*. This service is going to have public methods for getting information about the current user and for getting information about their contacts. Essentially, both of these consist of making a REST call to the live contacts service. The `GetUserInformation` method simply retrieves a URL by combining a URI template with the `locationid`. This returns an XML document serialized as a string.

```
[ServiceContract]
public interface IWindowsLiveIdService{
    [OperationContract]
    string GetUserInformation(string locationId, string delegationToken);

    [OperationContract]
    string GetContactsInformation(string locationId, string delegationToken);
}
```

Code snippet from IWindowsLiveIdService.cs

```
public class WindowsLiveIdService : IWindowsLiveIdService {
    public string GetUserInformation(string locationId, string delegationToken) {
        string uriTemplate =
 "https://livecontacts.services.live.com/@L@{0}/rest/LiveContacts/owner/";
        var xdoc = WindowsLiveContactAPIRequest(locationId,
                                               delegationToken, uriTemplate);
        return xdoc.ToString();
    }

    public string GetContactsInformation(string locationId, string delegationToken) {
        string uriTemplate =
  "https://livecontacts.services.live.com/@L@{0}/rest/LiveContacts/Contacts";
        var xdoc = WindowsLiveContactAPIRequest
                   (locationId,delegationToken,uriTemplate);
        var contacts = (from contact in xdoc.Descendants("Contact")
                        select contact).ToArray();
        foreach (var con in contacts) {
            RetrieveCID(locationId, delegationToken, con);
        }

        return xdoc.ToString();
    }

    private static void RetrieveCID(string locationId, string delegationToken,
                                    XElement con) {
        try {
            string uriTemplate =
  "https://livecontacts.services.live.com/@L@{0}/LiveContacts/Contacts/Contact(" +
  con.Element("ID").Value + ")/CID";
            var xdoc2 = WindowsLiveContactAPIRequest(locationId, delegationToken,
                                                     uriTemplate);
            var cid = xdoc2.Element("CID").Value;
            con.Add(new XElement("CID", cid));
        }
        catch (Exception) {
            // This will happen if there is no CID (ie no spaces page)
        }
    }

    private static XDocument WindowsLiveContactAPIRequest(string locationId,
```

```
                                                    string delegationToken,
                                                    string uriTemplate) {
            string uri = string.Format(uriTemplate, locationId);
            HttpWebRequest request = (HttpWebRequest)WebRequest.Create(uri);
            request.UserAgent = "Windows Phone 7 Sample";
            request.ContentType = "application/xml; charset=utf-8";
            request.Method = "GET";

            request.Headers.Add("Authorization",
                                "DelegatedToken dt=\"" + delegationToken + "\"");
            HttpWebResponse response = (HttpWebResponse)request.GetResponse();

            var xdoc = XDocument.Load(response.GetResponseStream());
            response.Close();
            return xdoc;
        }
    }
```

Code snippet from WindowsLiveIdService.cs

`GetContactsInformation` is slightly more complex since the initial REST request only provides half the information. In order to display the profile picture of a contact, you need to know what is known as their *CID*. This is not returned by default so the code iterates through each contact retrieving their CID with an additional REST call. As this might take a substantial amount of time for users with a large Contacts List, you may consider refactoring this and returning the list of contacts in pages, or performing this step asynchronously.

From the Windows Phone application, you need to add a service reference to this service; right-click on the project in the Solution Explorer, and select "Add Service Reference." Figure 13-12 illustrates how you should enter the address of the service and click Go. Once the service has been discovered, you should specify the Namespace and then click OK to proceed.

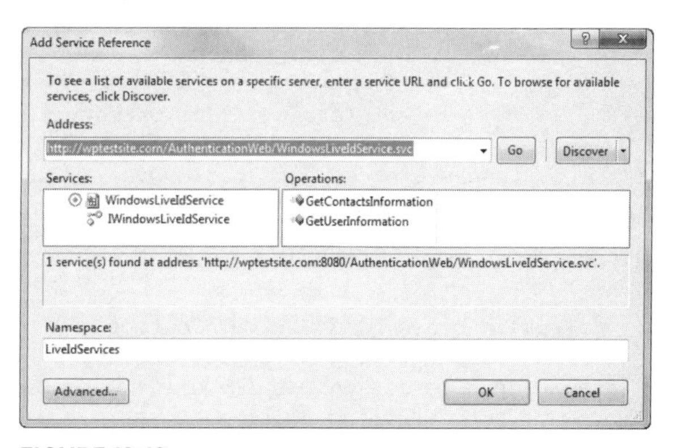

FIGURE 13-12

You will want to add some controls to display the information about the current user and their contacts. In this case, an `Image` and three `TextBlocks` are added to display the photo, first name, last name, and e-mail address of the current user. There is also a `ListBox` that will display the contacts with their profile images and display names:

Available for download on Wrox.com

```
<Grid.Resources>
    <DataTemplate x:Key="ContactItemTemplate">
        <Grid>
            <TextBlock TextWrapping="Wrap" Style="{StaticResource PhoneTextNormalStyle}"
                       Text="{Binding DisplayName}" Margin="55,0,0,0"/>
             <Image Height="50" VerticalAlignment="Top" HorizontalAlignment="Left"
```

```
                    Width="50" Source="{Binding Photo}"/>
            </Grid>
        </DataTemplate>
    </Grid.Resources>
    <Image x:Name="PhotoImage" HorizontalAlignment="Left" Height="100" Margin="25,25,0,0"
           VerticalAlignment="Top" Width="100" Source="{Binding Photo}"/>
    <TextBlock x:Name="FirstNameText" Margin="150,25,0,0" TextWrapping="Wrap"
               Text="{Binding FistName}" VerticalAlignment="Top"
               Style="{StaticResource PhoneTextNormalStyle}" HorizontalAlignment="Left"/>
    <TextBlock x:Name="LastNameText" Margin="150,50,0,0" TextWrapping="Wrap"
               Text="{Binding LastName}" VerticalAlignment="Top"
               Style="{StaticResource PhoneTextNormalStyle}" HorizontalAlignment="Left"/>
    <TextBlock x:Name="EmailText" Margin="150,75,0,0" TextWrapping="Wrap"
               Text="{Binding PrimaryEmail}" VerticalAlignment="Top"
               Style="{StaticResource PhoneTextNormalStyle}" HorizontalAlignment="Left"/>
    <ListBox x:Name="ContactsList" Margin="0,130,0,230"
             ItemTemplate="{StaticResource ContactItemTemplate}"/>
```

Code snippet from AuthenticationPage.xaml

To make setting the values of each of these controls easier, you should create two classes that will represent the information about the current user and each of their contacts:

Available for download on Wrox.com

```
public class WindowsLiveIdUser {
    public string ID { get; set; }
    public string FirstName { get; set; }
    public string LastName { get; set; }
    public string PrimaryEmail { get; set; }
    public BitmapImage Photo {
        get {
            if (string.IsNullOrEmpty(ID)) return null;
            return new BitmapImage(new Uri("http://" + ID +
  ".users.storage.live.com/MyData/MyProfile/GeneralProfile/ProfilePhoto"));
        }
    }
}

public class WindowsLiveIdContact {
    public string ID { get; set; }
    public string DisplayName { get; set; }
    public string CID { get; set; }
    public BitmapImage Photo {
        get {
            if (string.IsNullOrEmpty(CID)) return null;
            return new BitmapImage(new Uri("http://storage.live.com/users/" +
                   CID + "/myprofile/ExpressionProfile/ProfilePhoto"));
        }
    }
}
```

Code snippet from WindowsLiveIdClasses.cs

The only thing left to do is to call the service methods, populate instances of these classes, and bind them to the user interface. In the `ScriptNotify` method (most of which has been omitted for brevity), when there is a valid `ConsentToken` detected, the `LoadUserInformation` method is invoked. This starts a chain of actions that first invokes the `GetUserInformationAsync` method on the `WindowsLiveIdServiceClient`, and once a response is received invokes the `GetContactsInformationAsync` method. These are asynchronous service calls that raise an event when they complete. Each method returns with information about the user or their contacts as a string literal representation of an XML document:

Available for download on Wrox.com

```
LiveIdServices.WindowsLiveIdServiceClient liveIdClient;
public AuthenticationPage(){
    InitializeComponent();

    liveIdClient = new LiveIdServices.WindowsLiveIdServiceClient();
    liveIdClient.InnerChannel.OperationTimeout = new TimeSpan(0, 5, 0);
    liveIdClient.GetUserInformationCompleted +=
                                liveIdClient_GetUserInformationCompleted;
    liveIdClient.GetContactsInformationCompleted +=
                                liveIdClient_GetContactsInformationCompleted;
}

private void AuthenticationBrowser_ScriptNotify(object sender, NotifyEventArgs e){
    ...

    if (!string.IsNullOrEmpty(ConsentToken)){
        LoadUserInformation();
    }
}

private void LoadUserInformation(){
    liveIdClient.GetUserInformationAsync(LocationId, ConsentToken);
}

void liveIdClient_GetUserInformationCompleted(object sender,
                            LiveIdServices.GetUserInformationCompletedEventArgs e){
    var xdoc = XDocument.Parse(e.Result);

    var user= new WindowsLiveIdUser(){
                ID = LocationId,
                FirstName = xdoc.Element("Owner").Element("Profiles")
                                .Element("Personal").Element("FirstName").Value,
                LastName = xdoc.Element("Owner").Element("Profiles")
                               .Element("Personal").Element("LastName").Value,
                PrimaryEmail = xdoc.Element("Owner").Element("Emails")
                                   .Elements("Email").First().Element("Address").Value
            };
    this.Dispatcher.BeginInvoke(() =>
            {
                this.LayoutRoot.DataContext = user;
            });

    liveIdClient.GetContactsInformationAsync(LocationId, ConsentToken);
```

```
    }

    void liveIdClient_GetContactsInformationCompleted(object sender,
                            LiveIdServices.GetContactsInformationCompletedEventArgs e){
        var xdoc = XDocument.Parse(e.Result);
        var contacts = (from contact in xdoc.Descendants("Contact")
                        select new WindowsLiveIdContact(){
                                ID=contact.Element("ID").Value,
                                DisplayName=contact.Element("Profiles")
                                                .Element("Personal")
                                                .Element("DisplayName").Value,
                                CID=(contact.Element("CID")!=null?
                                                contact.Element("CID").Value:"")
                            }).ToArray();
        this.Dispatcher.BeginInvoke(() =>
                {
                    this.ContactsList.ItemsSource = contacts;
                });
    }
```

Code snippet from AuthenticationPage.xaml.cs

The XML documents are parsed and converted into objects that are then set as the data source for the relevant user interface controls. Figure 13-13 illustrates how the user's information, photo, and a list of contacts would be displayed.

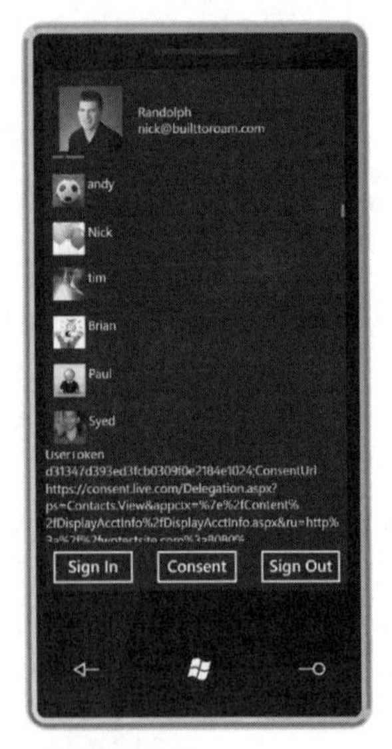

FIGURE 13-13

In order to display the photos for each contact in the list there is additional work done by the WCF Service before it returns the contact data. If the user has a large number of contacts this may take a long time to complete, resulting in the WCF Service timing out. It is recommended that this service be restructured to cater for this scenario before publishing your application.

SUMMARY

In this chapter, you've seen how to detect and handle different networking conditions. You've learned how to render Web content in your application using the `WebBrowser` control. Lastly, you saw how you can combine the `WebBrowser` control with a simple website to allow you to authenticate users using Live ID.

14

Consuming the Cloud

WHAT'S IN THIS CHAPTER

- Making basic web requests
- Using compression to reduce the size of requests
- Calling WCF services
- Working with OData services in XML and JSON

While many Windows Phone services, such as location and push notification, rely on the cloud to do some or all of their processing, it's also likely that you will want to connect to other remote services. These may be services offered by your organization either on premise or in the cloud, or they may be third-party offerings that have a cloud-based API for accessing them.

In this chapter, you will learn how to integrate with Windows Communication Foundation (WCF) and Simple Object Access Protocol (SOAP) services. You will also learn how you can reduce your message overhead through the use of RESTful (Representational State Transfer) services.

HTTP REQUEST

When you need to access content from the Web, whether it be a service or static content such as images or documents, Windows Phone gives you a couple of options as to how you access this content. However, they all boil down to performing an HTTP request. Windows Phone 7 does not provide support for raw Transmission Control Protocol (TCP) or User Datagram Protocol (UDP) socket-level communication, which means that HTTP requests are the lowest level of communication you can use. In this section, you'll learn how to use both the WebClient and HttpWebRequest classes to access content across the Web. Following this, you'll see how you can use WCF to provide a higher-level wrapper for making service requests.

One thing that you will notice about using either the WebClient or HttpWebRequest class is that all network calls — for example, to download or upload content — are asynchronous. Traditionally, you've had the choice between making synchronous or asynchronous calls, with the latter being avoided owing to additional complexity. However, using asynchronous calls when doing long-running, or high-latency, tasks is preferable as it reduces the possibility of building an unresponsive user interface (UI). Given Silverlight's heritage as a web technology, and thus heavily reliant on possibly high-latency network calls, and the need to keep the framework as small as possible, it made sense to only provide asynchronous networking calls.

WebClient

If you simply want to download content from a specific URL, then the WebClient class is by far the simplest way to go. To download the content, you create an instance of the WebClient, attach an event handler that will get invoked when the download is complete, and then start the download. In the following example, an XML file is downloaded using the DownloadStringAsync method. There are two event handlers attached to the WebClient instance, one to get progress information, the other a notification when the download has completed. The second parameter to the DownloadStringAsync method is an arbitrary object or identifier that can be accessed in the event handlers to identify which download the event is for. In this way, a single event handler can be reused among multiple instances of the WebClient class:

Available for download on Wrox.com

```
WebClient client = new WebClient();
public MainPage(){
    InitializeComponent();

    client.DownloadProgressChanged += client_DownloadProgressChanged;
    client.DownloadStringCompleted += client_DownloadStringCompleted;
    client.OpenReadCompleted += client_OpenReadCompleted;
}

private void WebClientButton_Click(object sender, RoutedEventArgs e){
    client.DownloadStringAsync(
                   new Uri("http://localhost/ServicesApplication/rssdump.xml"),
                   "sample rss");
}

void client_DownloadProgressChanged(object sender, DownloadProgressChangedEventArgs e){
    if (e.UserState as string == "sample rss"){
        this.DownloadProgress.Value = e.ProgressPercentage;
    }
}

void client_DownloadStringCompleted(object sender, DownloadStringCompletedEventArgs e){
    this.DownloadedText.Text = e.Result;
}
```

Code snippet from MainPage.xaml.cs

Whenever an event is raised by the `WebClient`*, you will notice that updates are made directly on the user interface. While* `WebClient` *operations, such as* `DownloadStringAsync`*, are performed on a background thread, the events they generate are automatically raised on the same thread that the* `WebClient` *instance was created on. In this case, because it was created on the UI thread, the events will be raised on that thread, removing the requirement to use* `Dispatcher.BeginInvoke` *to wrap any UI updates.*

The `DownloadStringAsync` method is good for downloading content that you know can be represented as a string, for example, XML or text documents. If you need to download binary content — for example, an image — you can use the `OpenReadAsync` method, which opens a binary stream so that you can progressively read the downloaded content:

```
private void WebClientButton2_Click(object sender, RoutedEventArgs e){
    client.OpenReadAsync(new Uri("http://localhost/ServicesApplication/desert.jpg"),
                         "my picture");
}

private void client_OpenReadCompleted(object sender, OpenReadCompletedEventArgs e){
    var strm = e.Result;
    var img = new BitmapImage();
    img.SetSource(strm);
    this.SampleImage.Source = img;
}
```

Code snippet from MainPage.xaml.cs

One thing to be aware of is that when you use the `OpenReadAsync` method, the `DownloadProgressChanged` event is not raised. If you want to update a progress bar with the download status, you will have to do this yourself. For example, the following code divides the content into 100 sections and reports download progress when each chunk has been read from the stream:

```
private void client_OpenReadCompleted(object sender, OpenReadCompletedEventArgs e){
    var strm = e.Result;

    var ms = new MemoryStream((int)strm.Length);
    var buffer = new byte[strm.Length / 100];
    var cnt = 0;
    var progress = 0;
    while (strm.Position < strm.Length){
        cnt = strm.Read(buffer, 0, buffer.Length);
        ms.Write(buffer, 0, cnt);

        progress++;
```

```
            this.Dispatcher.BeginInvoke(() =>{
                        this.DownloadProgress.Value = progress;
                    });
        }
        ms.Seek(0, SeekOrigin.Begin);
        var img = new BitmapImage();
        img.SetSource(ms);
        this.Dispatcher.BeginInvoke(() =>{
                this.SampleImage.Source = img;
            });
    }
```

Code snippet from MainPage.xaml.cs

Figure 14-1 illustrates the use of both `DownloadStringAsync` and `OpenReadAsync` to download an XML document (middle image) and an image (rightmost image), respectively. The leftmost image illustrates the progress bar indicating the download status.

FIGURE 14-1

Uploading Content

The `WebClient` can also be used to upload content using either the `UploadStringAsync` (for string data) or `OpenWriteAsync` (for binary data). Before you see an example of how this works, you'll create a simple service in an ASP.NET web application that can receive the content being uploaded. Start by creating a new WCF Service called SimpleService and defining the contract and implementation as follows:

Available for download on Wrox.com

```
[ServiceContract]
public interface ISimpleService{
    [WebInvoke]
    [OperationContract]
    bool FileUpload(Stream input);
}
```

Code snippet from ISimpleService.cs

```
<%@ ServiceHost Language="C#" Debug="true"
                Service="ServicesApplication.SimpleService"
                CodeBehind="SimpleService.svc.cs"
                Factory= "System.ServiceModel.Activation.WebServiceHostFactory"%>
```

Code snippet from SimpleService.svc

```
public class SimpleService : ISimpleService{
    public bool FileUpload(Stream input){
       var reader = new StreamReader(input);
       var txt = reader.ReadToEnd();
       if (txt.Length > 0){
           return true;
       }
       else{
           return false;
       }
    }
}
```

Code snippet from SimpleService.cs

In addition to the usual WCF `ServiceContract` and `OperationContract` attributes, this service has a `WebInvoke` attribute applied to the `FileUpload` method. This indicates that this method can be invoked via a REST service call. Typically, service calls are wrapped in a SOAP envelope, which WCF parses and then strips before sending the payload along to the appropriate method. In mobile applications it is not uncommon for the SOAP envelope to consume more space than the payload itself. REST provides a much lighter mechanism encoding the method to be invoked within the service's URL itself. To enable the REST end point, you should update the web.config file with the following system.serviceModel section, which defines an endpoint that uses the webHttpBinding. If this section is missing from the web.config file, you can insert it just before the closing configuration tag:

Available for download on Wrox.com

```
<system.serviceModel>
    <behaviors>
        <endpointBehaviors>
            <behavior name="WebBehavior">
```

```
                <webHttp />
            </behavior>
        </endpointBehaviors>
    </behaviors>
    <services>
        <service name="ServicesApplication.SimpleService">
            <endpoint address="pox" binding="webHttpBinding"
                      contract="ServicesApplication.ISimpleService"
                      behaviorConfiguration="WebBehavior" />
        </service>
    </services>
</system.serviceModel>
```

Code snippet from web.config

In order to invoke the `FileUpload` method via REST, you need to make an http POST to the following URL:

```
http://localhost/ServiceApplication/SimpleService.svc/FileUpload
```

There are four components to this URL:

1. The host, which in this case is localhost as the ASP.NET web application is running via IIS
2. The virtual directory, which in this case is the name of the ASP.NET web application, ServiceApplication. You can change this by changing the name of the virtual directory within IIS.
3. The base address of the service, which is SimpleService.svc corresponding to the service name. If you wish to remove the .svc extension, you can define a URL Rewrite rule (download URL Rewrite via the Microsoft Web Platform Installer for IIS 7). Add the following code to the Web.Config file for your ASP.NET web application which defines a URL rewrite to change SimpleService to SimpleService.svc.

```
<system.webServer>
  <rewrite>
    <rules>
      <rule name="RemoveSvc" stopProcessing="true">
        <match url="^SimpleService/(.*)$"/>
        <action type="Rewrite"
            url="SimpleService.svc/{R:1}" />
      </rule>
    </rules>
  </rewrite>
</system.webServer>
```

Code snippet from web.config

After making this change the new URL is:

```
http://localhost/ServiceApplication/SimpleService/FileUpload
```

4. The last part is the method to be invoked, which in this case is the `FileUpload` method.

Invoking this method from your Windows Phone application is just a matter of calling the `OpenWriteAsync` method and then writing content to the request stream that is accessible via the `OpenWriteCompleted` event handler:

```
private void WebClientButton3_Click(object sender, RoutedEventArgs e){
    client.OpenWriteAsync(
            new Uri("http://localhost/ServicesApplication/SimpleService/FileUpload"),
            "POST");
}

private void client_OpenWriteCompleted(object sender, OpenWriteCompletedEventArgs e){
    using (var strm = e.Result){
        var bytesToWrite= RawContent();
        strm.Write(bytesToWrite,0,bytesToWrite.Length);
        strm.Flush();
    }
}

private byte[] RawContent(){
    var content="Test content to be uploaded";
    var raw = Encoding.UTF8.GetBytes(content);
    return raw;
}
```

Code snippet from MainPage.xaml.cs

In this case, the use of the `OpenWriteAsync` method is overkill since the content being uploaded is actually a string. This could be replaced by using the `UploadStringAsync` method as follows:

```
client.UploadStringAsync(
            new Uri("http://localhost/ServicesApplication/SimpleService/FileUpload"),
            "POST", "Test content to be uploaded");
```

As with `DownloadStringAsync`, there are events for `UploadProgressChanged` and `UploadStringCompleted` that allow your application to indicate upload progress and completion to the user.

HttpWebRequest

The `WebClient` class is useful for doing simple HTTP operations such as downloading a file. However, in cases in which you need more control over the HTTP request, you should use the `HttpWebRequest` class. This allows you to set headers, specify the content type of the request, and stream data into the request. Using the `HttpWebRequest` class is not too dissimilar from the `WebClient`. The `WebClient` uses an event callback to indicate that either a download has completed or that the application can begin reading. When you use the `HttpWebRequest`, you supply a callback method as part of the request. This method gets invoked when the response is ready to be processed. The following code downloads an XML file using an `HttpWebRequest` instance instead of a `WebClient`:

Available for download on Wrox.com

```
private void HttpWebRequestButton_Click(object sender, RoutedEventArgs e){
    var req = HttpWebRequest.Create(
                        new Uri("http://localhost/ServicesApplication/rssdump.xml"))
                                                                as HttpWebRequest;
    req.BeginGetResponse(HttpWebRequestButton_Callback, req);
}

private void HttpWebRequestButton_Callback(IAsyncResult result){
    var req = result.AsyncState as HttpWebRequest;
    var resp = req.EndGetResponse(result);
    var strm = resp.GetResponseStream();
    var reader = new StreamReader(strm);

    this.Dispatcher.BeginInvoke(() =>{
            this.DownloadedText.Text = reader.ReadToEnd();
            this.TextViewer.Visibility = System.Windows.Visibility.Visible;
        });
}
```

Code snippet from MainPage.xaml.cs

The callbacks for the `HttpWebRequest` *are done on a background thread. This means that you don't need to interrupt the user interface to process the response. However, it does mean that if you want to update any UI element, you need to switch back to the UI thread using* `Dispatcher.BeginInvoke`*, as in this example.*

Uploading Content

As with the `WebClient`, you can use the `HttpWebRequest` object to upload content. Instead of calling `BeginGetResponse` immediately, you need to call `BeginGetRequestStream`. The callback for this method will allow you to write data to the request stream that will be uploaded. Once you've completed writing to the request stream, then you call `BeginGetResponse`. When the response is available, the appropriate callback will be invoked:

Available for download on Wrox.com

```
private void HttpWebRequestButton2_Click(object sender, RoutedEventArgs e){
    HttpWebRequest req = HttpWebRequest.Create(
            new Uri("http://localhost/ServicesApplication/SimpleService/FileUpload"))
                                                                as HttpWebRequest;
        req.Method = "POST";
        req.BeginGetRequestStream(HttpWebRequestButton2_RequestCallback, req);
}

private void HttpWebRequestButton2_RequestCallback(IAsyncResult result){
    var req = result.AsyncState as HttpWebRequest;
```

```
        using(var strm = req.EndGetRequestStream(result)){
                var bytesToWrite = RawContent();
                strm.Write(bytesToWrite, 0, bytesToWrite.Length);
                strm.Flush();
        }

        req.BeginGetResponse(HttpWebRequestButton_Callback, req);
    }
```

Code snippet from MainPage.xaml.cs

Cookies

Where the `HttpWebRequest` becomes more useful is if you have to perform a sequence of requests. Such an example might be for a pair of service methods, wherein the first is used to authenticate the user before the second fetches the required data. A common solution in this scenario is for the authentication service method to return a cookie that must then be passed along with all other service method requests. The `HttpWebRequest` facilitates this through a `CookieContainer` that is updated upon each response to define the cookies for any subsequent requests. Let's take a look at the method for authenticating the user:

Available for download on Wrox.com

```
[WebGet(UriTemplate="SignIn/{username}/{password}")]
[OperationContract]
bool SignIn(string username, string password);
```

Code snippet from ISimpleService.cs

```
public bool SignIn(string username, string password){
    if (string.IsNullOrEmpty(username) || string.IsNullOrEmpty(password)) return false;
    // Add additional logic to authenticate the user here
    WebOperationContext.Current.OutgoingResponse.Headers["Set-Cookie"] =
                        "username=" + username + "&password=" + password + "; path=/";
    return true;
}
```

Code snippet from SimpleService.svc.cs

This example does little more than verify that the username and password have been specified. A more complete implementation would verify the specified credentials against a database or similar data store. If you look at the interface definition, you will notice that this uses `WebGet` instead of `WebInvoke`. This indicates that this is an HTTP GET operation and that the format of the URL should be the method name, `SignIn`, followed by the two parameters separated by a "/" symbol. Assuming that a valid username and password are supplied, a cookie is attached to the outgoing response. Typically, you'd both want to add more logic to authenticate the user and hash the username/password combination so that it's not easily identifiable.

The second method in this sequence looks at the incoming request and determines if the authentication cookie has been specified. If they have been specified, the `FileUpload` is allowed to proceed:

Available for download on Wrox.com

```
[WebInvoke]
[OperationContract]
bool FileUploadCookieRequired(Stream input);
```

Code snippet from ISimpleService.cs

```
public bool FileUploadCookieRequired(Stream input){
    var authHeader = WebOperationContext.Current.IncomingRequest.Headers["Cookie"];
    if (string.IsNullOrEmpty(authHeader)) return false;
    var cookies = authHeader.Split('&');
    var authCookiesFound = (from cookie in cookies
                            where cookie.StartsWith("username") ||
                                  cookie.StartsWith("password")
                            select cookie).Count() == 2;
    if (!authCookiesFound) return false;
    return FileUpload(input) ;
}
```

Code snippet from SimpleService.svc.cs

To manage this cookie within your Windows Phone application, all you need to do is create a `CookieContainer` that can store any relevant cookies between consecutive `HttpWebRequest` calls:

Available for download on Wrox.com

```
private void HttpWebRequestButton3_Click(object sender, RoutedEventArgs e){
    string username = "Nick";
    string password = "MyPassword";
    HttpWebRequest req = HttpWebRequest.Create(
                 new Uri("http://localhost/ServicesApplication/SimpleService/SignIn/"
                                              + username + "/" + password))
                                                     as HttpWebRequest;
    req.CookieContainer = new CookieContainer();
    req.BeginGetResponse(SignIn_Callback, req);
}

private void SignIn_Callback(IAsyncResult result) {
    var req = result.AsyncState as HttpWebRequest;
    var resp = req.EndGetResponse(result);
    resp.Close();

    HttpWebRequest newReq = HttpWebRequest.Create(
new Uri("http://localhost/ServicesApplication/SimpleService/FileUploadCookieRequired"))
                                                          as HttpWebRequest;
     newReq.CookieContainer = req.CookieContainer;
```

```
        newReq.Method = "POST";
        newReq.BeginGetRequestStream(HttpWebRequestButton2_RequestCallback, newReq);
    }
```

Code snippet from MainPage.xaml.cs

The first `HttpWebRequest` calls the `SignIn` method, passing the username and password via the URL. An instance of the `CookieContainer` is attached to the first request in order to capture the cookie returned in the response from this method. The `CookieContainer` instance is then passed to the second `HttpWebRequest` object so that the cookie will be sent back to the server as part of the request.

One point to remember here is that this is just a simple scenario to demonstrate how you can share cookies between requests. As the example currently stands, credentials are transmitted in plaintext, and there is little protection to stop someone faking an authentication cookie. If you were going to use this in a production system, you would need to either make sure that the credentials are encrypted or use an SSL (Secure Sockets Layer) channel to communicate over.

Credentials

Unless you're accessing public data, you're going to need to provide some form of identification. In the previous section, you saw a contrived example for authenticating against a server. However, the most common forms of authentication you are likely to come across within Windows Phone 7 development are Basic, Digest, Forms, and Windows/NTLM. There is built-in support within both the `WebClient` and `HttpWebRequest` classes for supplying credentials for Basic authentication within your request. Digest, Forms, and Windows/NTLM Authentication are not supported out-of-the-box. The following code accesses content in a folder that has been set to use Basic Authentication. You can use IIS Manager to configure the authentication required for individual folders within your web application.

Available for download on Wrox.com

```
private void WebClientButton2_Click(object sender, RoutedEventArgs e){
    client.Credentials = new NetworkCredential("Nick", "MyPassword");
    client.OpenReadAsync(
            new Uri("http://localhost/ServicesApplication/BasicAuth/desert.jpg"));
}

private void HttpWebRequestButton4_Click(object sender, RoutedEventArgs e){
    HttpWebRequest req = HttpWebRequest.Create(
                new Uri("http://localhost/ServicesApplication/BasicAuth/rssdump.xml"))
                                                        as HttpWebRequest;
    req.Credentials = new NetworkCredential("Nick", "MyPassword");
    req.BeginGetResponse(HttpWebRequestButton_Callback, req);
}
```

Code snippet from MainPage.xaml.cs

Compression

So far, all of the requests have been using uncompressed data. In building a Windows Phone application, you have to be aware of network availability and bandwidth costs. This is especially true if your application needs to transfer data over cellular-based networks, which are often an order of magnitude slower than networks you may be used to in a home or work PC-based environment. If you need to transfer a large quantity of information between your application and a remote service, you should consider compressing the data.

You can leverage the compression support within IIS 7 to automatically compress data on the server side before it is transferred via HTP, significantly reducing the size and cost of data being downloaded to your application. IIS supports both static and dynamic compression of content. The former is applied to content types, such as images, that do not change over time and that the compressed data can be cached, whereas the latter is applied to dynamic content as it is served by IIS. In both cases, content is only compressed when a request to IIS is made that includes the Accept-Encoding header, set to "gzip" or "deflate" depending on the type of compression. This indicates that the client can accept compressed content.

The first thing to do is to ensure that you have enabled compression within IIS. Static Compression is installed by default. However, if you want to compress data coming from a service, you should also install the Dynamic Content Compression module available via the Microsoft Web Platform Installer (available from `www.microsoft.com/web/downloads/platform.aspx`) as shown in Figure 14-2.

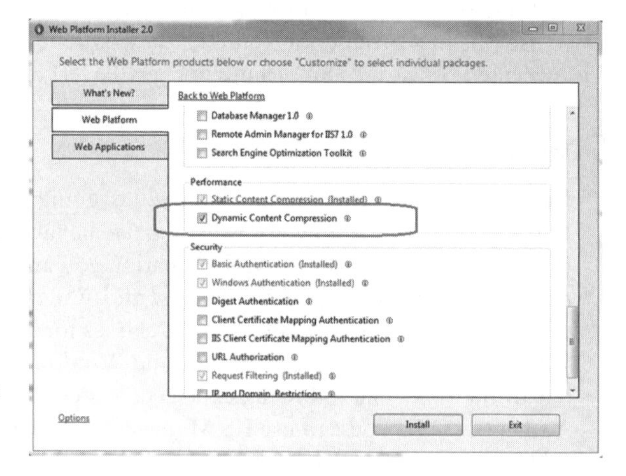

FIGURE 14-2

Once the Compression module has been installed, open IIS Manager and navigate to the virtual directory that you want to enable compression for. Select the Compression feature and ensure that both checkboxes are enabled, as shown in Figure 14-3.

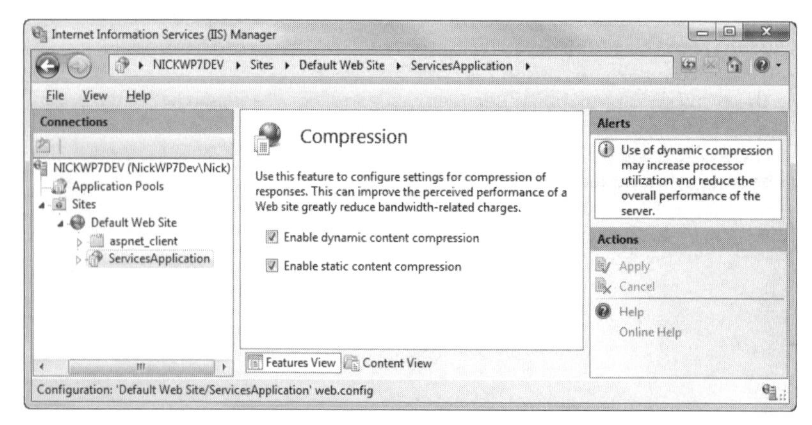

FIGURE 14-3

By default, this only enables compression for text and x-javascript MIME (Multipurpose Internet Mail Extensions) types. For completeness, you'll enable dynamic compression for all SOAP, JSON, and REST services. This won't affect any of your existing services or clients as compression is a feature the client must opt into. Any existing clients will continue to receive uncompressed data. To include additional MIME types for dynamic compression, select the root server node (not a virtual directory or website) in IIS Manager and select "Configuration Editor" from the Features view. From the dropdown box at the top of the screen, select "system.webServer/httpCompression" and then select the `dynamicTypes` property to edit. This will display a dialog similar to that shown in Figure 14-4.

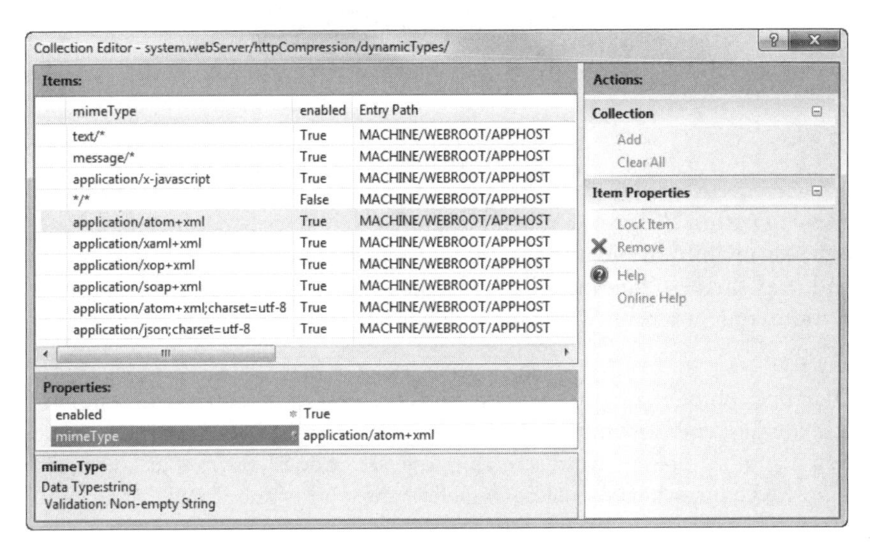

FIGURE 14-4

Make sure that all the different MIME types are present as they will ensure that dynamic content compression is enabled for SOAP, JSON, and REST service calls. You can also add the image/jpg MIME type to the `staticTypes` property to enable static compression for JPEG images. Once you have done this, make sure you apply the changes and then "Restart IIS." You must do this; otherwise, you will continue to get uncompressed data.

To ensure that it is working, you can use a tool like Fiddler (`www.fiddler2.com`) to validate that the data is returned compressed. Run Fiddler and select the Request Builder tab. Make sure the method type is set to GET and enter the URL for an image contained within your web application (e.g., **`http://localhost/ServicesApplication/desert.jpg`**). Into the "Request Headers" box, enter **Accept-Encoding: gzip, deflate** and click the Execute button. The first time you do this, you may get the response shown on the right of Figure 14-5, which is uncompressed. If you then repeat the request, you should see the response shown on the left of Figure 14-5, which indicates that the response is compressed. If you click on the yellow Notice bar, Fiddler will decompress the image for you, again showing the response in the right image.

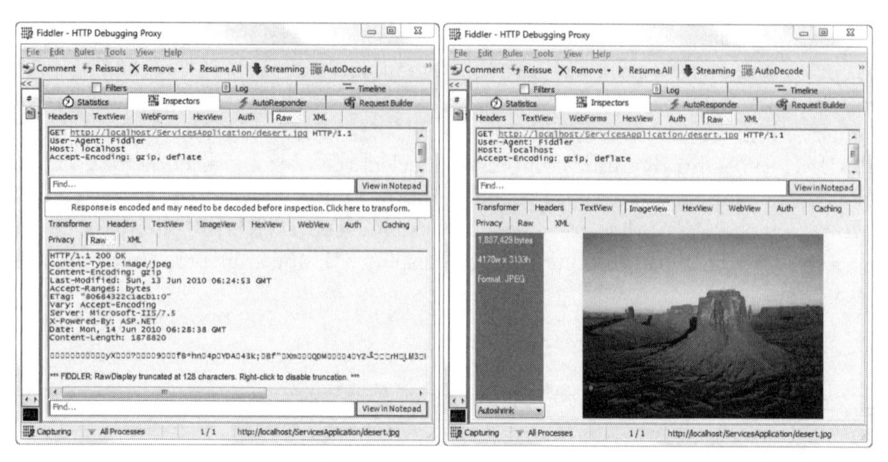

FIGURE 14-5

The reason for you not getting compression the first time the request is made is to do with the way that static compression within IIS works. When a resource is first accessed, it is immediately returned uncompressed. Meanwhile, asynchronously in the background, the resource is compressed and cached. Any further requests will receive the compressed version. This process only works well with static content where two or more requests to the same URI are expected to return identical responses.

On the other hand, Dynamic compression doesn't cache the compressed version of the content. This means that each incoming request will take slightly longer to process because of the response needing to be regenerated and compressed from scratch. However, it does ensure that all requests are serviced with compressed data, which may be of benefit if the data is quite large.

IIS will serve compressed data when it receives a request with the Accept-Encoding header set. However, it is not possible to set this header when making a request from a Windows Phone application as it is one of the restricted headers. A solution is to specify a custom header value in the request from the Windows Phone application, and have IIS translate it into the appropriate Accept-Encoding header value before processing the request. This can be done by defining a URL Rewrite rule that will detect the custom header (in this case a header called Compress-Data with a value of "true") and rewrite the request to include the Accept-Encoding header with a value of "gzip,deflate."

By default the URL Rewrite module is not permitted to set the Accept-Encoding header (referred to as a server side variable when working with rewrite rules). To override this behavior you have to permit the rewrite section of the web.config file to override the set of server side variables that can be modified. Locate and open the IIS configuration file, `applicationHost.config` (typically found at `C:\Windows\System32\inetsrv\config`). Modify the allowedServerVariables section within the rewrite sectionGroup to set the overrideModeDefault to **`Allow`**.

```
<configuration>
    <configSections>
        <sectionGroup name="system.webServer">
            <sectionGroup name="rewrite">
                <section name="allowedServerVariables" overrideModeDefault="Allow" />
```

Next, you just need to add the following rewrite rule to the system.webServer section of the web.config file of your ASP.NET web application.

```
<system.webServer>
    <rewrite>
      <allowedServerVariables>
        <add name="HTTP_ACCEPT_ENCODING" />
      </allowedServerVariables>
      <rules>
        <rule name="RewriteWithAcceptEncoding"
              patternSyntax="Wildcard" stopProcessing="false">
          <match url="*" />
          <conditions>
            <add input="{HTTP_COMPRESS_DATA}" pattern="true" />
          </conditions>
          <action type="None" />
          <serverVariables>
            <set name="HTTP_ACCEPT_ENCODING" value="gzip,deflate" />
          </serverVariables>
        </rule>
      </rules>
    </rewrite>
  </system.webServer>
```

Code snippet from web.config

From your Windows Phone application in order to request compressed data you just need to include the Compress-Data header with a value of true. Such a request will be intercepted by the RewriteWithAcceptEncoding rule that identifies the Compress-Data header (note that this is specified as the HTTP_COMPRESS_DATA input value - the rule is for the header name to be converted to uppercase, "-" is replaced with "_" and the "HTTP_" prefix appended to indicate it is a header). It then sets the Accept-Encoding header (again the header name is converted to server variable name HTTP_ACCEPT_ENCODING) to "gzip, deflate", which will in turn cause IIS to return compressed data.

Unlike the server, where there is built-in support for compression, Windows Phone does not automatically handle decompressing HTTP responses. To decompress the response, you will need to either write your own decompression library or use a third-party library. You're going to use SharpZipLib (`http://sharpdevelop.net/OpenSource/SharpZipLib/`). Unfortunately, out-of-the-box there are no Windows Phone projects for SharpZipLib, so you will need to create your own Windows Phone class library and include the source files. You will need to define the following additional compilation constants, `NETCF_2_0 NETCF` (see the Build tab of the project Properties window), and then fix any remaining build errors, of which there will be a few minor ones that are easily resolved. Once completed, you can add a reference to this project to your Windows Phone application.

In order for your Windows Phone application to tell the server that it is willing to accept compressed HTTP responses, you need to include the Compress-Data header set to `true` within the HTTP request. Setting this header is only a request, so once a response is received, you must check if it was actually compressed. This can be done by checking if the response includes the Content-Encoded header set to GZIP, which matches the compression type you requested. Depending on whether this header is set will determine whether the content needs to be decompressed or not.

```
private void HttpWebRequestButton5_Click(object sender, RoutedEventArgs e){
    HttpWebRequest req = HttpWebRequest.Create(
                    new Uri("http://localhost/ServicesApplication/desert.jpg"))
                                                            as HttpWebRequest;
    req.Headers["Compress-Data"] = "true";
    req.BeginGetResponse(HttpWebRequestButton_CompressedCallback, req);
}

private void HttpWebRequestButton_CompressedCallback(IAsyncResult result){
    var req = result.AsyncState as HttpWebRequest;
    var resp = req.EndGetResponse(result);
    var strm = resp.GetResponseStream();

    var ms = new MemoryStream();
    if ((resp.Headers["Content-Encoding"] + "").Contains("gzip")){
        using (var gzip = new GZipInputStream(strm)){
            gzip.PipeTo(ms);
        }
    }
    else {
        strm.PipeTo(ms);
    }
    ms.Seek(0, SeekOrigin.Begin);

    this.Dispatcher.BeginInvoke(() =>
            {
                var img = new BitmapImage();
                img.SetSource(ms);
                this.SampleImage.Source = img;
            });
}
```

Code snippet from Utilities.cs

This code uses an extension method, `PipeTo`, to transfer the contents of one stream into another stream:

```
public static class Utilities{
    public static void PipeTo(this Stream inputStream, Stream outputStream){
        var buffer = new byte[1000];
        var bytesRead = 0;
        while ((bytesRead = inputStream.Read(buffer, 0, buffer.Length)) > 0){
            outputStream.Write(buffer, 0, bytesRead);
        }
    }
}
```

Code snippet from Utilities.cs

If you run this code a couple of times, you should see that for the first one, or possibly two, times the content is downloaded uncompressed. Thereafter it will be downloaded compressed. IIS static compression sets an expiry on cached content, so after a period of idle time the content will be flushed, resulting in it being downloaded uncompressed the next time. If you want to ensure that the content is always compressed, consider changing the image/jpeg MIME type to dynamic content compression within IIS Manager.

WCF/ASMX SERVICES

`WebClient` and `HttpWebRequest` both provide a simple way to upload and download content from remote servers. This simplicity fades if you attempt to communicate with SOAP-based services such as an ASMX Web Service or a WCF Service. Luckily, Visual Studio gives you the ability to automatically generate proxy classes that can wrap up the complexities of calling services which utilize these higher-level specifications.

Service Configuration

By default, ASMX Web Services don't require any changes to make them accessible from a Windows Phone application. However, the same may not necessarily be true for WCF-based services. Owing to the constraints of the .NET Compact Framework, however, it is not possible to use some of the more sophisticated WCF bindings such as `wsHttpBinding`. This means that if you have a WCF Service that you want to access from your Windows Phone application, you will have to ensure that the configuration for that service uses the `basicHttpBinding`.

As you saw earlier in the chapter, you can also access WCF Services that use the `webHttpBinding`. *However, you can't use the Visual Studio "Add Service Reference" functionality to wrap calls to such a service. Instead, you have to use the* `HttpWebRequest` *or* `WebClient` *classes to manually issue service requests.*

You're going to create a simple service that returns a list of contacts from an online data store. Start by adding a new WCF Service called *ContactService.svc* to an ASP.NET web application. If you are creating this service in .NET Framework v4, then there will be no entry in your web.config file. This might come as a surprise, but the new simplified configuration system means that, by default, when you place a WCF Service within a virtual directory, that directory becomes the base address of the service and `basicHttpBinding` is automatically assigned to the HTTP protocol. This means that out-of-the-box a new WCF Service running on .NET Framework 4.0 is accessible to Windows Phone-based applications.

If, however, you need to tailor the configuration of your WCF Service — for example, to increase the maximum request or response size — or your WCF Service is running on an older version of the .NET Framework, you will need to know how to configure your service to use the basicHttpBinding. The easiest way to do this is with the WCF Service Configuration Editor, which is accessible via the Tools menu shown in the left image of Figure 14-6.

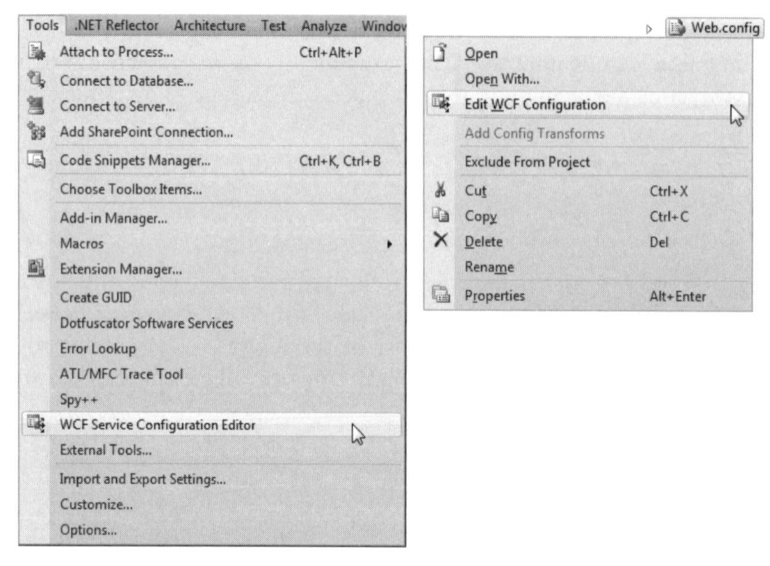

FIGURE 14-6

Once you have opened the Editor for the first time, you will notice that a new item appears in the right-click context menu for the web.config file within Solution Explorer. Selecting "Edit WCF Configuration" from this right-click menu is the quickest way to open the web.config file in the Editor. After you've opened the web.config file in the Editor, you should see two main panes, as shown in Figure 14-7.

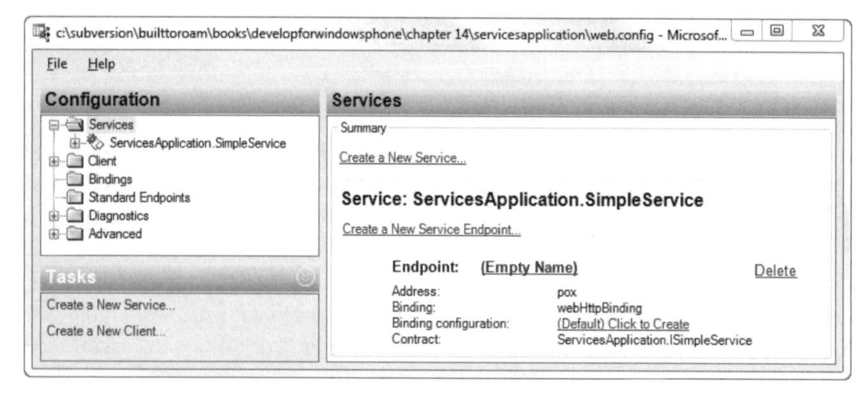

FIGURE 14-7

In this case, the REST service you created earlier is listed under the Services node in the Configuration pane but not the ContactService. If you are working with an older version of the .NET Framework, there may be a node for the ContactService. If it is missing, you can add ContactService to the list of Services by selecting "Create a New Service" from the Services pane and following the New Service Element Wizard that appears. Follow these steps to set up the configuration for your service:

1. Enter **ServicesApplication.ContactService** as the service type. If you click the Browse button, you can navigate to the bin folder, followed by double-clicking the `ServicesApplication` assembly.
2. The contract should be correctly selected as "ServicesApplication.IContactService."
3. The communication mode should remain as HTTP.
4. As you want to ensure accessibility from Windows Phone applications, you need to continue with the Basic Web Services interoperability mode.
5. Leave the Address field empty (ignoring the warning prompt).

Once you have confirmed the details of your new service configuration, you should be returned to the Configuration Editor with the newly created service end point selected, as shown in Figure 14-8. If you already had a service configuration for ServicesApplication.ContactService, you need to make sure that the Binding for your end point is set to `basicHttpBinding`.

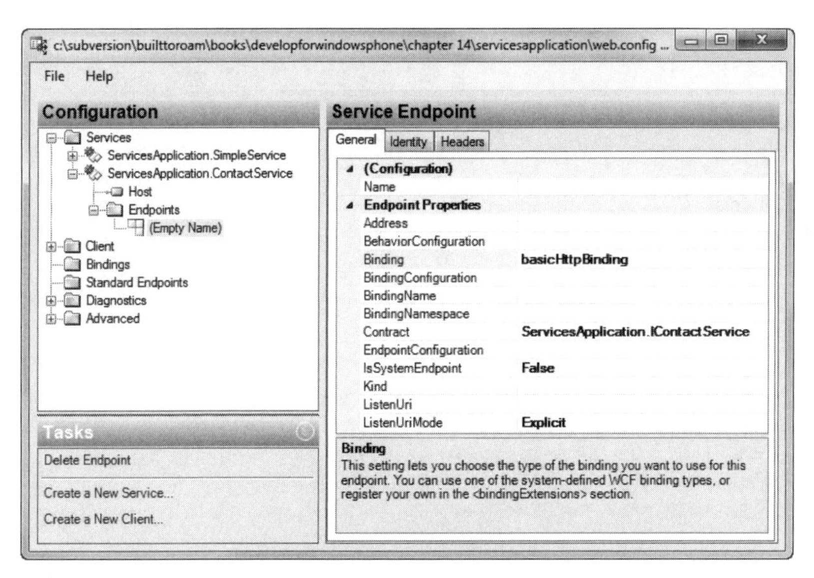

FIGURE 14-8

Add Service Reference

With your server-side service all set up to use `basicHttpBinding`, you're now ready to create the proxy classes in your Windows Phone application. Right-click on the project in Solution Explorer and select “Add Service Reference.” This will display the “Add Service Reference” dialog shown in Figure 14-9. Enter the address of the service into the Address field and click Go. Once the service has been discovered, the Services and Operations pane will display information about the service.

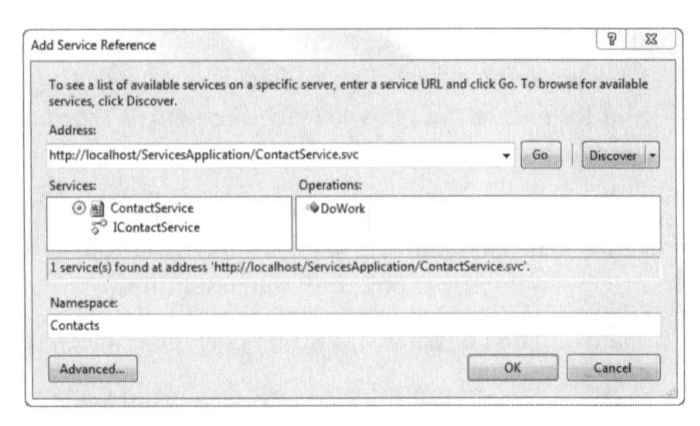

FIGURE 14-9

Enter a Namespace, in this case **Contacts**, and click OK to add the service reference. You should see a Service References folder appear in your Windows Phone project with a Contacts node beneath it.

Service Implementation and Execution

Currently, as configured by the WCF service project template, the ContactService has a single method, `DoWork`, that doesn’t do anything. Let’s change this method to `RetrieveContacts` and have it return an array of `Contact` objects. You’ll also create a second method called `SaveContact` that will accept a contact as a parameter. The interface and implementation are as follows:

```
[ServiceContract]
public interface IContactService{
    [OperationContract]
    Contact[] RetrieveContacts();

    [OperationContract]
    public void SaveContact(Contact contactToSave);
}
```

Code snippet from IContactService.cs

```
public class ContactService : IContactService{
    private static List<Contact> Contacts = new List<Contact>(){
        new Contact(){Id=Guid.NewGuid(), Name="Joe",
                      EmailAddress="Joe@FredsFarm.com"},
        new Contact(){Id=Guid.NewGuid(), Name="Barny",
                      EmailAddress="Barny@FredsFarm.com"}
    };

    public Contact[] RetrieveContacts(){
        return Contacts.ToArray();
    }

    public void SaveContact(Contact contactToSave){
        // Remove any contacts that have the same Id
```

```
            Contacts.RemoveAll((contact) => contact.Id == contactToSave.Id);

            // Save the new contact by adding it to the list
            Contacts.Add(contactToSave);
        }
    }
```

Code snippet from ContactService.svc.cs

After making these changes to the ContactService, you will need to update the generated proxy classes created in your Windows Phone application. Visual Studio can facilitate this by right-clicking on the appropriate node under the Service References node in Solution Explorer and selecting "Update Service Reference."

TESTING WCF SERVICES

Once you start to build more complex services, they become increasingly hard to debug and diagnose. When developing Windows Phone applications, it can be difficult to know whether the issue is with the client-side or server-side code. For these types of reasons, Visual Studio comes with a WCF Test Client that can be used to test your service methods and isolate any client code from the equation.

Figure 14-10 illustrates the WCF Test Client showing an invocation of the `SaveContacts` method.

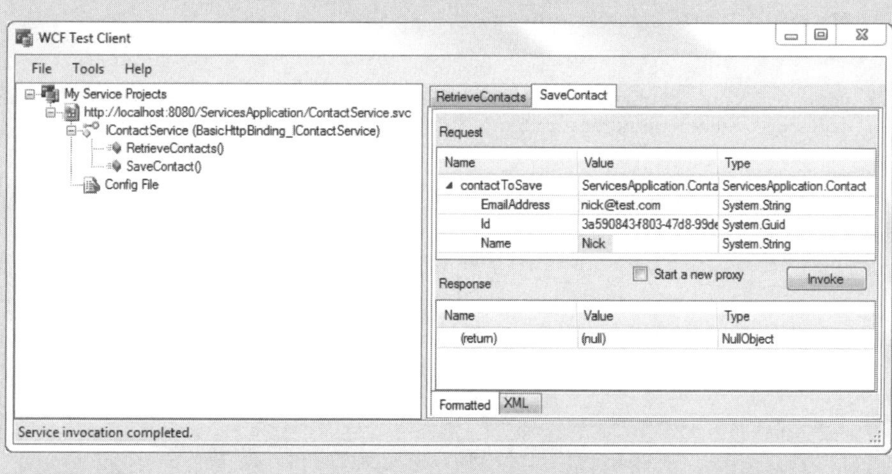

FIGURE 14-10

> The WCF Test Client may not start automatically when you run the project containing the WCF Service you wish to test. In this case it can be manually started from `C:\Program Files\Microsoft Visual Studio 10.0\Common7\IDE\WcfTestClient.exe`. Once started you can direct it to the service you wish to test by selecting File ➪ Add Service and entering the URL for the WCF Service.

The code for invoking these services within a Windows Phone application is relatively straightforward and is similar in structure to working with the `WebClient`. After an instance of the `ContactServiceClient` proxy class has been created, it can be used to invoke both the `RetrieveContacts` and `SaveContact` methods. These methods both have corresponding events that are raised when the server response for the method invocation is available:

```
Contacts.ContactServiceClient client = new Contacts.ContactServiceClient();
public ServicesPage(){
    InitializeComponent();
    client.RetrieveContactsCompleted += client_RetrieveContactsCompleted;
    client.SaveContactCompleted += client_SaveContactCompleted;
}

private void GetContactsButton_Click(object sender, RoutedEventArgs e){
    client.RetrieveContactsAsync();
}

void client_RetrieveContactsCompleted(object sender,
                                      Contacts.RetrieveContactsCompletedEventArgs e){
    ContactsList.ItemsSource = e.Result;
}

private void AddContactButton_Click(object sender, System.Windows.RoutedEventArgs e){
    var contact = new Contacts.Contact(){
        Id = Guid.NewGuid(),
        Name = this.NameText.Text,
        EmailAddress = this.EmailText.Text
    };
    client.SaveContactAsync(contact,contact);
}

void client_SaveContactCompleted(object sender,
                                 AsyncCompletedEventArgs e){
    var contact = e.UserState as Contacts.Contact;
    var list = this.ContactsList.ItemsSource
                   as ObservableCollection<Contacts.Contact>;
    list.Add(contact);
}
```

Code snippet from ServicesPage.xaml.cs

Figure 14-11 illustrates what the page looks like after the `RetrieveContacts` method has been invoked. The returned contacts are displayed in a listbox. A new contact can be added by pressing the "Add Contact" button after entering a name and e-mail address.

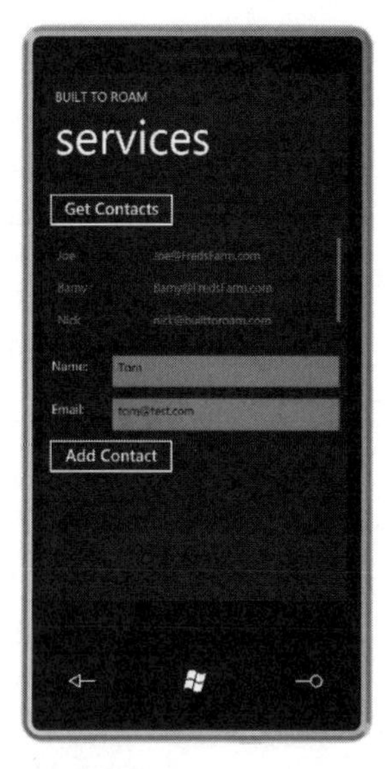

FIGURE 14-11

As with the `WebClient`, it is worth noting that while responding to an event raised by the `ContactServiceClient` class there is no need to switch back to the UI thread using `Dispatcher.BeginInvoke` as this has already been taken care of. On the flipside, be aware that whatever post processing you do perform will be done on the UI thread, which has the potential of causing the user experience to lock up. If you need to do any significant work to the data before displaying it, you should consider using a background thread to do this.

Custom Headers

There are occasions when you need to specify custom HTTP headers that should be included with each service request. To do this, you need to create an `OperationContextScope` and apply a new `HttpRequestMessageProperty` to the outgoing message. The following example illustrates this technique by setting the Authorization header, as if you were using Basic Authentication to secure your service:

```
using (new OperationContextScope(client.InnerChannel)){
    HttpRequestMessageProperty prop = new HttpRequestMessageProperty();
    string credentials = Convert.ToBase64String(Encoding.UTF8.GetBytes
                                        ("Nick" + ":" + "MyPassword"));
    prop.Headers["Authorization"] = "Basic " + credentials;
    OperationContext.Current.OutgoingMessageProperties.Add(
                                 HttpRequestMessageProperty.Name, prop);
    client.RetrieveContactsAsync();
}
```

Credentials

In the case of credentials, it is not necessary to use the custom HTTP header demonstrated above; instead, there is a `ClientCredentials` property on the generated proxy client that you can use to specify the username and password that should be used to authenticate against the server:

```
client.ClientCredentials.UserName.UserName = "Nick";
client.ClientCredentials.UserName.Password = "MyPassword";
```

Accessing secure services is covered in more detail in Chapter 18 on security.

WCF DATA SERVICES

Developers have long been frustrated by having to continually author CRUD code. This refers to code that "Creates, Reads, Updates, and Deletes" items within a database. Such code may operate directly against the database, or it may be in the form of proxy code written on the client or server to facilitate CRUD operations from a client application. WCF Data Services provides an extensible tool for publishing data using a REST-based interface so that client applications can consume and update it without the need to develop custom CRUD-style code.

OData with WCF Data Services

Although SOAP-based services offer many advantages owing to their self-descriptive nature, they also add significant overhead to the size of the messages being sent back and forth. Most desktop applications can get away with this because they typically have a high-speed connection through to the target server. With a Windows Phone application, where connectivity and bandwidth are both limited, it makes more sense to use a technology that has minimal overheads. WCF Data Services publishes and consumes data using the Open Data Protocol (OData) web protocol (`www.odata.org`). The default format for data is XML, but as you will see later on, you can also request data from WCF Data Services in JSON, making it extremely compact and easy to work with within your application.

In this section, you're going to use a sample patient database. You'll see how WCF Data Services can be used to publish out the set of existing patients and receive updates from the Windows Phone application.

ADO.NET Entity Data Model

In order to create a WCF Data Service, you first need to define a data context. The *data context* provides the mapping between the object model exposed by the WCF Data Service and the backend database. In this case, add a new ADO.NET Entity Data Model called *PatientModel.edmx* to your ASP.NET web application. This will display the Entity Data Model Wizard, which will prompt you to select a database and tables that you want to include in the model. Figure 14-12 illustrates the simple data model that is included in the entity model.

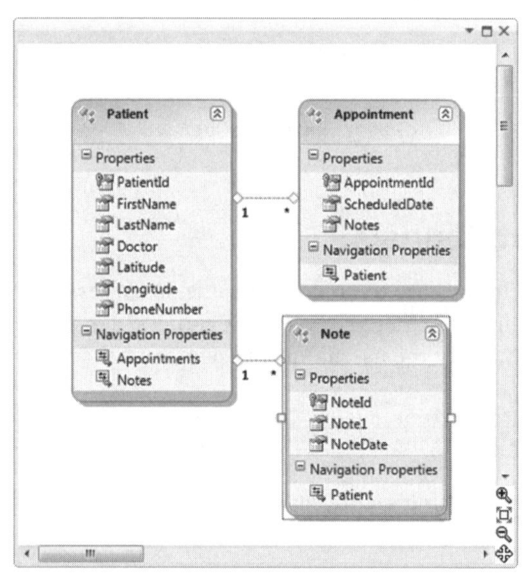

FIGURE 14-12

Once you have the entity model of your database, you need to add a new WCF Data Service to the project. Call it **PatientDataService.svc**. This takes the data accessed through the entity model and makes it accessible via the REST end point. The only changes from the WCF Data Service template you need to make are to specify the data context, which in this case is PatientEntities, and describe which entities are accessible. In this case, you want all entities to be both readable and writeable by clients. This can be specified by adding a new access rule for `"*"` with `EntitySetsRights.All`:

```
public class PatientDataService : DataService<PatientEntities>{
    public static void InitializeService(DataServiceConfiguration config){
        config.SetEntitySetAccessRule("*", EntitySetRights.All);
        config.DataServiceBehavior.MaxProtocolVersion = DataServiceProtocolVersion.V2;
    }
}
```

Code snippet from PatientDataService.svc.cs

When you run your application and navigate to the newly created WCF Data Service, you should see a list of entities that are exposed by the data service. In this case, you have Patients, Notes, and Appointments. If you now append one of these entities to the end of the URL of the data service, you should see a list of those entities. Figure 14-13 illustrates the start of the list of Patients in the test database.

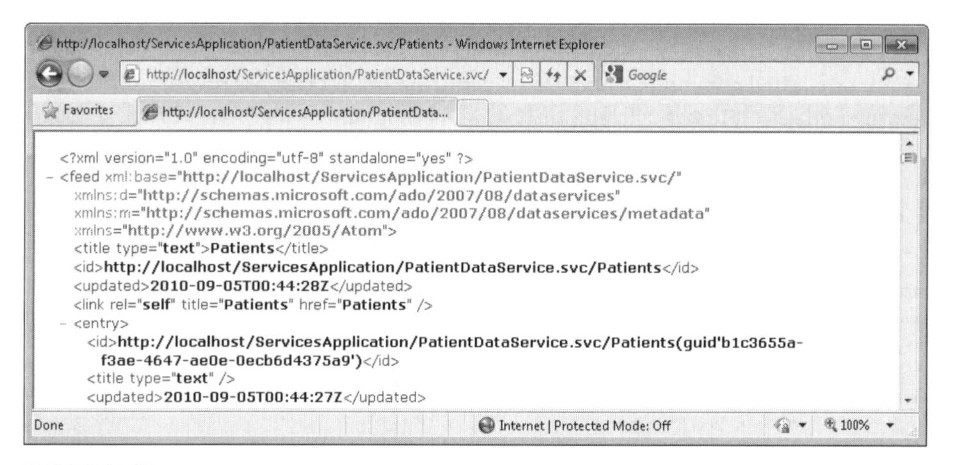

FIGURE 14-13

Now that the WCF Data Service has been created and exposes the contents of our database via a REST-based API, you are ready to use this from within our Windows Phone application. Rather than having to use the `HttpWebRequest` class and crafting the required URLs and XML parsing code manually, you can make use of a command-line tool called `DataSvcUtil` to automatically generate a set of proxy classes. Open a command prompt and navigate to the folder of your

Windows Phone application. Run the following command to create the WCF Data Service proxy classes (see Figure 14-14):

```
"%windir%\Microsoft.NET\Framework\v4.0.30319\DataSvcUtil.exe" /version:2.0
 /dataservicecollection /language:CSharp /out:PatientData.cs
 /uri:http://localhost/ServicesApplication/PatientDataService.svc
```

Then, within your Windows Phone project, right-click within Solution Explorer and select "Add Existing Item." Locate the newly created PatientData.cs file and click Add. Compiling the project at this stage will result in errors due to the generated wrapper relying on the OData Client Library, which is available from `www.microsoft.com/downloads`. Download and install the library, and then add a reference to it within your Windows Phone project. Within the "Add Reference" dialog, you will need to switch to the Browse tab and navigate to the folder where you installed the library. Select "System.Data.Services.Client.dll," and then click OK.

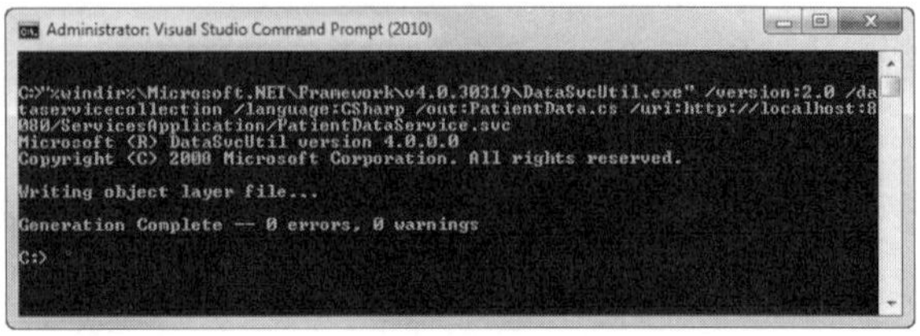

FIGURE 14-14

Queries

Working with WCF Data Services is slightly different from working with the more traditional SOAP services, which are orientated around method calls. Instead, a WCF Data Service is built around the concept of querying and selecting data through LINQ (Language Integrated Query) expressions. Let's start with a simple query to return all patients:

```
PatientEntities entities = new PatientEntities(
            new Uri("http://localhost/ServicesApplication/PatientDataService.svc"));

public DataServicesPage(){
    InitializeComponent();
}

private void GetPatientsButton_Click(object sender, RoutedEventArgs e){
    var query = entities.Patients;
    query.BeginExecute(Patients_Callback, query);
}

private void Patients_Callback(IAsyncResult result){
    var query = result.AsyncState as DataServiceQuery<Patient>;
    var patients = query.EndExecute(result).ToArray();

    this.Dispatcher.BeginInvoke(()=>{
        PatientsList.ItemsSource = patients;
    });
}
```

Code snippet from DataServicesPage.xaml.cs

The first thing you will notice is that using a WCF Data Service proxy class has similarities to working with an `HttpWebRequest` object. When you submit a query via the `BeginExecute` method, you supply a callback to be invoked once the server's response is obtained. Within this callback you have to use `Dispatcher.BeginInvoke` if you need to perform any updates to the user interface.

Before you look at some examples of the types of query expressions you can write, you should note that queries will eventually be converted into a unique URL used for a REST request. In the above example, the generated URL is the same as that displayed in Figure 14-13:

```
http://localhost/ServicesApplication/PatientDataService.svc/Patients
```

Let's start by adding a filter to only list patients with the first name of *Tom*. As you can see, you can use a LINQ expression to filter the `Patients` property of the `PatientEntities` object:

Available for download on Wrox.com

```
private void GetFilteredPatientsButton_Click(object sender, RoutedEventArgs e){
    var query = from p in entities.Patients
                where p.FirstName == "Tom"
                select p;

    var dsquery = query as DataServiceQuery<Patient>;
    dsquery.BeginExecute(Patients_Callback, dsquery);
}
```

Code snippet from DataServicesPage.xaml.cs

Note that the typecast to `DataServiceQuery` is required; otherwise, the `BeginExecute` method isn't accessible. By default, LINQ queries return an `IQueryable` object, but in this case, you're working with a WCF Data Service, so performing the typecast enables you to access functionality, such as `BeginExecute`, which is only applicable in a WCF Data Service context. As you would imagine, executing this query will return a data set containing a list of patients with a first name of Tom. You can verify this by examining the URL for this query, which clearly shows the application of the filter condition:

```
http://localhost/ServicesApplication/PatientDataService.svc/Patients()?$filter=FirstName
 eq 'Tom'
```

When performing a query, you may like to order the resulting data set:

```
var query = from p in entities.Patients
            orderby p.FirstName descending
            select p;

//http://localhost/ServicesApplication/PatientDataService.svc/Patients()?$orderby=
 FirstName desc
```

By default, a WCF Data Service query against the `patient` entity will only return details of the `patient` object itself. The ADO.NET Entity Data Model shown in Figure 14-12 indicates that

Patients also have Notes and Appointments. Within a query, you can use the `Expand` method to indicate that the query should deep-load one, or more, related entities:

```
var query = from p in entities.Patients.Expand("Appointments")
            select p;
//http://localhost/ServicesApplication/PatientDataService.svc/Patients()?$expand=
 Appointments

var query = from p in entities.Patients.Expand("Notes")
            select p;
//http://localhost/ServicesApplication/PatientDataService.svc/Patients()?$expand=
 Appointments

var query = from p in entities.Patients.Expand("Notes,Appointments")
            select p;
//http://localhost/ServicesApplication/PatientDataService.svc/Patients()?$expand=Notes,
 Appointments
```

If the appointment entity had a related entity script, you could do a further deep load to return all scripts written in appointments for the selected patients using the following query:

```
var query = from p in entities.Patients.Expand("Appointments/Scripts")
            select p;

//http://localhost/ServicesApplication/PatientDataService.svc/Patients()?$expand=
 Appointments/Scripts
```

Custom Methods and Stored Procedures

Whereas the default behavior of WCF Data Services is adequate for accessing data using basic queries, there are some limitations. For example, you would have difficulty executing a query to retrieve all patients that have at least one appointment scheduled, or all patients within a certain distance of a location. More complex queries such as these could be done entirely on the Windows Phone client side, by downloading the list of all patients and associated entities and then filtering them locally. This would, however, incur a significant overhead as all data would need to be transferred to the client's device. A better alternative may be to perform a custom SQL query on the server side to return only the required rows to the client. For this example, you're going to use a stored procedure on the SQL Server database called `PatientsWithAppointments` defined as follows:

```
CREATE PROCEDURE PatientsWithAppointments
AS
BEGIN
    select *
    from Patient p
    where (select COUNT(*) from Appointment apt where p.PatientId=apt.PatientId)>0
END
```

In order to use this stored procedure, you need to add it as a function within your ADO.NET Entity Data Model. Once you have created the stored procedure within the database, open up your

Entity Data Model within your ASP.NET web application and select "Update Model from Database." Step through the Wizard, ensuring that it locates and adds the `PatientsWithAppointments` method. In the Model Browser Tool window, navigate to the PatientsWithAppointments node that is located under PatientModel.Store ➪ Stored Procedure. Right-click on this node and select "Add Function Import." Figure 14-15 illustrates the "Add Function Import" dialog. Make sure that the correct stored procedure name is selected and that the return type is set to Entities of type Patient, then click OK.

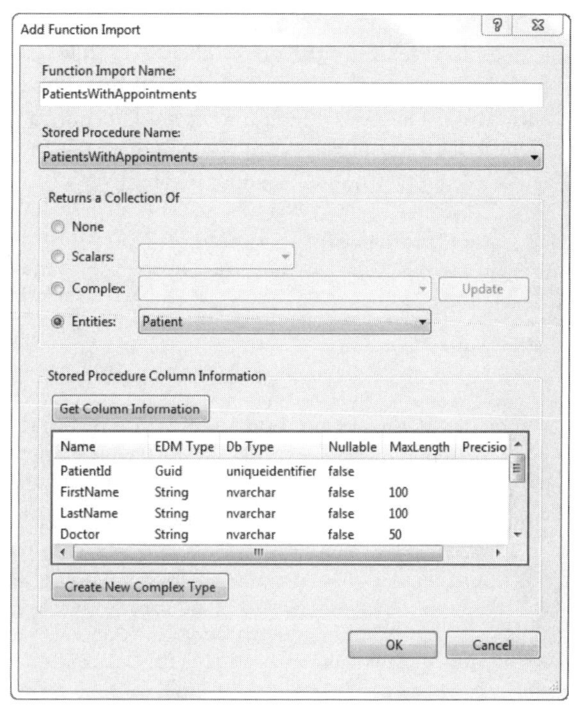

FIGURE 14-15

With the stored procedure accessible via the ADO.NET Entity Data Model, you now need to expose it via the WCF Data Service. This can be done by defining a service method with the `WebGet` attribute. Here you're defining two such methods to perform the more complex queries discussed previously. Note that the first makes use of the stored procedure you've just defined to return patients with at least one appointment. The second accepts two parameters that define the center of a search region. Patients within a certain distance to the specified location are returned. You've also extended the InitializeService method to allow access to the newly created service operations.

```
[WebGet]
public IQueryable<Patient> PatientsWithAppointments(){
    return this.CurrentDataSource.PatientsWithAppointments().AsQueryable<Patient>();
}

[WebGet]
public IQueryable<Patient> PatientsInRegion(double latitude, double longitude){
    return from p in this.CurrentDataSource.Patients
           where Math.Abs(p.Latitude - latitude) < 0.1 &&
                 Math.Abs(p.Longitude - longitude) < 0.1
           select p;
}
public static void InitializeService(DataServiceConfiguration config){
    config.SetEntitySetAccessRule("*", EntitySetRights.All);
    config.SetServiceOperationAccessRule("*", ServiceOperationRights.All);
    config.DataServiceBehavior.MaxProtocolVersion = DataServiceProtocolVersion.V2;
}
```

Code snippet from PatientDataServices.svc.cs

You are now ready to query these methods from the Windows Phone application. There is no need to regenerate the proxy classes as there is no strongly typed support for custom methods. In order to invoke these custom methods, you can use either the `CreateQuery` or `BeginExecute` method. The `BeginExecute` method accepts a URI as a parameter, allowing you to specify the URI that maps to your custom method. In the following example, the URI points to the `PatientsWithAppointments` method:

Available for download on Wrox.com

```
entities.BeginExecute<Patient>(
                new Uri("http://localhost/ServicesApplication/PatientDataService.svc/" +
                    "PatientsWithAppointments"), Patients_Callback,null);
```

Code snippet from DataServicesPage.xaml.cs

The `CreateQuery` method enables you to build up the query URI programmatically by passing in the name of the method you wish to invoke and the names and values of any parameters. Here you're calling the `PatientsInRegion` method, passing in values for the latitude and longitude. The generated URI will include the parameters within the query string.

Available for download on Wrox.com

```
var query = entities.CreateQuery<Patient>("PatientsInRegion")
                    .AddQueryOption("latitude", -37.998352)
                    .AddQueryOption("longitude", 145.083596);
query.BeginExecute(Patients_Callback, query);

//http://localhost/ServicesApplication/PatientDataService.svc/PatientsInRegion()?
 latitude=-37.998352&longitude=145.083596
```

Code snippet from DataServicesPage.xaml.cs

Updates, Inserts, and Deletes

Working with data returned from a WCF Data Service is relatively straightforward and to a certain extent can be treated as if you are working against an ADO.NET Entity Data Model directly on the server. If you modify an entity on the client, changes won't be committed until you call `UpdateObject` and then invoke `BeginSaveChanges`.

Available for download on Wrox.com

```
private void SavePatientButton_Click(object sender, RoutedEventArgs e){
    var p = this.PatientsList.SelectedItem as Patient;
    p.FirstName = "New Name";
    entities.UpdateObject(p);
    entities.BeginSaveChanges(SaveComplete_Callback, null);
}

private void SaveComplete_Callback(IAsyncResult result){
    var resp = entities.EndSaveChanges(result);
}
```

Code snippet from DataServicesPage.xaml.cs

To insert a new entity, you just need to create a new instance of the `entity` class, setting the appropriate properties. For each entity type there will be a corresponding `add` method, in this case, `AddToPatients`. To submit the changes to the server, invoke the `BeginSaveChanges` method:

Available for download on Wrox.com

```
var p = new Patient(){
    PatientId = Guid.NewGuid(),
    FirstName = "Nick",
    LastName = "Randolph",
    PhoneNumber = "00 00 00 0000",
    Doctor = "Fred",
    Latitude = -32.0,
    Longitude = 151.0
};
entities.AddToPatients(p);
entities.BeginSaveChanges(SaveComplete_Callback, null);
```

Code snippet from DataServicesPage.xaml.cs

Deleting an entity is done via the `DeleteObject` method, again followed by `BeginSaveChanges`:

Available for download on Wrox.com

```
private void SavePatientButton_Click(object sender, RoutedEventArgs e){
    var p = this.PatientsList.SelectedItem as Patient;
    entities.DeleteObject(p);
    entities.BeginSaveChanges(SaveComplete_Callback, null);
}
```

Code snippet from DataServicesPage.xaml.cs

JSON

One interesting aspect of WCF Data Services is that your data is not only published in a structured XML format, but also in a JSON-based format. This is a much more compact notation and can significantly reduce the amount of bandwidth consumed by large sets of data. This is an ideal format for mobile-based applications. Table 14-1 illustrates the difference in size between the XML and JSON formats. As you can see the JSON format comes in well below half that of the XML representation of the same data set.

TABLE 14-1: Data Comparison

XML	JSON
HTTP/1.1 200 OK Cache-Control: no-cache Content-Length: 1810 Content-Type: application/atom+xml; charset=utf-8 Server: Microsoft-IIS/7.5 DataServiceVersion: 1.0; X-AspNet-Version: 4.0.30319 X-Powered-By: ASP.NET	HTTP/1.1 200 OK Cache-Control: no-cache Content-Length: 788 Content-Type: application/json;charset=utf-8 Server: Microsoft-IIS/7.5 DataServiceVersion: 1.0; X-AspNet-Version: 4.0.30319 X-Powered-By: ASP.NET

(continued)

TABLE 14-1 *(continued)*

XML


```
Date: Wed, 18 Aug 2010 05:35:15 GMT

<?xml version="1.0" encoding="utf-8"
standalone="yes"?>
<entry xml:base="http://localhost/
ServicesApplication/PatientDataService.
svc/" xmlns:d="http://schemas.microsoft.com/
ado/2007/08/dataservices" xmlns:m="http://
schemas.microsoft.com/ado/2007/08/dataservices/
metadata" xmlns="http://www.w3.org/2005/Atom">
  <id>http://localhost/ServicesApplication/
PatientDataService.svc/Patients(guid'b1c3655a-
f3ae-4647-ae0e-0ecb6d4375a9')</id>
  <title type="text"></title>
  <updated>2010-08-18T05:35:15Z</updated>
  <author>
    <name />
  </author>
  <link rel="edit" title="Patient" href="Patient
s(guid'b1c3655a-f3ae-4647-ae0e-0ecb6d4375a9')" />
  <link rel="http://schemas.microsoft.com/
ado/2007/08/dataservices/related/Appointments"
type="application/atom+xml;type=feed"
title="Appointments" href="Patients(guid'b1c3655
a-f3ae-4647-ae0e-0ecb6d4375a9')/Appointments" />
  <link rel="http://schemas.microsoft
.com/ado/2007/08/dataservices/related/
Notes" type="application/atom+xml;type=feed"
title="Notes" href="Patients(guid'b1c3655a-f3ae-
4647-ae0e-0ecb6d4375a9')/Notes" />
  <category term="PatientModel.Patient"
scheme="http://schemas.microsoft.com/
ado/2007/08/dataservices/scheme" />
  <content type="application/xml">
    <m:properties>
      <d:PatientId m:type="Edm.Guid">b1c3655a-
f3ae-4647-ae0e-0ecb6d4375a9</d:PatientId>
      <d:FirstName>Frankz</d:FirstName>
      <d:LastName>Goh</d:LastName>
      <d:Doctor>Nick</d:Doctor>
      <d:Latitude m:type="Edm.Double">-
37.998352</d:Latitude>
      <d:Longitude m:type="Edm.
Double">145.083596</d:Longitude>
      <d:PhoneNumber>+1 425 001 0001</d:
PhoneNumber>
      <d:LastUpdated m:type="Edm.DateTime">2010-
06-27T12:53:47.51</d:LastUpdated>
      <d:Deleted m:type="Edm.Boolean">true</d:
Deleted>
    </m:properties>
  </content>
</entry>
```


Size: 1810 bytes

JSON


```
Date: Wed, 18 Aug 2010 05:40:39 GMT

{
"d" : {
"__metadata": {
"uri": "http://localhost/
ServicesApplication/PatientDataService
.svc/Patients(guid'b1c3655a-f3ae-
4647-ae0e-0ecb6d4375a9')", "type":
"PatientModel.Patient"
}, "PatientId": "b1c3655a-f3ae-4647-
ae0e-0ecb6d4375a9", "FirstName":
"Frankz", "LastName": "Goh", "Doctor":
"Nick", "Latitude": -37.998352,
"Longitude": 145.083596, "PhoneNumber":
"+1 425 001 0001", "LastUpdated": "\/
Date(1277643227510)\/", "Deleted":
true, "Appointments": {
"__deferred": {
"uri": "http://localhost/
ServicesApplication/PatientDataService
.svc/Patients(guid'b1c3655a-f3ae-4647-
ae0e-0ecb6d4375a9')/Appointments"
}
}, "Notes": {
"__deferred": {
"uri": "http://localhost/
ServicesApplication/PatientDataService
.svc/Patients(guid'b1c3655a-f3ae-4647-
ae0e-0ecb6d4375a9')/Notes"
}
}
}
}
```


Size: 788 bytes

Unfortunately, the proxy classes don't support working with JSON-formatted data, which means that you will need to create your own requests using the `HttpWebRequest` class, and you will need to manually parse the response into objects that you can work with. Creating the request is similar to what you saw earlier in this chapter, where you simply specified the URI that you wanted to access. The only difference is that you need to specify the Accept HTTP header so that the server knows to send JSON, rather than XML, in the response:

Available for download on Wrox.com

```
private void GetPatientsInJsonButton_Click(object sender, RoutedEventArgs e){
    var request = HttpWebRequest.Create(
          "http://localhost/ServicesApplication/PatientDataService.svc/Patients")
                                                            as HttpWebRequest;
    request.Accept = "application/json";
    request.BeginGetResponse(JsonPatients_Callback, request);
}
```

Code snippet from DataServicesPage.xaml.cs

Although you could manually parse the returned JSON data, there is a much easier way in the form of the `DataContractJsonSerializer` class found within the `System.ServiceModel.Web` assembly. This will parse a JSON stream and return the objects found within it. For it to do this, it needs to know what type of objects can be found within the stream. The following class declarations mirror the structure of the JSON returned by the server:

Available for download on Wrox.com

```
public class PatientData{
    public JsonPatient[] d { get; set; }
}

public class JsonPatient{
    public MetaData __metadata { get; set; }
    public string PatientId { get; set; }
    public string FirstName { get; set; }
    public string LastName { get; set; }
    public string Latitude { get; set; }
    public string Longitude { get; set; }
    public string PhoneNumber { get; set; }
    public DeferredItem Appointments { get; set; }
    public DeferredItem Notes { get; set; }
}

public class MetaData{
    public string uri { get; set; }
    public string type { get; set; }
}

public class DeferredItem{
    public Deferred __deferred { get; set; }
```

```
    }

    public class Deferred{
        public string uri { get; set; }
    }
```

Code snippet from DataServicesPage.xaml.cs

With these class declarations you can extract an object graph from the response using the DataContractJsonSerializer:

```
    private void JsonPatients_Callback(IAsyncResult result)
    {
        var request = result.AsyncState as HttpWebRequest;
        var response = request.EndGetResponse(result);

        var deserializer = new DataContractJsonSerializer(typeof(PatientData));
        var data = deserializer.ReadObject(response.GetResponseStream()) as PatientData;

        this.Dispatcher.BeginInvoke(() =>
        {
            this.PatientsList.ItemsSource = data.d;
        });
    }
```

Code snippet from DataServicesPage.xaml.cs

You can see in Figure 14-16 how the JSON appears in raw format (left image) and when the parsed object is set as the source for the listbox (right image).

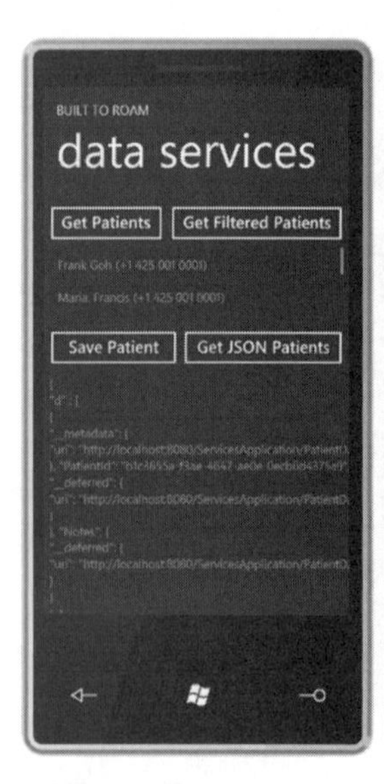

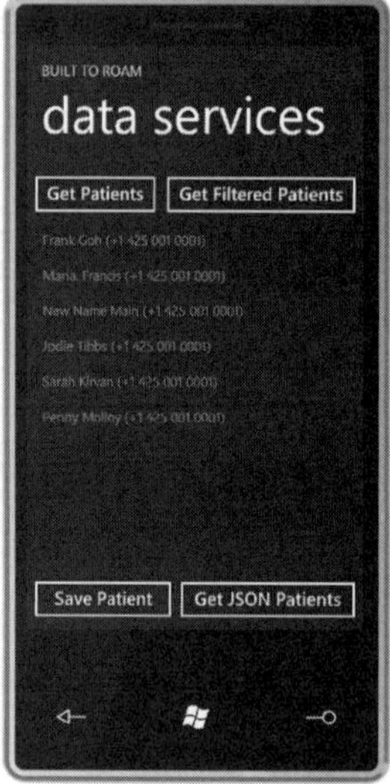

FIGURE 14-16

The `DataContractJsonSerializer` *is in the* `System.Runtime.Serialization.Json` *namespace. In order to use this class you will need to add a reference to* `System.ServiceModel.Web.dll` *to your application.*

SUMMARY

In this chapter, you've seen how to work with both the `WebClient` and `HttpWebRequest` classes in order to access remote data within your Windows Phone application. Although they provide the lowest-level mechanism for working with remote data, they are difficult to work with when dealing with structured data exposed via SOAP- or REST-based APIs. Both WCF and ASMX Web Services provide a high-level abstraction to invoke remote services, whereas WCF Data Services provides a great starting point for any Windows Phone application wishing to perform CRUD-style operations against a remote data source. Using the JSON content type coupled with IIS compression, this can yield a bandwidth optimized solution.

15

Data Visualization

WHAT'S IN THIS CHAPTER

- Understanding how data binding works
- How binding mode affects the binding direction
- How to use data binding in your application design
- Working with design time data
- Connecting element properties using data binding

Very few applications exist that don't make use of data in some form. In this chapter, you will learn how data binding and visualization are important in building a rich Windows Phone application. You will see how data binding is a powerful way to declaratively connect user interface (UI) elements to your data model and how you can use Expression Blend to configure data binding visually.

DATA BINDING

Probably the second most time-consuming aspect of building an application, after writing CRUD (Create, Read, Update, and Delete) code to interact with a database, is building the user interface — more specifically, displaying data, accepting and processing user input. The concept of *data binding* means that you can define a relationship between a UI element (also known as the *target*) and a data provider (also known as the *source*). Being able to define such a relationship both reduces the amount of code you need to write and encourages a clear separation of the data and presentation aspects of your application.

In this chapter, you're going to work with a library of books and illustrate how you can use data binding and the design capabilities of Expression Blend to build a rich user interface with only minimal lines of code. Let's start by taking a quick look at the structure of our data model, which consists of a book that has multiple authors, as shown in Figure 15-1.

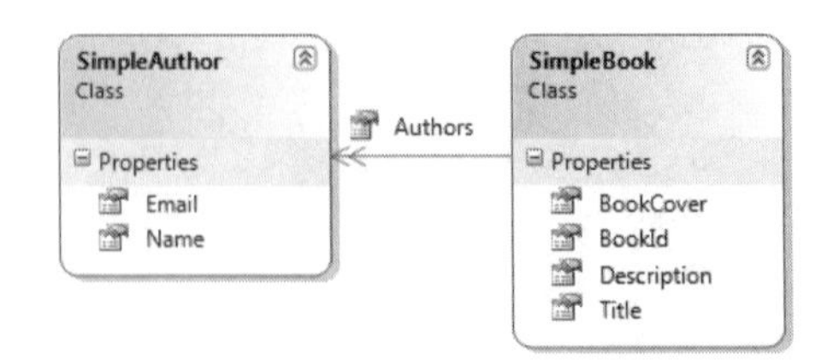

FIGURE 15-1

Rather than taking a theoretical approach to discussing data binding, we're going to jump right in and show how you can use data binding within your application. With this in mind, let's create a basic user interface that will display a single book. In this case, you're only going to be displaying the cover, title, and description:

```
<Grid x:Name="ContentGrid" Grid.Row="1">
    <Image x:Name="BookCoverImage" HorizontalAlignment="Left" Height="100"
           Margin="27,49,0,0" VerticalAlignment="Top" Width="100"/>
    <TextBlock x:Name="BookTitleText" Margin="143,50,0,0" TextWrapping="Wrap"
               VerticalAlignment="Top"/>
    <TextBlock x:Name="BookDescriptionText" Margin="143,110,0,0" TextWrapping="Wrap"
               VerticalAlignment="Top" Height="150"/>
</Grid>
```

Code snippet from MainPage.xaml

DataContext

If you were to display a book without data binding, you would typically write code that looks similar to the following, in order to set the `Source` and `Text` properties of the `Image` and `TextBlock` controls, respectively:

```
this.BookCoverImage.Source = myBook.BookCover;
this.BookTitleText.Text = myBook.Title;
this.BookDescriptionText.Text = myBook.Description;
```

With data binding, you can define a relationship between the data source, which in this case is an instance of the `SimpleBook` class, and the UI elements. The relationship can be declaratively defined within your XAML-based user interface as the following demonstrates:

```
<Grid x:Name="ContentGrid" Grid.Row="1">
    <Image x:Name="BookCoverImage" HorizontalAlignment="Left" Height="100"
           Margin="27,49,0,0" VerticalAlignment="Top" Width="100"
           Source="{Binding BookCover}" />
     <TextBlock x:Name="BookTitleText" Margin="143,50,0,0" TextWrapping="Wrap"
               VerticalAlignment="Top"
               Text="{Binding Title}" />
     <TextBlock x:Name="BookDescriptionText" Margin="143,110,0,0" TextWrapping="Wrap"
               VerticalAlignment="Top" Height="150"
               Text="{Binding Title}" />
</Grid>
```

Code snippet from MainPage.xaml

In the declaration for the Image and the two TextBlock controls, a new attribute has been added to set the Source and Text properties, respectively. Rather than setting them to an explicit value, we've used a notation (viz., a set of curly braces, { }) to indicate that the value has to be computed. In this case, the Binding keyword also indicates that the value is to be determined through the use of data binding. This is followed by the property on the data source that the element should be bound to. In summary, the Source property on the Image is the data binding target, while the BookCover property is the data source.

Binding statements within XAML are what introduce the separation of concern between your user interface and the data or model layers. With data binding, the designer is free to replace a TextBlock with a Button or other control and continue to bind it to the book's Title property. You could, in fact, decide to add entirely new controls within the XAML that the source code, and hence the developer, is completely unaware of. To make similar changes to the non-data binding–based example, the designer would have to interact with the developer as each change in the UI was made.

At this point you might be wondering how the data binding system knows what source object it should be referring to when you bind to the BookCover property. The source object is more commonly referred to as the *data context*, and there is a corresponding property on the FrameworkElement class called DataContext. Since every UI control ultimately derives from FrameworkElement, this means that each control has a DataContext property allowing you to set the data context for that element.

Within our current XAML we haven't defined the data context for our UI elements. This is permitted by the data binding system. If the DataContext property is null, there is nothing to bind to, so any binding statements are ignored.

In order to get your user interface to display the details of a SimpleBook object called myBook, you need to specify the data context for each control:

```
this.BookCoverImage.DataContext = myBook;
this.BookTitleText.DataContext = myBook;
this.BookDescriptionText.DataContext = myBook;
```

This doesn't look like we've achieved much more than the ability to declare the data binding using markup. If the designer decides to add or remove a control that requires data binding, you need to get the developer involved to make a minor code change. Luckily you can resolve this situation. One of the neatest features of the DataContext property is that if it has not been explicitly set (i.e., the property has the value null), it will inherit the data context of its parent. For example, in this case, you can refactor the code to set the DataContext on the ContentGrid. The Image and the two TextBlocks are contained within the ContentGrid element so they will inherit the DataContext assigned to the ContentGrid, and thus data binding will access the appropriate properties in order to set the Source and Text properties:

```
this.ContentGrid.DataContext = myBook;
```

It's worth noting that the DataContext property can be inherited over multiple levels. For example, you could also set the DataContext property on the PhoneApplicationPage and have it inherited through to the Image and TextBlock elements. The important thing to note is that if the

DataContext of a specific control is null, the framework will walk up the logical tree until it finds a suitably set DataContext.

XAML

```
<phone:PhoneApplicationPage ...
    Loaded="PhoneApplicationPage_Loaded">
    <Grid x:Name="LayoutRoot" ... >
        <Grid x:Name="ContentGrid" ... >
            <Image x:Name="BookCoverImage" Source="{Binding BookCover}"/>
            <TextBlock x:Name="BookTitleText" Text="{Binding Title}"/>
            <TextBlock x:Name="BookDescriptionText" Text="{Binding Description}"/>
```

Code snippet from MainPage.xaml

C#

```
public partial class MainPage : PhoneApplicationPage{
    SimpleBook myBook;
    private void PhoneApplicationPage_Loaded(object sender, RoutedEventArgs e) {
        myBook = CreateBook();
        this.DataContext = myBook;
    }
}
```

Code snippet from MainPage.xaml.cs

BindingMode

So far you've seen how you can create a data binding between a property on the data source and a property on the target UI element. When the data context is set, the current value of the data source property is used to set the property on the target element. However, there is more to data binding than simply copying a value from a property on the data source object to a property on the target element when the user interface is being initialized.

As an example, what should happen if the value of a data source property changes during runtime? Should the corresponding property on the target element update, or should it continue to display the original value? Alternatively, what if the layout binds the book's Title property to a TextBox control instead of a TextBlock? This would allow the user to alter the text, but should any change be pushed back to the data source object?

The data binding system allows for the behavior in both of these scenarios to be defined by specifying the binding mode. The Mode property is an enumeration value that can be set within the binding declaration and has three possible values:

- OneTime — The property on the target element is set to the current value of the property on the data source object when the binding is established.
- OneWay — The property on the target element is set when the binding is established and subsequently updated whenever the property on the data source changes. This is the default value.

- `TwoWay` — As with `OneWay`, the property on the target element is updated when the source property changes. In addition, the property on the source is updated whenever the property on the target element is changed.

Let's update the XAML to explicitly define the binding mode:

Available for download on Wrox.com

XAML

```
<Grid x:Name="ContentGrid" Grid.Row="1">
    <Image x:Name="BookCoverImage" HorizontalAlignment="Left" Height="100"
           Margin="27,49,0,0" VerticalAlignment="Top" Width="100"
           Source="{Binding BookCover, Mode=OneTime}" />
    <TextBlock x:Name="BookTitleText" Margin="143,50,0,0" TextWrapping="Wrap"
               VerticalAlignment="Top"
               Text="{Binding Title, Mode=OneWay}" />
    <TextBlock x:Name="BookDescriptionText" Margin="143,110,0,0" TextWrapping="Wrap"
               VerticalAlignment="Top" Height="60"
               Text="{Binding Description, Mode=OneWay}" />
    <TextBox Height="32" Margin="143,193,0,0" Name="TitleText"
             VerticalAlignment="Top"
             Text="{Binding Title, Mode=TwoWay}"/>
    <Button Content="Update" Height="70" HorizontalAlignment="Left"
            Margin="6,266,0,0" Name="UpdateDescriptionButton"
            VerticalAlignment="Top"
            Width="160" Click="UpdateDescriptionButton_Click" />
</Grid>
```

Code snippet from MainPage.xaml

C#

```
private void UpdateDescriptionButton_Click(object sender, RoutedEventArgs e){
    myBook.Description = "This is a new description";
}
```

Code snippet from MainPage.xaml.cs

The image for the book is not updatable, so the binding is set to `OneTime`. However, the two `TextBlocks` are both set to `OneWay` to allow the interface to be updated automatically whenever the data context object changes. Lastly, we've added a `TextBox` that is `TwoWay` bound to the `Title` property on the source, and a `Button`, which will be used to update the `Description` property programatically.

If you run this within the debugger, you will discover that the Update button does, indeed, set the `Description` property on the `myBook` object, and changing the text in the `TextBox` updates the `Title` property. However, both actions have no effect on the two `TextBlocks`, even though they are data bound to the very same source properties. This is because there is no way for the data binding system to know when one or more of the properties on the source object has changed. This is the role of the `INotifyPropertyChanged` interface.

The `INotifyPropertyChanged` interface is a relatively simple interface that is used to signal that a property on an object has changed values. If an object that implements `INotifyPropertyChanged` is set as a `DataContext` from an element, the Framework will take advantage of this to determine when user interface controls must be synchronized with the current state of the object.

In other words, you have to implement an interface that will notify listeners (the data binding infrastructure) that a property has changed. It's a relatively straightforward interface with only a single event, called PropertyChanged. The following is an implementation for the `SimpleBook` class:

Available for download on Wrox.com

```
public class SimpleBook : INotifyPropertyChanged {
    public Guid BookId { get; set; }
    private string title;
    public string Title {
        get{
            return title;
        }
        set {
            if (title == value) return;
            title = value;
            RaisePropertyChanged("Title");
        }
    }

    private string description;
    public string Description {
        get{
            return description;
        }
        set {
            if (description == value) return;
            description = value;
            RaisePropertyChanged("Description");
        }
    }
    public BitmapImage BookCover { get; set; }
    public SimpleAuthor[] Authors { get; set; }

    public event PropertyChangedEventHandler PropertyChanged;
    private void RaisePropertyChanged(string propertyName) {
        if (PropertyChanged != null) {
            PropertyChanged(this, new PropertyChangedEventArgs(propertyName));
        }
    }
}
```

Code snippet from SimpleBook.cs

In this example, both the `Title` and `Description` properties have been expanded out to use a backing variable. If the value is being changed, the `RaisePropertyChanged` method is called, which, in turn, ensures that there are listeners before raising the `PropertyChanged` method. The important thing to note is that the `propertyName` argument supplied must match the name of the property that has changed values. For example, within the `Title` property setter the `propertyName` supplied is *Title* and indicates to any listeners that if they are data bound to the `Title` property, they should update their corresponding property with the new value of the `Title` property.

PROPERTY CODE SNIPPET

Clearly, implementing the `INotifyPropertyChanged` interface requires additional code to be written for every property. Rather than having to type the same code structure each time, you can create a Visual Studio code snippet to generate the code for you. The easiest way to create a snippet is to use the Snippet Editor (`http://snippeteditor.codeplex.com/`). Figure 15-2 illustrates the Snippet Editor for a new snippet called *propdata*, which will reduce the amount of code you have to type for each property you define:

```
private $type$ $field$;
public $type$ $property$ {
    get { return $field$;}
    set {
        if($field$==value)return;
        $field$ = value;
        RaisePropertyChanged("$property$");
    }
}
$end$
```

After creating and saving this snippet (you should save it to the \Documents\Visual Studio 2010\Code Snippets\Visual C#\My Code Snippets folder), you can use it immediately within Visual Studio. In a code file, type **propdata** and press Tab to insert the snippet. Press Tab to iterate through the three placeholders for the type, field name, and property name.

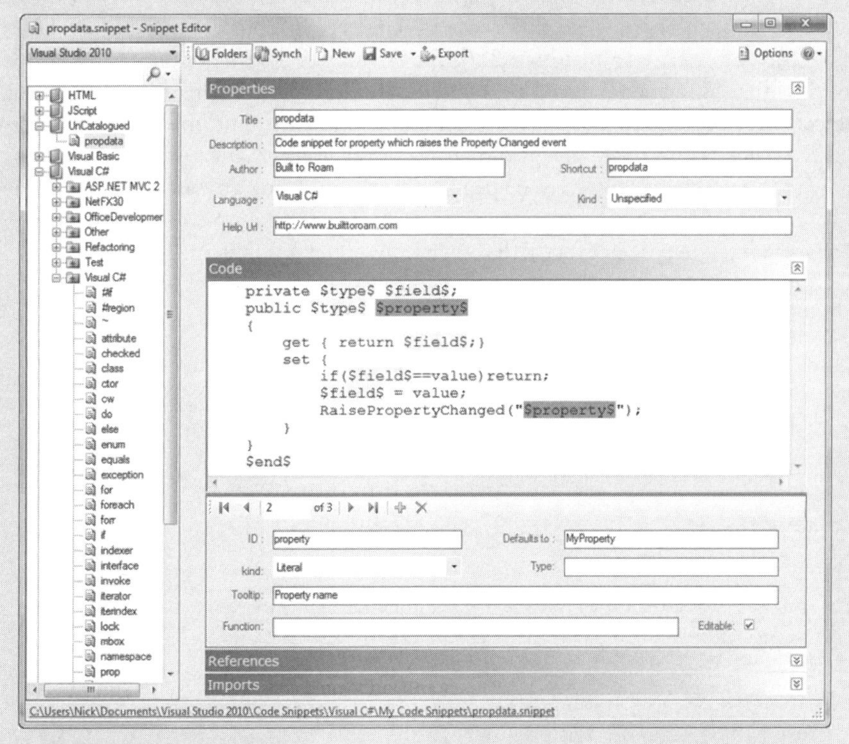

FIGURE 15-2

Value Converters

Sometimes when you use data binding, you will want to bind to a property on the target object that has a different data type or representation to the property on the source object. In this case, you can use a value converter to convert the source value into a suitable value to bind to the target property. To demonstrate this, let's add a `PublishedDate` property to the `SimpleBook` class. This is a nullable `DateTime` field so that you can represent books that have been and are yet to be published. However, on the user interface, you're simply going to display a checkbox to indicate if the book has been published or not.

XAML

```
<CheckBox Content="Published" HorizontalAlignment="Left" Margin="0,150,0,0"
          VerticalAlignment="Top" IsChecked="{Binding PublishedDate}"
          IsEnabled="False"/>
```

Code snippet from MainPage.xaml

C#

```
public class SimpleBook: INotifyPropertyChanged{
        ...
        public DateTime? PublishedDate { get; set; }
```

Code snippet from SimpleBook.cs

In this case, you're binding the `PublishedDate` property to the `IsChecked` property of the `CheckBox` element. This is a bool, so there is a clear difference in data types. What you need is a value converter that can translate the `DateTime?` value into a Boolean. A value converter, in the context of data binding, is any class that implements the `IValueConverter` interface. In theory, an `IValueConverter` can convert objects between any two types based on the type of the input object and the destination data type (specified using the `targetType` argument). However, in practice, value converters are usually implemented to convert between two specific data types. They are also typically named appropriately to indicate the type of conversion they perform. For example, here we're defining the `DateToBoolConverter`, which does exactly that — it converts a date into a Boolean value:

```
public class DateToBoolConverter : IValueConverter {
    public object Convert(object value, Type targetType,
                          object parameter, CultureInfo culture) {
        if (value is DateTime){
            if ((DateTime)value < DateTime.Now) {
                return true;
            }
        }
        return false;
    }

    public object ConvertBack(object value, Type targetType,
                              object parameter, CultureInfo culture) {
        throw new NotImplementedException();
    }
}
```

Code snippet from DateToBoolConverter.cs

What's interesting to note about the `IValueConverter` interface is that it is designed to be bi-directional. There is a `Convert` and a `ConvertBack` method, which when implemented correctly should be the reverse of each other. However, it is also quite common for a developer to implement only the `Convert` method. This is permitted if the value converter is only used in `OneTime` or `OneWay` data bindings. The `ConvertBack` method is used to convert the property on the data binding target back to the data type expected by the property on the data source, so it would only be invoked if the data binding mode was `TwoWay`.

Before you can use the `DateToBoolConverter` in a data binding, you need to create an instance that can be referenced within XAML. In this case, you're going to create an instance in the resources collection of the `ContentGrid`. To use this value converter, you just need to extend the binding syntax to specify the converter object to utilize:

```
<Grid.Resources>
    <local:DateToBoolConverter x:Key="DateToBoolConverter"/>
</Grid.Resources>
<CheckBox Content="Published" HorizontalAlignment="Left" Margin="-6,148,0,0"
          VerticalAlignment="Top" IsEnabled="False"
          IsChecked="{Binding PublishedDate,
                      Converter={StaticResource DateToBoolConverter}}" />
```

Code snippet from MainPage.xaml

In addition to the `CheckBox` that indicates whether a book has been published, you also want to add some more visual effects to the book display. If a book isn't published, the book cover image should be dimmed. You can do this by adjusting the `Opacity` of the image to allow the dark background to permeate through, effectively darkening the image. To do this, you can use another custom value converter that converts a `DateTime?` value into an opacity level. Rather than hard-coding the opacity to return when the book hasn't been published, you will pass this value in as an argument to the value converter's `Convert` method. In the following code, the parameter is cast to a double that is returned if the book isn't published:

```
public class DateToOpacityConverter:IValueConverter {
    public object Convert(object value, Type targetType,
                          object parameter, CultureInfo culture) {
        var defaultOpacity = 0.5;
        double.TryParse((string)parameter, out defaultOpacity);
        if (value is DateTime) {
            if ((DateTime)value < DateTime.Now)
            {
                return 1.0;
            }
        }
        return defaultOpacity;
    }

    public object ConvertBack(object value, Type targetType,
```

```
                                    object parameter, CultureInfo culture) {
            throw new NotImplementedException();
        }
    }
```

Code snippet from DateToOpacityConverter.cs

When you use this value converter within a XAML data binding statement, you can also specify the `ConverterParameter`. The value you specify will get passed into the value converter as a string that can then be parsed into the correct data type:

Available for download on Wrox.com

```
<Grid.Resources>
    <local:DateToOpacityConverter x:Key="DateToOpacityConverter"/>
</Grid.Resources>
<Image x:Name="BookCoverImage" HorizontalAlignment="Left" Height="136"
       Margin="6,6,0,0"
       VerticalAlignment="Top" Width="139"
       Source="{Binding BookCover, Mode=OneTime}"
       Opacity="{Binding PublishedDate,
                 ConverterParameter=0.6,
                 Converter={StaticResource DateToOpacityConverter}}"/>
```

Code snippet from MainPage.xaml

DESIGNING WITH DATA

We're going to take a small intermission from data binding to look at some of the design capabilities of Expression Blend for working with data. We'll break this down further by looking at two scenarios that may drive which features within Blend you use. The first looks at how you can use Blend to create and work with sample data to build up your user interface. The second works on the assumption that you already have some design-time data that you can work with.

Sample Data

You're going to continue the example of working with a list of books but come at it from a different angle. Imagine that you are creating a simple interface that will allow the user to scroll through a list of books. They should be able to select a book and view more information about the book.

In order to get a feel for what the interface will look like, you're going to need some sample data to work with. This is especially true before the project's developer has had time to build out a fully functional data access layer. The first step is to use the sample data feature of Blend to create appropriate data to represent the information about a book. Open the solution you've been working on in Blend, and open the Data window (leftmost image of Figure 15-3). From the "Create Sample Data" dropdown (second icon from right in center image of Figure 15-3), select "New Sample Data." This will create a default set of sample data consisting of a collection of entities made up of two properties. Edit these properties by selecting the property name and renaming them to **Title** and **Description**, as shown in the rightmost image of Figure 15-3.

FIGURE 15-3

The next thing to do is to change the data type to be a string. Select the icon alongside the property to drop down the Data Type Selector, shown in the leftmost image of Figure 15-4. For the `Description` property, set the "Max word count" to 83 (the actual number is not meaningful, just a large number to illustrate sample data generation). On the Collection node, click the plus sign to drop down the menu for adding new properties (center image in Figure 15-4). Add simple properties for both `BookCover`, set the `Type` to `Image`, and `PublishedData`, and leave the `Type` as `String` but set the `Format` to `Date`. Lastly, select "Add Collection Property" and name the property **Authors**. Add `Email` and `Name` properties, setting the `Type` to `String` and selecting "Email Address and Name" for the `Format`. The resulting data source should look similar to the rightmost image of Figure 15-4.

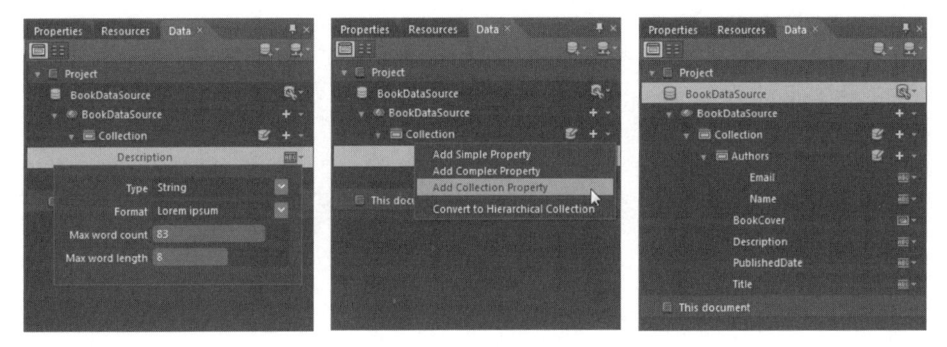

FIGURE 15-4

This process has created a set of sample data that you can use within your application. The data is stored within a XAML file with the same name as the data source, within the SampleData folder. If you want to adjust the sample data, you can either edit this file directly, or you can click the "Edit Sample Values" icon. This allows you to edit properties for the entities via a simple spreadsheet-type interface. Figure 15-5 illustrates the "Edit Sample Values" dialog for the collection of books.

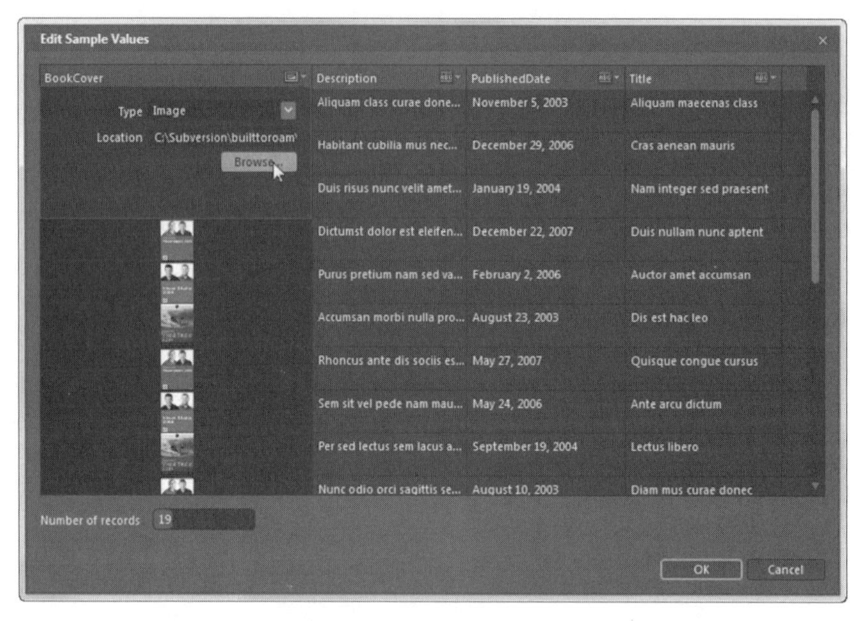

FIGURE 15-5

In Figure 15-5 you'll notice that we've selected a location for the images to be sourced from. This is useful if you have some stock images that you want to appear in your application during design time to make it appear more realistic. The dialog also allows you to modify the number of records that are in the sample data source.

Now that you have your sample data source, it's time to make use of it to build up your user interface. Select the `BookCover` and `Title` properties from the Data window and drag them across onto the main area of the page. You should see a prompt indicating that this will create a `ListBox`, as shown in the left image of Figure 15-6. When you release the mouse button, a `ListBox` should be created as in the right image of Figure 15-6 (you may need to reset the `Margin` property of the created `ListBox` for it to take up the whole ContentGrid area).

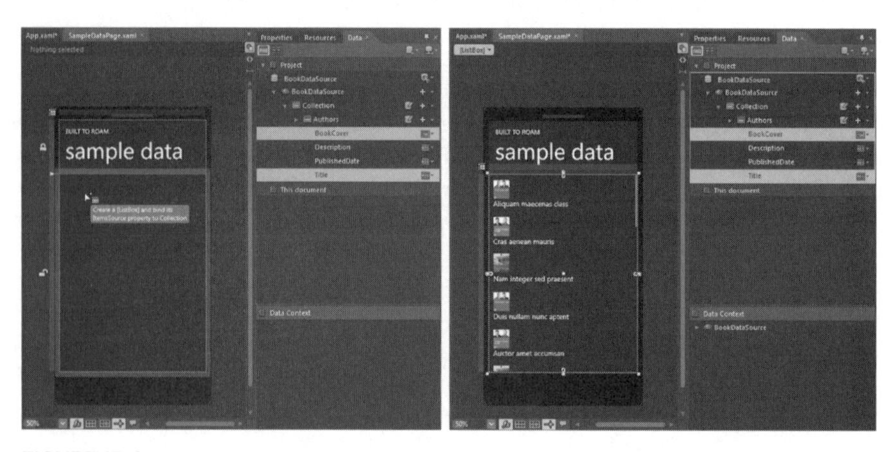

FIGURE 15-6

In the right image of Figure 15-6, the newly created `ListBox` is currently selected. From the Data window you can see that its data context is set to the `BookDataSource`. This is also indicated in the lower panel of the Data window entitled *Data Context*, as well as being evident from the yellow border around the `BookDataSource` in the top panel. If you change to the Properties window, you will notice that the `ItemsSource` property for the `ListBox` has a yellow border and the Advanced Options square, shown on the right side of Figure 15-7, is also colored yellow. This indicates that this property is data bound. Clicking the "Advanced Options," followed by "Data Binding" opens the Create Data Binding window shown on the left of Figure 15-7. Here you can select what the property is bound to. In this case, it has been configured to be the main collection of the `BookDataSource`.

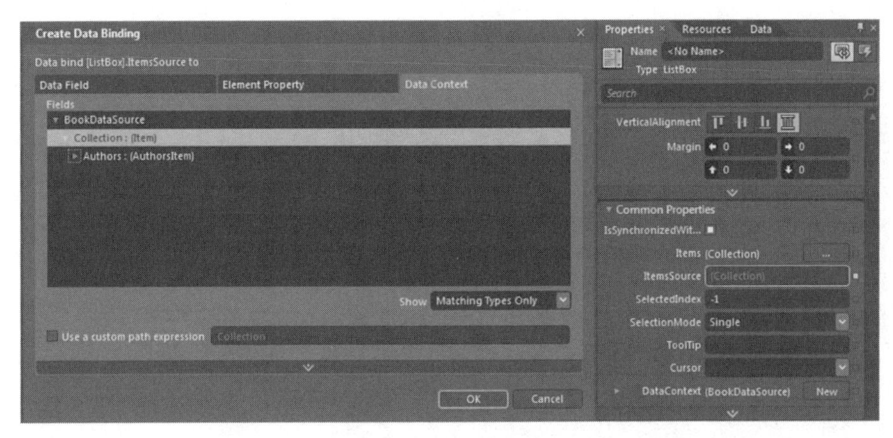

FIGURE 15-7

We'll come back to the Create Data Binding window later as it can be used to control binding direction as well as specify the value converter to use. Close the window to return to the Layout Designer.

In the Objects and Timeline window, right-click the listbox and select Edit Additional Templates ➪ Edit Generated Items (Item Template) ➪ Edit Current. You will notice that the tree structure changes to illustrate the visual representation of each item within the listbox (i.e., the Item Template). Figure 15-8 illustrates that each item is made up of a `StackPanel` that contains an `Image` and a `TextBlock`.

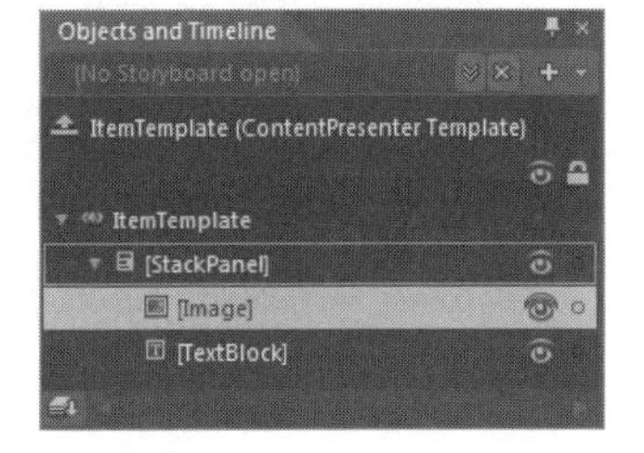

FIGURE 15-8

In this case, you want to change this to use a `Grid` instead of a `StackPanel`. Right-click the `StackPanel` and select Change Layout Type ➪ Grid. Rearrange the `Image` and `TextBlock` so that they are alongside each other. If you look at the `Source` property of the `Image` and the `Text` property of the `TextBlock`, you should see that they are both data bound. This is indicated again by the yellow border. If you click Advanced Options ➪ Data Binding, you will see that they are bound to the `BookCover` and `Title` properties of the sample data, respectively.

Now it's time to display the details of the selected book. Start by resizing the listbox to fill approximately half of the content grid. From the Data window, click the "Details Mode" icon, second in from the left at the very top of the window. This changes the behavior for when you drag items from the Data window onto the page layout. Previously you were in List mode, so dragging items onto the layout would create a listbox to display them in. In Details view, a small form will instead be generated to allow data entry or display of a single record. Select the `BookCover`, `Title`, and `Description` fields, and drag them into the empty space on the page. You will see that in addition to creating an `Image` and two `TextBlock` elements to represent the desired properties, Blend has also created corresponding labels. Before you rearrange these items, right-click each `TextBlock` element in the Objects and Timeline window and select Edit Style ➪ Apply Resource ➪ PhoneTextNormalStyle, as shown in Figure 15-9, to select the style.

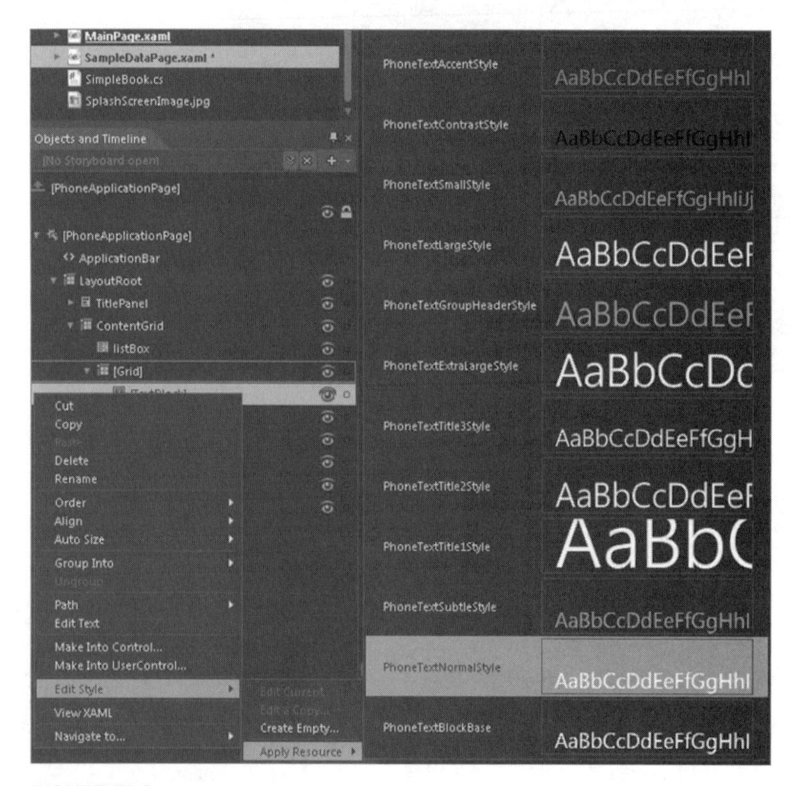

FIGURE 15-9

Now you can rearrange your layout to best display the Image, Title, and Description. As well as relying on the automatic data binding that is configured when you drag fields onto the layout from the Data window, it is possible to manually configure data bindings. We will demonstrate this by connecting a `CheckBox` control to the `PublishedData` field.

From the Assets window, drag a CheckBox into the same area as the other book detail elements and set the Content to "Published." Click the "Advanced Options" button next to the `IsChecked` property. Choose "Data Binding . . . " Then, select Select "All Properties" from the Show dropdown, followed by selecting "`PublishedDate`," as shown in Figure 15-10.

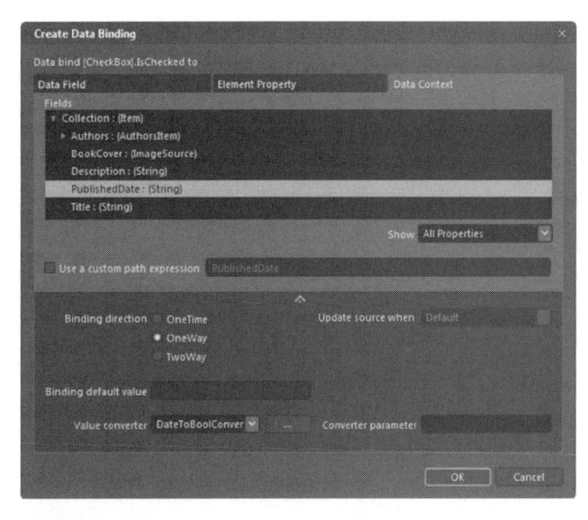

FIGURE 15-10

As was discussed previously, `PublishedDate` isn't the same data type as `IsChecked`, which means that you need to use a value converter. Select the `DataToBoolConverter` you created earlier by clicking the Ellipses button and selecting the appropriate type. Completing the details of the "Create Data Binding" dialog shown in Figure 15-10 is essentially the step that is performed automatically when you drag fields from the Data window onto your layout.

You might be wondering how the details elements you have just added to the page are related to the selected item from the `ListBox` created earlier. You may have noticed that when you added the details elements, they were all placed within a `Grid`. This is a useful way to group elements together, particularly if you want to set the `DataContext` property of each element to the same value. If you look at the `DataContext` property for the `Grid`, you will see that it is itself data bound. However, rather than being bound to a property of a data source, the `DataContext` is bound to the `SelectedItem` property of the `ListBox`. This is known as *Element Property Binding* and will be covered in more detail later on.

When we first created the sample data source and started dragging elements onto the page, we glossed over exactly how the data source was being loaded and wired to the page. If you look in the SampleData\BookDataSource folder, you will see that it has created several files to represent the sample data. The main file is BookDataSource.xaml, which you can open to view and even edit the XAML used to represent the sample data. Here's an excerpt from the top of this file showing a collection of books, which, in turn, have a nested collection of authors:

```
<SampleData:BookDataSource
   xmlns:SampleData="clr-namespace:Expression.Blend.SampleData.BookDataSource"
   xmlns="http://schemas.microsoft.com/winfx/2006/xaml/presentation">
  <SampleData:BookDataSource.Collection>
    <SampleData:Item Title="Aliquam maecenas class"
                     Description="Aliquam class curae nisi leo nisl ..."
                     PublishedDate="November 5, 2003"
                     BookCover="/BindMeABook;
    component/SampleData/BookDataSource/BookDataSource_Files/ProVS2005_Cover.jpg">
      <SampleData:Item.Authors>
        <SampleData:AuthorsItem FullName="Aaberg, Jesper"
                                Email="someone@example.com" />
        <SampleData:AuthorsItem FullName="Adams, Ellen"
                                Email="user@adventure-works.com" />
```

The sample data file, BookDataSource.xaml, holds the XAML representation of the sample data, but it won't automatically get loaded unless it is referenced somewhere within the application. If you look in app.xaml, you will notice that there is such a reference, thus creating an instance of the sample data set that can be referenced anywhere in the application by the `BookDataSource` key:

```
<Application
    ...
    x:Class="BindMeABook.App">
    <Application.Resources>
        <SampleData:BookDataSource x:Key="BookDataSource" d:IsDataSource="True"/>
```

Of course, the last thing to do is to connect this instance of the sample data to the data context of the page. This is done by setting the `DataContext` on the `LayoutRoot Grid` control of the page:

```
<phone:PhoneApplicationPage
    ...
    x:Class="BindMeABook.SampleDataPage">
    <Grid x:Name="LayoutRoot" Background="{StaticResource PhoneBackgroundBrush}"
          DataContext="{Binding Source={StaticResource BookDataSource}}">
```

Looking further down the XAML for the page, you will notice that the `ListBox` is bound to a property called `Collection`. As there was only one list of books, we didn't rename the default collection in the sample data. You should take this opportunity to return to the Data window in Blend and change the name from *Collection* to **Books**. This will automatically update all references in the page. The resulting `ListBox` element should look similar to the following:

```
<ListBox x:Name="listBox" ItemTemplate="{StaticResource ItemTemplate}"
         ItemsSource="{Binding Books}" Height="298" VerticalAlignment="Top"
         ScrollViewer.HorizontalScrollBarVisibility="Disabled"/>
```

Code snippet from SampleDataPage.xaml

Having the sample data wired up is useful for designing the layout in Blend. You can even run the application and see that the data is loaded and the user interface is functional. However, as you build out the rest of your application, you will want to replace the sample data with your real data source. Although this is desirable at runtime, you may still want to keep the sample data so that you have a rich design-time experience. To do this, you can simply disable the sample data at runtime. From the Data window, click the icon next to the `BookDataSource` and uncheck the "Enable When Running Application" item in the dropdown menu. Now when you run your application, you won't see any data, which is as you would expect since you haven't wired up any real data. You still have design-time support in Blend because it uses a small trick in the XAML file to specify a design-time `DataContext`. If you look at the `LayoutRoot` element, you will notice a subtle change in the property that is being data bound to. Instead of specifying the `DataContext` property, which would set the runtime data source, it now references `d:DataContext`, which is an attribute that is specific to the design experience within Blend (or another designer such as Visual Studio):

```
<phone:PhoneApplicationPage
    ...
    xmlns:d="http://schemas.microsoft.com/expression/blend/2008"
    x:Class="BindMeABook.SampleDataPage">
```

```
<Grid x:Name="LayoutRoot" Background="{StaticResource PhoneBackgroundBrush}"
      d:DataContext="{Binding Source={StaticResource BookDataSource}}">
```

Code snippet from SampleDataPage.xaml

The last step, of course, is to wire up real data. Here you are simply setting a new `BookViewModel` instance as the data source. The `BookViewModel` contains a single property, `Books`, that holds the array of books that you want to display. Note that the structure mirrors that of the sample data you created in Blend — the source has a property `Books`, made up of entities that have properties such as `Title`, `BookCover`, and `Description`:

```
void SampleDataPage_Loaded(object sender, RoutedEventArgs e){
    this.DataContext = new BookViewModel();
}

public class BookViewModel{
    public SimpleBook[] Books { get; set; }

    public BookViewModel()
    {
        Books = CreateBooks();
    }

    private SimpleBook[] CreateBooks(){ ... }
}
```

Code snippet from SampleDataPage.xaml.cs

SAMPLE DATA TYPES

One interesting aspect of data binding is that it's often referred to as *string-binding*. This is because all the data bindings are set up using string literals within the XAML file. You would have thought that this would be a recipe for disaster as we have always been taught not to use string literals wherever possible. This danger is still present; for example, if you were to change the `Title` property to `BookTitle`, then the data bindings that depend on this property would no longer work. Luckily, both Visual Studio and Expression Blend have reached a point where they can detect some of these issues, providing you build warnings to indicate where properties can't be found.

The plus side of data binding being defined using strings is that there is no hard coding of data types. In this example, you created a set of sample data within Blend, then wired up real data based on your existing `SimpleBook` class. The use of data binding makes it possible to interchange data sources so long as the data types have the same data structure.

Running the application, you should see a programmatically generated list of books displayed at the top of the page. When you select an item from the `ListBox`, you should see the details appear in the lower half of the page, as shown in Figure 15-11.

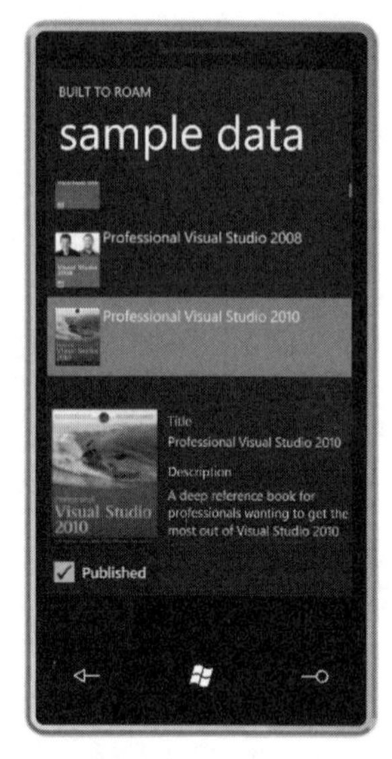

FIGURE 15-11

Design-Time Data

In some cases while you are building a Windows Phone application, you will already have some data available that you will want to use at design time. This might be a static XML file or it might be code that creates instances of your data classes. In this section, you'll look at how you can take that data and use it at design time, switching it out for real data at runtime.

Model View ViewModel (MVVM)

Before we go any further, let's take a look at a pattern that is known as Model View ViewModel (MVVM). For those familiar with other architecture patterns such as MVC (Model-View-Controller) or MVP (Model View Presenter), MVVM isn't too different. You have the *Model*, which defines the data or information that makes up your application. You also have the *View*, which in the case of a Windows Phone application will most likely correlate to a Silverlight-based page. Lastly, you have what is known as the *ViewModel*, which you can think of as the current state of the View. Unlike the Model, which is solely focused on keeping track of the application data, the ViewModel is closely associated with a particular View. The ViewModel is often structured in a manner that follows the visual layout of data on the page and has properties that correlate to properties of elements on the page. The idea is that the View can be bound to the ViewModel using data binding, rather than having to write code to continually route logic and data between the Model and the View. As an example, a `TextBlock` within the View may data bind to a `ViewModel` property called `Author` that concatenates individual `FirstName` and `LastName` fields from the Model. A future change in how the data is represented in the Model wouldn't affect the View(s), as the `ViewModel`'s `Author` property would simply be updated.

Creating Your ViewModel

There isn't much more to MVVM than that, and rather than have you read through any more theory on MVVM, let's build a simple user interface that allows editing of the details of a book, this time starting with a `ViewModel` that we'll create in code. Since the `ViewModel` is closely associated with the layout of a View, it is common for them to be named accordingly. In this case, the View is going to be a `PhoneApplicationPage` called DesignTimePage.xaml. The corresponding `ViewModel` will be a regular C# class called `DesignTimeViewModel`:

```
public class DesignTimeViewModel{}
```

The layout is going to include the book cover, title, description, whether the book has been published, and the list of authors. Our Model in this case is the `Book` class, which, in turn, contains an array of `Author` objects:

Available for download on Wrox.com

```
public class Book {
    public Guid BookId { get; set; }
    public string Title { get; set; }
    public string Description { get; set; }
    public BitmapImage BookCover { get; set; }
    public DateTime? PublishedDate { get; set; }
    public Author[] Authors { get; set; }
}

public class Author{
    public string FullName { get; set; }
    public string Email { get; set; }
}
```

Code snippet from Book.cs

In the opening section of this chapter, the `SimpleBook` class was modified to implement the `INotifyPropertyChanged` interface directly. This is not a good idea because it tightly couples your View (i.e., the page) to the Model, making it hard to change either without breaking the other. It also assumes that you are able to make changes to each of the properties within the Model that you want to data bind against (i.e., allowing changes in the data to propagate to the user interface). If the Model is being generated by a tool, based on a service interface or database structure, this may be impossible, or require significant work each time the Model is regenerated.

A better scenario is for `INotifyPropertyChanged` to be implemented by the `ViewModel`. Within the `ViewModel` you create properties that correlate to properties of the visual elements that make up the View. When a data bound control within the View changes, it will update the underlying `ViewModel` property. These changes can, in turn, be passed through to the Model either as they happen or based on some event (e.g., the user selecting Save). Let's take a look at what the `DesignTimeViewModel` would look like:

Available for download on Wrox.com

```
public class DesignTimeViewModel : INotifyPropertyChanged{
    private string title;
    public string Title{
        get{ return title; }
        set{
            if (title == value) return;
            title = value;
            RaisePropertyChanged("Title");
        }
    }

    private string description;
    public string Description{
        get{ return description; }
        set{
            if (description == value) return;
            description = value;
            RaisePropertyChanged("Description");
        }
    }
```

```
        private BitmapImage bookCover;
        public BitmapImage BookCover{
            get{ return bookCover; }
            set{
                if (bookCover == value) return;
                bookCover = value;
                RaisePropertyChanged("BookCover");
            }
        }

        private bool isPublished;
        public bool IsPublished{
            get{ return isPublished; }
            set{
                if (isPublished == value) return;
                isPublished = value;
                RaisePropertyChanged("IsPublished");
            }
        }

        private ObservableCollection<Author> authors = new ObservableCollection<Author>();
        public ObservableCollection<Author> Authors{
            get { return authors; }
        }

        public event PropertyChangedEventHandler PropertyChanged;
        private void RaisePropertyChanged(string propertyName){
            if (PropertyChanged != null){
                PropertyChanged(this, new PropertyChangedEventArgs(propertyName));
            }
        }
    }
```

Code snippet from DesignTimeViewModel.cs

You will notice that the `DesignTimeViewModel` looks remarkably similar to the `Book` class. However, there are a few subtle differences. Firstly, instead of a `PublishedDate`, the `DesignTimeViewModel` has an `IsPublished` property. This makes it easier to bind the `IsChecked` property of the `CheckBox` as it no longer requires the use of a value converter. As you would expect, each property raises the `PropertyChanged` event so that the user interface can update when the `ViewModel` changes.

The other difference is that instead of an `Author` array, it exposes an `ObservableCollection` of type `Author`. The `ObservableCollection` class provides a base implementation for the `INotifyCollectionChanged` interface, which raises events whenever the collection is modified. We'll come back to the `DesignTimeViewModel` later on, but let's jump across to Blend and wire up the user interface.

One thing you will notice is that it is very rare to have to write code that sets properties on visual elements. If you find yourself writing such code, you should step back and look around to see if there is a way that you can do it declaratively. This isn't to say that writing code in the code-behind file for a page is bad or should be avoided. There are times when the easiest or simplest way is to write a single line of code in the code-behind file. However, if it can be done declaratively within XAML, it will enable designers to visualize and interact with the change as part of the design experience within Blend or Visual Studio.

Creating a Data Source

From the Data window in Blend, select "Create Object Data Source" from the rightmost icon in the toolbar at the top of the window, as shown in the leftmost image of Figure 15-12. In the "Create Object Data Source" dialog, select the `DesignTimeViewModel` and click OK.

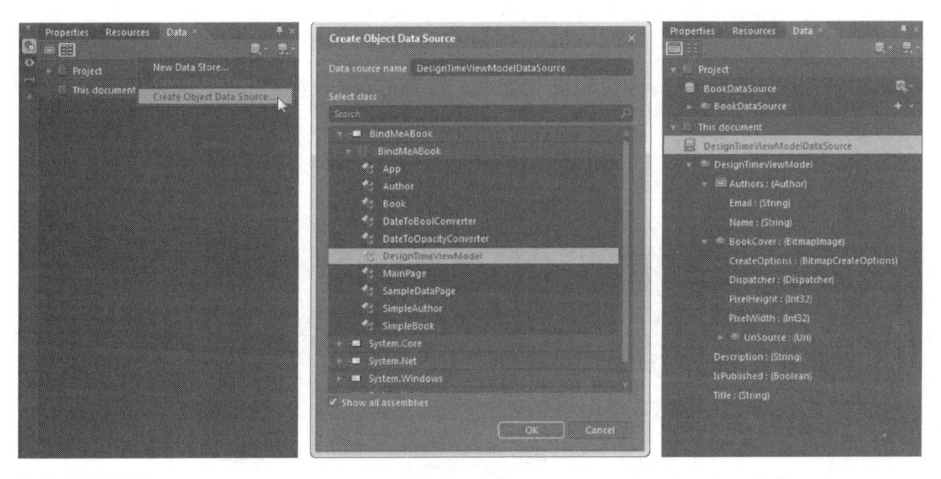

FIGURE 15-12

What you have just done is to add an instance of the `DesignTimeViewModel` class to the current page. The instance is displayed in the "This document" section of the Data window, as shown in the rightmost image of Figure 15-12. If you look at the XAML for the page, you will see that there is a `DesignTimeViewModel` element within the Resources collection for the page:

```
<phone:PhoneApplicationPage
    ...
    xmlns:local="clr-namespace:BindMeABook"
    x:Class="BindMeABook.DesignTimePage">
    <phone:PhoneApplicationPage.Resources>
        <local:DesignTimeViewModel x:Key="DesignTimeViewModelDataSource"
                                   d:IsDataSource="True"/>
    </phone:PhoneApplicationPage.Resources>
```

At run time an instance of the `DesignTimeViewModel` will be created and added to the resources available to the page. You can extend the XAML to set properties and even add `Author` instances to the Authors collection:

```
<phone:PhoneApplicationPage.Resources>
    <local:DesignTimeViewModel x:Key="DesignTimeViewModelDataSource"
                               d:IsDataSource="True"
        Title="Professional Visual Studio 2010"
        Description="A deep reference book for professionals
                     wanting to get the most out of Visual Studio 2010"
        IsPublished="true" >
        <local:DesignTimeViewModel.BookCover>
            <BitmapImage UriSource="/Images/ProVS2010_Cover.jpg" />
        </local:DesignTimeViewModel.BookCover>
        <local:DesignTimeViewModel.Authors>
            <local:Author FullName="Nick Randolph" Email="nick@builttoroam.com" />
        </local:DesignTimeViewModel.Authors>
    </local:DesignTimeViewModel>
</phone:PhoneApplicationPage.Resources>
```

Building the User Interface

From the Data window, select items from under the DesignTimeViewModelDataSource node and drag them onto the page. Figure 15-13 illustrates the `BookCover` that has already been placed on the page and the `Title` that is currently being added to the page.

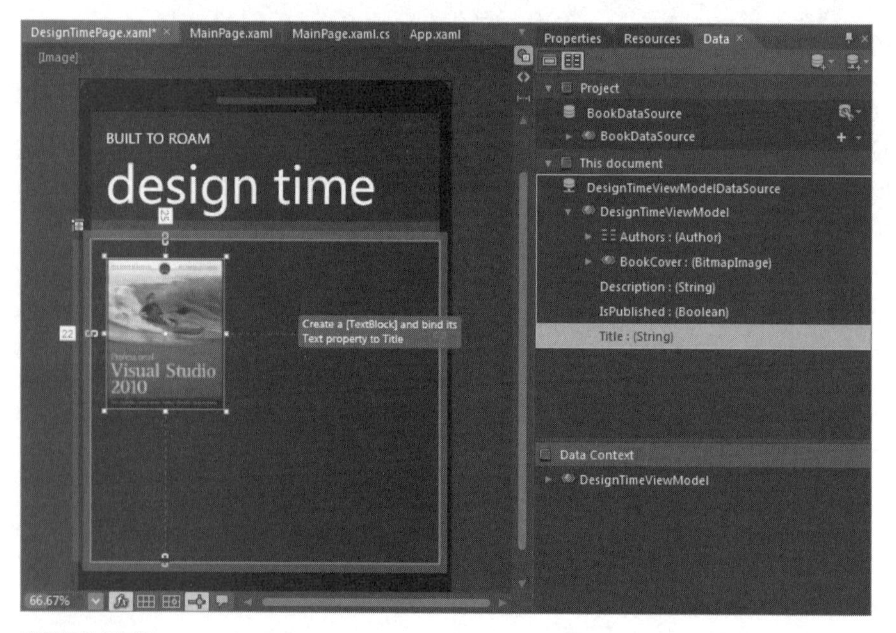

FIGURE 15-13

Note that as you drag elements onto the page, Blend attempts to guess the type of visual element to create. In the case of string properties, this is a `TextBlock`. However, we want to create an interface where the user can modify the properties of the book. Instead of dragging the properties from the Data window, first, drag the appropriate element onto the page from the Assets window. For example, for the `Title` property we want a `TextBox`, so drag one of these from the Assets window onto the page. To bind the `TextBox` to the `Title` property, you can then drag the Title node from the Data window onto the `TextBox`. Blend is quite good at predicting the property and mode of binding that you want. For example, in this case, it will bind the `Title` property to the `Text` property of the `TextBox`. However, to be safe, it's recommended that you hold down the Shift key when you drag the node onto the visual element. This will display the "Create Data Binding" dialog, shown in Figure 15-14, which allows you to verify the property on the visual element to bind to as well as the mode of the binding. The only limitation with this dialog is that it doesn't allow you to specify value converters. If you've designed your `ViewModel` to work with your View, this should rarely be an issue as the data type of your `ViewModel` properties should match the corresponding property on the View.

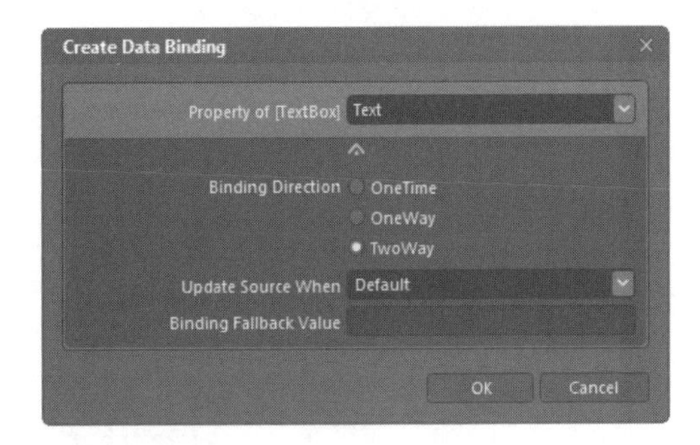

FIGURE 15-14

To add the list of authors to the page, click the "List Mode" icon in the top-left corner of the Data window. Then drag the Authors node onto the page. This will automatically create a listbox populated with the authors' names and e-mail addresses. You can then edit the Item Template and adjust the layout of these items individually, as done earlier in the chapter.

You are now ready to run your Windows Phone application. When the `DesignTimePage` loads, you can edit the `Title` and `Description` fields. If you set a break point in the corresponding properties in the `ViewModel`, you will see that the two-way binding is indeed working.

Separating Design and Runtime Data

The instance of the `DesignTimeViewModel` that you created in Expression Blend is great for building your layout. However, at runtime, you will want to populate the `ViewModel` with real data. By declaring the `ViewModel` explicitly in XAML, it will be created at runtime by calling the zero-parameter constructor. As such, there is no opportunity for you to populate it with real data before it is displayed. It would be great if it were possible to pass in a parameter such as the ID of the book to be displayed, or a reference to a helper class that can load the book information into the `ViewModel`. Unfortunately, this is not possible to do within XAML.

There is a way to address this issue with what's known as the *locator pattern*. Essentially, instead of creating an instance of the `ViewModel` within the page, you set the `DataContext` of the page to be a reference to a property on a centrally accessible locator object. This property will return an instance of the `ViewModel` that has been loaded with the necessary information. This is starting to sound a little theoretical again, so let's look at how this plays out in an example.

We're going to start with a very simple ViewModel locator that simply creates and retains a single instance of the DesignTimeViewModel. When the DesignTimeViewModel is created, it is initialized through the LoadFromBook method that sets the various properties to be equal to the values in the supplied book. There is also a read-only property, BookViewModel, which is used to make the instance accessible so that it can be set as the DataContext on a page:

Available for download on Wrox.com

```
public class ViewModelLocator{
    private static DesignTimeViewModel BookViewModelInstance;

    public ViewModelLocator(){
        var book = new Book(){
                BookId = Guid.NewGuid(),
                Title = "Professional Visual Studio 2010",
                Description = "A deep reference book for professionals wanting to
                               get the most out of Visual Studio 2010",
                BookCover = new BitmapImage(new Uri("/Images/ProVS2010_Cover.jpg",
                                                    UriKind.Relative)),
                PublishedDate = new DateTime(2011, 4, 10),
                Authors = new Author[]{new Author(){FullName="Nick Randolph",
                    Email="nick@builttoroam.com"},
                    new Author(){FullName="David Gardner"},
                    new Author(){FullName="Michael Minutillo"},
                    new Author(){FullName="Chris Anderson"}
                }
            };

        BookViewModelInstance = new DesignTimeViewModel();
        BookViewModelInstance.LoadFromBook(book);
    }

    public DesignTimeViewModel BookViewModel{
        get{
            return BookViewModelInstance;
        }
    }
}
```

Code snippet from ViewModelLocator.cs

To use a locator within your Windows Phone application, you first need to create an instance of it. Since you want the locator to be accessible anywhere within your application, it makes sense to declare it within the app.xaml file:

Available for download on Wrox.com

```
<Application
    ...
    x:Class="BindMeABook.App"
    xmlns:local="clr-namespace:BindMeABook">
    <Application.Resources>
        <local:ViewModelLocator x:Key="Locator" />
```

Code snippet from App.xaml

Since this instance can be referenced anywhere within the application, you can update the DesignTimePage.xaml to use this object as the page's `DataContext`:

```
<phone:PhoneApplicationPage
    ...
    xmlns:local="clr-namespace:BindMeABook"
    x:Class="BindMeABook.DesignTimePage"
    DataContext="{Binding Path=BookViewModel, Source={StaticResource Locator}}" >
```

Code snippet from App.xaml

Now when the application loads, an instance of the locator is created and added to the application resources collection. When the `DesignTimePage` loads, it references the locator, requesting the value of the `BookViewModel` property, which it then sets to be its `DataContext`. You can run the application to see this in effect. To get a feel for how things load, place break points in the constructor for the `BookLocator` and in the `BookViewModel` property.

All that you've actually done so far is move the logic that sets the property values for the `DesignTimeViewModel` from within the page's XAML to the constructor of the `BookLocator`. What we really want to do is to create sample data during design time and to load real data at run time. This can be done by testing to see whether the application is running in a designer or not. The `DesignerProperties.IsInDesignTool` property can be queried to detect if the code is running in the designer or running on an actual device (or emulator):

```
Book book;
if (DesignerProperties.IsInDesignTool){
    book = new Book(){
              BookId = Guid.NewGuid(),
              Title = "Professional Visual Studio 2010",
              Description = "A deep reference book for professionals wanting to
                             get the most out of Visual Studio 2010",
              BookCover = new BitmapImage(new Uri("/Images/ProVS2010_Cover.jpg",
                                                  UriKind.Relative)),
              PublishedDate = new DateTis ime(2011, 4, 10),
              Authors = new Author[]{new Author(){FullName="Nick Randolph",
              Email="nick@builttoroam.com"},
              new Author(){FullName="David Gardner"},
              new Author(){FullName="Michael Minutillo"},
              new Author(){FullName="Chris Anderson"}}};
}
else{
        book = LoadBook();
}
```

This is rather clumsy and will rapidly lead to a mess of code to support design and runtime data sources. To refactor this code and make it more scalable, we're going to use the repository pattern. When the locator is created, it determines whether to load the design or runtime repository. These repositories both implement the same interface, `IRepository`, which means that they can easily be interchanged. You will also need to modify the constructor of the `DesignTimeViewModel` to accept an `IRepository` instance.

Start with the `IRepository` and the two implementations:

Available for download on Wrox.com

```
public interface IRepository{
    Book BookToDisplay();
}
```

Code snippet from IRepository.cs

```
public Book BookToDisplay(){
    var book = new Book(){
                BookId = Guid.NewGuid(),
                Title = "Professional Visual Studio 2010",
                Description = "A deep reference book for professionals wanting to
                           get the most out of Visual Studio 2010",
                BookCover = new BitmapImage(new Uri("/Images/ProVS2010_Cover.jpg",
                                                UriKind.Relative)),
                PublishedDate = new DateTime(2011, 4, 10),
                Authors = new Author[]{new Author(){FullName="Nick Randolph",
                    Email="nick@builttoroam.com"},
                    new Author(){FullName="David Gardner"},
                    new Author(){FullName="Michael Minutillo"},
                    new Author(){FullName="Chris Anderson"}
                }};
    return book;
}
```

Code snippet from DesignTimeRepository.cs

```
Public Book BookToDisplay(){
    var book = new Book();
    // Load book information from isolated storage or from a cloud service
    return book;
}
```

Code snippet from RuntimeRepository.cs

Now, update the `DesignTimeViewModel` and the `BookLocator` constructors:

Available for download on Wrox.com

```
private IRepository Repository { get; set; }
public DesignTimeViewModel(IRepository repository){
    this.Repository = repository;
    LoadFromBook(this.Repository.BookToDisplay());
}
```

Code snippet from DesignTimeViewModel.cs

```
public ViewModelLocator(){
    IRepository repository;
    if (DesignerProperties.IsInDesignTool){
```

```
            repository = new DesignTimeRepository();
        }
        else{
            repository = new RuntimeRepository();
        }

        BookViewModelInstance = new DesignTimeViewModel(repository);
    }
```

Code snippet from ViewModelLocator.cs

Now you have sufficiently separated the logic for creating your design-time and runtime data. You can extend the `IRepository` interface to include any methods that are required by this and other view models. All you have to do is provide implementations for both design time and run time.

MVVM Light Toolkit

The MVVM Light Toolkit (`http://mvvmlight.codeplex.com/`) provides a number of components that can assist you in working with the MVVM pattern in your Windows Phone application.

Base View Model

You will have noticed that many classes implement the `INotifyPropertyChanged` interface and contain a `RaisePropertyChanged` method that is used to raise the `PropertyChanged` event. Instead of implementing these in every `ViewModel` you create, the MVVM Light Toolkit provides a base class that implements the `INotifyPropertyChanged` interface and provides a `RaisePropertyChanged` method and other base functionality that you may find useful along the way.

To use the Toolkit, you need to download it and follow the installation instructions on the website. Once it is installed, you need to reference `GalaSoft.MvvmLight.WP7.dll` and `GalaSoft.MvvmLight.Extras.WP7.dll`. With these assemblies referenced, it's just a matter of modifying the `DesignTimeViewModel` to inherit from `ViewModelBase` and removing both the `PropertyChanged` event and the `RaisePropertyChanged` method:

Available for download on Wrox.com

```
using GalaSoft.MvvmLight;

public class DesignTimeViewModel : ViewModelBase{
    ...
}
```

Code snippet from DesignTimeViewModel.cs

Command Pattern

One of the other areas where the MVVM Light Toolkit is useful is in implementing the Command Pattern within your Windows Phone application. Unfortunately, it is not possible to data bind an event raised by a visual element, for example, a button click, to a method on the underlying `ViewModel`. However, it is possible to trigger a behavior based on an event and then have that behavior call the appropriate method.

As we've done previously, let's work through an example. In this case, were going to add a very simple method to the `ViewModel` called `ClearTitle` that simply clears the Title text for the book being displayed:

```
public class DesignTimeViewModel : ViewModelBase{
    public RelayCommand ClearTitleCommand { get; private set; }

    public DesignTimeViewModel(IRepository repository){
        ...
        ClearTitleCommand = new RelayCommand(ClearTitle);
    }

    public void ClearTitle(){
        this.Title = "";
    }
```

Code snippet from DesignTimeViewModel.cs

In addition to specifying the `ClearTitle` method, we've created a `RelayCommand` instance that points to this method. This is required in order to be able to bind to the command.

Now let's return to Blend and add a Clear button near the `Title TextBox`. From the Behaviors node on the Assets window, drag the `EventToCommand` behavior onto the newly created Clear button. You will see a new EventToCommand node appear in the Objects and Timeline window, and you'll notice the properties for this behavior in the Properties window. The `EventName` will already be set to the `Click` event, and the default value for `SourceName` is `[Parent]`, as shown in Figure 15-15. This indicates that this behavior will be triggered on the `Click` event of its parent, which, of course, is the Clear button.

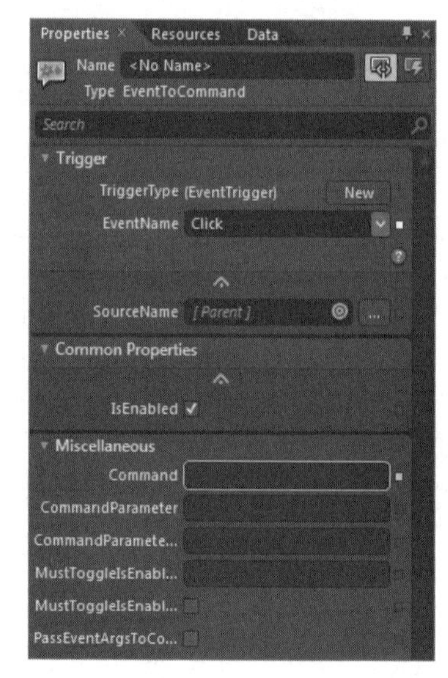

FIGURE 15-15

Click the "Advanced Options" next to the Command property and select the "ClearTitleCommand" node. This effectively sets the `ClearTextCommand` as the command that will be invoked when the behavior is triggered.

That's all that's needed to bind a command to an event that is raised by a visual element. However, there is one more point of interest about the use of commands. The `Clear` command should only be invoked when the `Title` isn't empty. The constructor for the `RelayCommand` supports a second argument that is a predicate function that should return `true` if the command is permittable. You can update the `DesignTimeViewModel` accordingly:

```
public class DesignTimeViewModel : ViewModelBase{
    public RelayCommand ClearTitleCommand { get; private set; }

    private string title;
    public string Title{
        get{
            return title;
        }
        set{
            if (title == value) return;
            title = value;
            RaisePropertyChanged("Title");
            RaisePropertyChanged("CanClearTitle");
        }
    }

    public DesignTimeViewModel(IRepository repository){
        ...
        ClearTitleCommand = new RelayCommand(ClearTitle , ()=>CanClearTitle);
    }

    public void ClearTitle(){
        this.Title = "";
    }

    public bool CanClearTitle{
        get{
            return !string.IsNullOrEmpty(this.Title);
        }
    }
```

Code snippet from DesignTimeViewModel.cs

It may seem weird that `CanClearTitle` is declared as a property when the `RelayCommand` requires a predicate function. The reason for this is that this property has dual functions. It is used to determine whether the `ClearTitle` method can be invoked, but it will also be used to enable and disable the Clear button through data binding. For this to work, when the `Title` is cleared, or, in fact, any time the `Title` is modified, the `CanClearTitle` property needs to be queried to determine if the button should be enabled or disabled. This is why there is a second call to the `RaisePropertyChanged` method, passing in `CanClearTitle`.

When you run the application, you should be able to click the Clear button and for it to both clear the `Title TextBox` and disable the Clear button. If you enter text into the `TextBox`, you should notice that the Clear button is re-enabled.

Element and Resource Binding

One aspect of binding that we haven't touched on is the ability to bind to items other than custom C# objects created programmatically. It is also possible to bind to resources and properties exposed by other visual elements within the page. You have, in fact, seen both of these before. For example, when the BookLocator was added to the app.xaml, it was accessed from within a page via a resource binding. Similarly, earlier in the chapter, the Grid that contained the details for the selected book used an element binding to data bind to the SelectedItem property of the ListBox.

Let's first look at how resource data binding works. Add a TextBlock to a page and click the "Advanced Options" icon next to the Style property. From the pop-up list select "Local Resources." This will display all the resources that are available within the current scope and match the type of the property (in this case, the Style property). We've selected the PhoneTextNormalStyle so that the TextBlock reflects the normal style of text on the Windows Phone, as shown in Figure 15-16.

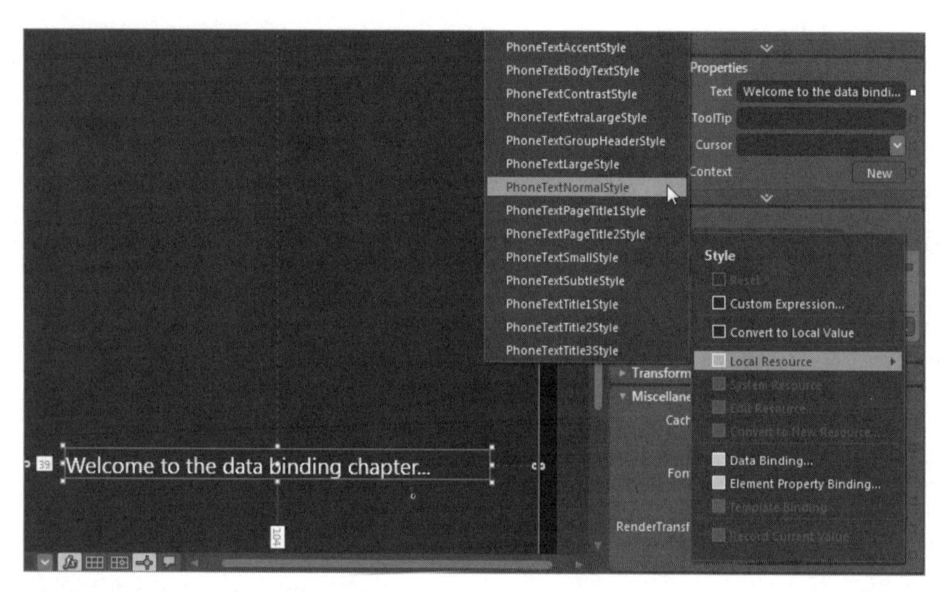

FIGURE 15-16

If you look at the XAML for the TextBlock, you will see that the Style property value uses the same curly { } braces that you've seen all the way through this chapter. However, the difference is that instead of it being a Binding, it's a StaticResource:

```
<TextBlock HorizontalAlignment="Left" Margin="39,0,0,104" TextWrapping="Wrap"
           Text="Welcome to the data binding chapter..." VerticalAlignment="Bottom"
           Width="396" Style="{StaticResource PhoneTextNormalStyle}"/>
```

A StaticResource is one that is loaded once and remains fixed until discarded. Windows Phone doesn't have the notion of a DynamicResource, unlike its bigger brother WPF. Here the StaticResource is being looked up by the key PhoneTextNormalStyle. This resource exists in the global resources dictionary available to all Windows Phone applications. Every page within a Silverlight application has access to all resources defined in the global resources dictionary, those defined within app.xaml and those that are defined within the page. Subsequently, nested elements within a page can access resources defined in these locations, or any other parent element.

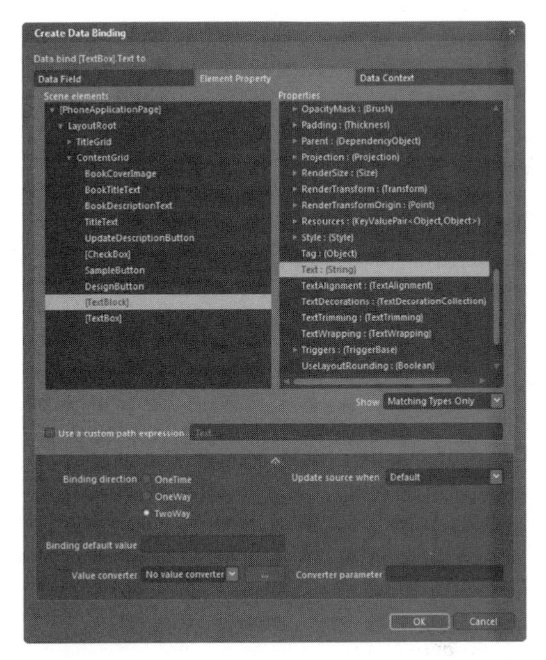

FIGURE 15-17

Element-binding allows one element to bind to a property on another element. Here you're going to use a very simple example of having a `TextBlock` (you can use the one created earlier in this section) and a `TextBox`. The `Text` property on the `TextBox` will be bound to the `Text` property of the `TextBlock` using a `TwoWay` binding, as shown in Figure 15-17. When setting up this binding, make sure you select the Element Property tab in the "Create Data Binding" dialog.

The corresponding XAML for the `TextBox` is as follows.

Available for download on Wrox.com

```
<TextBox Margin="39,0,8,8" TextWrapping="Wrap"
         Text="{Binding Text, ElementName=textBlock, Mode=TwoWay}"
         VerticalAlignment="Bottom"/>
```

Code snippet from MainPage.xaml

SUMMARY

In this chapter, you've learned how to use both Visual Studio and Expression Blend to design and work with data in your application. Being able to effectively switch between design-time and runtime data is essential to working in a team where there are both developers and designers. The ability to separate responsibilities between these two roles is one of the best reasons to use the MVVM and data binding throughout your Windows Phone application.

16

Storing and Synchronizing Data

WHAT'S IN THIS CHAPTER

- Saving application settings to isolated storage
- Reading and writing to isolated storage
- How to cache and persist objects
- Strategies for synchronizing data

In the second part of working with data, this chapter delves into how you can persist data between application sessions. Rather than having to fetch data from an online source each time your application is run, you can periodically synchronize data into an application-specific data repository known as *Isolated Storage*.

Throughout this chapter you will learn how to wrap the saving and subsequent loading of data from Isolated Storage. You will also learn about the new OData (Open Data Protocol) standard and how the use of this standard can help build a synchronization framework that can improve the efficiency of your application.

ISOLATED STORAGE

The only form of persistent on-device storage that you can access from your Windows Phone application is what is known as *Isolated Storage*. As its name implies, this is storage that your application can read and write to that is isolated from other applications on the device. Your application is the only application that can read and write to the Isolated Storage allocated to your application. This prevents applications from sharing or interfering with data belonging to other applications.

If you are used to being able to read and write directly to the file system, then working with Isolated Storage may take a little getting used to. As a developer you no longer have complete access to the file system. Instead, you are allocated a single folder into which you can read and write files. Isolated Storage for a Windows Phone application is similar to Isolated Storage for a Silverlight application for the desktop. There is one difference in that a desktop Silverlight application has a quota, which defines the maximum amount of storage the application can use without requesting further space through the `IncreaseQuotaTo` method. But this quota is not enforced by Windows Phone, allowing an application to consume as much storage space as available.

ApplicationSettings

There are two main ways to use Isolated Storage. The first is to simply store data as a key-value pair, whereas the second is similar to more traditional file system access, where you can read and write to one or more file streams.

Quite often you will want to persist user information between sessions of the application. For example, you may not want the user to have to type in his or her username every time they run your application, or you may like to store an authentication token retrieved from a cloud-based service. To easily save such values between application runs, you can store it in ApplicationSettings. This is a dictionary of objects, identified by a string key and accessible as a singleton within the `IsolatedStorageSettings` class:

Available for download on Wrox.com

```
// Save the username
IsolatedStorageSettings.ApplicationSettings["UserName"] = "Nick";

// Retrieve the username
var userName = IsolatedStorageSettings.ApplicationSettings["UserName"].ToString();
```

Code snippet from MainPage.xaml.cs

You can also store more complex objects:

Available for download on Wrox.com

```
public class Credentials {
    public string UserName { get; set; }
    public string Domain { get; set; }
}
var user = new Credentials(){UserName="Nick_Randolph",Domain="BuiltToRoam"};

// Save the user
IsolatedStorageSettings.ApplicationSettings["LastUsersCredentials"] = user;

// Retrieve the user
var selectedPerson =
     IsolatedStorageSettings.ApplicationSettings["LastUsersCredentials"]
                                                     as Credentials;
MessageBox.Show(selectedPerson.UserName);
```

Code snippet from MainPage.xaml.cs

Data-Binding

Rather than explicitly saving and retrieving values throughout your application, you may want to wrap access to the ApplicationSettings object. This will allow you to centralize the list of keys used to read and write settings, as well as providing a strongly typed interface through which settings can be accessed. The following `Settings` class illustrates how you can build a wrapper that defines all the keys that are used to store and retrieve data, as constants. There is also a helper method, `RetrieveSetting`, which either returns the stored item or the default value, which in most cases is just `null`:

Available for download on Wrox.com

```
public class Settings {
    private const string ApplicationTitleKey = "ApplicationTitle";
    private const string LastUsersCredentialsKey = "LastUsersCredentials";

    private T RetrieveSetting<T>(string settingKey) {
        object settingValue;
        if (IsolatedStorageSettings.ApplicationSettings.TryGetValue(settingKey,
                                                             out settingValue)){
            return (T)settingValue;
        }
        return default(T);
    }

    public string ApplicationTitle {
        get {
            return RetrieveSetting<string>(ApplicationTitleKey);
        }
        set {
            IsolatedStorageSettings.ApplicationSettings[ApplicationTitleKey] = value;
        }
    }

    public Credentials LastUsersCredentials {
        get {
            return RetrieveSetting<Credentials>(LastUsersCredentialsKey);
        }
        set {
            IsolatedStorageSettings
                    .ApplicationSettings[LastUsersCredentialsKey] = value;
        }
    }
}
```

Code snippet from ApplicationSettings.cs

For each data item that is saved to Isolated Storage using ApplicationSettings, the `Settings` wrapper class exposes a corresponding property. This makes it easy to read and write settings from within your code and helps avoid the common problem of typos or case mismatches in string arguments.

You might be wondering why these methods and properties aren't static, given that they are used to retrieve values from the ApplicationSettings dictionary, which is a singleton object. The answer to this lies in the ability to create an instance of the `Settings` class declaratively in App.xaml. As the App.xaml file is only ever processed once, when the application starts up, it makes sense to create a single instance of the `Settings` class, which can then be referenced anywhere within the application:

Available for download on Wrox.com

```
<Application
    x:Class="ManageYourData.App"
    ...
    xmlns:local="clr-namespace:ManageYourData">
    <Application.Resources>
        <local:Settings x:Key="DataSettings" />
    </Application.Resources>
```

Code snippet from App.xaml

This snippet creates an instance of the `Settings` class and adds it to the application resource dictionary using a key of `DataSettings`. Alternatively, you could have easily created this in code and exposed it as a property of the application within the app.xaml.cs file. However, defining the instance of the `Settings` class in XAML means that you can more easily reference it from within the XAML for other pages and controls within your application. For example, if you wanted to data-bind the `ApplicationTitle` property to a `TextBox` you can do this all within the page XAML.

Available for download on Wrox.com

```
<TextBox DataContext="{StaticResource DataSettings}" Height="72"
         HorizontalAlignment="Left" Margin="166,0,0,0" Name="TextToSave"
         VerticalAlignment="Top" Width="314"
         Text="{Binding ApplicationTitle, Mode=TwoWay}" />
```

Code snippet from MainPage.xaml

When the text in the `TextBox` is changed, it will update the `ApplicationTitle` property on the `Settings` instance defined in the application resource dictionary with the key `DataSettings`. This, in turn, will persist the new value into the ApplicationSettings dictionary within Isolated Storage.

The default behavior of ApplicationSettings is for any changes to be written to Isolated Storage when the application closes. However, there is a possibility that if the application terminates unexpectedly, this information won't be written correctly, if at all. You can force changes to be written immediately to Isolated Storage by invoking the `Save` *method:*

```
IsolatedStorageSettings.ApplicationSettings.Save();
```

IsolatedStorageFileStream

Although storing objects by key is quick and easy, there are times when you need more control over how the data is being stored. The second way to work with Isolated Storage is to treat it as if it were a regular file system directory filled with one or more files that are accessed via the `IsolatedStorageFileStream` class. In order to create a file, you can simply create a new instance of the `IsolatedStorageFileStream`, specifying the name of the file you want to open; the `FileMode`, which in this case is to `Create` the file; and the user store for the application. You can think of the *user store* as being the root directory or folder of the application, under which you can create files and directories to organize your data:

Available for download on Wrox.com

```
private void CreateFile_Click(object sender, RoutedEventArgs e){
    using(var myFileStream = new IsolatedStorageFileStream("test.txt",FileMode.Create,
                                 IsolatedStorageFile.GetUserStoreForApplication()))
    using(var writer = new StreamWriter(myFileStream)){
        writer.WriteLine("This is some text that I want to write to a file");
    }
}
```

Code snippet from MainPage.xaml.cs

You will notice from this code snippet that what you get is a stream that you can write to. In this case, the stream is wrapped in a `StreamWriter`, which makes it easier to write textual-formatted data. The use of `using` statements ensures that both the writer and the stream are flushed and closed once all the information has been written.

To read information back out of the Isolated Storage file, you simply need to create another `IsolatedStorageFileStream` instance, supplying the same filename. Notice that the `FileMode` has been set to `Open` because you want to read what is in the current file, rather than creating a new file:

Available for download on Wrox.com

```
private void OpenButton_Click(object sender, RoutedEventArgs e){
    using(var myFileStream = new IsolatedStorageFileStream("test.txt",FileMode.Open,
                                 IsolatedStorageFile.GetUserStoreForApplication()))
    using(var reader = new StreamReader(myFileStream)){
        var text= reader.ReadToEnd();
        MessageBox.Show(text);
    }
}
```

Code snippet from MainPage.xaml.cs

This time the stream has been wrapped in a `StreamReader` to make it easier to read textual information. Again, the use of `using` statements ensures that both the reader and the stream are closed once they're finished with.

The GetUserStoreForApplication method returns an IsolatedStorageFile instance. You can think of this as a File Manager through which you can create, list, and delete both files and directories. The following code illustrates how you can detect if a file exists and how you can delete it if it does exist:

Available for download on Wrox.com

```
private void DeleteButton_Click(object sender, RoutedEventArgs e){
    IsolatedStorageFile directory = IsolatedStorageFile.GetUserStoreForApplication();()
    if (directory.FileExists("test.txt")){
        directory.DeleteFile("test.txt");
    }
    else{
        MessageBox.Show("File doesn't exist");
    }
}
```

Code snippet from MainPage.xaml.cs

The IsolatedStorageFile instance supports methods such as CreateDirectory, DeleteDirectory, and DirectoryExists for working with directories, and CreateFile, DeleteFile, and FileExists for working with files.

DATA-CACHING

Isolated Storage is well suited to applications that only need to store limited amounts of information via use of the ApplicationSettings key-value store, or applications that want low-level control on how the data is stored. However, neither of these options provides an easy-to-use object or relational repository that can be used throughout your application. If you were building for other platforms, you might consider using SQL Server Compact or even SQL Server Express as a data repository for your application. At this stage, there is no built-in support for any database technology within the Windows Phone development tools. In this section, you'll see how you can build a repository that can be used to persist and synchronize data with a backend server.

Object Cache

Before you start building a persistent store, you'll start with an in-memory object cache. Whether you are downloading content from a remote service (e.g., a Web Service or an RSS feed) or loading data from Isolated Storage, you are going to want to track the data in memory for the life of the application to avoid constantly serializing and deserializing it. You can easily create global variables in the App.xaml.cs that can be referenced anywhere within the application, but this is rather clumsy and can quickly become hard to manage and work with. An alternative is to create an *object cache* that can be used to track any number of objects. The object cache can be created as a singleton in the App.xaml, making it accessible from anywhere within your application either as a StaticResource that can be used for data-binding in XAML, or via code.

To start with, let's define an interface called `IObjectCache` that will define the methods that you want our object cache to expose. This interface inherits from the interface `IInitializeAndCleanup`, which defines the methods `Initialize` and `Cleanup` that are used to set up and tear down the object cache. These methods have been separated into their own interface so that they can be reused independently:

```
public interface IObjectCache : IInitializeAndCleanup {
    List<IObjectKeyMap> ObjectKeyMappings { get; }

    bool Exists<T>(T item) where T : class, IEntityBase;

    IEnumerable<T> FindByKey<T>(object key) where T : class, IEntityBase;

    IEnumerable<T> Select<T>() where T : class, IEntityBase;

    void Insert<T>(T item) where T : class, IEntityBase;

    void Delete<T>(T item) where T : class, IEntityBase;
}
```

Code snippet from IObjectCache.cs

```
public interface IInitializeAndCleanup {
    void Initialize();
    void Cleanup();
}
```

Code snippet from IInitializeAndCleanup.cs

The `IObjectCache` interface defines a property, `ObjectKeyMappings`, which is a list of object-to-key mappings. The `IObjectKeyMap` interface stores the `Type` of an entity and the name of the property on that `Type` that returns a unique key identifying an entity of that `Type`. Take the example of the class called `Car` with a property `Id` that returns a unique integer that can be used to identify the `Car`. In this case, an `IObjectKeyMap` would be added to the `ObjectKeyMappings` list that has the `EntityType` property set to `Car` and the `KeyPropertyName` property set to `Id`:

```
public interface IObjectKeyMap {
    Type EntityType { get; set; }
    string KeyPropertyName { get; set; }
}
```

Code snippet from IObjectKeyMap.cs

You will also notice that most of the methods defined on the `IObjectCache` interface are generic methods that have constraints requiring the `Type` parameter, `T`, to be a class that implements the `IEntityBase` interface. Initially, the `IEntityBase` interface is an empty interface that inherits from `INotifyPropertyChanged`. You will see when you start persisting and synchronizing entities why it is important for entities to implement `INotifyPropertyChanged`. The `IEntityBase` interface will be extended later to include additional properties.

Available for download on Wrox.com

```
public interface IEntityBase : INotifyPropertyChanged{ }
```

Code snippet from IEntityBase.cs

The other methods defined on the `IObjectCache` interface will be covered as you go through the implementation. Let's start by creating our concrete `ObjectCache` class that inherits from `IObjectCache`. The following snippet shows the public and private properties, along with the constructor:

Available for download on Wrox.com

```
public class ObjectCache : IObjectCache{
    public List<IObjectKeyMap> ObjectKeyMappings { get; private set; }
    public ObjectCache() {
        ObjectKeyMappings = new List<IObjectKeyMap>();
    }

    private Dictionary<Type, PropertyInfo> EntityKeyPropertyCache { get; set; }
    private Dictionary<Type, Func<object, object>> EntityKeyFunctions { get; set; }
    private Dictionary<Type, List<IEntityBase>> InMemoryCache { get; set; }
}
```

Code snippet from ObjectCache.cs

The `EntityKeyPropertyCache` and `EntityKeyFunctions` properties are used to track the key mapping properties and functions used to determine the key for a particular entity, and the `InMemoryCache` is used to track all entities added to the object cache.

Before you can start adding or querying objects in the cache, you need to set up the cache so that it is ready to accept new entities. In the `Initialize` method, the various dictionaries need to be instantiated. For efficiency reasons, you will cache within the EntityKeyFunctions dictionary a function that returns the current value of a given entities key property. This is done by creating a small lambda expression that will return the property value for any given entity via use of a `PropertyInfo` instance found via reflection. For performance both the `PropertyInfo` and key mapping function are cached. At this stage, there are no cleanup tasks required:

Available for download on Wrox.com

```
public void Initialize() {
    InMemoryCache = new Dictionary<Type, List<IEntityBase>>();
    EntityKeyFunctions = new Dictionary<Type, Func<object, object>>();
    EntityKeyPropertyCache = new Dictionary<Type, PropertyInfo>();

    foreach (var registeredType in ObjectKeyMappings){
        var entityType = registeredType.EntityType;
        var cache = new List<IEntityBase>();
        InMemoryCache[entityType] = cache;
        EntityKeyPropertyCache[entityType] =
           entityType.GetProperty(registeredType.KeyPropertyName);
```

```
            EntityKeyFunctions[entityType] =
               (entity) => EntityKeyPropertyCache[entity.GetType()].GetValue(entity, null);
        }
    }

    public void Cleanup(){ }
```

Code snippet from ObjectCache.cs

Accessing entities that have been added to the `ObjectCache` is done through a method called `Select`. This is a generic method in which the `Type` parameter is used to determine the class of objects that are to be returned. Rather than accessing the `InMemoryCache` property directly, you'll create a wrapper generic method, `EntityCache`, that uses the `Type` parameter to return the corresponding `List` from the InMemoryCache dictionary. The `Select` method converts the returned `List<IEntityBase>` into a read-only `IEnumerable<T>`, which can then be used in a LINQ (Language Integrated Query) expression or, as you will see later, bound directly to the user interface (UI):

```
private List<IEntityBase> EntityCache<T>() where T : class, IEntityBase{
    var entityType = typeof(T);
    List<IEntityBase> cache = InMemoryCache[entityType];
    return cache;
}

public IEnumerable<T> Select<T>() where T : class, IEntityBase{
    var cacheList = EntityCache<T>();
    return cacheList.AsReadOnly().OfType<T>();
}
```

Code snippet from ObjectCache.cs

You've heard a bit about the key mapping functions that are defined for each entity so as to determine a unique key for an entity being added to the object cache, but where do these get used? There are a couple of methods that use the key mapping functions; let's start with the `FindByKey` method. For convenience, we've added a helper method, `KeyFunction`, that is used to safely retrieve the key mapping function for an entity that matches the `Type` parameter, `T`:

```
private Func<object, object> KeyFunction<T>() {
    Func<object, object> keyFunction;
    if (!EntityKeyFunctions.TryGetValue(typeof(T), out keyFunction)) {
        keyFunction = (object keyItem) => (object)keyItem;
    }
    return keyFunction;
}
```

Code snippet from ObjectCache.cs

The FindByKey method performs a LINQ expression yielding any entities where the key mapping function returns a match to the supplied key. In the previous example of a Car with an Id property as the key property, if a key value of 5 was used, the FindByKey method would return all Car entities in the ObjectCache with an Id of 5:

```
public IEnumerable<T> FindByKey<T>(object key) where T : class, IEntityBase{
    var keyFunction = KeyFunction<T>();
    var matches = from x in Select<T>()
                  where keyFunction(x).Equals(key)
                  select x;
    return matches;
}
```

Code snippet from ObjectCache.cs

If you wanted to test if there was an entity in the ObjectCache with a particular key, you could call the Exists method. This method will return true if there is an entity already in the ObjectCache with the specified key value. Note that you will use FirstOrDefault, rather than Count, because Count always requires the full list to be enumerated, whereas FirstOrDefault will stop enumerating the list of entities as soon as a matching entity has been matched:

```
public bool Exists<T>(T item) where T : class, IEntityBase {
    var matches = FindMatches(item).FirstOrDefault();
    return matches != null;
}

private IEnumerable<T> FindMatches<T>(T item) where T : class, IEntityBase {
    var keyFunction = KeyFunction<T>();
    var key = keyFunction(item);
    return FindByKey<T>(key);
}
```

Code snippet from ObjectCache.cs

In this case, you want our ObjectCache to contain only a single entity for any given key. Thus, before an entity is inserted, it is tested to see if it is unique:

```
public void Insert<T>(T item) where T : class, IEntityBase {
    if (Exists(item)) {
        throw new EntityExistsException();
    }
    EntityCache<T>().Add(item);
}
```

Code snippet from ObjectCache.cs

Deleting items from the ObjectCache is simply a matter of removing the entity from the corresponding entity list:

```
public void Delete<T>(T item) where T : class, IEntityBase {
    EntityCache<T>().Remove(item);
}
```

Code snippet from ObjectCache.cs

Since the ObjectCache tracks references to objects in memory, there is no need to have a method that is called when the object changes. You'll see later that the INotifyPropertyChanged interface is used to detect when objects do change so that these changes can be persisted.

Earlier you saw that the key mapping functions were created based on a list of IObjectKeyMap entries. Of course, you can't create instances of an interface, so you need a concrete class, ObjectKeyMap, that implements this interface:

```
public class ObjectKeyMap : IObjectKeyMap {
    [TypeConverter(typeof(StringToTypeConverter))]
    public Type EntityType { get; set; }
    public string KeyPropertyName { get; set; }
}
```

Code snippet from ObjectKeyMap.cs

The EntityType property has a TypeConverter attribute associated with it. This allows for type conversion between the name of a class as a string and the corresponding Type object. This is required so that you can create instances of the ObjectKeyMap in XAML. By default, Silverlight is unable to convert between a string and the Type that it refers to, when setting property values from XAML:

```
public class StringToTypeConverter : TypeConverter {
    public override bool CanConvertFrom(ITypeDescriptorContext context,
                                        Type sourceType){
        return sourceType.IsAssignableFrom(typeof(string));
    }

    public override object ConvertFrom(ITypeDescriptorContext context,
                                       CultureInfo culture, object value){
        var typeName = value as string;
        if (String.IsNullOrEmpty(typeName)){
            return null;
        }
        var type = Type.GetType(typeName);
        return type;
    }
}
```

Code snippet from StringToTypeConverter.cs

You now have a base implementation of an ObjectCache that you can use within your application. In this case, you're going to extend the ObjectCache into an application-specific cache called

CarStore with an additional property, Cars, that wraps a call to Select with the Type parameter set to Car. The Car class implements IEntityBase and has properties for (int)Id, (string)Make, (string)Model, and (int)Year:

Available for download on Wrox.com

```
public class CarStore : ObjectCache
{
    public Car[] Cars
    {
        get {
            return Select<Car>().ToArray();
        }
    }
}
```

Code snippet from CarStore.cs

Using the CarStore within your application is as simple as creating an instance of the CarStore in the App.xaml file. In this case, the ObjectCache and associated interfaces have been placed in a separate assembly with the namespace BuiltToRoam.Data. The CarStore and Car classes are defined in the application project with the namespace ManageYourData:

Available for download on Wrox.com

```
<Application
    x:Class="ManageYourData.App"
    ...
    xmlns:data="clr-namespace:BuiltToRoam.Data;assembly=BuiltToRoam.Data"
    xmlns:local="clr-namespace:ManageYourData">
    <Application.Resources>
        <local:CarStore x:Key="Cache">
            <data:ObjectCache.ObjectKeyMappings>
                <data:ObjectKeyMap
                        EntityType="ManageYourData.Car,ManageYourData"
                        KeyPropertyName="Id" />
            </data:ObjectCache.ObjectKeyMappings>
        </local:CarStore>
    </Application.Resources>
```

Code snippet from App.xaml

The CarStore instance has been created in the application resource dictionary with a key of Cache. This key can be used to reference the CarStore instance from anywhere within the application. Before you can use the object cache, you need to initialize it. This can be done in the Application_Launching and Application_Activated event handlers within the App.xaml.cs file. At this point, the CarStore instance can also be assigned to a CLR property making it easier to reference in code. For completeness, make sure that the Cleanup method is called in the Application_Closing and Application_Deactivated event handlers:

Available for download on Wrox.com

```
public CarStore Store { get; private set; }

private void Application_Launching(object sender, LaunchingEventArgs e) {
    Store = this.Resources["Cache"] as CarStore;
```

```
        Store.Initialize();
    }

    private void Application_Activated(object sender, ActivatedEventArgs e){
        Store = this.Resources["Cache"] as CarStore;
        Store.Initialize();
    }

    private void Application_Closing(object sender, ClosingEventArgs e) {
        Store.Cleanup();
    }

    private void Application_ Deactivated(object sender, ClosingEventArgs e) {
        Store.Cleanup();
    }
```

Code snippet from App.xaml.cs

You might be wondering why you went to all the effort of creating the `CarStore` instance in XAML if you were just going to access it via a property defined in the `App` class. The reason is that it makes it much easier to do data-binding directly to items stored in the object cache. A `ListBox` has been added to the `ContentGrid` that has the `CarStore` instance (referenced by the `Cache` key) as its `DataContext`. Each item in the `ListBox` is arranged using a `StackPanel` containing three `TextBlock` elements. These elements are bound to the `Year`, `Model`, and `Make` properties of the `ListBox` items. The items are loaded from the `Cars` property on the `CarStore`:

```
<Grid x:Name="ContentGrid" Grid.Row="1" DataContext="{StaticResource Cache}">
    <ListBox Name="CarsList" ItemsSource="{Binding Cars}">
        <ListBox.ItemTemplate>
            <DataTemplate>
                <StackPanel Orientation="Horizontal" d:LayoutOverrides="Height">
                    <TextBlock TextWrapping="Wrap" Text="{Binding Year}"
                               Style="{StaticResource PhoneTextNormalStyle}"
                               Margin="10,0,0,0"/>
                    <TextBlock TextWrapping="Wrap" Text="{Binding Model}"
                               Style="{StaticResource PhoneTextNormalStyle}"
                               Margin="10,0,0,0"/>
                    <TextBlock TextWrapping="Wrap" Text="{Binding Make}"
                               Style="{StaticResource PhoneTextNormalStyle}"
                               Margin="10,0,0,0"/>
                </StackPanel>
            </DataTemplate>
        </ListBox.ItemTemplate>
    </ListBox>
</Grid>
```

Code snippet from CarsPage.xaml

Although being able to data-bind directly to the object cache is incredibly useful, there are times when you will want to be able to reference the object store from code. The following example shows how you can reference the `CarStore` instance in code and add a Car using the `Insert` method:

Available for download on Wrox.com

```
private void CarsButton_Click(object sender, RoutedEventArgs e) {
    var store = (Application.Current as App).Store;
    store.Insert(new Car { Id = 2, Make = "Ford", Model = "Fiesta", Year = 1998 });
    store.Insert(new Car { Id = 3, Make = "Honda", Model = "Civic", Year = 2001 });
}
```

Code snippet from MainPage.xaml.cs

You can also create your own properties in the `CarStore` that return subsets of the data held in the object cache. For example, you could create a property called `ModernCars` that only returns cars where the year of manufacture was in the last 10 years:

Available for download on Wrox.com

```
public Car[] ModernCars {
    get {
        return (from car in Select<Car>()
                where car.Year >= DateTime.Now.AddYears(-10).Year
                select car).ToArray();
    }
}
```

Code snippet from CarStore.cs

Persistent Storage

The `ObjectCache` is useful for tracking objects during a single run of an application. In order for objects to be tracked over multiple runs of an application, it is necessary to persist them to Isolated Storage. Rather than incorporating the logic to read and write to Isolated Storage into the `ObjectCache`, you're going to define an interface that includes operations to `Select`, `Insert`, `Delete`, and `Update` objects. Like the `IObjectCache` interface, the `IStorageProvider` interface inherits from `IInitializeAndCleanup` to manage the lifetime of the storage provider. It also includes an `IEntityConfiguration` list that is used to define the types that will be stored:

Available for download on Wrox.com

```
public interface IStorageProvider : IInitializeAndCleanup{
    List<IEntityConfiguration> StorageConfigurations { get; }

    IEnumerable<T> Select<T>() where T : class;

    void Insert(object item);

    void Delete(object item);

    void Update(object item);
}
```

Code snippet from IStorageProvide.cs

```
public interface IEntityConfiguration{
    Type EntityType { get; set; }
}
```

Code snippet from IEntityConfiguration.cs

The benefit of defining the `IStorageProvider` interface is that you can provide your own implementation without modifying the rest of the application logic. In this chapter, You'll learn about an implementation that serializes objects to Isolated Storage in JSON.

IStorageProvider: JSON Serialization

The first implementation of the `IStorageProvider` interface uses the `DataContractJsonSerializer` to persist objects. The `SimpleStorageProvider` class has a `StoreName` property, which defines a folder within Isolated Storage. Within this folder, there will be a single file for each type of object to be persisted. This is a simple example of how you can build your own storage provider and has not been optimized for storing large quantities of objects:

Available for download on Wrox.com

```
public class SimpleStorageProvider : IStorageProvider {
    public List<IEntityConfiguration> StorageConfigurations { get; private set; }
    public string StoreName { get; set; }

    public SimpleStorageProvider() {
        StorageConfigurations = new List<IEntityConfiguration>();
    }

    public Dictionary<Type, IInitializeAndCleanup> Contexts { get; set; }
}
```

Code snippet from SimpleStorageProvider.cs

The declaration of the `SimpleStorageProvider` includes a dictionary called *Contexts* that will contain a `TypeStorageContext` for each `Type` of object to be persisted. Note that as the `TypeStorageContext` class has a `Type` parameter, it is not possible to specify the second `Type` parameter of the dictionary. Instead, since the `TypeStorageContext` class implements the `IInitializeAndCleanup` interface, this is supplied instead. Before completing the implementation of the `SimpleStorageProvider`, let's first look at how the `TypeStorageContext` works:

Available for download on Wrox.com

```
public class TypeStorageContext<T> : IInitializeAndCleanup {
    private string FileName { get; set; }
    private IsolatedStorageFileStream FileStream { get; set; }
    private DataContractJsonSerializer Serializer { get; set; }

    public TypeStorageContext(string fileName) {
        FileName = fileName;
    }
}
```

Code snippet from TypeStorageContext.cs

The constructor of the `TypeStorageContext` has a single argument, `fileName`, that defines the file in Isolated Storage into which all objects of `Type T` will be serialized. When the `Initialize` method is called, both the `FileStream` and `Serializer` objects are created. It's important to make sure that the `FileStream` is closed during `Cleanup` to avoid file corruption or lost data:

```
public void Initialize() {
    IsolatedStorageFile isf = IsolatedStorageFile.GetUserStoreForApplication();
    Serializer = new DataContractJsonSerializer(typeof(T));
    FileStream = isf.OpenFile(FileName,
                              System.IO.FileMode.OpenOrCreate,
                              System.IO.FileAccess.ReadWrite);
}

public void Cleanup() {
    FileStream.Close();
}
```

Code snippet from TypeStorageContext.cs

The `DataContractJsonSerializer` can't be used to read and write directly to the `FileStream`. Instead, you need to use an intermediary `MemoryStream` into which the serialized object can be placed. In the case of the `Write` method, the object is serialized to the `MemoryStream`, and then the corresponding byte array is written to the `FileStream`:

```
private void Write<T>(T item) where T : class
{
    using (var ms = new MemoryStream())
    {
        Serializer.WriteObject(ms, item);
        ms.Flush();
        var bytes = ms.ToArray();
        var lengthBytes = BitConverter.GetBytes(bytes.Length);
        FileStream.Write(lengthBytes, 0, 4);
        FileStream.Write(bytes, 0, bytes.Length);
    }
    FileStream.Flush();
}
```

Code snippet from TypeStorageContext.cs

The `Read` method reverses this by reading the byte array and storing this within a `MemoryStream` instance from which the object can then be deserialized:

```
private T Read<T>() where T : class
{
    var lenghtBytes = new byte[4];
    FileStream.Read(lenghtBytes, 0, 4);
    var bytes = new byte[BitConverter.ToInt32(lenghtBytes, 0)];
    FileStream.Read(bytes, 0, bytes.Length);
    using (var ms = new MemoryStream(bytes))
```

```
        {
            var instance = Serializer.ReadObject(ms) as T;
            return instance;
        }
    }
```

Code snippet from TypeStorageContext.cs

With these two primitive methods available, you can now define public methods to `Select` and `Insert` objects. The `Select` method returns an iterator that will progressively return objects from the `FileStream` until the end of the file is reached. Each new object being inserted is simply appended to the end of the existing `FileStream` contents within the `Insert` method:

```
    public IEnumerable<T> Select<T>() where T : class {
        while (FileStream.Position < FileStream.Length){
            var instance = Read<T>();
            yield return instance;
        }
    }

    public void Insert<T>(T item) where T : class
    {
        var entity = item as SimpleEntityBase;
        entity.EntityId = Guid.NewGuid();

        FileStream.Seek(0, System.IO.SeekOrigin.End);
        Write(item);
    }
```

Code snippet from TypeStorageContext.cs

You will notice in the `Insert` method that the object being inserted is cast to a `SimpleEntityBase`. In addition to implementing `IEntityBase`, required for a number of the methods on the `ObjectCache`, the `SimpleEntityBase` class exposes a property called `EntityId`. This is a `Guid` that is used to uniquely identify the entity when it is saved to Isolated Storage:

```
    public class SimpleEntityBase : IEntityBase {
        public Guid EntityId { get; set; }

        public event PropertyChangedEventHandler PropertyChanged;

        protected void RaisePropertyChanged(string propertyName) {
            if (PropertyChanged != null) {
                PropertyChanged(this, new PropertyChangedEventArgs(propertyName));
            }
        }
    }
```

Code snippet from SimpleEntityBase.cs

To implement the Delete method, you simply need to iterate through the objects that have been saved in the FileStream looking for an object that has the same EntityId as the object to be deleted. Once you have located that object, to delete it you simply need to shift the remaining objects in the FileStream so that they overwrite the object to be deleted:

```
public void Delete<T>(T item) where T : class{
    var entity = item as SimpleEntityBase;
    var searchId = entity.EntityId;
    FileStream.Seek(0, System.IO.SeekOrigin.Begin);
    var lastEntityStartIndex = FileStream.Position;
    while (FileStream.Position < FileStream.Length){
        var nextEntity = Read<T>() as SimpleEntityBase;
        if (nextEntity.EntityId == searchId){
            var buffer = new byte[1000];
            int count;
            var entitySize = FileStream.Position - lastEntityStartIndex;
            while ((count=FileStream.Read(buffer, 0, buffer.Length))>0){
                FileStream.Seek(-(entitySize+count), System.IO.SeekOrigin.Current);
                FileStream.Write(buffer, 0, count);
                FileStream.Seek(entitySize + count, System.IO.SeekOrigin.Current);
            }
            FileStream.SetLength(FileStream.Length - entitySize);
        }
        lastEntityStartIndex = FileStream.Position;
    }
}
```

Code snippet from TypeStorageContext.cs

Lastly, you need to update the length of the FileStream to ensure that the last entity is not duplicated.

The Update method is a combination of deleting the existing object and then writing the updated object to the end of the FileStream. Since the new object might have a different length, when serialized to JSON it is important not to simply overwrite the original object in the FileStream:

```
public void Update<T>(T item) where T : class{
    Delete(item);

    FileStream.Seek(0, System.IO.SeekOrigin.End);
    Write(item);
}
```

Code snippet from TypeStorageContext.cs

Now let us return to the SimpleStorageProvider and complete its implementation. You need to implement the IInitializeAndCleanup interface first by creating the Initialize and Cleanup methods. The Initialize method ensures that a directory exists for the store, based on the StoreName property. It then iterates through each of the configurations and creates an appropriate

TypeStorageContext for each entity Type. Note again the need to use reflection because of the limitations around Type parameters for generic types:

Available for download on Wrox.com

```
public void Initialize(){
    IsolatedStorageFile isf = IsolatedStorageFile.GetUserStoreForApplication();
    isf.CreateDirectory(StoreName);

    Contexts = new Dictionary<Type, IInitializeAndCleanup>();
    foreach (var config in StorageConfigurations) {
        string fileName = String.Format(@"{0}\{1}", StoreName,config.EntityType.Name);

        var storeType = typeof(TypeStorageContext<>).MakeGenericType(config
.EntityType);
        var store = Activator.CreateInstance(storeType, fileName)
                                                    as IInitializeAndCleanup;
        store.Initialize();
        Contexts[config.EntityType] = store;
    }
}

public void Cleanup() {
    foreach (var config in StorageConfigurations) {
        (Contexts[config.EntityType] as IInitializeAndCleanup).Cleanup();
    }
}
```

Code snippet from SimpleStorageProvider.cs

It is important that the Cleanup method call the corresponding Cleanup method on each of the TypeStorageContext instances. This will close and release the FileStream that each context maintains for reading and writing to the file in Isolated Storage. What's left is to implement the methods to Select, Insert, Delete, and Update objects. For each of these, it's simply a matter of locating the right TypeStorageContext, using the CurrentStore method, and then calling the corresponding method:

Available for download on Wrox.com

```
private TypeStorageContext<T> CurrentStore<T>() {
    return Contexts[typeof(T)] as TypeStorageContext<T>;
}

public IEnumerable<T> Select<T>() where T : class {
    var store = CurrentStore<T>();
    return store.Select<T>();
}

public void Insert<T>(T item) where T : class {
    var store = CurrentStore<T>();
    store.Insert(item);
}

public void Delete<T>(T item) where T : class {
    var store = CurrentStore<T>();
```

```
        store.Delete(item);
    }

    public void Update<T>(T item) where T : class {
        var store = CurrentStore<T>();
        store.Update(item);
    }
```

Code snippet from SimpleStorageProvider.cs

You now have an implementation of `IStorageProvider` that can be used to persist objects to Isolated Storage. What you now have to do is to integrate that into the `ObjectCache` so that objects are persisted in addition to being tracked in memory.

Persisting Objects

If you recall, the `ObjectCache` is based on the `IObjectCache` interface. At this stage, neither the interface nor the implementation knows anything about persisting objects. Start by thinking about how you want the in-memory cache to work. Do you want it to persist objects to disk every time there is a change? Or do you want to be able to make changes and then persist the updates when you have completed all the changes? You can actually accommodate both of these options by defining a `CacheAction` enumeration. This defines a set of flag values that will be used to determine which actions trigger items to be persisted:

Available for download on Wrox.com

```
[Flags]
public enum CacheAction{
    None = 0,
    Insert = 1,
    Delete = 2,
    Update = 4
}
```

Code snippet from IObjectCache.cs

The `IObjectCache` interface needs to be updated to include a reference to an implementation of `IStorageProvider`. Changes are persisted using the storage provider when either the `Persist` method is called or one of the actions defined by the `CacheMode` property is performed:

Available for download on Wrox.com

```
public interface IObjectCache : IInitializeAndCleanup {
    ...
    IStorageProvider Store { get; set; }
    CacheAction CacheMode { get; set; }
    void Persist();
}
```

Code snippet from IObjectCache.cs

Now onto the implementation within the `ObjectCache` class — the starting point, of course, is to include the additional properties required by the changes to the `IObjectCache` interface:

```
public IStorageProvider Store { get; set; }
public CacheAction CacheMode { get; set; }
```

Code snippet from ObjectCache.cs

You're also going to include a dictionary that will indicate whether all the objects of a specific type have been loaded from the `IStorageProvider` into memory. This helps facilitate lazy loading of entities.

When you have a large number of objects of a specific type, you may not want to load them all into memory at once. This implementation simply uses a Boolean value to indicate whether entities have been loaded. To make this more efficient, you may want to only load a subset of values at any time. You should update this implementation to use your own logic to track which objects have been loaded into memory.

There is also a list of object action pairs that makes up the History of changes that have been made to the `ObjectCache`. This is used to record the changes if the `CacheMode` is not configured to persist all changes as they occur:

```
private Dictionary<Type, bool> EntitiesLoaded { get; set; }
private List<KeyValuePair<IEntityBase, CacheAction>> History { get; set; }
```

Code snippet from ObjectCache.cs

The `Initialize` and `Cleanup` methods require minor additions to initialize the storage provider and create and populate the additional dictionary and list:

```
public void Initialize() {
    if (Store != null) {
        Store.Initialize();
    }

    History = new List<KeyValuePair<IEntityBase, CacheAction>>();
    InMemoryCache = new Dictionary<Type, List<IEntityBase>>();
    EntitiesLoaded = new Dictionary<Type, bool>();
    EntityKeyFunctions = new Dictionary<Type, Func<object, object>>();
    EntityKeyPropertyCache = new Dictionary<Type, PropertyInfo>();

    foreach (var registeredType in ObjectKeyMappings) {
        var entityType = registeredType.EntityType;
        var cache = new List<IEntityBase>();
        EntityKeyPropertyCache[entityType] =
           entityType.GetProperty(registeredType.KeyPropertyName);
        EntityKeyFunctions[entityType] =
           (entity) => EntityKeyPropertyCache[entity.GetType()].GetValue(entity, null);
```

```
            InMemoryCache[entityType] = cache;
            EntitiesLoaded[entityType] = false;
        }
    }

    public void Cleanup() {
        if (Store != null) {
            Store.Cleanup();
        }
    }
```

Code snippet from ObjectCache.cs

You also need to update the `EntityCache` method so that the first time it is called for a `Type` of entity, all existing objects of that `Type` are loaded from persistent storage into memory:

Available for download on Wrox.com

```
    private List<IEntityBase> EntityCache<T>() where T : class, IEntityBase {
        var entityType = typeof(T);
        List<IEntityBase> cache = InMemoryCache[entityType];

        if (!EntitiesLoaded[typeof(T)]) {
            foreach (var entity in Store.Select<T>().OfType<IEntityBase>()){
                cache.Add(entity);
            }
            EntitiesLoaded[typeof(T)] = true;
        }

        return cache;
    }
```

Code snippet from ObjectCache.cs

To determine whether an action should be persisted immediately, there is an added helper method called `ShouldPersist`. This accepts a `CacheAction` to test against the `CacheMode` value. If that action has been included in the `CacheMode` value, it will return `true`, indicating that the action should be immediately persisted using the `IStorageProvider` implementation:

Available for download on Wrox.com

```
    private bool ShouldPersist(CacheAction mode) {
        return (this.CacheMode & mode) > 0;
    }
```

Code snippet from ObjectCache.cs

The `ShouldPersist` method is used in the `Insert` and `Delete` methods to determine whether to pass the object to the `IStorageProvider` implementation or add an entry to the History list:

Available for download on Wrox.com

```
    public void Insert<T>(T item) where T : class, IEntityBase {
        if (!IsUnique(item)) {
            throw new EntityExistsException();
        }
```

```
        EntityCache<T>().Add(item);
        if (ShouldPersist(Data.CacheAction.Insert)) {
            Store.Insert(item);
        }
        else {
            History.Add(new KeyValuePair<IEntityBase, CacheAction>
                                        (item, Data.CacheAction.Insert));
        }
        item.PropertyChanged += item_PropertyChanged;
    }

    public void Delete<T>(T item) where T : class, IEntityBase {
        EntityCache<T>().Remove(item);
        if (ShouldPersist(Data.CacheAction.Insert)) {
            Store.Update(item);
        }
        else {
            History.Add(new KeyValuePair<IEntityBase, CacheAction>
                                        (item, Data.CacheAction.Delete));
        }
        item.PropertyChanged -= item_PropertyChanged;
    }
```

Code snippet from ObjectCache.cs

The last statement in both the `Insert` and `Delete` methods involves adding or removing an event handler on the `PropertyChanged` event. This is used to ensure that any future modifications made to an object that has been added to the `ObjectCache` are tracked and properly persisted. Remember that this wasn't required previously since modifying an object that had been added to the in-memory `ObjectCache` also modified the actual object in the cache (they're two references to the same object).

```
    void item_PropertyChanged(object sender, PropertyChangedEventArgs e) {
        if (ShouldPersist(Data.CacheAction.Update)){
            Store.Update(sender);
        }
        else {
            History.Add(new KeyValuePair<IEntityBase, CacheAction>
                                        (sender as IEntityBase, Data.CacheAction.Update));
        }
    }
```

Code snippet from ObjectCache.cs

The last method to implement is the `Persist` method, which is used to explicitly flush any changes recorded into the History list to persistent storage:

```
    public void Persist() {
        while (History.Count > 0) {
            var action = History[0];
            switch (action.Value) {
```

```
                case Data.CacheAction.Insert:
                    Store.Insert(action.Key);
                    break;
                case Data.CacheAction.Delete:
                    Store.Delete(action.Key);
                    break;
                case Data.CacheAction.Update:
                    Store.Update(action.Key);
                    break;

            }
            History.RemoveAt(0);
        }
    }
```

Code snippet from ObjectCache.cs

If you recall from earlier, you created an instance of the `CarStore` (which inherits from `ObjectCache`) in App.xaml that could be accessed anywhere within the application. In order to use either of the `IStorageProvider` implementations, they also need to be defined within the App.xaml file. The following XAML creates an instance of the SimpleStorageProvider class IStorageProvider. The `Store` property on the `CarStore` instance defines which implementation is being used; in this case, it is set to the `SimpleCarsData` key, which is, of course, the `SimpleStorageProvider` implementation:

Available for download on Wrox.com

```
<Application
    x:Class="ManageYourData.App"
    ...
    xmlns:data="clr-namespace:BuiltToRoam.Data;assembly=BuiltToRoam.Data"
    xmlns:simple="clr-namespace:BuiltToRoam.Data.SimpleStorage;
                  assembly=BuiltToRoam.Data.SimpleStorage"
    xmlns:local="clr-namespace:ManageYourData">
    <Application.Resources>
        <simple:SimpleStorageProvider StoreName="Cars" x:Key="SimpleCarsData">
            <simple:SimpleStorageProvider.StorageConfigurations>
                <data:EntityConfiguration
                      EntityType="ManageYourData.Car,ManageYourData" />
            </simple:SimpleStorageProvider.StorageConfigurations>
        </simple:SimpleStorageProvider>
        <local:CarStore x:Key="Cache"
                        Store="{StaticResource SimpleCarsData}"
                        CacheMode="Insert,Delete,Update">
            <data:ObjectCache.ObjectKeyMappings>
                <data:ObjectKeyMap
                      EntityType="ManageYourData.Car,ManageYourData"
                      KeyPropertyName="Id" />
            </data:ObjectCache.ObjectKeyMappings>
        </local:CarStore>
    </Application.Resources>
```

Code snippet from App.xaml

The CacheMode has been set to Insert, Delete, and Update. In other words, when any action occurs, the change will immediately be persisted. You will notice that the SimpleStorageProvider instance has a single entry in the StorageConfigurations node. An instance of the EntityConfiguration class is used that provides a minimal implementation of the IEntityConfiguration interface:

Available for download on Wrox.com

```
public class EntityConfiguration : IEntityConfiguration {
    [TypeConverter(typeof(StringToTypeConverter))]
    public Type EntityType { get; set; }
}
```

Code snippet from EntityConfiguration.xaml

To illustrate the persistent storage in action, let's extend the code from earlier for working with the CarStore:

Available for download on Wrox.com

```
private void CarsButton_Click(object sender, RoutedEventArgs e) {
    var store = (Application.Current as App).Store;
    var car1 = new Car { Id = 1, Make = "Holden", Model = "Frontera", Year = 1995 };
    store.Insert(car1);
    var car2 = new Car { Id = 2, Make = "Ford", Model = "Fiesta", Year = 1998 };
    store.Insert(car2);
    var car3 = new Car { Id = 3, Make = "Honda", Model = "Civic", Year = 2001 };
    store.Insert(car3);

    car1.Model = "New Fiesta";

    store.Delete(car2);

    this.NavigationService.Navigate(new Uri("/CarsPage.xaml", UriKind.Relative));
}
```

Code snippet from MainPage.xaml.cs

In this example, you are creating three cars that are added to the object cache. car1 is then updated, and car2 is deleted. Because of the CacheMode value, the update and deletions are persisted immediately. If you run this application for a second time and click the button, you will notice that the EntityExistsException is raised as there is already an entity matching the Id of car1. Since pressing the CarsButton repeatedly causes this exception, you may want to add an additional button that just loads up the CarsPage without attempting to insert any Cars:

Available for download on Wrox.com

```
private void ViewCarsButton_Click(object sender, RoutedEventArgs e){
    this.NavigationService.Navigate(new Uri("/CarsPage.xaml", UriKind.Relative));
}
```

Code snippet from MainPage.xaml.cs

Synchronization

Now that you have a mechanism for storing data on the device, you need to consider how you are going to get data onto the device in the first place. In some cases, you may want to access a Web Service or a data feed and simply cache the information you receive in the object cache. However, the challenge then becomes: How do you know when that data needs to be refreshed? Alternatively, if the user is entering data on the device, you need to determine when this information should be sent to the server. Lastly, you should consider what happens when data changes on both the device and the server at the same time.

This topic is generally referred to as *synchronization* and is not an easy challenge to solve. Windows Mobile has had a strong legacy of support for Merge Replication and other synchronization frameworks that made it easy to synchronize data between SQL Server Compact on the device and a backend SQL Server database. Although these frameworks aren't available for Windows Phone, it is relatively easy to roll your own synchronization framework that leverages the object cache you've just covered when combined with a backend data service. As with the object cache, you'll start by defining an interface for the synchronization service. You'll then move on to looking at a base implementation that uses WCF Data Services to synchronize data. You can, of course, create your own implementation that connects to alternative data sources.

Let's start by updating the `IObjectCache` interface and adding a `Sync` action to the `CacheAction` enumeration. When the `Sync` `CacheAction` is specified, any changes that are persisted will also be synchronized to the server:

Available for download on Wrox.com

```
[Flags]
public enum CacheAction{
    None = 0,
    Insert = 1,
    Delete = 2,
    Update = 4,
    Sync = 8
}

public interface IObjectCache : IInitializeAndCleanup {
    ...
    ISyncProvider SyncService { get; set; }
    void Sync<T>() where T : class, IEntityBase;
}
```

Code snippet from IObjectCache.cs

The `IObjectCache` interface now defines a property called `SyncService` that is an implementation of the `ISyncProvider` interface. It also includes a method named `Sync` that will synchronize all elements of the `Type T` with the backend data service. Before you look at the implementation of this method, let's look at the `ISyncProvider` interface in more detail:

```
public interface ISyncProvider : IInitializeAndCleanup {
    List<IEntityConfiguration> SyncConfigurations { get; }

    void RetrieveChangesFromServer<T>(DateTime lastUpdated,
                                      Action<T[]> ChangedEntitiesFromServer)
                                                  where T : class, IEntityBase;

    void SaveChangesToServer<T>(KeyValuePair<T, CacheAction>[] EntitiesToSave,
             Action<KeyValuePair<T, CacheAction>[]> SaveChangesToServerCompleted)
                                                    where T : class, IEntityBase;
}
```

Code snippet from ISyncProvider.cs

As you saw with the `IStorageProvider`, the `ISyncProvider` also has an IEntityConfiguration list that is used to define configuration information for each object type that is to be synchronized. Depending on the implementation, this may mean that different object types are synchronized to different servers or data sources, for example. It also inherits from `IInitializeAndCleanup` to allow the implementation to set up and tear down at appropriate times. The other two methods in the `ISyncProvider` interface are for receiving changes from the server and for saving any local changes back to the server. As you will see later, these methods are used together to perform synchronization of both server and client changes.

Let's look at these two methods in a little more detail. The `ReceiveChangesFromServer` method accepts a time stamp that indicates the time the last update was downloaded from the server. It is important to note that this should not be the time that an object was *created* using the Windows Phone application — it should be the time stamp for when the object was last *updated* on the server. The second parameter is a callback method that has a single parameter consisting of an array of objects. The `SaveChangesToServer` method is almost the reverse. It accepts a list of objects and actions to perform (e.g., insert, update, or delete) and a callback that will be invoked when the objects have been saved.

You'll come back to the `ISyncProvider` interface after looking at how the synchronization process works within the `ObjectCache`. To start with, the `ObjectCache` needs to be updated to include the new property and methods of the `IObjectCache`. In addition to adding the `SyncService` property, a dictionary of time stamps is added to track when different object types were last synchronized. Both the `Initialize` and `Cleanup` methods are updated to set up and tear down the `ISyncProvider` implementation:

```
public ISyncProvider SyncService { get; set; }
private Dictionary<Type, DateTime> LastSynced { get; set; }

public void Initialize(){
    if (Store != null){
        Store.Initialize();
```

```
        }

        if (SyncService != null){
            SyncService.Initialize();
        }

        History = new List<KeyValuePair<IEntityBase, CacheAction>>();
        InMemoryCache = new Dictionary<Type, List<IEntityBase>>();
        EntitiesLoaded = new Dictionary<Type, bool>();
        EntityKeyFunctions = new Dictionary<Type, Func<object, object>>();
        EntityKeyPropertyCache = new Dictionary<Type, PropertyInfo>();
        LastSynced = new Dictionary<Type, DateTime>();

        foreach (var registeredType in ObjectKeyMappings){
            var entityType = registeredType.EntityType;
            var cache = new List<IEntityBase>();
            InMemoryCache[entityType] = cache;
            EntitiesLoaded[entityType] = false;
            LastSynced[entityType] = DateTime.MinValue;
            EntityKeyPropertyCache[entityType] =
             entityType.GetProperty(registeredType.KeyPropertyName);
            EntityKeyFunctions[entityType] =
             (entity) => EntityKeyPropertyCache[entity.GetType()].GetValue(entity, null);
        }
    }

    public void Cleanup(){
        if (SyncService != null){
            SyncService.Cleanup();
        }

        if (Store != null){
            Store.Cleanup();
        }
    }
```

Code snippet from ObjectCache.cs

When objects are loaded into memory from Isolated Storage, it is necessary to determine their last updated date by querying the `LastUpdated` property of each object:

Available for download on Wrox.com

```
    private List<IEntityBase> EntityCache<T>() where T : class, IEntityBase {
        var entityType = typeof(T);
        List<IEntityBase> cache = InMemoryCache[entityType];

        if (!EntitiesLoaded[typeof(T)]) {
            var lastUpdated = DateTime.MinValue;
            foreach (var entity in Store.Select<T>().OfType<IEntityBase>()){
                cache.Add(entity);
                 lastUpdated = (entity.LastUpdated > lastUpdated ? entity.LastUpdated :
                                                                         lastUpdated);
```

```
            }
            LastSynced[typeof(T)] = lastUpdated;
            EntitiesLoaded[typeof(T)] = true;
        }

        return cache;
    }
```

Code snippet from ObjectCache.cs

This requires an update to the `IEntityBase` interface to include the `LastUpdated` property. Previously, you didn't need to track when an object was last updated, nor did you have to track whether the object had been deleted or changed in the Windows Phone. As far as the `ObjectCache` was concerned, once the changes had been persisted, that was all that mattered. However, now you need to track this information so that you know which objects need to be further synchronized with the server:

Available for download on Wrox.com

```
public interface IEntityBase : INotifyPropertyChanged {
    bool Deleted { get; set; }
    DateTime LastUpdated { get; set; }
    bool IsChanged { get; set; }

    void UpdateFrom(IEntityBase newEntity);
}
```

Code snippet from IEntityBase.cs

Because you now need to track whether an object has been deleted, rather than simply deleting it, you need to update the `Delete` method of the `ObjectCache`. Now instead of deleting the object, the `Deleted` property is set to `true`, and the `Update` method is invoked on the `IStorageProvider` implementation:

Available for download on Wrox.com

```
public void Delete<T>(T item) where T : class, IEntityBase {
    EntityCache<T>().Remove(item);
    item.Deleted = true;
    if (ShouldPersist(Data.CacheAction.Delete)) {
        Store.Update(item);
    }
    else {
        History.Add(new KeyValuePair<IEntityBase, CacheAction>
                                (item, Data.CacheAction.Delete));
    }
    item.PropertyChanged -= item_PropertyChanged<T>;
}
```

Code snippet from ObjectCache.cs

We've also added an `UpdateFrom` method to the `IEntityBase` interface. This is required so that objects that already exist in the object cache can be updated with the contents of an object with an identical key that has been fetched from the server. As an example, the `Car` entity class could implement the `UpdateFrom` method as follows:

```
public class Car:SimpleEntityBase {
    ...
    public override void UpdateFrom(IEntityBase newEntity) {
        var car = newEntity as Car;
        this.Make = car.Make;
        this.Model = car.Model;
        this.Year = car.Year;

        base.UpdateFrom(newEntity);
    }
}
```

Code snippet from Car.cs

The `Sync` method itself is relatively simple as it just needs to invoke the `RetrieveChangesFromServer` method on the `ISyncProvider` implementation. However, to make sure that the correct time stamp is used, there is a call to the `EntityCache` method to load objects of the appropriate `Type` into memory:

```
public void Sync<T>() where T : class, IEntityBase {
    // Make sure all entities are loaded from the persistent storage
    var cache = EntityCache<T>();

    DateTime lastupdated;
    if (!LastSynced.TryGetValue(typeof(T), out lastupdated)) return;
    SyncService.RetrieveChangesFromServer<T>(lastupdated,
                                             ChangedEntitiesFromServer);
}
```

Code snippet from ObjectCache.cs

Once the `ISyncProvider` has retrieved the changes from the server, the `ChangedEntitiesFromServer` method is invoked. This method has two parts: In the first, the entities returned from the server are used to update the object cache. If there is a match with an existing entity, that entity is either updated or deleted (if the entity from the server is marked as *Deleted*). Alternatively, if there is no match, the entity is added to the object cache, but only if it is not marked as *Deleted*. There is no point tracking objects that have been deleted on the server in the object cache.

```
private void ChangedEntitiesFromServer<T>(T[] entities) where T : class, IEntityBase
{
    DateTime lastupdated = LastSynced[typeof(T)];

    foreach (var entity in entities) {
        lastupdated = (entity.LastUpdated > lastupdated) ?
                      entity.LastUpdated : lastupdated;
```

```
            var existing = this.FindMatches(entity).FirstOrDefault();
            if (existing != null){
                if (!entity.Deleted){
                    existing.UpdateFrom(entity);
                }
                else{
                    this.Delete(entity);
                }
                existing.IsChanged = false;
            }
            else if(!entity.Deleted) {
                entity.IsChanged = false;
                this.Insert(entity);
            }
        }

        this.Persist();
        LastSynced[typeof(T)] = lastupdated;

        var entitiesToSave = (from entity in Select<T>()
                              where entity.IsChanged == true
                              select new KeyValuePair<T, CacheAction>(
                                  entity,
                                  (entity.LastUpdated <= DateTime.MinValue ?
                                                 CacheAction.Insert :
                                                (entity.Deleted ?
                                                        CacheAction.Delete :
                                                        CacheAction.Update))
                                  )).ToArray();
        SyncService.SaveChangesToServer<T>(entitiesToSave, ChangesSaved);
    }
```

Code snippet from ObjectCache.cs

The second part of the method gathers any local changes and calls the `SaveChangesToServer` method. Note that the action to perform for each object that is marked as changed is determined based on the `LastUpdated` and `Deleted` properties. If the object doesn't exist on the server — in other words, it was recently created in the Windows Phone application — the `LastUpdated` property will be set to `DateTime.MinValue`.

The `ChangesSaved` method is invoked when the synchronization is complete. It makes sure that the `LastUpdated` time is updated to reflect when the most recent change was made to objects in the object cache:

Available for download on Wrox.com

```
    private void ChangesSaved<T>(KeyValuePair<T, CacheAction>[] savedEntities)
                                                where T : class, IEntityBase {
        var lastUpdated = LastSynced[typeof(T)];
        foreach (var entity in savedEntities) {
            if (lastUpdated < entity.Key.LastUpdated) {
                lastUpdated = entity.Key.LastUpdated;
            }
        }
```

```
        LastSynced[typeof(T)] = lastUpdated;
        this.Persist();
    }
```

Code snippet from ObjectCache.cs

ISyncProvider: WCF Data Service

The example implementation of the `ISyncProvider` interface is going to synchronize with a WCF Data Service. For this you will create a simple service that exposes Car records from a SQL Server database. You have a database that contains a table called `Car`, as shown in Figure 16-1.

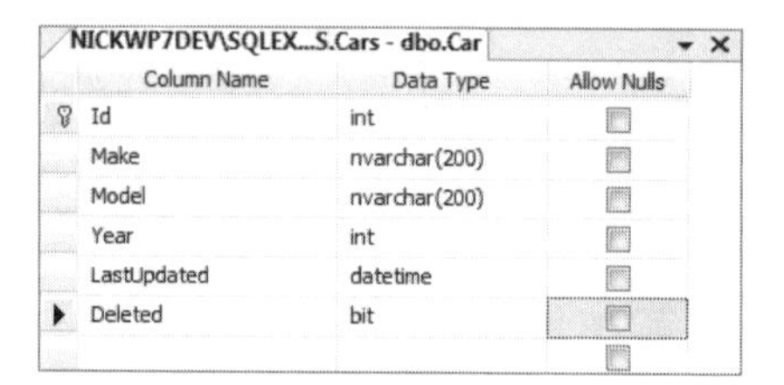

Column Name	Data Type	Allow Nulls
Id	int	☐
Make	nvarchar(200)	☐
Model	nvarchar(200)	☐
Year	int	☐
LastUpdated	datetime	☐
Deleted	bit	☐

FIGURE 16-1

In order to expose this as a WCF Data Service, you need to create a new project called *CarServices*, based on the ASP.NET web application project template (make sure you select ".NET Framework v4"). Into this project, add a new item called **CarModel.edmx** based on the ADO.NET Entity Data Model item template. This will prompt you to define the connection string to the database, as well as selecting the tables to include in the data model. Specify the connection string to the `Cars` database and make sure that the `Car` table is included. Then accept the default values to create the data model.

The next step is to create another item called **CarService.svc**, this time based on the WCF Data Service item template. Update the boilerplate code by specifying the `CarEntities` class as the `DataService` context and allowing all entity operations on the `Cars` entity set. Your code should match the following `CarService` class:

Available for download on Wrox.com

```
public class CarService : DataService< CarsEntities> {
    public static void InitializeService(IDataServiceConfiguration config) {
        config.SetEntitySetAccessRule("*", EntitySetRights.All);
    }
}
```

Code snippet from CarService.svc.cs

If you set CarService.svc to be the start page (right-click the file in Solution Explorer and select "Set as Start Page") and then run the web application, you should be able to navigate to the Cars folder (e.g., `http://localhost:7208/CarService.svc/Cars`) and see a list of cars that are in the database.

One of the important things that need to happen when objects are saved to the database is that the `LastUpdated` property must be set to the current server time. There are several ways to do this using features of SQL Server (e.g., triggers), but in this case, you're going to override the behavior of ADO.NET Entity Framework to update the `LastUpdated` property at the point where objects are being saved. You'll also intercept any attempts to delete an object. Instead of deleting the object, the `Deleted` property will be set to `true` and the new object state to saved. To do this, you need to extend the Entity Model so that each entity type implements the IEntityTracking interface:

Available for download on Wrox.com

```
public interface IEntityTracking {
    DateTime LastUpdated { get; set; }
    bool Deleted { get; set; }
}
```

Code snippet from IEntityTracking.cs

```
public partial class Car : IEntityTracking { }
```

Code snippet from CarModel.custom.cs

Notice that the Car class is a partial class that is defined in a separate file from the rest of the entity. This means that you can easily update your Entity Model from the database schema without worrying about it overwriting your custom code. As the Car class already has properties for LastUpdated and Deleted (based on the database schema), there is no need to implement the IEntityTracking interface explicitly.

To update the LastUpdated property on entities being saved, you need to override the SaveChanges method on the CarsEntities class. In this method, any Modified, Added, or Deleted changes are retrieved. If the change is to delete an item, then the Deleted property is set to true and the state is changed to Modified. The LastUpdated property is set to the current server time, and the changes are then saved:

Available for download on Wrox.com

```
public partial class CarsEntities{
    public override int SaveChanges(SaveOptions options){
        var changes = this.ObjectStateManager.GetObjectStateEntries(
                                                  EntityState.Modified |
                                                  EntityState.Added |
                                                  EntityState.Deleted);
        foreach (var change in changes){
            var entity = change.Entity as IEntityTracking;
            if (entity != null){
                if (change.State == EntityState.Deleted){
                    change.ChangeState(EntityState.Modified);
                    entity.Deleted = true;
                }
                entity.LastUpdated = DateTime.Now;
            }
        }

        return base.SaveChanges(options);
    }
}
```

Code snippet from CarModel.custom.cs

This wraps up the service implementation. Let's return to the WCF Data Service implementation of the ISyncProvider interface. You'll start by extending the EntityConfiguration class by adding properties specific to using WCF Data Services. The WCFSyncConfiguration class includes properties that define the base URL (e.g., http://localhost:7208/CarService.svc), the entity set name (e.g., Cars), the name of the properties that tracks when the entity was last updated (e.g., LastUpdated), and the name of the property that is the entity key:

```
public class WCFSyncConfiguration : EntityConfiguration {
    public string BaseUrl { get; set; }

    public string EntitySetName { get; set; }

    public string LastUpdated { get; set; }

    public string KeyMapping { get; set; }
}
```

Code snippet from WCFSyncConfiguration.cs

The ISyncProvider implementation, WCFSyncService, contains an IEntityConfiguration list. This will be populated with WCFSyncConfiguration instances declared in XAML as you have done previously for the IStorageProvider implementations. WCFSyncService also contains several other dictionaries that are there to optimize the synchronization process by pre-calculating URL and key mapping functions:

```
public class WCFSyncService: ISyncProvider {
    public List<IEntityConfiguration> SyncConfigurations { get; private set; }

    public WCFSyncService() {
        SyncConfigurations = new List<IEntityConfiguration>();
    }

    private Dictionary<Type, string> ServerChangesLinks { get; set; }
    private Dictionary<Type, string> SaveChangesLinks { get; set; }
    private Dictionary<Type, string> InsertLinks { get; set; }
    private Dictionary<Type, PropertyInfo> EntityKeyPropertyCache { get; set; }
    private Dictionary<Type, Func<object, object>> EntityKeyFunctions { get; set; }
}
```

Code snippet from WCFSyncService.cs

Since ISyncProvider inherits from IInitializeAndCleanup, the first set of methods in WCFSyncService are the Initialize and Cleanup methods. The Initialize method creates the dictionaries responsible for caching the WCF Data Service URLs for requesting changes and saving changes, as well as for caching key mapping functions. It then iterates through the configurations and creates appropriate URLs and mapping functions for each object type being synchronized:

```
public void Initialize() {
    ServerChangesLinks = new Dictionary<Type, string>();
    SaveChangesLinks = new Dictionary<Type, string>();
    InsertLinks = new Dictionary<Type, string>();
    EntityKeyFunctions = new Dictionary<Type, Func<object, object>>();
    EntityKeyPropertyCache = new Dictionary<Type, PropertyInfo>();
    foreach (var syncType in SyncConfigurations.OfType<WCFSyncConfiguration>()){
        ServerChangesLinks[syncType.EntityType] = syncType.BaseUrl + "/" +
                                                  syncType.EntitySetName +
                                                  "?$filter=" + syncType.LastUpdated +
                                                  "%20gt%20datetime'{0}'";
        SaveChangesLinks[syncType.EntityType] = syncType.BaseUrl + "/" +
                                                syncType.EntitySetName + "({0})";
        InsertLinks[syncType.EntityType] = syncType.BaseUrl + "/" +
                                                syncType.EntitySetName;
        EntityKeyPropertyCache[syncType.EntityType] = syncType.EntityType.GetProperty
                                                              (syncType.KeyMapping);
        EntityKeyFunctions[syncType.EntityType] =
       (entity) => EntityKeyPropertyCache[syncType.EntityType].GetValue(entity, null);
    }
}

public void Cleanup() { }
```

Code snippet from WCFSyncService.cs

It is worth looking at the format for the URLs for requesting server changes, saving changes, and for inserting new objects. Essentially, these follow a fixed structure based on the way WCF Data Services exposes sets of entities. The following URL can be broken down into the base URL (`http://localhost:7208/CarService.svc`), the entity set name (`Cars`), the name of the property that determines when the entity was last updated (`LastUpdated`), and, lastly, the time stamp to compare against. In this case, the use of the greater than operator (`gt`) indicates that when a GET action is performed on this URL, all the `Car` entities that were updated after the specified date will be returned. Clearly, the last updated date will vary after the initial sync, resulting in only the changed entities being downloaded from the server:

```
http://localhost:7208/CarService.svc/Cars?$filter=LastUpdated%20gt%20datetime'0001-
01-01T00:00:00.0000000'
```

In order to save changes to an entity that already exists on the server, the changed entity is sent to the following URL via a `PUT` action. Again, the format of the URL can be broken into the base URL, the entity set name, and lastly the ID of the entity to be updated (`100`):

```
http://localhost:7208/CarService.svc/Cars(100)
```

The last URL format is for inserting new entities. This is just a matter of sending the new entity via a `POST` action. The format of the URL is just the base URL plus the entity set name:

```
http://localhost:7208/CarService.svc/Cars
```

Now that you know what the format of the URLs to retrieve and save information to WCF Data Services is, you can use this information to complete the implementation of the `RetrieveChangesFromServer` and `SaveChangesToServer` methods. Actually, these methods are relatively simple as they create an instance of a wrapper class and then invoke a method to perform the core logic. This is done to track the callback that needs to be invoked when the operation completes:

```
public void RetrieveChangesFromServer<T>(DateTime lastUpdated,
                                         Action<T[]> ChangedEntitiesFromServer)
                                                  where T : class, IEntityBase {
    string url = ServerChangesLinks[typeof(T)];
    var datetimeText = lastUpdated.ToString("o");
    var uri = string.Format(url, datetimeText);
    var wrapper = new ServerChangesRequestWrapper<T>() {
                                 RequestUrl = uri,
                                 ChangedEntitiesFromServer = ChangedEntitiesFromServer };
    wrapper.RetrieveChangesFromServer();
}

public void SaveChangesToServer<T>(KeyValuePair<T, CacheAction>[] EntitiesToSave,
                 Action<KeyValuePair<T, CacheAction>[]> SaveChangesToServerCompleted)
                                                   where T : class, IEntityBase {
    string url = SaveChangesLinks[typeof(T)];
    var wrapper = new SaveChangesRequestWrapper<T>(){
                               SaveChangesUrl = url,
                               InsertUrl = InsertLinks[typeof(T)],
                               EntitiesToSave = EntitiesToSave,
                               SaveChangesToServerCompleted = SaveChangesToServerCompleted,
                               KeyMapping = EntityKeyFunctions[typeof(T)]
                          };
    wrapper.SaveChangesToServer();
}
```

Code snippet from WCFSyncService.cs

The `ServerChangesRequestWrapper` is split into two parts. The first method is the `RetrieveChangesFromServer` method, which is the method that is invoked from the `WCFSyncService` class. This method is responsible for preparing the request and setting the Accept header to ensure that the response is JSON. The second method is the `SyncCallback` method, which is invoked when the request returns. In this case, the `DataContractJsonSerializer` is used to parse the response and return an `EntitySet` of Type `T`. The `EntitySet` class is a simple wrapper around an array of objects of Type `T`. These are the objects that have changed on the server since the last updated time. This array is then supplied as the parameter to the `ChangedEntitiesFromServer` callback that gets handled by the `ObjectCache`:

```
private class ServerChangesRequestWrapper<T> {
    public string RequestUrl { get; set; }
    public Action<T[]> ChangedEntitiesFromServer { get; set; }

    public void RetrieveChangesFromServer() {
        var request = HttpWebRequest.Create(RequestUrl) as HttpWebRequest;
```

```
        request.Accept = "application/json";
        request.BeginGetResponse(SyncCallback, request);
    }

    private void SyncCallback(IAsyncResult result) {
        var request = result.AsyncState as HttpWebRequest;
        var response = request.EndGetResponse(result);

        var deserializer = new DataContractJsonSerializer(typeof(EntitySet<T>));
        var data = deserializer.ReadObject(response.GetResponseStream())
                                as EntitySet<T>;

        var entities = (from entity in data.d.results.OfType<T>()
                        select entity).ToArray();

        ChangedEntitiesFromServer(entities);
    }
}
```

Code snippet from WCFSyncService.cs

```
public class EntitySet<T> {
    public T[] d { get; set; }
}
```

Code snippet from JsonHelper.cs

The `SaveChangesRequestWrapper` is slightly more complex since the objects that have changed within the Windows Phone application have to be serialized to JSON and then added as the payload to the request. In this example, each object is sent via an individual request so there is an `entityIndex` that tracks which object is being sent. The `SaveChangesToServer` method prepares the request by first determining the request URL (note that in the case of updates, this needs to include the ID of the object being updated), then determining the method to perform (i.e., `PUSH`, `PUT`, or `DELETE` for Insert, Update, or Delete, respectively), and, lastly, writing the serialized object to the request stream:

Available for download on Wrox.com

```
private class SaveChangesRequestWrapper<T> where T : class, IEntityBase {
    public string SaveChangesUrl { get; set; }
    public string InsertUrl { get; set; }
    public KeyValuePair<T, CacheAction>[] EntitiesToSave { get; set; }
    public Action<KeyValuePair<T, CacheAction>[]> SaveChangesToServerCompleted
                                                                { get; set; }
    public Func<object, object> KeyMapping { get; set; }

    private DataContractJsonSerializer serializer;
    private int entityIndex;

    public void SaveChangesToServer()
    {
```

```
if (EntitiesToSave == null || EntitiesToSave.Length == 0)
{
    SaveChangesToServerCompleted(EntitiesToSave);
    return;
}

if (serializer == null)
{
    serializer = new DataContractJsonSerializer(typeof(T));
}
var entityToSave = EntitiesToSave[entityIndex];
var key = KeyMapping(entityToSave.Key);
string keyString;
if (key is string)
{
    keyString = "'" + key.ToString() + "'";
}
else if (key is Guid)
{
    keyString = "guid'" + key.ToString() + "'";
}
else
{
    keyString = key.ToString();
}
var uri = string.Format(SaveChangesUrl, keyString);
if (entityToSave.Value == CacheAction.Insert)
{
    entityToSave.Key.LastUpdated = DateTime.Now.ToUniversalTime();
    uri = InsertUrl;
}
var request = HttpWebRequest.Create(uri) as HttpWebRequest;
request.Accept = "application/json";
request.ContentType = "application/json";
switch (entityToSave.Value)
{
    case CacheAction.Insert:
        request.Method = "POST";
        break;
    case CacheAction.Update:
        request.Method = "PUT";
        break;
    case CacheAction.Delete:
        request.Method = "DELETE";
        break;
}

request.BeginGetRequestStream((callback) =>
{
    var req = callback.AsyncState as HttpWebRequest;
    using (var strm = req.EndGetRequestStream(callback))
    {
        var ms = new MemoryStream();
        serializer.WriteObject(ms, entityToSave.Key);
```

```
                ms.Flush();
                ms.Seek(0, SeekOrigin.Begin);
                var reader = new StreamReader(ms);
                var txt = reader.ReadToEnd();
                System.Diagnostics.Debug.WriteLine(txt);

                serializer.WriteObject(strm, entityToSave.Key);
            }
            request.BeginGetResponse(SyncCallback, request);
        }, request);

    }

    private void SyncCallback(IAsyncResult result)
    {
        var request = result.AsyncState as HttpWebRequest;
        var response = request.EndGetResponse(result);

        if (response.ContentLength > 0)
        {
            var deserializer = new DataContractJsonSerializer(typeof(Entity<T>));
            var data = deserializer.ReadObject(response.GetResponseStream())
                                  as Entity<T>;
            EntitiesToSave[entityIndex].Key.LastUpdated = data.d.LastUpdated;
            EntitiesToSave[entityIndex].Key.Deleted = data.d.Deleted;
            EntitiesToSave[entityIndex].Key.IsChanged = false;
        }

        entityIndex++;
        if (entityIndex < EntitiesToSave.Length)
        {
            SaveChangesToServer();
        }
        else
        {
            SaveChangesToServerCompleted(EntitiesToSave);
        }
    }
}
```

Code snippet from WCFSyncService.cs

```
public class Entity<T>{
    public T d { get; set; }
}
```

Code snippet from JsonHelper.cs

The `SyncCallback` method is invoked when the request has completed. Depending on the method being performed, there may be a response that will contain the updated object. This is deserialized and used to update the relevant properties on the existing object. If there are more objects to be sent to the server, the next object is sent by calling the `SaveChangesToServer` method. Otherwise, the `SaveChangesToServerCompleted` callback is invoked, which returns to the `ObjectCache`.

For a `Car` object to be successfully sent to the server, there are some minor additions to both the `Car` class and its base class, `SimpleEntityBase`. By default, all public properties will be serialized. However, there are properties, such as `EntityId` and `IsChanged`, that shouldn't be sent to the server because these are only relevant to the object cache within the Windows Phone application. To override the default behavior, you can specify which properties are serialized by using the `DataMember` attribute. This needs to be used on both the `Car` and `SimpleEntityBase` classes, and you also need to add the `DataContract` attribute to both class declarations:

```
[DataContract]
public class Car : SimpleEntityBase {
    [DataMember]
    public int Id { ... }

    [DataMember]
    public int Year { ... }

    [DataMember]
    public string Model { ... }

    [DataMember]
    public string Make { ... }

    public override void UpdateFrom(IEntityBase newEntity) { ... }
}
```

Code snippet from Car.cs

```
[DataContract]
public class SimpleEntityBase : IEntityBase {
    public Guid EntityId { ... }
    public bool IsChanged { ... }

    [DataMember]
    public bool Deleted { ... }

    [DataMember]
    public DateTime LastUpdated { ... }

    public virtual void UpdateFrom(IEntityBase newEntity) { }

    public event PropertyChangedEventHandler PropertyChanged;

    protected void RaisePropertyChanged(string propertyName) { ... }
}
```

Code snippet from SimpleEntityBase.cs

The last thing to do is to update the App.xaml to include an instance of the `WCFSyncService`.

Available for download on Wrox.com

```
<Application
    x:Class="ManageYourData.App"
    ...
    xmlns:data="clr-namespace:BuiltToRoam.Data;assembly=BuiltToRoam.Data"
    xmlns:simple="clr-namespace:BuiltToRoam.Data.SimpleStorage;
assembly=BuiltToRoam.Data.SimpleStorage"
    xmlns:wcf="clr-namespace:BuiltToRoam.Data.WCFDataService;
assembly=BuiltToRoam.Data.WCFDataService"
    xmlns:local="clr-namespace:ManageYourData">
    <Application.Resources>
        <simple:SimpleStorageProvider StoreName="Cars" x:Key="SimpleCarsData">
            <simple:SimpleStorageProvider.StorageConfigurations>
                <data:EntityConfiguration
                      EntityType="ManageYourData.Car,ManageYourData" />
            </simple:SimpleStorageProvider.StorageConfigurations>
        </simple:SimpleStorageProvider>
        <wcf:WCFSyncService x:Key="WCF">
            <wcf:WCFSyncService.SyncConfigurations>
                <wcf:WCFSyncConfiguration
                     BaseUrl="http://localhost:7208/CarService.svc"
                     EntitySetName="Cars" LastUpdated="LastUpdated"
                     EntityType="ManageYourData.Car,ManageYourData"
                     KeyMapping="Id" />
            </wcf:WCFSyncService.SyncConfigurations>
        </wcf:WCFSyncService>
        <local:CarStore x:Key="Cache" Store="{StaticResource SimpleCarsData}"
                        CacheMode="Insert,Delete,Update,Sync"
                        SyncService="{StaticResource WCF}">
            <data:ObjectCache.ObjectKeyMappings>
                <data:ObjectKeyMap
                      EntityType="ManageYourData.Car,ManageYourData"
                      KeyPropertyName="Id" />
            </data:ObjectCache.ObjectKeyMappings>
        </local:CarStore>
    </Application.Resources>
</Application>
```

Code snippet from App.xaml

You can also add a button that will allow the user to invoke synchronization of cars between the Windows Phone application and the server:

```
private void SyncButton_Click(object sender, RoutedEventArgs e){
    (Application.Current as App).Store.Sync<Car>();
}
```

Code snippet from MainPage.xaml.cs

This completes the overview of how you can use an object cache coupled with persistent storage and synchronization to provide data to your application. You may want to consider triggering the synchronization process automatically when objects in the Windows Phone application change, based on the `CacheMode` value.

SUMMARY

In this chapter, you've learned about reading and writing to Isolated Storage to save information between sessions of your application. You've seen how you can use both an in-memory and persistent object cache to allow data to be shared between pages within your application, and across multiple application runs. The object cache can be declared in XAML, making it easy to configure and data-bind to. Lastly, you saw how this can be extended to include support for synchronizing data to a data service, allowing centralized data to be viewed via the Windows Phone application regardless of whether there is an active connection available.

17

Frameworks

WHAT'S IN THIS CHAPTER

- Using the Managed Extensibility Framework to make your Windows Phone application composable and extensible
- Tracking application usage using the Microsoft Silverlight Analytics Framework
- Testing your application using the Silverlight Unit Testing Framework
- Automating the Windows Phone emulator

For small, lightweight Windows Phone applications there is no need to use any additional frameworks other than what comes out-of-the-box with Silverlight. However, as your application grows and you extend its capabilities, it will become important to have an application architecture that enables it to grow while still remaining robust.

In this chapter, you will learn about the Managed Extensibility Framework (MEF) for componentizing your application and the Microsoft Silverlight Analytics Framework (MSAF) for tracking application usage. You will also learn how to test your application as part of your development process.

MANAGED EXTENSIBILITY FRAMEWORK

The Managed Extensibility Framework (MEF) was introduced in the desktop version of .NET Framework v4.0 as a way to encourage developers to architect their applications into components that could be glued together at runtime. This not only promotes application componentization, but also allows applications to be extensible and more testable. Despite Windows Phone applications typically being significantly smaller in size and complexity than their desktop counterparts, they can still benefit from the use of MEF.

Import and Export

At the most basic level, MEF is about importing and exporting components. For example, if you have a class that provides logging functionality, you could export it by marking it with the `Export` attribute. Another component that wishes to log information could then import that class by declaring a property and marking it with the `Import` attribute. MEF handles connecting the two so that at runtime the component can log information using an instance of the class that exported the logging functionality.

Before we go on, let's see this example in action. The first thing to do is to add references to the MEF assemblies within your Windows Phone application.

At the time of writing, there isn't an official release of MEF for Windows Phone. However, the source code for the Microsoft Silverlight Analytics Framework (MSAF), discussed in the next section, includes a port of MEF for Windows Phone. Download the source code for MSAF from `http://msaf.codeplex.com` *and add references to the "ComponentModel" and "Composition.Initialization" projects to your application.*

Create a class called `SimpleLogger` in a new class library called *LoggerImplementation*. This class exposes a single method, `Log`, which simply writes the string to the debug console. This class is exposed to MEF using the `Export` attribute:

```
[Export]
public class SimpleLogger{
    public void Log(string information){
        Debug.WriteLine(information);
    }
}
```

Code snippet from SimpleLogger.cs

The `MainPage` of the application defines a property, `Logger`, which is marked with the `Import` attribute:

```
public partial class MainPage : PhoneApplicationPage{
    [Import]
    public SimpleLogger Logger { get; set; }

    public MainPage(){
        InitializeComponent();
        Logger.Log("Application Started");
    }
}
```

Code snippet from MainPage.xaml.cs

At runtime, MEF detects all properties that have been marked with the `Import` attribute. It then searches through the list of types that have been exported by being marked with the `Export` attribute. Where there is a match between the type exported and the type of the property marked for import, an instance of the type will be created and assigned to the property.

You might be wondering at this stage how MEF gets involved. To invite MEF to join the party, you need to invoke the `Initialize` method on the `CompositionHost` class. To ensure that this gets done prior to any component in the application being accessed, you need to invoke this at application startup. You can do this either by hooking both the `Launching` and `Activated` events on the application, or you can create an application service that implements `IApplicationService`. In this case, you'll do the latter as this will result in better code encapsulation and separation of concerns:

```
public class MEFApplicationService : IApplicationService {
    public MEFApplicationService() {
        CompositionHost.Initialize(
            new AssemblyCatalog(Application.Current.GetType().Assembly),
            new AssemblyCatalog(typeof(LoggerImplementation.SimpleLogger).Assembly)
        );
    }
    public void StartService(ApplicationServiceContext context) { }
    public void StopService() { }
}
```

Code snippet from MEFApplicationService.cs

The `Initialize` method accepts an arbitrary number of catalogs that define the locations that MEF should search when resolving types. The `AssemblyCatalog` class is used to provide static references to the assemblies to search, which in this case is the main application assembly and the `LoggerImplementation` assembly (within the main application project you will need to add a reference to the `LoggerImplementation` project).

Adding an instance of the MEFApplicationService to the Application.ApplicationLifetimeObjects collection in App.xaml will ensure that the `Initialize` method gets called to initialize MEF before any application code has a chance to run.

```
<Application
    x:Class="Frameworks.App"
    ...
    xmlns:local="clr-namespace:Frameworks">

    <Application.ApplicationLifetimeObjects>
        <local:MEFApplicationService/>
        ...
    </Application.ApplicationLifetimeObjects>
</Application>
```

Code snippet from App.xaml

The one thing that remains is to make a call to MEF to get it to satisfy all the imports on the MainPage:

Available for download on Wrox.com

```
public MainPage() {
    InitializeComponent();
    CompositionInitializer.SatisfyImports(this);

    Logger.Log("Application Started");
}
```

Code snippet from MainPage.xaml.cs

At this point, you might be wondering why you couldn't have just created an instance of the SimpleLogger class in the constructor. The short answer is that you could have easily done this. However, let's take a look at things from the point of view of what happens if you want to replace the type of logger you use. This would require us to go through and replace all references to the SimpleLogger class with a reference to the new logger.

Clearly, you could improve the design of your logging components by creating an interface, ILogger, which is implemented by both loggers. Notice that the Export attribute has been modified to include the ILogger type as a parameter. This indicates that the class should be exported as an ILogger (making it available for any Import that is looking for an ILogger implementation). This means that the type of the Logger property on MainPage can be ILogger.

Available for download on Wrox.com

```
namespace LoggerImplementation {
    [Export(typeof(ILogger))]
    public class SimpleLogger : ILogger {
        public void Log(string information) {
            Debug.WriteLine(information);
        }
    }
}
```

Code snippet from SimpleLogger.cs

```
namespace AnotherLoggerImplementation {
    [Export(typeof(ILogger))]
    public class IsolatedStorageLogger : ILogger {
        public void Log(string information) {
            IsolatedStorageSettings.ApplicationSettings
                                    ["LastLoggedInformation"] = information;
        }
    }
}
```

Code snippet from IsolatedStorageLogger.cs

```
public partial class MainPage : PhoneApplicationPage {
    [Import]
```

```
        public ILogger Logger { get; set; }
    }
```

Code snippet from MainPage.xaml.cs

To switch between the two `ILogger` implementations, all you need to do is to change which assembly is searched by MEF in the MEFApplicationService constructor:

```
    public MEFApplicationService() {
        CompositionHost.Initialize(
            new AssemblyCatalog(Application.Current.GetType().Assembly),
            new AssemblyCatalog(
        typeof(AnotherLoggerImplementation.IsolatedStorageLogger).Assembly));
    }
```

Code snippet from MEFApplicationService.cs

ImportMany

MEF can also be used to import multiple classes that all implement the same interface using the `ImportMany` attribute. The best way to illustrate this is with an example. Later in this chapter you'll learn about the Microsoft Silverlight Analytics Framework (MSAF) and how you can test your Windows Phone application. For each of these sections you're going to add at least one additional `PhoneApplicationPage` to the application, and you want to architect the Windows Phone application so that adding these additional pages is as easy and as isolated as possible.

To keep each of these sections separate, let's place them in their own assemblies. Create two new projects based on the Windows Phone Class Library project template, and name them **MSAFSample** and **TestingSample**. Open the MSAFSample project, and add a new `PhoneApplicationPage` called MSAFPage.xaml. In this case, rather than exporting the page itself, you're going to create an additional class that will be exported. This class will expose two properties called `Name` and `RootPage`. The former can be used to provide a friendly name for the component, while the latter will be used to navigate to the page. The advantage of this technique, over exporting the page directly, is that it makes it much easier to identify the page and present it within a dynamically generated user interface (UI).

```
namespace MSAFSample {
    public class SampleDeclaration : ISample {
        public string Name {
            get {
                return "MSAF Sample";
            }
        }

        public Uri RootPage {
            get {
                return new Uri("/MSAFSample;component/MSAFPage.xaml",
                               UriKind.Relative);
            }
```

```
        }
    }
}
```

Code snippet from SampleDeclaration.cs

You'll notice that the `SampleDeclaration` class implements the `ISample` interface (in the same way that `SimpleLogger` and `IsolatedStorageLogger` both implemented `ILogger`) but that there is no `Export` attribute declared. This is because the declaration of the `ISample` interface has an `InheritedExport` attribute applied to it. This means that any class that implements this interface will automatically be exported.

```
namespace SharedInterfaces {
    [InheritedExport(typeof(ISample))]
    public interface ISample {
        string Name { get; }
        Uri RootPage { get; }
    }
}
```

Code snippet from ISample.cs

In the TestingSample project, create two new `PhoneApplicationPage` instances named TestSamplePage.xaml and TestSample2Page.xaml. This project is going to expose two different implementations of the `ISample` interface, effectively exporting two samples that the user will be able to access.

```
namespace TestingSample {
    public class SampleDeclaration : ISample {
        public string Name {
            get {
                return "Testing Sample";
            }
        }

        public Uri RootPage {
            get {
                return new Uri("/TestingSample;component/TestSamplePage.xaml",
                               UriKind.Relative);
            }
        }
    }
}
```

Code snippet from SampleDeclaration.cs

```
namespace TestingSample {
    public class Sample2Declaration : ISample {
        public string Name {
            get {
```

```
                return "Testing Sample 2";
            }
        }

        public Uri RootPage {
            get {
                return new Uri("/TestingSample;component/TestSample2Page.xaml",
                               UriKind.Relative);
            }
        }
    }
}
```

Code snippet from Sample2Declaration.cs

Now that you have two class libraries that export three `ISample` implementations, you can modify our Windows Phone application to import them. To do this, you first need to create the property with the `Import` attribute in the `MainPage`:

Available for download on Wrox.com

```
public partial class MainPage : PhoneApplicationPage {
    [ImportMany]
    public IEnumerable<ISample> Samples { get; set; }
}
```

Code snippet from MainPage.xaml.cs

In this case, since you expect there to be multiple implementations of the `ISample` interface imported, it is necessary to make the property an `IEnumerable<ISample>` collection. The only thing left is to make sure that both assemblies are referenced in the `MEFApplicationService` constructor:

Available for download on Wrox.com

```
public MEFApplicationService(){
    CompositionHost.Initialize(
        new AssemblyCatalog(Application.Current.GetType().Assembly),
        new AssemblyCatalog(typeof(MSAFSample.MSAFPage).Assembly),
        new AssemblyCatalog(typeof(TestingSample.TestSamplePage).Assembly));
}
```

Code snippet from MEFApplicationService.cs

Although there are three implementations of the `ISample` interface, there are only two assembly references added to the AssemblyCatalog (three if you include the assembly that contains the application itself). This is because the `TestingSample` assembly includes both the `SampleDeclaration` and `Sample2Declaration` classes, and there is no need to reference the same assembly multiple times. The MEF framework will find all classes defined within each referenced assembly.

When you run the application, you want to display the list of available samples within a listbox. Then, when the user clicks a button, the selected sample should be displayed by opening the appropriate page.

```
public MainPage() {
    InitializeComponent();
    CompositionInitializer.SatisfyImports(this);
    Loaded += new RoutedEventHandler(MainPage_Loaded);
}

void MainPage_Loaded(object sender, RoutedEventArgs e) {
    // Only load the samples the first time this page is loaded
    if (this.SamplesList.Items.Count == 0) {
        foreach (var sample in Samples) {
            this.SamplesList.Items.Add(sample);
        }
    }
}

private void OpenSampleButton_Click(object sender, RoutedEventArgs e) {
    var sample = this.SamplesList.SelectedItem as ISample;
    if (sample == null)
        return;
    this.NavigationService.Navigate(sample.RootPage);
}
```

Code snippet from MainPage.xaml.cs

Figure 17-1 illustrates the application in action. The first image illustrates the list of samples available for the user to run. By selecting a sample and hitting the "Open Sample" button, the appropriate page is displayed. The middle image is the page for the MSAF example, while the rightmost image is the page for the second testing sample.

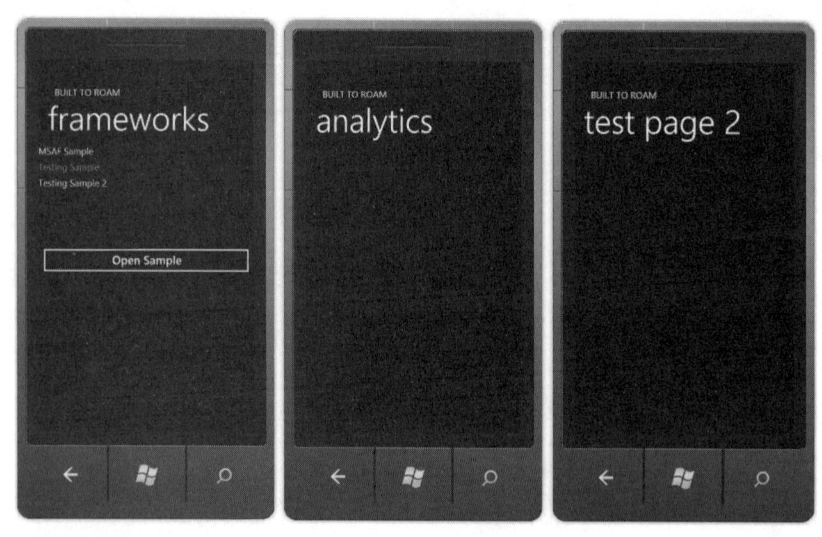

FIGURE 17-1

APPLICATION COMPOSITION

There are several frameworks that can help you build more complex and composable applications. By building your application in components and allowing them to communicate through a loosely coupled messaging system, the components can be developed in relative isolation and then mixed and recombined in various ways. Two such frameworks that are worth mentioning here are:

- **Composite WPF (Prism)** (`http://compositewpf.codeplex.com`) — The composite application guidance for WPF (and subsequently Silverlight for WP7) was created to help tackle the challenges faced by large monolithic applications. Building discrete, loosely coupled, semi-independent components and combining them into an application shell is seen as an effective way to address many of these challenges.
- **nRoute** (`http://nroute.codeplex.com`) — nRoute is another composite application framework, created to allow applications written for silverlight, WPF, and now Windows Phone, to use Model View View Model (MVVM) to build and integrate components into an application.

MICROSOFT SILVERLIGHT ANALYTICS FRAMEWORK

The Microsoft Silverlight Analytics Framework (MSAF) includes support for Windows Phone development. This gives you a standard approach to include usage tracking within your Windows Phone application that is somewhat independent of the actual tracking site or service you use. It's only somewhat independent as each service (e.g., Google Analytics or Omniture) requires a custom behavior to be created in order for it to be used with the framework. Luckily, the major tracking services already have behaviors available for use within your application.

To get started with tracking usage within your Windows Phone application, you need to download and install the latest version of the MSAF library from `http://msaf.codeplex.com`.

If you already have the source code downloaded from the previous section, this can be referenced instead.

We're going to continue the previous example, adding the MSAF to the MSAFSample project. Start by adding references to the following assemblies to both your application and the MSAFSample project:

- `ComponentModel` [MEF]
- `Composition.Initialization` [MEF]
- `Microsoft.WebAnalytics` [MSAF]
- `Microsoft.WebAnalytics.Behaviors` [MSAF]
- `Microsoft.WebAnalytics.Navigation` [MSAF]
- `System.Windows.Interactivity`

You'll notice from this list that the first two assemblies are the MEF libraries. MSAF uses MEF internally so you need to add these references, along with `System.Windows.Interactivity`.

The next step is to create an instance of the `WebAnalyticsService` class that implements both the `IApplicationService` and `IApplicationLifetime` interfaces. If you are simply adding the MSAF, you can create this as an instance within the ApplicationLifetimeObjects collection in the App.xaml.

```
<Application.ApplicationLifetimeObjects>
    <msaf:WebAnalyticsService/>
</Application.ApplicationLifetimeObjects>
```

However, in this case, since you're only using MSAF within the MSAFSample project, you will use a more componentized approach that leverages MEF. Instead of adding an instance of the `WebAnalyticsService` to the `ApplicationLifetimeObjects` collection in the app.xaml file, you're going to dynamically create an instance of the `WebAnalyticsService` and add it to the collection during the initialization of the application. Start by defining an interface named `IApplicationServiceFactory`:

Available for download on Wrox.com

```
namespace SharedInterfaces {
    [InheritedExport(typeof(IApplicationServiceFactory))]
    public interface IApplicationServiceFactory {
        IApplicationService Create();
    }
}
```

Code snippet from IApplicationServiceFactory.cs

Now, you need to add a simple implementation to the MSAFSample project that will return a new instance of the WebAnalyticsService:

Available for download on Wrox.com

```
namespace MSAFSample {
    public class WebAnalyticsServiceFactory : IApplicationServiceFactory {
        public IApplicationService Create() {
            return new WebAnalyticsService();
        }
    }
}
```

Code snippet from WebAnalyticsServiceFactory.cs

The last thing to do is to update the `MEFApplicationService` class that you created earlier. You start by adding a property of type `IEnumerable<IApplicationServiceFactory>`, which is essentially a list of the service factories that have been imported. At the end of the constructor this list is iterated over, creating the `IApplicationService` instances that are then added to the ApplicationLifetimeObjects collection.

Available for download on Wrox.com

```
public class MEFApplicationService : IApplicationService {
    [ImportMany]
    public IEnumerable<IApplicationServiceFactory> ServiceFactories { get; set; }

    public MEFApplicationService() {
        CompositionHost.Initialize(
            new AssemblyCatalog(Application.Current.GetType().Assembly),
            new AssemblyCatalog(typeof(MSAFSample.MSAFPage).Assembly),
            new AssemblyCatalog(typeof(
                                Microsoft.WebAnalytics.AnalyticsEvent).Assembly),
            new AssemblyCatalog(typeof(
                    Microsoft.WebAnalytics.Behaviors.TrackAction).Assembly));

        CompositionInitializer.SatisfyImports(this);
        foreach (var factory in ServiceFactories) {
            Application.Current.ApplicationLifetimeObjects.Add(factory.Create());
        }
    }

    public void StartService(ApplicationServiceContext context) { }

    public void StopService() { }
}
```

Code snippet from MEFApplicationService.cs

The other additions to the constructor are references to both the `WebAnalytics` and `WebAnalytics.Behaviors` assemblies within the AssemblyCatalog. This is to ensure that the MSAF library can be initialized correctly (remember that it uses MEF internally).

You're now at the point where you can begin tracking user activity within the MSAFSample project. To start with, you'll simply log activity to the console using the `ConsoleAnalytics` behavior. Open Expression Blend. Within the Assets window you should see an Analytics node under Behaviors, as shown in Figure 17-2. Drag an instance of the `ConsoleAnalytics` behavior across onto the LayoutRoot node within the Objects and Timeline window.

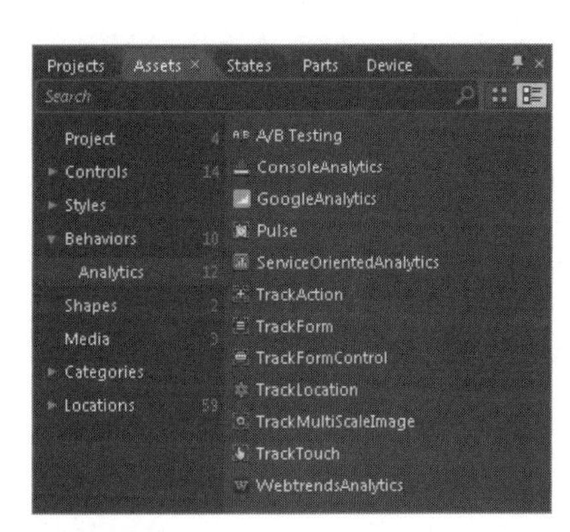

FIGURE 17-2

Adding the `ConsoleAnalytics` behavior to the page enabled tracking of events to the console. Now you need to define which events and information you want to track. In this case, you'll keep it simple and just report when a button is clicked on the page. To do this, add a button to the page, and then drag an instance of the `TrackAction` behavior from the Assets window onto the button. Figure 17-3 illustrates the Objects and Timeline window for the page.

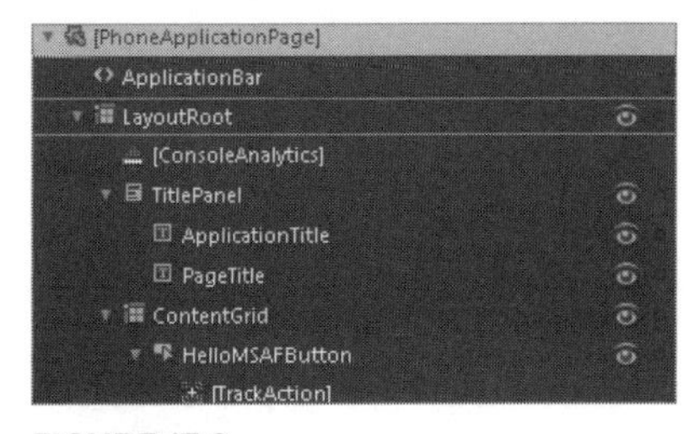

FIGURE 17-3

You need to configure the properties of the `TrackAction` behavior. In Figure 17-4, you can see that the `TrackAction` has been wired up to the `Click` event of the `HelloMSAFButton`. The `Category` and the `Value` properties have also been configured. These properties will be sent through to the analytics provider so that you can identify the event type and the specific button that was clicked.

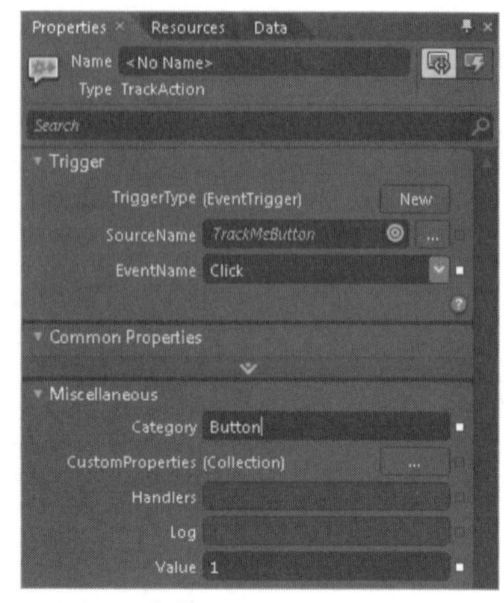

FIGURE 17-4

If you run the application, you will see that each time you click the `Button` event information will be sent to the debugger console window.

```
Click: TrackMeButton, 1,
Click: TrackMeButton, 1,
Click: TrackMeButton, 1,
Click: TrackMeButton, 1,
```

Sending usage information to the console window can help while debugging your Windows Phone application, but it doesn't really help monitor how your application is being used by users out in the wild. To do this, you can make use of one of the behaviors designed for use with a third-party analytics service. This will send the event tracking data to the third-party service so that you can use its tools to aggregate data and run reports to determine how your application is being used.

In this example, you'll add the `GoogleAnalytics` behavior to the application by dragging it from the Assets window onto the LayoutRoot node in the Objects and Timeline window. Unlike the `ConsoleAnalytics` behavior, which didn't have any properties to set, you need to configure the `GoogleAnalytics` behavior by specifying the Web Property ID that the tracking information will be recorded against, as shown in Figure 17-5. To get a Web Property ID, you simply need to sign up for a Google Analytics account at `www.google.com/analytics`.

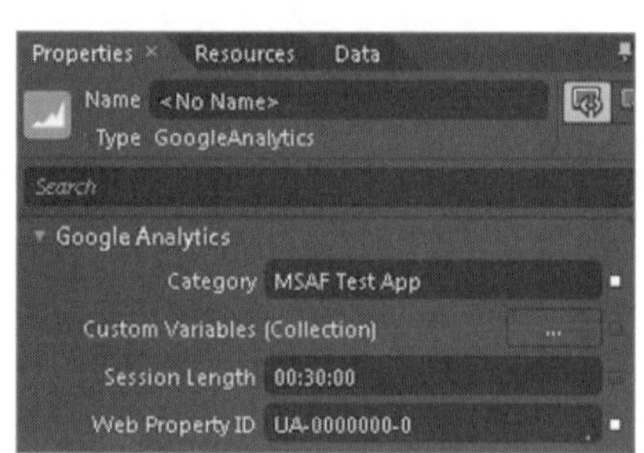

FIGURE 17-5

While you're making some changes to the application, you can also experiment with adding a `TrackLocation` behavior to the `LayoutRoot`. This will periodically record the geo-location of the device. Remember that if you use this behavior, you must prompt the user to advise them that you will be recording their location.

After your application has been run a few times, you can then log in to your Google Analytics account and monitor how your application is being used. Figure 17-6 illustrates the "Event Tracking Overview" area of Google Analytics for a test account.

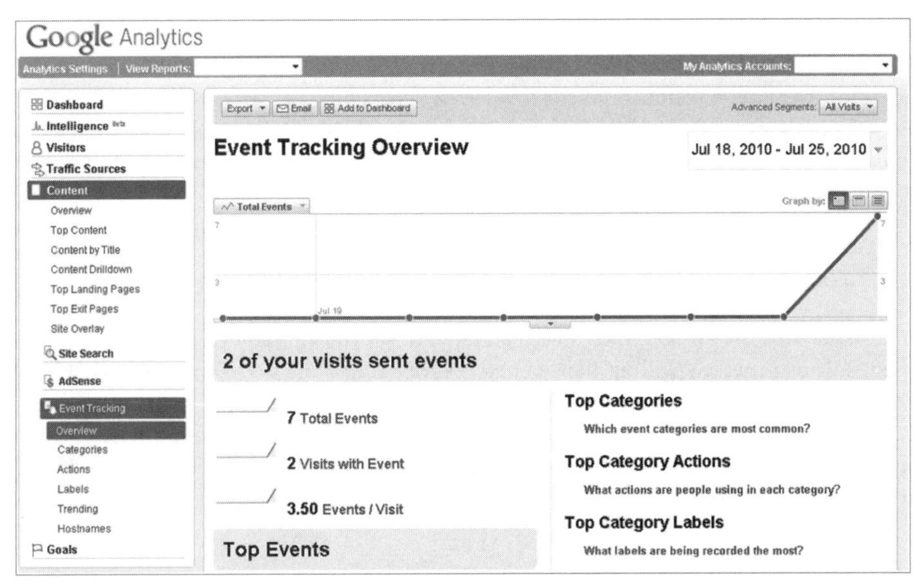

FIGURE 17-6

TESTING

One of the most important aspects in building a robust application is *testing*. Windows Phone applications are no exception. In fact, because of the unique application life cycle and device features, you will probably have to do significantly more testing to ensure correct behavior under all usage scenarios. Your testing matrix should involve unit testing and functional and usability testing, and you should remember to make sure that at least the final level of testing is done on a real device.

Unit Testing

One area of testing that you can automate is the execution of *unit tests*. You can either adopt a test-driven development (TDD) approach, in which case you write tests that fail and then fill in the code to make those tests pass, or you can simply write tests retrospectively to validate that your code executes as expected. In this section, we're going to illustrate how to use the Silverlight Unit Test Framework (SUTF) to build and execute unit tests for your application code.

Getting Started

We're going to continue on using the same Windows Phone application from the previous section. However, to avoid polluting the application itself with the testing framework and test cases, we'll create the unit tests in a separate project. We're going to be testing code that will reside in the TestingSample project (which, if you recall from earlier in the chapter, was created to illustrate the use of MEF), so we'll create a corresponding unit test project called *TestingSample.Tests*. One point to be aware of is that because SUTF has a visual output, you need to create the project based on the Windows Phone Application template, rather than the Windows Phone Class Library template.

The next thing to do is to add a reference to the SUTF assemblies. SUTF actually forms part of the Silverlight 4 Toolkit (`http://silverlight.codeplex.com`). However, for assemblies that are compatible with Windows Phone, you should consult Jeff Wilcox's blog at `www.jeff.wilcox.name`. He periodically posts updates to SUTF that have been built to work with Windows Phone applications. When you download SUTF you should add references to the following two assemblies within the TestingSample.Tests project:

```
Microsoft.Silverlight.Testing.dll
Microsoft.VisualStudio.QualityTools.UnitTesting.Silverlight.dll
```

To initialize and display the SUTF user interface, you need to set the Content of the application's `MainPage` to the result returned by calling the `UnitTestSystem.CreateTestPage` method. This will iterate through all classes within the current assembly looking for classes that have been marked with the `TestClass` attribute. Within such classes, it will then look for methods that have the `TestMethod` attribute. The names of these methods form the list of tests that will be invoked by the SUTF framework, with the eventual test results displayed in a tree-like structure:

Available for download on Wrox.com

```
public MainPage() {
    InitializeComponent();

    Content = UnitTestSystem.CreateTestPage();

    IMobileTestPage imtp = Content as IMobileTestPage;
    if (imtp != null) {
        BackKeyPress += (x, xe) => xe.Cancel = imtp.NavigateBack();
    }
}
```

Code snippet from MainPage.xaml.cs

The SUTF UI allows you to drill down into the results of each unit test to see what the causes of any failures are. To allow you to navigate back up the hierarchy, you need to override the functionality of the Back button to give the SUTF framework a chance to handle it internally. As you can see in the code, this is done by casting the Content of the page to an `IMobileTestPage` object and then wiring up an event handler for the `BackKeyPress` event.

Let's add some basic tests just to get started:

Available for download on Wrox.com

```
[TestClass]
public class MyFirstTests {
    [TestMethod]
    public void ShouldPass() {
        Assert.IsTrue(true);
    }

    [TestMethod]
    public void ShouldFail() {
        Assert.IsTrue(false);
    }
}
```

Code snippet from MyFirstTests.cs

This code illustrates a class, `MyFirstTests`, which has the `TestClass` attribute and contains two methods, both of which have the `TestMethod` attribute. As indicated by their names, the `ShouldPass` method will pass, while the `ShouldFail` will fail owing to the `IsTrue` assertion being incorrect. Figure 17-7 illustrates the visual feedback given when these tests are executed by SUTF.

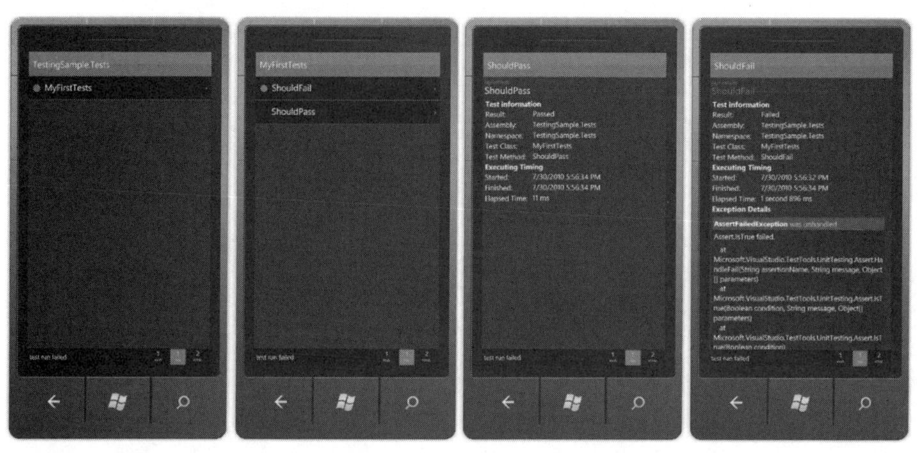

FIGURE 17-7

The leftmost image of Figure 17-7 shows the list of all classes that have been found with the `TestClass` attribute applied. Classes with a red dot beside them indicate those that have tests which have failed. Clicking the class name drills down to the list of methods within that class that have the `TestMethod` attribute (second image). Again, the red dot indicates methods that have failed. The last two images (on the right) illustrate the information gathered about a successful test and one that has failed.

Now that you've seen how SUTF works, let's create a test that actually validates some functionality. In the TestSample project, create a class called `Customer` that has a property named `FirstName`:

Available for download on Wrox.com

```
namespace TestingSample {
    public class Customer {
        public string FirstName { get; set; }
    }
}
```

Code snippet from Customer.cs

In the TestingSample.Tests project, add a corresponding test class — `CustomerTests` — and apply the `TestClass` attribute. The first thing you're going to test is that the `Customer` class implements the `INotifyPropertyChanged` interface. This is required if you want to wire up event handlers to detect when properties change value or if you want to use the `Customer` class for binding to the UI.

MODEL VERSUS VIEWMODEL

One thing to be aware of here is that it is good practice not to data-bind directly to the underlying Model (in this case, a `Customer` object). Instead, it is better to create a ViewModel that passes through the properties of the Customer that you wish to bind to:

```
public class CustomerViewModel {
    private Customer Customer { get; set; }
    public string FirstName {
        get { return Customer.FirstName; }
        set { Customer.FirstName = value; }
    }
}
```

This way the Model doesn't need to implement the `INotifyPropertyChanged` interface and can be a plain old CLR object (POCO). It also provides a helpful abstraction point. If the structure of the Model ever changes, you can handle any required translation within the ViewModel without affecting any existing bindings within the View layer. To keep this testing example simple, we'll assume that the `Customer` object is going to implement `INotifyPropertyChanged` directly.

```
[TestClass]
public class CustomerTests {
    [TestMethod]
    public void Implement_INotifyPropertyChanged_Test() {
        var customer = new Customer();
        Assert.IsInstanceOfType(customer, typeof(INotifyPropertyChanged));
    }
}
```

Code snippet from CustomerTests.cs

Running SUTF will demonstrate that this test case fails, since the `Customer` class does not currently implement the `INotifyPropertyChanged` interface. You can fix this by updating the `Customer` class to implement the interface:

```
public class Customer:INotifyPropertyChanged {
    public event PropertyChangedEventHandler PropertyChanged;

    public string FirstName { get; set; }
}
```

Code snippet from Customer.cs

Of course, simply implementing the interface by adding the `PropertyChanged` event to the `Customer` class doesn't meet the requirement of being able to detect when one of the property values changes. So, let's add two further tests that will initially fail and highlight the flaw in the implementation above. The first verifies that the `PropertyChanged` event fires correctly when the `FirstName` property is set, whereas the second verifies that the correct `PropertyName` value is passed through in the event args:

```
[TestMethod]
public void PropertyChangedEvent_FirstName_Test() {
    var customer = new Customer();
    var eventRaised = false;
    customer.PropertyChanged += (s, e) =>
        {
            eventRaised=true;
        };
    customer.FirstName = "Test Name";
    Assert.IsTrue(eventRaised);
}

[TestMethod]
public void PropertyChangedEvent_FirstName_PropertyNameCorrect_Test() {
    var customer = new Customer();
    var eventRaised = false;
    customer.PropertyChanged += (s, e) =>
        {
            eventRaised = e.PropertyName == "FirstName";
        };
    customer.FirstName = "Test Name";
    Assert.IsTrue(eventRaised);
}
```

Code snippet from CustomerTests.cs

Again, these test methods will fail when run against the current `Customer` implementation. You can make small adjustments to pass both these tests:

```
public class Customer : INotifyPropertyChanged {
    public event PropertyChangedEventHandler PropertyChanged;

    private string _FirstName;

    public string FirstName {
        get { return _FirstName; }
        set {
            PropertyChanged(this, new PropertyChangedEventArgs("FirstName"));
            _FirstName = value;
        }
    }
}
```

Code snippet from Customer.cs

Notice how each time you modify the implementation, you are doing so by writing the minimum amount of code necessary to get the current test cases to pass. This doesn't immediately lead to the best code, but it forces us to write test cases that cover all the scenarios. If you look at this implementation of the `FirstName` property, you'll see that there are potential issues with it:

- The `PropertyChanged` event is raised before setting the backing variable, meaning that if the event handler reads the property's value, it will obtain the old value, and not the newly set value.
- There is no check to see if the value of `FirstName` is changing, resulting in the `PropertyChanged` event being raised each time the property is set, even if the new value is the same as the old one.
- There is no check to make sure the `PropertyChanged` event isn't `null`, indicating that no one is currently listening for the event.

You can write individual test cases for each of these issues and progressively improve the `Customer` implementation. An advantage of this technique is that you will obtain an extensive list of test cases that will help validate that the class behaves correctly. It also has the inverse function of providing pseudo-documentation for how the class is expected to be used. If someone who hasn't seen the class wants to understand how it behaves, without looking at the code for the class (it might be in a compiled assembly that they can't reverse engineer), they can deduce how to interact with the class via the various unit tests that describe its correct behavior.

Test Attributes

There are several other attributes that you can use within your test cases to make the test reporting more descriptive and to enable more advanced functionality such as applying filtering to only run a subset of the tests.

Description

It is recommended that you name each test method according to the functionality that it is designed to test. That said, quite often it can be difficult to include enough information in the name to make it sufficiently clear what the test does. The `Description` attribute can be applied to a method to provide additional information:

Available for download on Wrox.com

```
[TestMethod]
[Description("Sample test that should PASS")]
public void ShouldPass() {
    Assert.IsTrue(true);
}
```

Code snippet from MyFirstTests.cs

Bug

The `Bug` attribute can be used to provide information as to why the test case was added. For example, in the following code, a bug was detected where the property value wasn't being set before the `PropertyChanged` event was fired. The code also indicates that the bug has been fixed, so currently the test case is expected to pass.

Available for download on Wrox.com

```
[TestMethod]
[Bug("Property value wasn't set when PropertyChanged event fired", Fixed = true)]
public void PropertyChangedEvent_FirstName_PropertyValueNotSet_Test() {
    var customer = new Customer();
    var eventRaised = false;
    customer.PropertyChanged += (s, e) => {
            eventRaised = customer.FirstName == "Test Name";
        };
    customer.FirstName = "Test Name";
    Assert.IsTrue(eventRaised);
}
```

Code snippet from CustomerTests.cs

WorkItem

For teams that use a work item tracking system such as Team Foundation Server (TFS), it can be useful to link each unit test with the work item that covered its creation. This can be done using the `WorkItem` attribute. This takes an integer value that can be used to correlate the test case with a work item:

Available for download on Wrox.com

```
[TestMethod]
[WorkItem(287)]
public void PropertyChangedEvent_FirstName_EventOnlyRaisedWhenValueChanges_Test() {
    var customer = new Customer();
    var eventRaised = 0;
    customer.PropertyChanged += (s, e) => {
            eventRaised++;
        };
    customer.FirstName = "Test Name";
    customer.FirstName = "Test Name";
    customer.FirstName = "Different Name";
    Assert.AreEqual(2, eventRaised,
        "PropertyChanged event should only be raised when the FirstName changes");
}
```

Code snippet from CustomerTests.cs

Tag

Sometimes you may only want to run a subset of the test cases. As your unit test library continues to grow, for example, it may take a significantly long period of time to execute the full set. You can annotate test classes and methods with one or more `Tag` attributes to allow you to group test methods. When a `Tag` attribute is applied to a class, it is the equivalent to applying that `Tag` to all test methods contained within that class. Figure 17-8 illustrates how you can specify which tests to run based on their `Tag`. In this case, all test methods with the "All Samples" tag will be run.

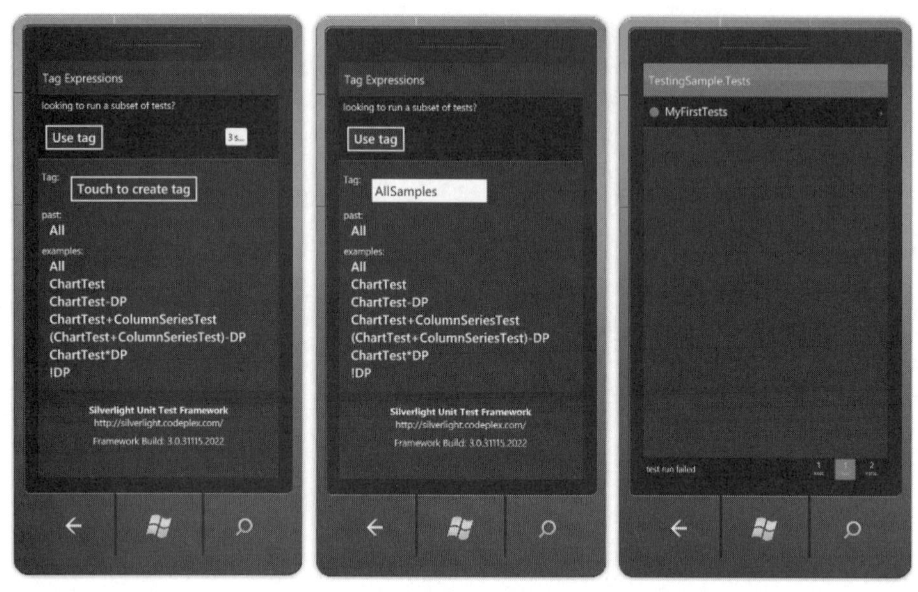

FIGURE 17-8

The interface for specifying the `Tag` values can accept relatively complex expressions. For example, "AllSamples – FailSample" would run test methods that have a `Tag` value of `AllSamples` but not those with a `Tag` value of `FailSample`. From the following code, only the `ShouldPass` test method would be run:

Available for download on Wrox.com

```
[TestClass]
[Tag("AllSamples")]
public class MyFirstTests {
    [TestMethod]
    public void ShouldPass() { ... }

    [TestMethod]
    [Tag("FailSample")]
    public void ShouldFail() { ... }
}
```

Code snippet from MyFirstTests.cs

Ignore

If, for whatever reason, you want one or more test methods to be skipped, you can apply the `Ignore` attribute to either a test class or individual test methods. This can be useful while external reasons, such as the unavailability of a service, are temporarily causing a unit test to fail.

ExpectedException

Sometimes the code you are testing may throw exceptions to indicate specific error conditions. It can be useful to have unit tests that verify that your application behaves correctly in the face of errors. Rather than using a try-catch block within your test method to verify that the correct exception has been thrown, you can annotate the method with the `ExpectedException` attribute. This asserts that the correct execution of the unit test should result in an exception of the specified type being thrown. If you specify a message, as in the following example, the message in the exception is matched against the message supplied:

```
[TestMethod]
[ExpectedException(typeof(ArgumentNullException), "Value can't be null")]
public void FirstName_NullValueGeneratesException_Test()
{
    var customer = new Customer();
    customer.FirstName = null;
}
```

Code snippet from CustomerTests.cs

Asynchronous

A lot of functionality within your Windows Phone application, particularly functionality requiring content to be downloaded, will be done asynchronously to prevent blocking the UI. Traditionally, asynchronous code can make testing your application quite difficult. Within the SUTF framework the `Asynchronous` attribute can be applied to a test method to allow such functionality to be easily tested. Take the following example, in which the `UpdateAsync` method takes approximately 2,000 milliseconds(ms) to complete (simulated here with a call to `Thread.Sleep`):

```
public event EventHandler UpdateComplete;
public bool IsChanged { get; private set; }

public void UpdateAsync() {
    ThreadPool.QueueUserWorkItem(InternalUpdate);
}

private void InternalUpdate(object state) {
    Thread.Sleep(2000);
    IsChanged = false;
    if (UpdateComplete != null) {
        UpdateComplete(this, EventArgs.Empty);
    }
}
```

Code snippet from CustomerTests.cs

To test this method, we can queue up a set of actions that will be performed asynchronously (this requires the `Asynchronous` attribute to be applied to the test method). We start by calling the `UpdateAsync` method to kick off the `asynchronous` method. Then we specify a conditional statement that causes the work flow to suspend until the `customerUpdated` flag is set (this is set to `true` in the `UpdateComplete` event handler). We then queue up the `Assert` to validate that the `IsChanged` property on the `Customer` is `false`. Lastly, we call `EnqueueTestComplete` to mark the end of the `asynchronous` test method.

Available for download on Wrox.com

```
[TestMethod]
[Asynchronous]
public void Update_Test()
{
    var customer = new Customer();
    var customerUpdated = false;
    customer.UpdateComplete += (s, e) =>
    {
        customerUpdated = true;
    };

    EnqueueCallback(() => customer.UpdateAsync());
    EnqueueConditional(() => customerUpdated);
    EnqueueCallback(() => Assert.IsFalse(customer.IsChanged));
    EnqueueTestComplete();
}
```

Code snippet from CustomerTests.cs

Timeout

Sometimes a unit test will fail and get stuck within an infinite loop or similar state that causes an excessive amount of time to be spent executing a test. This is especially true when using the `Asynchronous` attribute where a test case could lock indefinitely. (One possible scenario is for an `asynchronous` method to never cause a conditional statement enqueued via the `EnqueueConditional` method to become `true`.) To avoid breaking your test run, you can use the `Timeout` attribute to provide an upper time limit on the duration of a test method. The following limits the Update_Test unit test to a maximum of 4,000ms. If the unit test takes longer than 4,000ms, it is considered a failure, and execution moves on to the next unit test.

Available for download on Wrox.com

```
[TestMethod]
[Asynchronous]
[Timeout(4000)]
public void Update_Test() { ... }
```

Code snippet from CustomerTests.cs

TestInitialize and TestCleanup

Each time a test is run, you may want to establish an initial known state, or you may want to set up services that are going to be called by the test method. This can be done in a method marked with the `TestInitialize` attribute. Similarly, at the end of each test, a method with the `TestCleanup` attribute will be invoked to perform any cleanup activities. The methods with the `TestInitialize` and `TestCleanup` attributes will automatically be called before and after each test method located in the class they are contained within.

ClassInitialize and ClassCleanup

You may find that some states and services only need to be set up (and subsequently cleaned up) once all tests within a test class have been executed. In this case, instead of putting the setup (or teardown) code in a method with the `TestInitialize` (or `TestCleanup`) attribute, you can mark the methods with the `ClassInitialize` (or `ClassCleanup`) attribute.

There are also corresponding attributes that can be applied across the test assembly. The `AssemblyInitialize` and `AssemblyCleanup` attributes can be applied to methods that you want run before and after all tests within the same assembly.

Visual Studio Test Report

Although the visual output of SUTF is great during development and debugging, it isn't particularly good for regression or automated testing. It is hard to compare two or more test runs against each other to look for regressions or improvements. Luckily, SUTF is extensible and allows you to customize both the report format and where the report is written. Here we're going to look at two of these extension points in order to output a TRX file, which is the same report format that MS Test generates when you run a unit test within Visual Studio.

The first point where you can extend SUTF is by defining which log provider (or providers) the test information is written to. A log provider can be any class that implements the `LogProvider` interface and determines which test information is reported and the format it is written in. The default configuration of SUTF actually defines two log providers, one to write to the Visual Studio output window and another that writes test information to an XML document. The latter provider is useful as it generates an XML document that can easily be retrieved off the device and integrated into a test automation process.

The good news is that you typically don't need to implement your own `LogProvider`. Instead, you can rely on the default `VisualStudioLogProvider` to generate output in the TRX file format. However, what you do need to create is a `TestReportingProvider`. This provider is what writes the formatted test report information to a file or sends the output to an external report service.

One option would be to write the test report to Isolated Storage. However, the challenge then would be how to retrieve the report from Isolated Storage from a desktop machine. Although there is an automation library for working with the emulator, it doesn't permit the direct reading or writing to the Isolated Storage folder of an application.

The approach we're going to take is to send the generated report to a WCF Service. Let's start by creating a new application, called *TestReportRetriever*, based on the Windows ➪ Console Application template.

For this next section, you will need to run Visual Studio in Administrator mode. The WCF Service we're going to create will need to be able to register itself to receive requests on a custom port on the local computer. This task requires administrative rights. To start Visual Studio as an Administrator, right-click the Microsoft Visual Studio 2010 icon in the Start menu, and select "Run as Administrator."

In your console application, add a new item, called *TestReportService*, based on the WCF Service item template. This will actually generate three files: ITestReportService.cs, TestReportService.cs, and app.config. The service declaration (in ITestReportService.cs) will contain a single method,

`SaveReportFile`, that accepts two parameters, being the name of the log file and the contents generated by the test framework.

Available for download on Wrox.com

```
[ServiceContract]
public interface ITestReportService {
    [OperationContract]
    [WebInvoke(BodyStyle = WebMessageBodyStyle.Wrapped)]
    void SaveReportFile(string logName, string content);
}
```

Code snippet from ITestReportService.cs

You may get some build errors at this point because of some missing assembly references. On the Application tab of the project Properties page (right-click the console project in Solution Explorer and select Properties), make sure that the Target Framework is set to ".NET Framework 4" (not the Client Profile). Then add references to System.ServiceModel and System.ServiceModel.Web. The next thing to do is to update the service implementation in TestReportService.cs:

Available for download on Wrox.com

```
public class TestReportService : ITestReportService {
    public void SaveReportFile(string logName, string content) {
        var outputfile = Path.Combine(Properties.Settings.Default.BaseOutputPath,
                                      logName);
        if (File.Exists(outputfile)) {
            File.Delete(outputfile);
        }
        File.WriteAllText(outputfile, content);
        Console.WriteLine(content);
    }
}
```

Code snippet from TestReportService.cs

This code relies on an application setting called *BaseOutputPath*. (Go to the Settings tab of the project Properties page, create a new setting called *BaseOutputPath* of type string with scope of Application, and then give it a value that is the path to a folder on the computer.) The test results (i.e., the content argument) are then written to a file within this folder, as well as being written to the Console window.

You also need to update the app.config file to change the binding type to `basicHttpBinding` and increase the limit on the size of the content that can be sent to the service:

Available for download on Wrox.com

```
<system.serviceModel>
    <behaviors>
        <serviceBehaviors>
            <behavior name="">
                <serviceMetadata httpGetEnabled="true" />
                <serviceDebug includeExceptionDetailInFaults="true" />
            </behavior>
        </serviceBehaviors>
    </behaviors>
    <services>
        <service name="TestReportRetriever.TestReportService">
```

```
            <endpoint address="" binding="basicHttpBinding"
                      bindingConfiguration="reportBinding"
                      contract="TestReportRetriever.ITestReportService">
            </endpoint>
            <endpoint address="mex" binding="mexHttpBinding"
                      contract="IMetadataExchange" />
            <host>
                <baseAddresses>
                    <add baseAddress="http://localhost:8732/TestReportService/" />
                </baseAddresses>
            </host>
        </service>
    </services>
    <bindings>
        <basicHttpBinding>
            <binding name="reportBinding" maxBufferSize="1000000"
                     maxBufferPoolSize="1000000"
                     maxReceivedMessageSize="1000000">
            <readerQuotas maxStringContentLength="1000000"
                          maxArrayLength="1000000" />
            </binding>
        </basicHttpBinding>
    </bindings>
</system.serviceModel>
```

Code snippet from app.config

The last thing to do to complete the Console application is to start the TestReportService service whenever the application is started. Update the code in Program.cs to create and open a ServiceHost using the `TestReportService` type. The console application will wait for the user to press Enter before closing the service and terminating.

Available for download on Wrox.com

```
static void Main(string[] args){
    using (ServiceHost host = new ServiceHost(typeof(TestReportService))){
        host.Open();
        Console.WriteLine("The service is ready");
        Console.ReadLine();
        host.Close();
    }
}
```

Code snippet from Program.cs

In order to add a service reference to the TestingSample.Tests project, you will need to have an instance of the Console application running. Build the Console application, then navigate to the Bin\Debug folder of the Console application in Windows Explorer. Right-click TestReportRetriever.exe, and select "Run as Administrator." This should display a console similar to Figure 17-9, indicating that the service is running.

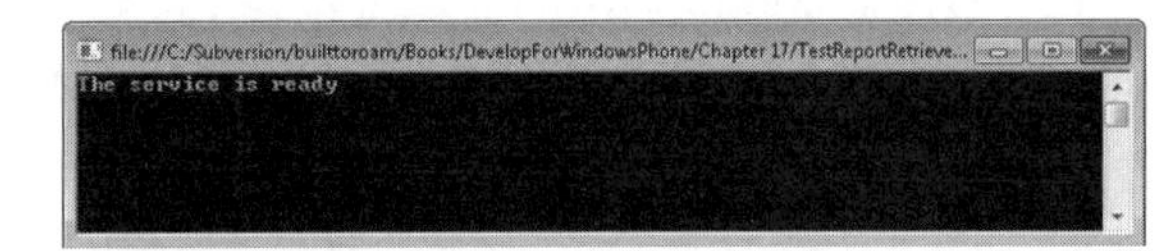

FIGURE 17-9

With the Console application running, right-click the TestingSample.Tests project, and select "Add Service Reference." In the Address field, enter **http://localhost:8732/TestReportService/** (note that this matches the value of the `baseAddress` element in the app.config file for the Console application) and click Go (Figure 17-10). When the service is discovered, give it a name, **Reporting**, and click OK. This will add the necessary proxy classes to your test project for you to be able to call the WCF Service.

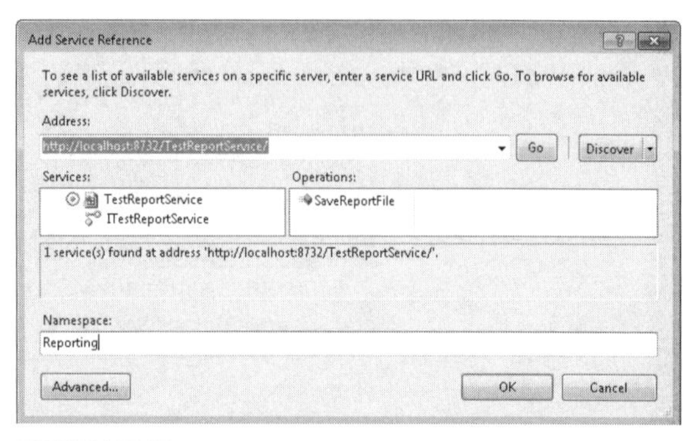

FIGURE 17-10

Next, add a new class called `ServiceReportingProvider` to the TestingSample.Tests project. The `ServiceReportingProvider` class should inherit from `TestReportingProvider`, and you should override the `WriteLog` method to invoke the `SaveReportFileAsync` method on an instance of the `TestReportServiceClient`:

Available for download on Wrox.com

```
public class ServiceReportingProvider : TestReportingProvider {
    TestReportServiceClient reportClient = new TestReportServiceClient();

    public ServiceReportingProvider(TestServiceProvider testService)
        : base(testService) { }

    public override void WriteLog(Action<ServiceResult> callback,
                                  string logName, string content) {
        this.IncrementBusyServiceCounter();
        reportClient.SaveReportFileCompleted += (s, e) => {
                this.DecrementBusyServiceCounter();
            };
        reportClient.SaveReportFileAsync(logName, content);

        base.WriteLog(callback, logName, content);
    }
}
```

Code snippet from ServiceReportingProvider.cs

The `SaveReportFileAsync` method is wrapped by calls to `IncrementBusyServiceCounter` and `DecrementBusyServiceCounter`. This keeps track of asynchronous calls to ensure that they are done in the correct order.

To configure SUTF to make use of the `ServiceReportingProvider` class to submit test reports, you need to modify the constructor for the `MainPage` in the TestingSample.Tests project. In this case, we start with the default settings for SUTF and register a new instance of the `ServiceReportingProvider` as the TestReporting service. These settings are then used when initializing SUTF by calling `CreateTestPage`.

```
public MainPage() {
    InitializeComponent();

    var settings = UnitTestSystem.CreateDefaultSettings();
    settings.TestService.RegisterService(TestServiceFeature.TestReporting,
                         new ServiceReportingProvider(settings.TestService));
    Content = UnitTestSystem.CreateTestPage(settings);
    IMobileTestPage imtp = Content as IMobileTestPage;
    if (imtp != null)
    {
        BackKeyPress += (x, xe) => xe.Cancel = imtp.NavigateBack();
    }
}
```

Code snippet from MainPage.xaml.cs

Configure Visual Studio to run both the TestingSample.Tests and TestReportRetriever projects by right-clicking the solution node, in Solution Explorer, and selecting “Set Startup Projects.” In the “Solution Properties Pages” dialog that is displayed, make sure the Common Properties ➪ Startup Project node is selected, and then select “Multiple startup projects” from the right-hand pane. In the list of projects, find the two projects and set the Action to Start. Also make sure that the TestReportRetriever project is started before the TestingSample.Tests by moving either project up or down in the Projects list. When you’re done, click OK to accept changes. You’re now ready to run the solution — this will start up the TestReportRetrieve project to receive the output of the tests, followed by the TestingSample.Tests project to run the tests within the Windows Phone 7 emulator. Figure 17-11 illustrates the content that has been sent to the `SaveReportFile` method in the TestReportRetriever Console application.

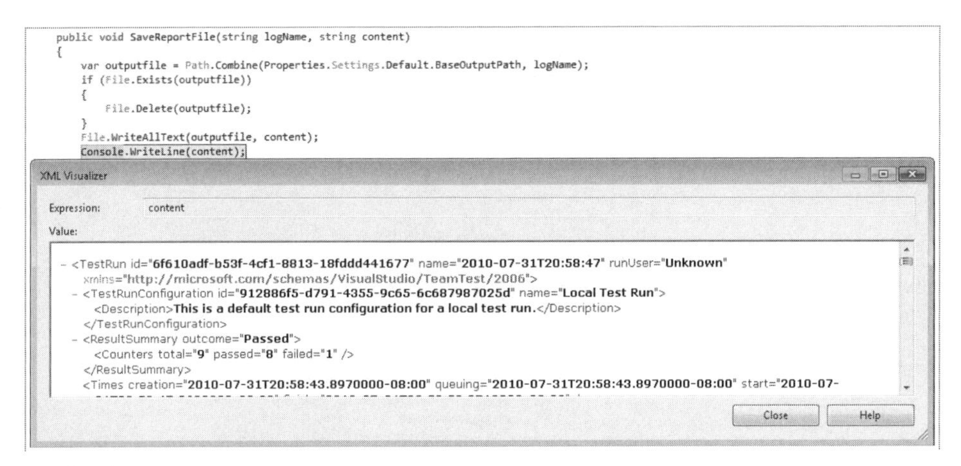

FIGURE 17-11

Figure 17-11 uses the XML Visualizer instead of the normal Text Visualizer to display the `content` *variable. When the application is paused and you hover over the* `content` *variable, you will see the debugging tooltip appear. If you click the Down button next to the magnifying glass in the tooltip, you will see a list of available visualizers. In this case, the XML Visualizer has been selected because it provides the most readable format for the test report XML.*

Emulator Automation

In the previous section, you saw how to write and execute test cases, as well as how to send the test report to an external WCF Service so that the report could be analyzed further. However, there is still a lot of manual intervention that is required in order to launch both the TestingSample.Tests application and the TestReportRetriever. This process can be automated further so that the emulator is launched (if necessary), the TestingSample.Tests application is installed and executed, and the test results are gathered, before the TestingSample.Tests application is uninstalled, returning the emulator to a clean state. Full automation of this process means that it can be integrated into a continuous build process allowing for regression testing to be performed.

To achieve this, we are going to extend the TestReportRetriever console application to connect to the emulator and install the TestingSample.Tests application. The first thing to do is to add a reference to the `Microsoft.SmartDevice.Connectivity` assembly within the TestReportRetriever application (you will have to browse to C:\Program Files\Common Files\microsoft shared\Phone Tools\CoreCon\10.0\Bin\Microsoft.Smartdevice.Connectivity.dll in order to locate this assembly). Installing and uninstalling an application on the emulator can be done via an instance of the `Device` class, as shown in the following code:

Available for download on Wrox.com

```
internal static void InstallAndLaunchApplication(this Device device,
                                           Guid applicationGuid,
                                           string iconFileName, string xapFileName) {
    CleanupApplication(device, applicationGuid);
    device.InstallApplication(applicationGuid, applicationGuid,
                              "NormalApp", iconFileName, xapFileName);
    var app = device.GetApplication(applicationGuid);
    app.Launch();
}

internal static void CleanupApplication(this Device device, Guid applicationGuid) {
    if (device.IsApplicationInstalled(applicationGuid)) {
        RemoteApplication application = device.GetApplication(applicationGuid);
        application.TerminateRunningInstances();
        application.Uninstall();
    }
}
```

Code snippet from Utilities.cs

The `applicationGuid` that needs to be supplied to both methods corresponds to the `ProductID` field found in the WMAppManifest.xml file for the application you want to install:

Available for download on Wrox.com

```
<?xml version="1.0" encoding="utf-8"?>
<Deployment xmlns="http://schemas.microsoft.com/windowsphone/2009/deployment"
            AppPlatformVersion="7.0">
    <App xmlns="" ProductID="{fcaf2b94-78de-417d-a41a-8aba3e7ea2e1}"
         Title="TestingSample.Tests" RuntimeType="Silverlight"
         Version="1.0.0.0" Genre="apps.normal"
         Author="TestingSample.Tests author"
         Description="Sample description" Publisher="TestingSample.Tests">
    ...
```

Code snippet from WMAppManifest.xml

In order to execute the `InstallAndLaunchApplication` method, you must first obtain an instance of the `Device` class. Within the `Main` method of the `Program` class, you can do this by first creating an instance of the `DatastoreManager`, which, in turn, is used to get a list of supported platforms and subsequently a `Device` instance:

Available for download on Wrox.com

```
DatastoreManager dsmgrObj = new DatastoreManager(1033);
Platform WP7SDK = dsmgrObj.GetPlatforms().Single(p => p.Name == "Windows Phone 7");
Device device = WP7SDK.GetDevices().Single(d => d.Name == "Windows Phone 7 Emulator");
```

Code snippet from Program.cs

Once the TestReportService has been started, you'll then connect to the emulator, install the application, and finally launch it.

Available for download on Wrox.com

```
try{
    device.Connect();
    device.InstallAndLaunchApplication(applicationGuid, iconFile, xapFile);
}
finally{
    device.Disconnect();
}
```

Code snippet from Program.cs

If you were to run the updated TestReportReceiver application as it currently stands, the emulator would be left in a state where the TestingSample.Test application was still installed and potentially still running. After the application has been launched, you want to wait until the tests have completed. To do this, you'll add an `AutoResetEvent` to the `Program` class and call `WaitOne` immediately after the `InstallApplication` method call. This will cause the console application to pause until signaled to continue. The full code for the `Program` class is as follows:

Available for download on Wrox.com

```
class Program{
    static AutoResetEvent endTestRunEvent = new AutoResetEvent(false);
    static void Main(string[] args) {

        // Constants defining the ProjectID and paths to the icon and xap file
```

```
            // to use when installing the application
            Guid applicationGuid = new Guid("fcaf2b94-78de-417d-a41a-8aba3e7ea2e1");
            string xapFile =
                 @"..\..\..\TestingSample.Tests\Bin\Debug\TestingSample.Tests.xap";
            string iconFile = @"..\..\..\TestingSample.Tests\ApplicationIcon.png";

            DatastoreManager dsmgrObj = new DatastoreManager(1033);
            Platform WP7SDK = dsmgrObj.GetPlatforms()
                                      .Single(p => p.Name == "Windows Phone 7");

            Device device = WP7SDK.GetDevices()
                                  .Single(d => d.Name == "Windows Phone 7 Emulator");

            using (ServiceHost host = new ServiceHost(typeof(TestReportService))) {
                host.Open();
                Console.WriteLine("The service is ready");

                Console.WriteLine("Connecting to Windows Phone 7 Emulator/Device...");
                Try {
                    device.Connect();
                    Console.WriteLine("Windows Phone 7 Emulator/Device Connected...");
                    device.InstallAndLaunchApplication(applicationGuid, iconFile, xapFile);
                    Console.WriteLine("Waiting for test run to complete");
                    endTestRunEvent.WaitOne(
                              Properties.Settings.Default.TestRunTimeoutInSeconds * 1000);
                    device.CleanupApplication(applicationGuid);
                }
                finally {
                    device.Disconnect();
                }
                host.Close();
            }
        }

        public static void EndTestRun() {
            endTestRunEvent.Set();
        }
    }
```

Code snippet from Program.cs

With this version of the TestReportRetriever application, the application doesn't wait for a user to press Enter in order for it to exit; instead, it will start the emulator and install the TestingSample. Tests application before waiting for `endTestRunEvent` to be set. This is done via the static `EndTestRun` method. All you need to do now is to modify the TestReportService to call `EndTestRun` when it has received the test report from SUTF.

The call to `endTestRunEvent.WaitOne` *references the TestRunTimeoutInSeconds application setting. This is to ensure that the console application does not become stuck and will eventually terminate regardless of whether SUTF has been able to successfully run all of the test cases. Of course, as your testing matrix grows, so too will the expected runtime of your unit test suite; make sure that you update the TestRunTimeoutInSeconds application setting to give a reasonable time for all tests to complete.*

Rather than simply terminating the TestReportRetriever application when the first test report is received via the `SaveReportFile` method of the TestReportService, we'll create a second method on the WCF Service called `ReportFinalResult`:

Available for download on Wrox.com

```
[ServiceContract]
public interface ITestReportService
{
    [OperationContract]
    [WebInvoke(BodyStyle = WebMessageBodyStyle.Wrapped)]
    void SaveReportFile(string logName, string content);

    [OperationContract]
    [WebInvoke(BodyStyle = WebMessageBodyStyle.Wrapped)]
    void ReportFinalResult(bool failure, int failures, int totalScenarios,
                           string message);
}
```

Code snippet from ITestReportService.cs

The implementation of this second method will simply call the `EndTestRun` method on the `Program` class. This will, in turn, set the `endTestRunEvent`, allowing the `TestReportRetriever` application to uninstall the application from the emulator and terminate gracefully:

Available for download on Wrox.com

```
public void ReportFinalResult(bool failure, int failures, int totalScenarios,
                              string message) {
    Program.EndTestRun();
}
```

Code snippet from TestReportService.cs

Now all that's left to do is to update the `ServiceReportingProvider` so that at the end of a complete test run it calls the new WCF Service method. Make sure that you update the service reference (run the TestReportRetriever application, and then, while it's still running, right-click the TestingSample.Tests ➪ Service References ➪ Reporting node (in Solution Explorer) and select the Update Service Reference option; then override the `ReportFinalResult` method within the `ServiceReportingProvider` class.

```
public override void ReportFinalResult(Action<ServiceResult> callback, bool failure,
                                       int failures, int totalScenarios,
                                       string message) {
    this.IncrementBusyServiceCounter();
    reportClient.SaveReportFileCompleted += (s, e) => {
            this.DecrementBusyServiceCounter();
        };

    reportClient.ReportFinalResultAsync(failure, failures, totalScenarios, message);
    base.ReportFinalResult(callback, failure, failures, totalScenarios, message);
}
```

Code snippet from ServiceReportingProvider.cs

As the TestReportRetriever console application will now launch the TestingSample.Tests project within the emulator, you can set the TestReportRetriever project to be the only startup project (right-click the project node in Solution Explorer and select "Set as Startup Project"). When you run the TestReportRetriever application, you will notice that the console is displayed, followed by the emulator (if it wasn't already running); then you will see the TestingSample.Tests application installed and run. Once the report has been received, the application will automatically be uninstalled, and the console application will exit. The test report generated can then be found within the folder you specified via the BaseOutputPath setting of the TestReportRetriever project.

The completed TestReportRetriever console application can be added as a step within your build process to automatically generate test reports after each check in or on a periodic basis to ensure that you are not introducing defects during the development phase of your Windows Phone application.

SUMMARY

In this chapter, you've seen several frameworks that you can use to assist you in building componentized and extensible Windows Phone applications. The use of a testing framework such as the Silverlight Unit Testing Framework can significantly improve the reliability of your code and can assist with the early identification and isolation of bugs introduced during the development of your application.

18

Security

WHAT'S IN THIS CHAPTER

- Understanding device and data security
- Using encryption to protect data on the device
- Authenticating the current user
- How to integrate with online platforms such as Twitter and Facebook

Increasingly, smart phones have become more like a mini-computer in your pocket rather than a simple device that can only make and receive phone calls. This movement has brought with it some major security challenges. As mobile phones travel with us, there is a much higher risk of them being stolen or simply misplaced. Their increasing storage and computing power means that they can also contain a large amount of personal information about ourselves, our friends, and our work. There is also the usual conflict between convenience and data security. You want the information on your phone to be readily accessible, yet you also want it to be protected. In this chapter, you'll learn how you can protect Windows Phone application data both on the device and on the wire. You'll also see how you can use several different authentication techniques to access remote data in a secure fashion.

ON THE DEVICE

There are a couple of aspects to protecting data that resides on the device. These include securing access to the device, being able to clear or reset the device in case it is lost, and securing the data in a format that can't easily be compromised.

Device Security

The best way to ensure that the data on your phone is secure is to protect the device itself from unwanted attention by making sure it is with you at all times. Clearly, there are times when this isn't possible, especially if the device gets stolen. In these cases, you need to ensure that there is adequate security to protect the data on the device.

The first step in protecting the data on the device is setting a password on the Lock screen. When the device hasn't been used for a specific period of time or the Power button is pressed, the Lock screen is displayed. If a password hasn't been set, the Lock screen is dismissed by scrolling it upward with a simple flick. Alternatively, if a password is required, the user will be prompted to enter it at this point; thus, this acts as the first line of defense. Figure 18-1 shows the sequence of steps (from left to right) to add a password and adjust the Lock screen time-out.

FIGURE 18-1

From the Settings screen, select "lock & wallpaper." This screen allows you to toggle whether a password is required when the Lock screen is dismissed.

Device Management

If, for whatever reason, your phone gets misplaced or stolen, the Windows Phone Live portal (`http://windowsphone.live.com`) allows you to locate, ring, lock, and even erase data stored on the device, as illustrated in Figure 18-2.

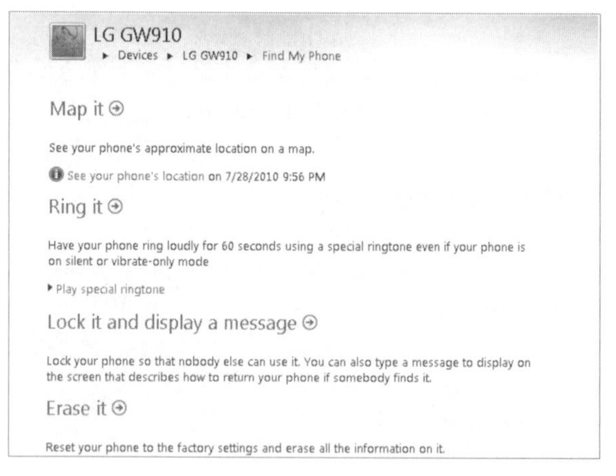

FIGURE 18-2

To remotely lock a device, navigate to the portal and sign in with your Windows Live ID. If you have multiple Windows Phones, you may have to select the device you want to manage. After selecting the device, select the "Lock it and display a message" option. This will ensure that wherever your device is, whether it has been misplaced or stolen, your personal data will be safe. You need to specify a 4-digit password that will need to be re-entered on the actual device in order to continue using it, as well as an optional message, as shown in Figure 18-3. If you suspect that you've misplaced the device, having it ring will allow you to locate it.

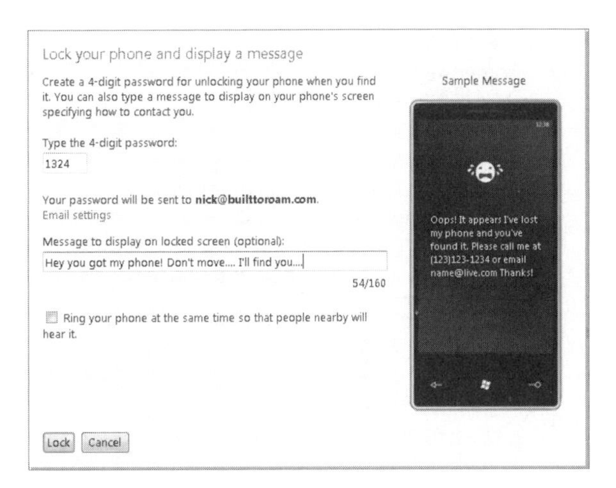

FIGURE 18-3

Once you're ready, you can click the Lock button and the device will be locked as soon as it can be contactable.

The next thing to do is to try to locate the device using the "Map it" function. Figure 18-4 illustrates a map showing the location of the phone. This will either trigger a memory as to where you might have left it, or it can be used by the police to track down the unsuspecting thief.

FIGURE 18-4

You can use the "Ring it" function from the portal without having to lock the device. This is useful if you can't locate your phone but you think it is nearby. The last option, "Erase it," can be used to remove all data from the phone. You should use this as a last resort. Once you have used this option, you will have to set up accounts, install applications, and sync content to your Windows Phone if you manage to re-locate the device. This feature, however, will ensure that you are able to remove access to any sensitive data stored on the device, even after it is in the hands of a thief.

Data Encryption

Let's assume that someone has been able to steal your Windows Phone and you haven't been able to lock the device or erase the information from it. You can still protect the data within your application by encrypting it. The following code uses the Advanced Encryption Standard (AES) to encrypt a string of data. There are a couple of points to note about using this algorithm. First, this uses a salt that is known only to this application on this device, which makes it much harder for someone to do a dictionary attack (in which they use a dictionary of pre-encrypted dictionary entries) to decrypt your data.

```
private const string SaltKey = "EncryptionSalt";
private static string Salt;
static EncryptionHelper(){
    if (!IsolatedStorageSettings.ApplicationSettings.TryGetValue(SaltKey, out Salt)){
        var rnd = new Random();
        var charList =
                   "abcdefghijklmnopqrstuvwxyzABCDEFGHIJKLMNOPQRSTUVWXYZ0123456789";
        for (int i = 0; i < 100; i++){
            Salt += charList[rnd.Next(0, charList.Length)];
        }
        IsolatedStorageSettings.ApplicationSettings[SaltKey] = Salt;
    }
}
```

Code snippet from EncryptionHelper.cs

The `Encrypt` method itself requires a password to be entered. This could also be randomly generated by the application and stored in Isolated Storage (like the previous code for generating and storing the salt). However, there is a risk that should the device by compromised, the password, salt, and encrypted data could be extracted. In this case, it would be relatively easy for someone to decrypt the data. For particularly sensitive data you may request that the user enter the password each time he or she wants to access the secured data. This would protect the data so that it couldn't be decrypted without the involvement of the user, but it would come at a cost of convenience as the user would have to enter the password every time.

```
public static string Encrypt(this string data, string password){
    byte[] utfdata = UTF8Encoding.UTF8.GetBytes(data);
    byte[] saltBytes = UTF8Encoding.UTF8.GetBytes(Salt);

    // The encryption algorithm
    using (var aes = new AesManaged()){
        Rfc2898DeriveBytes rfc = new Rfc2898DeriveBytes(password, saltBytes);

        aes.BlockSize = aes.LegalBlockSizes[0].MaxSize;
        aes.KeySize = aes.LegalKeySizes[0].MaxSize;
        aes.Key = rfc.GetBytes(aes.KeySize / 8);
        aes.IV = rfc.GetBytes(aes.BlockSize / 8);

        // Encryptor
        ICryptoTransform encryptTransf = aes.CreateEncryptor();

        // We're going to write the encrypted data to a memory stream
        using (var encryptStream = new MemoryStream())
        using (var encryptor = new CryptoStream(encryptStream,
                                                encryptTransf,
                                                CryptoStreamMode.Write)){
```

```
                encryptor.Write(utfdata, 0, utfdata.Length);
                encryptor.Flush();
                encryptor.Close();

                byte[] encryptBytes = encryptStream.ToArray();
                // Convert to Base64 - not required but good for testing as
                // it's easier to read
                string encryptedString = Convert.ToBase64String(encryptBytes);
                return encryptedString;
            }
        }
    }
```

Code snippet from EncryptionHelper.cs

The `Decrypt` method works in a similar fashion, but in reverse.

```
    public static string Decrypt(this string data, string password){
        byte[] encryptBytes = Convert.FromBase64String(data);
        byte[] saltBytes = Encoding.UTF8.GetBytes(Salt);

        // The encryption algorithm
        using (var aes = new AesManaged()){
            Rfc2898DeriveBytes rfc = new Rfc2898DeriveBytes(password, saltBytes);

            aes.BlockSize = aes.LegalBlockSizes[0].MaxSize;
            aes.KeySize = aes.LegalKeySizes[0].MaxSize;
            aes.Key = rfc.GetBytes(aes.KeySize / 8);
            aes.IV = rfc.GetBytes(aes.BlockSize / 8);

            // Decryptor
            ICryptoTransform decryptTrans = aes.CreateDecryptor();

            // We're going to write the decrypted data to a memory stream
            using (var decryptStream = new MemoryStream())
            using (var decryptor = new CryptoStream(decryptStream,
                                                    decryptTrans,
                                                    CryptoStreamMode.Write)){

                decryptor.Write(encryptBytes, 0, encryptBytes.Length);
                decryptor.Flush();
                decryptor.Close();

                byte[] decryptBytes = decryptStream.ToArray();
                string decryptedString = UTF8Encoding.UTF8.GetString(decryptBytes,
                                                          0, decryptBytes.Length);
                return decryptedString;
            }
        }
    }
```

Code snippet from EncryptionHelper.cs

Both the Encrypt and Decrypt methods are extension methods, meaning that they can be applied in a fluent manner to any string. This is illustrated in the following code that encrypts the text in the SourceText TextBox and then decrypts the text into the EncryptedText TextBox.

Available for download on Wrox.com

```
private const string Password = "SuperStrongPasswordThatOnlyIKnow";

private void EncryptButton_Click(object sender, RoutedEventArgs e)
{
    this.EncryptedText.Text= this.SourceText.Text.Encrypt(Password);
}

private void DecryptButton_Click(object sender, RoutedEventArgs e)
{
    this.DecryptedText.Text = this.EncryptedText.Text.Decrypt(Password);
}
```

Code snippet from MainPage.xaml.cs

Because the AES algorithm is symmetric, the contents of the DecryptedText TextBox should match the text that was originally entered in the SourceText TextBox.

OVER THE WIRE

When data is sent between your Windows Phone application and a remote server, there is always a risk that someone is listening and intercepting packets of data. The challenge is how to protect this data from as many forms of attacks as possible. There are two concerns that are of interest here: the use of encryption to ensure that listeners can't interpret the raw data and the authentication of the application with the remote server to ensure that you can verify with whom you are communicating.

Transport

There are many strategies for protecting data that is being transmitted between two points across the Internet. One option is to use a symmetric encryption method — such as AES, which you read about earlier in this chapter — to encrypt and decrypt data at each end. This requires the sender and receiver to arrange for the secure exchange of salt and password values to allow for the data to be correctly decoded. Another option is to rely on a Secure Socket Layer (SSL) connection between the two points. This is the most commonly used form of transport layer security, wrapping all communications in a layer of encryption where the necessary keys have been mutually agreed on.

From your Windows Phone application the use of SSL can be as easy as using HTTPS instead of HTTP to access a service using the HttpWebRequest or WebClient classes. For example, if you wanted to download a set of data using SSL, you could make a request to https://myremoteserver.com/mydataset. This would automatically invoke a secure channel through which the data can be downloaded.

If you are using WCF, then there is a little more work that you will have to do. On the server, you will need to configure both your WCF Service and IIS to use SSL. The following snippet is an example of the WCF configuration for a service called SSLService. By setting the security mode to "Transport," this WCF Service expects to be called over a secure channel.

```
<system.serviceModel>
    <services>
        <service behaviorConfiguration="SSLServiceBehaviors"
                 name="WebAuthSample.SSLService">
            <endpoint binding="basicHttpBinding"
                      bindingConfiguration="securedWithSSL"
                      contract="WebAuthSample.ISSLService" />
        </service>
    </services>
    <bindings>
        <basicHttpBinding>
            <binding name="securedWithSSL"  >
                <security mode="Transport" />
            </binding>
        </basicHttpBinding>
    </bindings>
    <behaviors>
        <serviceBehaviors>
            <behavior name="SSLServiceBehaviors">
                <serviceDebug includeExceptionDetailInFaults="false" />
            </behavior>
        </serviceBehaviors>
    </behaviors>
</system.serviceModel>
```

When you add this service to your Windows Phone application, you will notice that the security element is also added to the ClientConfig file. For example:

```
<system.serviceModel>
    <bindings>
        <basicHttpBinding>
            <binding name="BasicHttpBinding_ISSLService" maxBufferSize="2147483647"
                maxReceivedMessageSize="2147483647">
                <security mode="Transport" />
            </binding>
        </basicHttpBinding>
    </bindings>
    <client>
        <endpoint address="https://developmentserver/WebAuthSample/SSLService.svc"
                  binding="basicHttpBinding"
                  bindingConfiguration="BasicHttpBinding_ISSLService"
                  contract="SecuredService.ISSLService"
                  name="BasicHttpBinding_ISSLService" />
    </client>
</system.serviceModel>
```

With this configuration information you can invoke a WCF Service over SSL by simply specifying a secure endpoint (for example if it was hosted on IIS the endpoint address would start with HTTPS).

Authentication

If you are responsible for the development of both the Windows Phone application and the remote services it uses, then you will have to decide which method you are going to use to allow the services to authenticate your application, or perhaps more specifically, the user of your application. Alternatively, you may be simply consuming services from a third party, in which case the services may dictate a wide range of different authentication mechanisms. This section will cover some of the most widely used authentication mechanisms and how these can be implemented by your Windows Phone 7 application.

Basic

We'll start with HTTP Basic Authentication, which relies on a simple username and password combination. Basic Authentication is enabled on IIS via the Internet Information Services (IIS) Manager console (Start ➪ Control Panel ➪ Administrative Tools IIS Manager), as shown in Figure 18-5.

FIGURE 18-5

Select the virtual directory, or in this case the BasicAuth subfolder, and select Authentication from the Feature View pane. This will display the list of available authentication methods. In this case, Anonymous Authentication has been disabled and Basic Authentication has been enabled. If you now attempt to browse to a file within this folder, you will be prompted for a username and password.

IIS AUTHENTICATION TECHNIQUES

On some installations of IIS, Basic Authentication isn't installed by default. If this is the case, you can add it either by downloading and running the Web Platform Installer (`www.microsoft.com/web/downloads/platform.aspx`) or by going to Start ➪ Control Panel ➪ Programs and Features ➪ Windows Features.

The Web Platform Installer is a convenient way to install and configure the tools you need for building and deploying websites. The different authentication options for IIS can be configured by selecting the Web Platform ➪ Web Server (customize) menu item and scrolling down the list of features to the Security section. Check the box next to the appropriate authentication methods you want enabled.

In the "Windows Features" dialog, you can again control which authentication methods are available by expanding the tree to Internet Information Services ➪ World Wide Web Services ➪ Security. Again, check the box next to the authentication methods you want enabled.

From your Windows Phone application, calling a service that uses Basic Authentication is just a matter of setting the `Credentials` property on the `HttpWebRequest` object (this property also exists on the `WebClient` class), as in the following example:

```
private void BasicAuthButton_Click(object sender, RoutedEventArgs e){
    var request = HttpWebRequest.Create("http://localhost" +
                                         "/AuthSample/BasicAuth/BasicAuthText.txt");
    request.Credentials = new NetworkCredential("<username>", "<password>");
    request.BeginGetResponse(FileComplete, request);
}

void FileComplete(IAsyncResult result){
    var request = result.AsyncState as HttpWebRequest;
    var response = request.EndGetResponse(result);
    using(var strm = response.GetResponseStream())
    using (var reader = new StreamReader(strm)){
        var txt = reader.ReadToEnd();
        ...
    }
}
```

Code snippet from MainPage.xaml.cs

If you are consuming a WCF Service, then there are a couple of more things you need to do. In addition to enabling Basic Authentication on IIS, you also need to configure the WCF Service to expect the use of Basic Authentication. This is done by specifying the security element on the `basicHttpBinding`, setting the `mode` to "TransportCredentialOnly" and setting the `clientCredentialType` to "Basic."

```
<system.serviceModel>
    <services>
        <service behaviorConfiguration="BasicAuthBehavior"
                 name="AuthSample.BasicAuth.BasicAuthService">
            <endpoint binding="basicHttpBinding" bindingConfiguration="basicAuth"
                      contract="AuthSample.BasicAuth.IBasicAuthService" />
            <!--<endpoint address="mex" binding="mexHttpBinding"
                          contract="IMetadataExchange" />-->
        </service>
    </services>
    <bindings>
        <basicHttpBinding>
            <binding name="basicAuth">
                <security mode="TransportCredentialOnly">
                    <transport clientCredentialType="Basic" />
                </security>
            </binding>
        </basicHttpBinding>
    </bindings>
    <behaviors>
        <serviceBehaviors>
            <behavior name="BasicAuthBehavior">
                <serviceMetadata httpGetEnabled="true" />
```

```
                <serviceDebug includeExceptionDetailInFaults="true" />
            </behavior>
        </serviceBehaviors>
    </behaviors>
</system.serviceModel>
```

Code snippet from Web.config

When you add a reference to this service within your Windows Phone application, the ClientConfig file will automatically include the correct security node:

Available for download on Wrox.com

```
<system.serviceModel>
    <bindings>
        <basicHttpBinding>
            <binding name="BasicHttpBinding_IBasicAuthService"
                maxBufferSize="2147483647"
                maxReceivedMessageSize="2147483647">
                <security mode="TransportCredentialOnly" />
            </binding>
        </basicHttpBinding>
    </bindings>
    <client>
        <endpoint address="http://localhost/AuthSample/BasicAuth/BasicAuthService.svc"
            binding="basicHttpBinding"
            bindingConfiguration="BasicHttpBinding_IBasicAuthService"
            contract="BasicAuthSample.IBasicAuthService"
            name="BasicHttpBinding_IBasicAuthService" />
    </client>
</system.serviceModel>
```

Code snippet from ServiceReference.ClientConfig

If you disable Anonymous Authentication on the IIS directory where the WCF Service is contained, you must also remove the mex end point from the web.config file (in the example this element was commented out), otherwise the service will fail to start properly. This will cause an issue if you then attempt to use the "Add Service Reference" feature of Visual Studio. The work-around is to temporarily enable Anonymous Authentication while you add the service reference to your Windows Phone application. The security settings will still be set up correctly in the ClientConfig file since the WCF Service is still expecting to be called using Basic Authentication.

In code, all you need to do when calling a service that makes use of Basic Authentication is set the `UserName` and `Password` properties on the `ClientCredentials.UserName` object. The service can then be called as normal:

Available for download on Wrox.com

```
private void BasicCallServiceButton_Click(object sender, RoutedEventArgs e){
    var client = new BasicAuthSample.BasicAuthServiceClient();
    client.ClientCredentials.UserName.UserName = "<username>";
    client.ClientCredentials.UserName.Password = "<password>";
    client.DoWorkCompleted += BasicService_DoWorkCompleted;
    client.DoWorkAsync();
}

void BasicService_DoWorkCompleted(object sender, AsyncCompletedEventArgs e){...}
```

Code snippet from MainPage.xaml.cs

It's important to remember that with Basic Authentication the username and password are effectively sent in plaintext. This means that you should consider using this over SSL.

Forms (WCF Authentication Services)

An alternative to using Basic Authentication is to use Forms-based Authentication. This is the default option if you create a new web application using the ASP.NET Web Application project template in Visual Studio. For an existing ASP.NET Web Application, you can use Forms Authentication by enabling Forms Authentication in IIS and then adding the following element to the web.config file within your project:

Available for download on Wrox.com

```
<system.web>
    <authentication mode="Forms">
        <forms loginUrl="~/Account/Login.aspx" timeout="2880" />
    </authentication>
</system.web>
```

Code snippet from web.config

In this case, if a user isn't currently authenticated, the user will be directed to the Account/Login.aspx page in order to enter his or her credentials. Of course, you will need to configure the website so that the login page is accessible to unauthenticated users.

Forms Authentication works by issuing a token to users once they've logged in that is persisted as a cookie. Each time a request is made for a web page, the cookie is validated to ensure that the user is logged in. If this fails, the user is redirected to the `loginUrl` defined in the web.config file.

To use Forms Authentication to secure a WCF Service, you first have to configure the security for the folder that contains the WCF Service. Click the "ASP.NET Configuration" button from the toolbar on the Solution Explorer, shown in Figure 18-6.

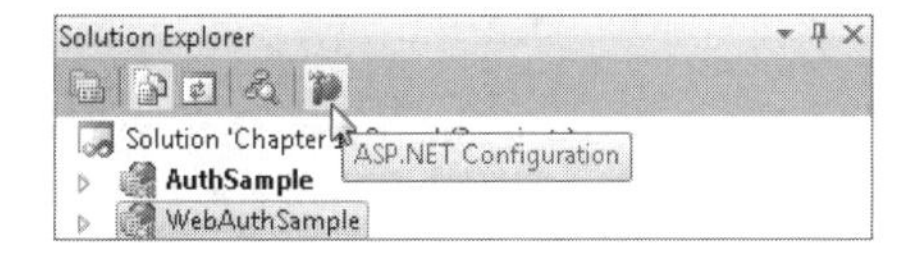

FIGURE 18-6

This will display the ASP.NET Web Application Administration portal for the project, shown in Figure 18-7.

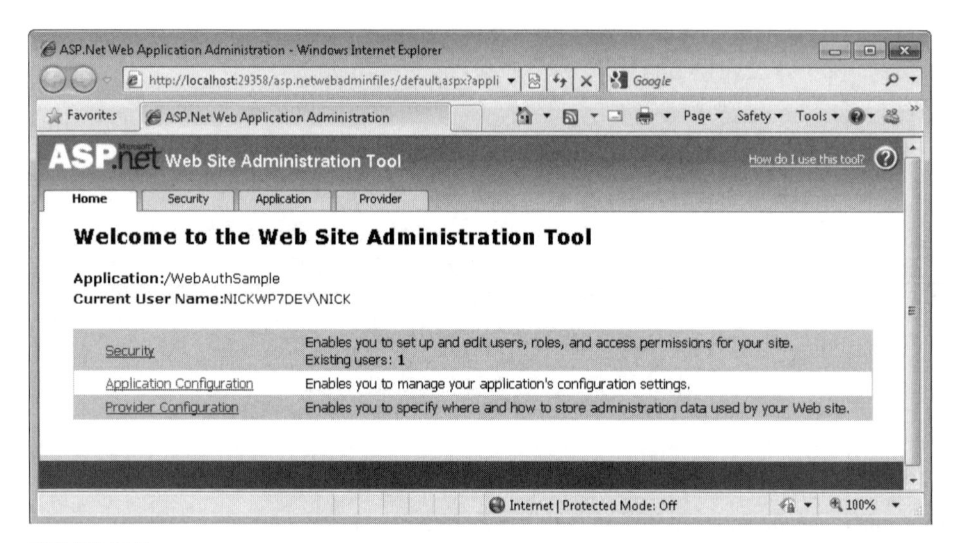

FIGURE 18-7

Click through to the Security tab. Start by clicking the "Enable Roles" link. Then go to "Create or Manage Roles" and create a role called **Managers**. Go back to the Security tab and click "Create user." Create two users, one that belongs to the Managers group (in this case, **AManager**) and one that doesn't (in this case, **NotAManager**).

The default access rule for all folders within your ASP.NET Web Application is to "Allow all authenticated users." You need to add additional rules to permit only Managers to access the secured folder (in this case, the FormsAuth folder). Go back to the Security tab and select "Manage access rules." Find the folder you want to configure the rules for and select "Add new access rule." Add the following rules:

Anonymous users	Deny
All users	Deny
Role [Managers]	Allow

If you look in the folder that you have just secured, you should see that a web.config file has been created that lists the authorization settings for this folder.

Available for download on Wrox.com

```
<system.web>
    <authorization>
        <deny users="?" />
        <deny users="*" />
        <allow roles="Managers" />
    </authorization>
</system.web>
```

Code snippet from web.config

The next thing you need to do is to expose Forms Authentication to your Windows Phone application. As the user won't be using a Web browser to navigate to your ASP.NET Web Application, you need to expose the Forms Authentication service as a WCF Service that can be called from your Windows Phone application. You do this by creating a service markup that exposes the built-in AuthenticationService. Add a file called *AuthenticationService.svc* to the root of your ASP.NET Web Application and add to it the following markup:

Available for download on Wrox.com

```
<%@ ServiceHost
  Language="C#"
  Service="System.Web.ApplicationServices.AuthenticationService"
  Factory="System.Web.ApplicationServices.ApplicationServicesHostFactory" %>
```

Code snippet from AuthenticationService.svc

The AuthenticationService also needs to be enabled by adding the following element to the web.config file:

Available for download on Wrox.com

```
<system.web.extensions>
    <scripting>
        <webServices>
            <authenticationService enabled="true" />
        </webServices>
    </scripting>
</system.web.extensions>
```

Code snippet from web.config

This will expose the AuthenticationService with a default end point that in itself is enough to be useful for validating credentials from the user. However, if you want to use the generated token to permit access to other services, you have to configure the service to allow the use of cookies. The following snippet from the web.config file configures both the AuthenticationService and the FormsAuthService (which is the WCF Service you want to call from your Windows Phone application) to allow the use of cookies:

Available for download on Wrox.com

```
<system.serviceModel>
    <services>
        <service behaviorConfiguration="DefaultBehavior"
                 name="System.Web.ApplicationServices.AuthenticationService">
```

```
            <endpoint binding="basicHttpBinding" bindingConfiguration="withCookies"
                      bindingNamespace="http://asp.net/ApplicationServices/v200"
                      contract="System.Web.ApplicationServices
.AuthenticationService" />
        </service>
        <service behaviorConfiguration="DefaultBehavior"
                 name="WebAuthSample.FormsAuth.FormsAuthService">
            <endpoint binding="basicHttpBinding" bindingConfiguration="withCookies"
                      contract="WebAuthSample.FormsAuth.IFormsAuthService" />
            <endpoint address="mex" binding="mexHttpBinding"
                      contract="IMetadataExchange" />
        </service>
    </services>
    <bindings>
        <basicHttpBinding>
            <binding name="withCookies" allowCookies="true" >
                <security mode="None" />
            </binding>
        </basicHttpBinding>
    </bindings>
    <behaviors>
        <serviceBehaviors>
            <behavior name="DefaultBehavior">
                <serviceMetadata httpGetEnabled="true" />
                <serviceDebug includeExceptionDetailInFaults="true" />
            </behavior>
        </serviceBehaviors>
    </behaviors>
    <serviceHostingEnvironment aspNetCompatibilityEnabled="true"  />
</system.serviceModel>
```

Code snippet from web.config

In this snippet, you also enable the ASP.NET Compatibility mode, which is required in order for Forms Authentication to be applied to the WCF Service. You also need to configure the WCF Service itself to use this Compatibility mode by specifying the `AspNetCompatibilityRequirements` attribute.

```
[AspNetCompatibilityRequirements(
          RequirementsMode = AspNetCompatibilityRequirementsMode.Allowed)]
public class FormsAuthService : IFormsAuthService{
    public void DoWork(){ ... }
}
```

Code snippet from FormsAuthService.svc.cs

You now need to add a service reference within your Windows Phone application to both the AuthenticationService.svc and the FormsAuthService.svc services (you will need to temporarily enable Anonymous access to the FormsAuth folder so that the service reference can be added). Figure 18-8 illustrates the "Add Service Reference" dialog for the AuthenticationService, which shows the operations available.

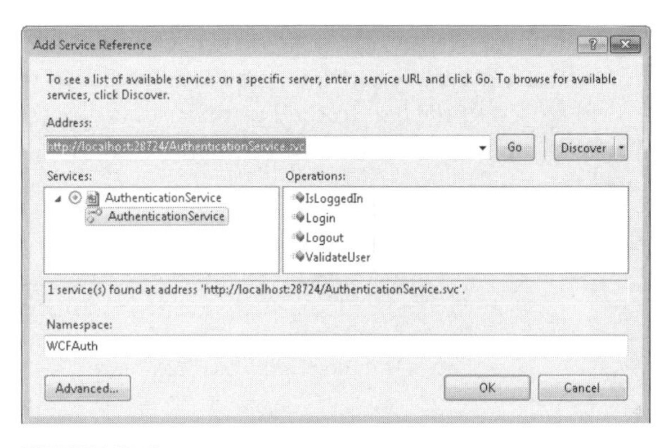

FIGURE 18-8

Once you have added these service references, don't forget to deny access to Anonymous users for the FormsAuth folder.

Adding the service references won't enable the use of cookies within the Windows Phone application. To do this, you need to set the `enableHttpCookieContainer` attribute to `true` for both services. This should give you a ClientConfig file that looks similar to the following:

Available for download on Wrox.com

```
<system.serviceModel>
    <bindings>
        <basicHttpBinding>
            <binding name="BasicHttpBinding_IFormsAuthService"
                     maxBufferSize="2147483647"
                     maxReceivedMessageSize="2147483647"
                     enableHttpCookieContainer="true">
                <security mode="None" />
            </binding>
            <binding name="BasicHttpBinding_AuthenticationService"
                     maxBufferSize="2147483647"
                     maxReceivedMessageSize="2147483647"
                     enableHttpCookieContainer="true">
                <security mode="None" />
            </binding>
        </basicHttpBinding>
    </bindings>
    <client>
        <endpoint address="http://localhost:28724/FormsAuth/FormsAuthService.svc"
            binding="basicHttpBinding"
            bindingConfiguration="BasicHttpBinding_IFormsAuthService"
            contract="FormsAuthSample.IFormsAuthService"
            name="BasicHttpBinding_IFormsAuthService" />
        <endpoint address="http://localhost:28724/AuthenticationService.svc"
            binding="basicHttpBinding"
            bindingConfiguration="BasicHttpBinding_AuthenticationService"
            contract="WCFAuth.AuthenticationService"
            name="BasicHttpBinding_AuthenticationService" />
    </client>
</system.serviceModel>
```

Code snippet from ServiceReference.ClientConfig

Making a call to the FormsAuthService is now a two-step process. The first step is to authenticate the user by calling the `Login` method on the AuthenticationService. This will return a value indicating whether the user was successfully logged on or not. If the user has been successfully authenticated, the `CookieContainer` attached to the `AuthenticationServiceClient` instance can be reused in the call to the FormsAuthService. In doing this, the token that was returned as a cookie from the `Login` method will be submitted alongside the `DoWork` request.

```
AuthenticationServiceClient authclient = new AuthenticationServiceClient();

private void FormsCallServiceButton_Click(object sender, RoutedEventArgs e){
    authclient.CookieContainer = new CookieContainer();
    authclient.LoginCompleted += authclient_LoginCompleted;
    authclient.LoginAsync("AManager", "1234qwer", "", true);
}

void authclient_LoginCompleted(object sender, LoginCompletedEventArgs e){
    var client = new FormsAuthSample.FormsAuthServiceClient();
    client.CookieContainer = authclient.CookieContainer;
    client.DoWorkCompleted += FormsService_DoWorkCompleted;
    client.DoWorkAsync();
}

void FormsService_DoWorkCompleted(object sender, AsyncCompletedEventArgs e){...}
```

Code snippet from MainPage.xaml.cs

One thing you may find useful is the ability to extract information about who is calling the WCF Service. This can be done by interrogating the `HttpContext.Current.User.Identity` instance, shown in Figure 18-9. This can be used to carry out work or return data that is specific to that user.

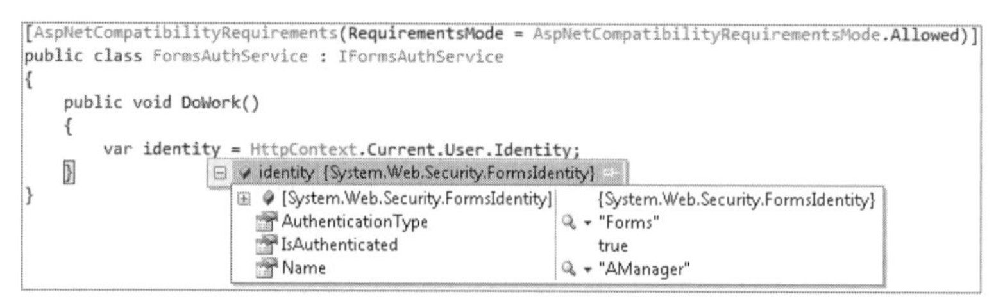

FIGURE 18-9

You can also use Forms Authentication with the `HttpWebRequest` class by specifying the `CookieContainer` before making the request:

```
void authfileclient_LoginCompleted(object sender, LoginCompletedEventArgs e){
    var request = HttpWebRequest.Create(
       "http://localhost:28724/FormsAuth/FormsAuthText.txt") as HttpWebRequest;
    request.CookieContainer = authclient.CookieContainer;
    request.BeginGetResponse(FormsFileComplete, request);
}
```

Code snippet from MainPage.xaml.cs

One thing to note about this code snippet is that the result of the `Create` method is cast to an `HttpWebRequest`. This is because the `CookieContainer` is not defined on the `WebRequest` base class, which is what the `Create` method returns by default.

It is recommended that you use an SSL connection for all Forms-based Authentication and service calls. This will help prevent credentials being hijacked and avoid username and password details being transmitted in plaintext.

OAuth 1.0 (Twitter)

So far you've been working with the authentication schemes that are built into IIS and ASP.NET. These are great if you control both the client (i.e., the Windows Phone application) and the server (i.e., the ASP.NET Web Application). However, if you don't — which is the case if you wish to integrate with any of the existing social networking sites — then you will most likely have to use a third-party authentication scheme. One such scheme that is getting a lot of attention recently is OAuth (Open Authorization). This section will cover how to authenticate against Twitter as an example of using OAuth 1.0. Then in the next section, you'll use the much simpler OAuth 2.0 to authenticate against Facebook.

We'll start by describing the user's experience with signing in with their Twitter account. When the user clicks the Authenticate button in the leftmost image of Figure 18-10, a `WebBrowser` control is displayed and immediately navigated to the Twitter authentication page (second image). This page advises the user that he is about to give your application access to his Twitter account.

FIGURE 18-10

After the user has signed in (third image from the left), he will be navigated to a final page that includes an authorization PIN (rightmost image). At this point, the Windows Phone application will detect that this page has been loaded and will automatically extract the PIN and use it to retrieve an access token from Twitter that will be used with each subsequent request. In some cases, the user may not see this last page as the application automatically hides the `WebBrowser` control when the final page is detected.

Let's walk through the code to do this. The first step is to define a class, `OAuthRequest`, which will be used to hold all the relevant URLs. This will help make this code more reusable for other sites that use OAuth. The `RequestUri`, `AuthorizeUri`, and `AccessUri` are defined by Twitter for use by any application wishing to authenticate using OAuth. The `ConsumerKey` and `ConsumerSecret` are strings specific to your application. When you sign up for an application account with Twitter, you will receive this pair of strings, which are then used in preparing the various OAuth requests.

Available for download on Wrox.com

```
public class OAuthRequest {
    public string RequestUri { get; set; }
    public string AuthorizeUri { get; set; }
    public string AccessUri { get; set; }

    public string ConsumerKey { get; set; }
    public string ConsumerSecret { get; set; }

    public string Method { get; set; }

    public string NormalizedUri { get; set; }
    public string NormalizedParameters { get; set; }

    public string VerifierPin { get; set; }

    public string Token { get; set; }
    public string TokenSecret { get; set; }

    public IDictionary<string, string> Parameters { get; private set; }

    public OAuthRequest()
    {
        Parameters = new Dictionary<string, string>();
    }
}
```

Code snippet from OAuthRequest.cs

Step 1: Retrieve a Request Token from Twitter

This step involves generating a request signature, preparing the request to the `RequestUri` (including setting the Authorization header), and then processing the response. In this case, the response includes a request token that will form part of the `AuthorizeUri`:

Available for download on Wrox.com

```
public void RetrieveRequestToken(Action Callback) {
    this.GenerateSignature(this.RequestUri);

    var request = HttpWebRequest.Create(this.NormalizedUri);
```

```
        request.Method = this.Method;
        request.Headers[HttpRequestHeader.Authorization] =
                                  this.GenerateAuthorizationHeader();

        request.BeginGetResponse((result) =>
            {
                var req = result.AsyncState as HttpWebRequest;
                var resp = request.EndGetResponse(result) as HttpWebResponse;

                using (var strm = resp.GetResponseStream())
                using (var reader = new StreamReader(strm)){
                    var responseText = reader.ReadToEnd();
                    this.ParseKeyValuePairs(responseText);
                }
            Callback();

        }, request);
    }
```

Code snippet from OAuthRequest.cs

Step 2: Navigate to the AuthorizeUri

The `Callback` method passed into the `RetrieveRequestToken` method in Step 1 navigates the `WebBrowser` control to the `AuthorizeUri` so that the user can log in:

```
RetrieveRequestToken(() =>{
    browser.Dispatcher.BeginInvoke(() => {
        browser.Navigate(new Uri(this.AuthorizeUri));
    });
});
```

Code snippet from OAuthRequest.cs

Step 3: User Signs into Twitter

Step 4: WebBrowser Navigated Event Is Detected and the PIN Is Extracted.

```
public void Authenticate(WebBrowser browser,
                         Action<IDictionary<string, string>> callback) {
    var baseAuthorizeUri = this.AuthorizeUri;
    browser.Navigated += (s, e) => {
        if (e.Uri.AbsoluteUri.ToLower().StartsWith(baseAuthorizeUri)) {
            if (!e.Uri.Query.Contains("oauth_token")) {
                var htmlString = browser.SaveToString();

                var authPinName = "oauth_pin>";
                var startDiv = htmlString.IndexOf(authPinName) + authPinName.Length;
                var endDiv = htmlString.IndexOf("<", startDiv);
```

```
                    var pin = htmlString.Substring(startDiv, endDiv - startDiv);

                    this.VerifierPin = pin;
                    this.RetrieveAccessToken(callback);
                }
            }
        };

        RetrieveRequestToken(() => {
            browser.Dispatcher.BeginInvoke(() => {
                browser.Navigate(new Uri(this.AuthorizeUri));
            });
        });
    }
```

Code snippet from OAuthRequest.cs

Step 5: Access Token Is Retrieved

As with Step 1, this step involves generating a request signature, preparing the request, this time to the `AccessUri` (including setting the Authorization header), and then processing the response. In this case, the response includes an access token that can be used for making further calls to Twitter.

```
    public void RetrieveAccessToken(Action<IDictionary<string, string>> Callback) {
        this.GenerateSignature(this.AccessUri);

        var request = HttpWebRequest.Create(this.NormalizedUri);
        request.Method = this.Method;
        request.Headers[HttpRequestHeader.Authorization] =
                            this.GenerateAuthorizationHeader();
        request.BeginGetResponse((result) => {
            var req = result.AsyncState as HttpWebRequest;
            var resp = request.EndGetResponse(result) as HttpWebResponse;

            Dictionary<string, string> responseElements;

            using (var strm = resp.GetResponseStream())
            using (var reader = new StreamReader(strm))
            {
                var responseText = reader.ReadToEnd();
                responseElements = this.ParseKeyValuePairs(responseText);
            }

            Callback(responseElements);

        }, request);
    }
```

Code snippet from OAuthRequest.cs

When the access token has been extracted, the final `Callback` method is invoked. This method is specified as the second command to the `Authenticate` method seen in Step 3. Putting this all together, you can now invoke this process from your Windows Phone application:

Available for download on Wrox.com

```
private OAuthRequest orequest;
private void AuthenticateButton_Click(object sender, RoutedEventArgs e) {
    orequest = new OAuthRequest() {
        RequestUri = "http://api.twitter.com/oauth/request_token",
        AuthorizeUri = "http://api.twitter.com/oauth/authorize",
        AccessUri="http://api.twitter.com/oauth/access_token",
        Method = "POST",
        ConsumerKey = "<your consumer key>",
        ConsumerSecret = "<your consumer secret>"
    };

    orequest.Authenticate(this.AuthenticationBrowser, AuthenticationComplete);
}

private void AuthenticationComplete(IDictionary<string, string> responseElements) {
    this.Dispatcher.BeginInvoke(() => {
        this.AuthenticationBrowser.Visibility = System.Windows.Visibility.Collapsed;
    });
}
```

Code snippet from TwitterPage.xaml.cs

In this case, the callback specified for the `Authenticate` method simply hides the `WebBrowser` control. The responseElements dictionary will contain any additional information returned by Twitter about the user who has just signed in.

This example omits the details of how to generate the signatures and authorization header information. Instead, full working source code is available for download.

OAuth 2.0 (Facebook)

The last form of authentication we're going to look at is OAuth 2.0, which is much simpler than OAuth 1.0 and relies heavily on the use of SSL instead of the generation of signatures. We'll start by defining a class called `OAuth2`:

Available for download on Wrox.com

```
public class OAuth2{
    public string LocalRedirectUrl { get; set; }
    public string AuthorizeBaseUrl { get; set; }
    public string AccessTokenBaseUrl { get; set; }
    public string PageType { get; set; }

    public string ClientId { get; set; }
    public string ClientSecret { get; set; }
}
```

Code snippet from OAuth2.cs

As with the previous Twitter/OAuth 1.0 example, when you sign up for an account with Facebook, you will be told what the various URLs are that are used by their implementation of OAuth 2.0. You'll also be given a `ClientId` and `ClientSecret` that will be used to prepare requests.

Step 1: Prepare Authorization URL

This is the URL that the WebBrowser will navigate to in order for the user to sign into Facebook. Unlike in OAuth 1.0, where a request token was retrieved, in this case, the URL is generated by the client. The URL will use SSL to ensure that any data transmitted is encrypted:

```
var authorizeUrl = string.Format(AuthorizeUrl,
                        AuthorizeBaseUrl, ClientId, LocalRedirectUrl,PageType);
```

Code snippet from OAuth2.cs

Step 2: Navigate to Authorization URL

```
browser.Navigate(new Uri(authorizeUrl));
```

Code snippet from OAuth2.cs

Step 3: User Signs into Facebook Account

Step 4: Navigated Event Is Intercepted

In the `Navigated` event handler, the URI is checked to see if it matches the `LocalRedirectUrl`. When this is matched, the relevant token information is extracted out of the URI query parameters. This information is then used to navigate the WebBrowser to a different URL that will retrieve the access token:

```
public void Authenticate(WebBrowser browser, Action callback) {
    browser.Navigated += (s, e) => {
        if (e.Uri.AbsoluteUri.ToLower().StartsWith(LocalRedirectUrl)) {
            ExtractCode(e.Uri);

            var accessTokenUrl = string.Format(AccessTokenUrl,
                              AccessTokenBaseUrl, ClientId,LocalRedirectUrl,
                              PageType, ClientSecret, Code);
            browser.Navigate(new Uri(accessTokenUrl));
        }
        else if (e.Uri.AbsoluteUri.ToLower().StartsWith(AccessTokenBaseUrl)) {
            var contents = browser.SaveToString();
            ExtractAccessToken(contents);
            callback();
        }

    };

    var authorizeUrl = string.Format(AuthorizeUrl,AuthorizeBaseUrl,
                                    ClientId, LocalRedirectUrl,PageType);
```

```
        browser.Navigate(new Uri(authorizeUrl));
    }

    private void ExtractCode(Uri uri) {
        var code = uri.Query.Trim('?');
        var bits = code.Split('=');
        this.Code = bits[1];
    }
```

Code snippet from OAuth2.cs

Step 5: Extract Access Token

In the code from the previous step, the second check on the URI is for the `AccessTokenBaseUrl`. If this is located, the access token is extracted from the query parameters:

```
private void ExtractAccessToken(string browserContents) {
    browserContents = HttpUtility.HtmlDecode(browserContents);
    var start = browserContents.IndexOf("access_token");
    var end = browserContents.IndexOf("</PRE>");
    var paramlist = browserContents.Substring(start, end - start);
    foreach (var param in paramlist.Split('&')){
        var bits = param.Split('=');
        switch (bits[0]){
            case AccessTokenKey:
                this.AccessToken = bits[1];
                break;
            case ExpiresKey:
                this.Expires = bits[1];
                break;
        }
    }
}
```

Code snippet from OAuth2.cs

Putting this all together, we can invoke the authentication process by simply calling the `Authenticate` method on an instance of the `OAuth2` class. The first parameter is the `WebBrowser` control that the user will interact with in order to sign into Facebook, and the second parameter is the callback method that will be invoked when the user has been successfully authenticated:

```
OAuth2 facebook = new OAuth2() {
    ClientId = "<your client id>",
    ClientSecret = "<your client secret>",
    LocalRedirectUrl = "http://www.facebook.com/connect/login_success.html",
    AuthorizeBaseUrl = "https://graph.facebook.com/oauth/authorize",
    AccessTokenBaseUrl = "https://graph.facebook.com/oauth/access_token",
    PageType = "web_server"
};
```

```
private void AuthenticateButton_Click(object sender, RoutedEventArgs e) {
    facebook.Authenticate(this.AuthenticationBrowser, AuthComplete);
}

void AuthComplete() {
    var request = HttpWebRequest.Create
     ("https://graph.facebook.com/me?access_token=" + facebook.AccessToken);
    request.BeginGetResponse(ProfileResponse, request);
}

void ProfileResponse(IAsyncResult result) {
    var request = result.AsyncState as HttpWebRequest;
    var response = request.EndGetResponse(result);
    using (var strm = response.GetResponseStream())
    using (var reader = new StreamReader(strm))
    {
        var txt = reader.ReadToEnd();
    }
}
```

Code snippet from FacebookPage.xaml.cs

Once the access token has been retrieved, it can be used to call other Facebook APIs. In this case, it's calling the API to retrieve the profile of the authenticated user. The variable `txt` would contain a JSON string similar to the following:


```
{"id":"6923674","name":"Nick Randolph","first_name":"Nick","last_name":"Randolph","
link":"http:\/\/www.facebook.com\/profile.php?id=6923174","hometown":{"id":11037276
8552,"name":"Sydney, New South Wales, Australia"},"education":[{"school":{"id":1117
48757717,"name":"University of Western Australia"},"degree":{"id":1120461219902,"na
me":"BE\/BCom"},"concentration":[{"id":112242345470433,"name":"IT"},{"id":112344032
3522,"name":"Accounting"},{"id":1124252114657,"name":"Finance"}]},{"school":{"id":1
082269677771,"name":"Hollywood Senior High School"},"year":{"id":1123924286252,"nam
e":"1995"}}],"gender":"male","email":"nick@builttoroam.com","timezone":10,"locale":
"en_US","verified":true,"updated_time":"2010-05-08T01:46:41+0000"}
```


SUMMARY

In this chapter, you have seen several ways to secure the data both within your Windows Phone application and on the wire when sending data to and from a remote server. You've also seen how you can use a variety of different strategies to authenticate the user and your application against the remote server.

19

Gaming with XNA

WHAT'S IN THIS CHAPTER

- Understanding the game loop
- Adding and rendering content
- Accepting input using the accelerometer, keyboard, and touch
- Using 3D models and shapes to create a scene

In the opening chapter of this book, you learned that there are two frameworks that you can use to develop Windows Phone applications. So far you have taken a deep dive into the world of Silverlight development for Windows Phone applications. However, for more game-oriented projects, the XNA Framework may be a more suitable development platform. Unlike Silverlight, which is primarily driven by XAML markup and events, XNA starts and ends with a *game loop*.

This chapter provides an overview of the XNA Framework and its associated development tools. You'll learn how to load and display sprites, as well as learn how to use transforms and lighting within your game. This is a teaser, and if you want to really get into building games for Windows Phone, you should look to some of the other great online resources dedicated to covering XNA development.

GETTING STARTED

In the previous chapters of this book, you have been building applications for Windows Phone using Silverlight, which combines a sophisticated layout and rendering system with an event-driven approach to logic. With Silverlight you can use Expression Blend to design the layout of the application and then wire up the event logic using Behaviors or more traditional event handlers in the code-behind file. Such applications are typically driven by their user interaction, which fits in well with the event-driven model of Silverlight.

On the other hand, most games continually update the screen in order to move or animate elements around the screen, and to adjust lighting and other visual and audio effects. Although it is possible to build a game in Silverlight, the event-driven model can often become a hindrance. The alternative is to use the XNA Framework. Instead of being event-driven, XNA uses the concept of a *game loop* to schedule updating and drawing of the screen.

Before you go further into the structure of a game loop, let's start building your first game using Visual Studio. Although there is no support for creating XNA-based games within Expression Blend, you may like to look into other products within the Expression Suite, such as Expression Design, as they can assist with the creation of game content such as images and 3D (three-dimensional) models. From Visual Studio's "New Project" dialog, shown in Figure 19-1, select the "XNA Game Studio 4.0" node. You will see that there are several different project templates allowing you to create games for Windows Phone, Xbox, and Windows. For the time being, select the Windows Phone Game (4.0) project template, give it the name **DontTouchTheWalls**, and click OK. You're going to be building a simple game in which the goal is to prevent the ball from touching the walls by rotating the device.

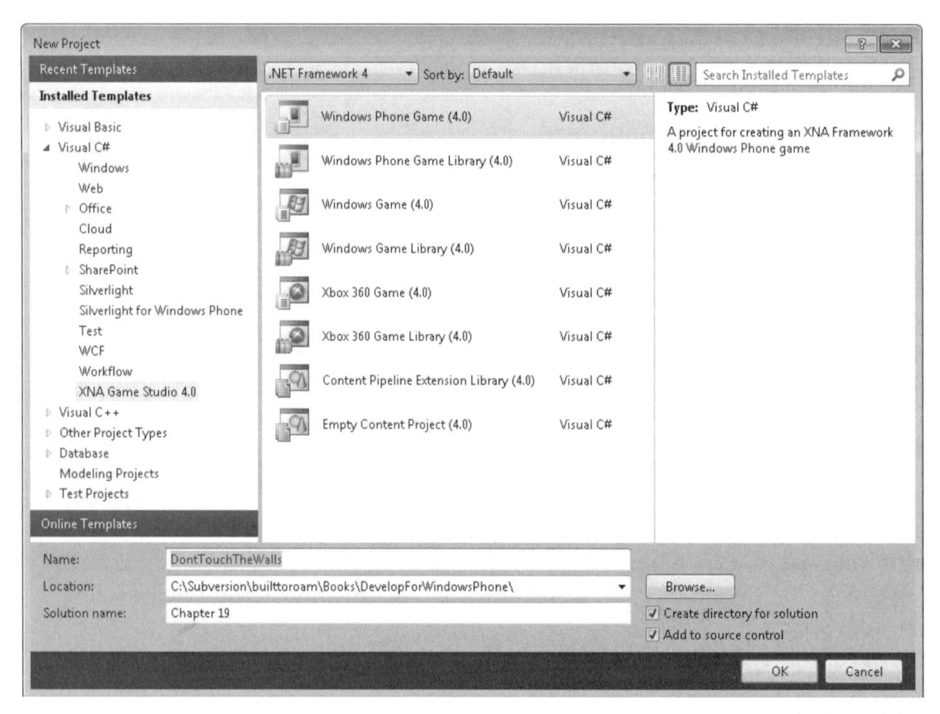

FIGURE 19-1

After clicking OK, you will see that a solution with two projects has been created. The first is the main game project, DontTouchTheWalls, and contains all the logic for loading and running your game. The second is a content project that will be used to hold all the resources (also referred to as *assets*) — such as images, sounds, fonts, and 3D models — that your game requires.

The entry point for the game is the `Game1` class, which inherits from the XNA `Game` class. There is no reason that this class has to be called `Game1`, so you'll start by renaming the class to

DontTouchWallsGame (if you rename the file in Solution Explorer, Visual Studio will assist you with renaming the class).

Double-click the DontTouchTheWalls project node in Solution Explorer to open the project Properties page shown in Figure 19-2. Here you can set properties that control how the game will appear on the Windows Phone. The Game thumbnail property determines the image that will appear in the Games hub along with the title of the game. As with Silverlight applications, the Tile title and Tile image will be used to determine how the tile is displayed when the game is pinned to the Start.

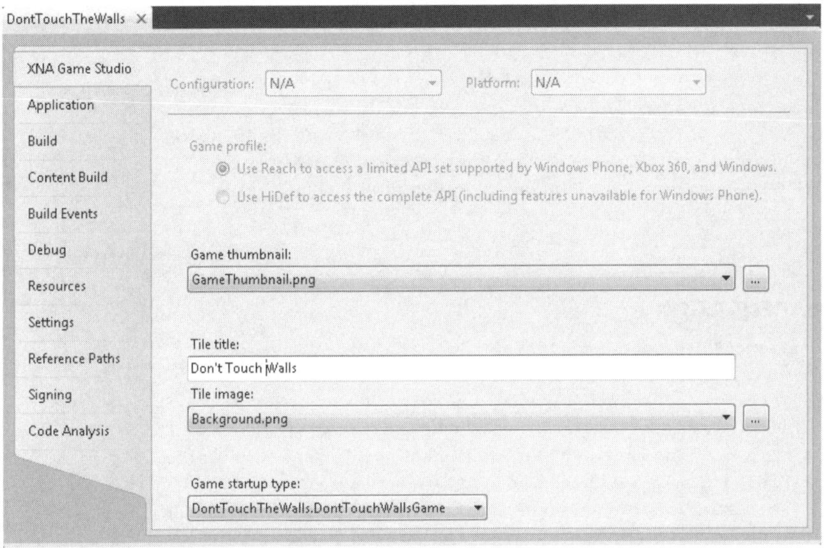

FIGURE 19-2

If you press F5 to run the game, you will see it launch in the emulator, displaying a blue screen — the default game source code doesn't do anything other than display a solid-color background. You can press the Back button to exit the application. From the Start screen if you go to the Games hub, you will see your game listed, as shown in Figure 19-3. Any game that the user installs on her Windows Phone will appear in the Games hub, and this is the games equivalent to the Applications list.

Users can click games listed within the hub to launch the game. The hub also includes information about the user's Xbox Live account such as their avatar and list of achievements.

FIGURE 19-3

GAMES HUB FOR APPLICATIONS

There is no reason why you can't build a normal application using the XNA Framework. Clearly, if you do this, you don't want your application to be listed within the Games hub. You can adjust the `Genre` element within the WMAppManifest.xml file to indicate the type of application the package contains and hence where it should be displayed.

GENRE: GAME

```
<?xml version="1.0" encoding="utf-8" ?>
<Deployment
    xmlns="http://schemas.microsoft.com/windowsphone/2009/deployment"
    AppPlatformVersion="7.0">
    <App xmlns="" ProductID="{2a650825-bb07-462b-9f17-a3072b2d587d}"
         Title="Don't Touch Walls" RuntimeType="XNA"
         Version="1.0.0.0" Genre="Apps.Games"
         Author="Nick Randolph" Description=""
         Publisher="Built to Roam">
```

GENRE: APPLICATION

```
<?xml version="1.0" encoding="utf-8" ?>
<Deployment
    xmlns="http://schemas.microsoft.com/windowsphone/2009/deployment"
    AppPlatformVersion="7.0">
    <App xmlns="" ProductID="{2a650825-bb07-462b-9f17-a3072b2d587d}"
         Title="Don't Touch Walls" RuntimeType="XNA"
         Version="1.0.0.0" Genre="Apps.Normal"
         Author="Nick Randolph" Description=""
         Publisher="Built to Roam">
```

When you ran the game, you would have seen that the background color didn't quite fill the screen. In fact, there are two reasons for this. First, as with Silverlight applications, XNA games do not run in Full-Screen mode by default. This can easily be rectified by setting the `IsFullScreen` property to `true` on the `GraphicsDeviceManager` instance that is created in the constructor of the `DontTouchWallsGame` class. Setting `IsFullScreen` to `true` will hide the System Tray that would otherwise appear at the top of the screen.

At this point, you've eliminated the space reserved for the System Tray, but there will still be a border along the edges of the screen. The second reason for the background not filling the screen is that the buffer that is used to accelerate drawing to the screen may be the wrong dimensions. To fix this, set the `PreferredBackBufferWidth` and `PreferredBackBufferHeight` properties of the `GraphicsDeviceManager` class. These changes are reflected in the following code snippet for the `DontTouchWallsGame` constructor:

Available for download on Wrox.com

```
public DontTouchWallsGame() {
    graphics = new GraphicsDeviceManager(this);

    // Set the Windows Phone screen resolution and full screen
    graphics.PreferredBackBufferWidth = 480;
    graphics.PreferredBackBufferHeight = 800;
    graphics.IsFullScreen = true;

    Content.RootDirectory = "Content";

    // Frame rate is 30 fps by default for Windows Phone.
    TargetElapsedTime = TimeSpan.FromSeconds(1/30.0);
}
```

Code snippet from DontTouchWallsGame.cs

Figure 19-4 illustrates three screenshots of the game running unaltered (leftmost image), with the buffer size corrected (middle image), and then in Full-Screen mode (rightmost image). There are subtle width and height differences in the size of the game in the three images.

FIGURE 19-4

Before you look at how to start building your game, and specifically the game loop, it is worth taking a quick look at the structure of the solution created for your game. The main game project, DontTouchTheWalls, doesn't look that dissimilar to the project structure of a Silverlight Windows

Phone application. As shown in Figure 19-5, in the Properties folder there are the AppManifest.xml, AssemblyInfo.cs, and WMAppManifest.xml files, which are used to define the contents of the game and whether the application should be displayed in the Applications or Games list on the device. Unlike a Silverlight application, where the entry point is contained within the App.xaml file, which in turn loads MainPage.xaml as the default page, an XNA game starts, and ends, within the game class, in this case `DontTouchWallsGame`.

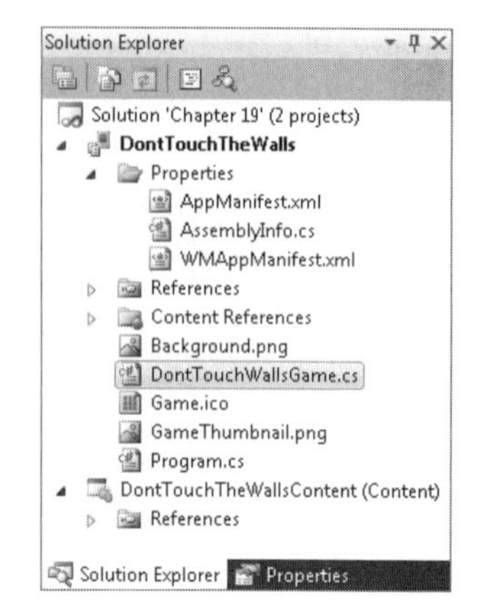

FIGURE 19-5

The second project, DontTouchTheWallsContent, is designed to hold any resources such as images, sounds, fonts, and other visual effects that your game may require. You'll come back to this later in the chapter in order to load resources to be used within the game.

One of the most significant advantages of using the XNA Framework to develop a game is that your game can target multiple platforms. Using the same code base, you can create a game that will work on Windows Phone, Xbox, and Windows. To illustrate how easy this is, right-click the game project, DontTouchTheWalls, and select "Create Copy of Project for Windows" from the context menu. This will create an additional project, "Windows Copy of DontTouchTheWalls," that contains all the same files as the DontTouchTheWalls project. In fact, this isn't just a copy of the project; it is an active link to the original project. This means that if a file is added to one project, it will appear in the other, and vice versa.

After creating the Windows project, you may notice that you have some build errors within Program.cs. This file provides the entry point for games running on Windows or the Xbox. The build errors will be related to the renaming of the default game class, `Game1`, to DontTouchWallsGame. Update the code in Program.cs to load the correct game by changing `Game1` to `DontTouchWallsGame`, as in the following code:

Available for download on Wrox.com

```
#if WINDOWS || XBOX
    static class Program
    {
        /// <summary>
        /// The main entry point for the application.
        /// </summary>
        static void Main(string[] args)
        {
            using (DontTouchWallsGame game = new DontTouchWallsGame())
            {
                game.Run();
            }
        }
    }
#endif
```

Code snippet from Program.cs

You will notice that this code uses conditional compilation symbols so that the `Program` class is only compiled when the game is being built for Windows or Xbox. At points within your game, you may need to write code that is only applicable to one or more platforms, for example, to load different content or adjust to differences in screen size. You can use the symbols `WINDOWS`, `XBOX`, or `WINDOWS_PHONE` to conditionally include a section of code for a given target device type.

If you open the project properties window for the Windows project, you will notice that the "Game profile" section is enabled, Figure 19-6. This was disabled for the Windows Phone project because it doesn't support HiDef so there is nothing to configure. However, for the Windows profile, you may want to set whether your game requires access to the high-definition APIs. Although it would seem that you would want to support the highest definition possible in your game, you should be aware that this may prevent some users from playing your game if their hardware doesn't support high-definition.

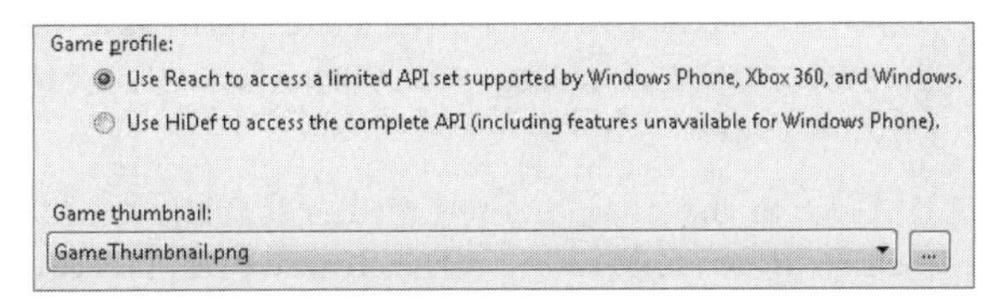

FIGURE 19-6

With these adjustments made, you can go ahead and run the Windows copy of the game (you may need to set the Windows project as the startup project by right-clicking the project and selecting "Set as Startup Project"). The game will run in full screen, again filling the screen with the default blue background color. To exit the application, you will need to switch back to Visual Studio and stop the running process.

Game Loop

The core of any XNA game is the *game loop*, which is made up of successive calls to the `Update` and `Draw` methods on the `game` class. As their names would imply, the `Update` method is where any of the game's logic should be executed, and the `Draw` method is where the screen contents are drawn onto the screen. Nearly all the code to respond to user input, manipulate object, check for collisions, and even control audio playback should be done within the `Update` method. This ensures that the `Draw` method is only responsible for refreshing the contents of the screen based on the current state of the game.

The speed at which the main game loop iterates controls the responsiveness of the input controls, but also has an effect on the amount of processing and battery power the game utilizes. There are two approaches to control the timing of the game loop that result in either a fixed or variable time interval between `Update` methods.

Fixed Time Step

The default approach uses a fixed time interval between successive calls to the `Update` method. In the constructor you saw earlier there was a line that sets the `TargetElapsedTime` property to a 30th of a second. This will give an approximate frame rate of 30 frames per second.

Figure 19-7 illustrates the life cycle of a game that uses a fixed-time-step game loop. For the moment, ignore the Initialize, Load Content, and Unload Content steps, as these will be covered in the section on loading content later in the chapter. The important part of Figure 19-7 is the Update–Draw loop.

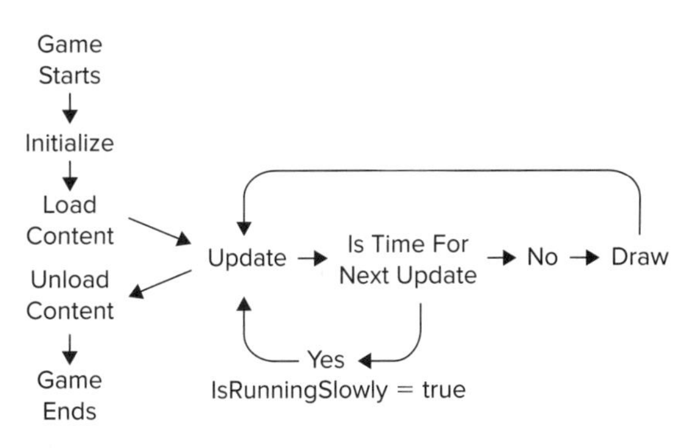

FIGURE 19-7

Notice that after the `Update` method has completed the system checks to see whether it is time for the `Update` method to be invoked again. If the `Update` method takes longer to execute than the `TargetElapsedTime`, the system will skip the call to the `Draw` method. If this happens, the `IsRunningSlowly` flag (within the `GameTime` object that is an argument to the `Update` method) is set to `true`.

Skipping the `Draw` method means that the speed of any animations, or game logic, will execute as close as possible to the target frame rate. However, it means that the user will not see the visual change, as the last frame will be stuck on the screen. Within your `Update` method you should inspect the `IsRunningSlowly` property and if it is set to true, skip any non-essential logic so that the `Update` method can complete as quickly as possible to give the Draw method an opportunity to execute before it is time for the next update.

A couple of points are worth noting about using a fixed-time-step game loop: First, the `TargetElapsedTime` is just that — it is the target for the elapsed time between calls to the `Update` method. If the Update–Draw loop takes less than the `TargetElapsedTime`, the game will just idle until it is time to call the `Update` method again. If the `Update` method takes longer than the `TargetElapsedTime`, the call to `Draw` is dropped, and the `Update` method is called again. The second point is really an extension of this scenario, in that while calls to the `Draw` method may be skipped, the `Update` method will be called the correct number of times. This can be important in some games that rely on the `Update` method to increment counters or move objects over time.

Variable Time Step

Using a variable-time-step game loop results in alternating calls to the `Update` and `Draw` methods, irrespective of how long each takes to execute, as illustrated in Figure 19-8. The `ElapsedGameTime` property on the `GameTime` object can be queried to determine the time interval between calls to the `Update` method. If the game is taking too long to complete the `Update` method, you may want to short-circuit some of the non-essential logic to improve responsiveness of the game.

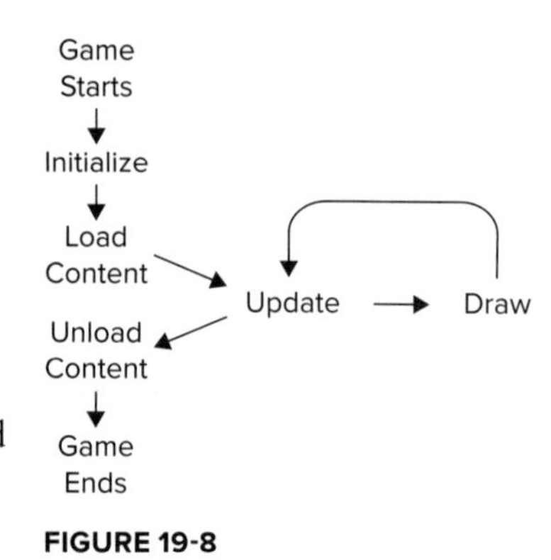

FIGURE 19-8

To set up the game loop to use a variable time step, you simply need to set the `IsFixedTimeStep` property on the `game` class to `false`. Since the `Update` method is called as soon as the `Draw` method has completed, there is no need to specify a `TargetElapsedTime`.

Game Life Cycle

The game loop is the core to any XNA game. However, there are other aspects to the life cycle of a game that you should be aware of.

Loading and Unloading Content

When a game starts, the first point where you can start initializing the game is within the `Initialize` method. Here you can initialize any game variables that will be used to track the state of the game. The `Initialize` method then invokes the `LoadContent` method, which is where you should load any graphical content that will be used by your game.

Windows Phone requires that all applications and games start with a specific period of time (as defined by the application policy guidelines available from `http://developer.windowsphone.com`). As such, if your game has a lot of graphical content, you want to load them as they are required, rather than at startup. This will help ensure that your game loads quickly.

Exiting the Game

There is no built-in support for exiting an XNA game. The Windows Phone Game (4.0) template that you used to create the game includes a simple mechanism for ensuring that the user can exit the game. In the `Update` method, the state of the Buttons collection is queried to determine if the user has pressed the Back button:

Available for download on Wrox.com

```
if (GamePad.GetState(PlayerIndex.One).Buttons.Back == ButtonState.Pressed)
    this.Exit();
```

Code snippet from DontTouchWallsGame.cs

If the Back button is being pressed, the `Exit` method is invoked to end the current instance of the game.

Application Events

The behavior and life cycle of XNA games on Windows Phone are similar to those of applications built with Silverlight. When the game or application goes into the background, they receive a `Deactivated` event. This is the first and only opportunity for the game to persist the current game state and halt any background processes. The following code illustrates how you can wire up event handlers for the phone application events:

Available for download on Wrox.com

```
public DontTouchWallsGame()
{
    graphics = new GraphicsDeviceManager(this);

    //Set the Windows Phone screen resolution
    graphics.PreferredBackBufferWidth = 480;
    graphics.PreferredBackBufferHeight = 800;
    graphics.IsFullScreen = true;

#if WINDOWS_PHONE
    PhoneApplicationService.Current.Launching += Current_Launching;
    PhoneApplicationService.Current.Deactivated += Current_Deactivated;
    PhoneApplicationService.Current.Activated += Current_Activated;
#endif
    Content.RootDirectory = "Content";

    // Frame rate is 30 fps by default for Windows Phone.
    TargetElapsedTime = TimeSpan.FromSeconds(1/30.0);
}
```

```
#if WINDOWS_PHONE
void Current_Launching(object sender, LaunchingEventArgs e) {...}

void Current_Activated(object sender, ActivatedEventArgs e) {...}

void Current_Deactivated(object sender, DeactivatedEventArgs e) {...}
#endif
```

Code snippet from DontTouchWallsGame.cs

It is important to remember that this code is specific to the Windows Phone version of the game, since neither the Windows nor the Xbox version of the .NET platform have the same events. When the user returns to a Windows Phone game, by pressing the Back button, the game will be activated again. If the game has been tombstoned, this may involve restarting the game prior to the activated event being raised.

When you are persisting the current game state in the `Deactivate` method, you have two options. You can either manually handle reading and writing to Isolated Storage (see Chapter 16), or you can use the State dictionary within the current `PhoneApplicationService` instance. The following code saves a simple string that could represent the current game state. You might also want to retrieve information from the State dictionary in the `Launching` event for the game state that needs to be persisted across multiple game instances.

Available for download on Wrox.com

```
void Current_Activated(object sender, Microsoft.Phone.Shell.ActivatedEventArgs e) {
    var gameState = PhoneApplicationService.Current.State["GameState"] as string;
}

void Current_Deactivated(object sender, Microsoft.Phone.Shell.DeactivatedEventArgs e) {
    PhoneApplicationService.Current.State["GameState"] = "Running";
}
```

Code snippet from DontTouchWallsGame.cs

RENDERING

OK, so enough talk about the structure of a game — let's get into building out the components of this game. The game play is relatively simple: The user needs to keep the ball away from the walls by changing the angle of the device. When the user hits a wall, the game will display the length of time for which they were able to keep the ball away from the wall.

Content

The first thing to do is to display an image of a ball. For this you'll use an image file, called ball.png, which has a transparent background. XNA uses the concept of a *Content Pipeline*, whereby designers can generate content, such as graphics, audio, and 3D models, using their tool of choice without any regard for how it will be consumed within the game. To include the content in the XNA game, a content importer is used to import the content into a format that is understood by one of XNA's default content processors. Alternatively, a content processor may be used to convert the content into a managed-code object that can be directly used within the XNA game. XNA ships with several importers and processors that handle a large number of the standard content types for images, audio, and 3D models.

To add an image to the XNA game, right-click the content project, DontTouchWallsGameContent, and select Add ➪ Existing Item. Locate the image you want to add, in this case, ball.png, and click Add. If the image isn't already in the project folder, it will be copied there and included in the project, as shown in Figure 19-9.

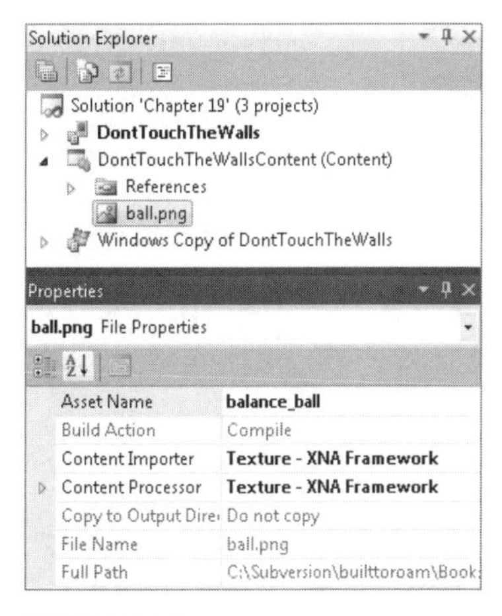

FIGURE 19-9

In the Properties window of Figure 19-9, you can see that the Asset Name has been updated to "balance_ball." This is the name that will be used within the game to access this asset. You can also see that the Content Importer and Content Processor properties are both set to "Texture - XNA Framework." This implies that the image will be loaded as a texture and that both the importer and processor are part of the XNA Framework.

Within the game you need to load this content into a local variable so that it can be used to draw the ball on the screen during the `Draw` method. The game class, `DontTouchWallsGame`, inherits the `Content` property from the `Game` base class, which returns an instance of the `ContentManager` class. The generic `Load` method found within the `ContentManager` object can then be used to load the image into memory. In this case, you know that the image will be loaded into a `Texture2D` object in preparation for it being drawn to the screen.

Available for download on Wrox.com

```
Texture2D balanceBall;
protected override void LoadContent() {
    spriteBatch = new SpriteBatch(GraphicsDevice);

    balanceBall = this.Content.Load<Texture2D>("balance_ball");
}
```

Code snippet from DontTouchWallsGame.cs

Sprites

The term *sprite* is typically used to refer to an image that represents a smaller part of a larger scene. For example, most games are composed of hundreds, or thousands, of sprites, each being positioned on the screen so as to build up a more realistic scene. *Spritesheets* are a concept widely used by web developers to improve performance. The *Spritesheet* is a single image made up of a large number of smaller images that are used throughout a website. Most browsers will download the Spritesheet once and reuse it to display all the smaller images, significantly reducing the download time and number of transfers required for the site.

In an XNA game, a *Sprite* is essentially any image that makes up the scene to be drawn. For example, the ball image that was just loaded in the `LoadContent` method is a Sprite that can be drawn to the screen as part of a SpriteBatch:

```
protected override void Draw(GameTime gameTime) {
    GraphicsDevice.Clear(Color.CornflowerBlue);

    this.spriteBatch.Begin();
    this.spriteBatch.Draw(this.balanceBall,
                          new Rectangle(100, 100, 100, 100), Color.White);
    this.spriteBatch.End();


    base.Draw(gameTime);
}
```

Code snippet from DontTouchWallsGame.cs

The ball image will be drawn at position (100,100) with a width and height of 100, as shown in Figure 19-10. Note that the coordinate system is based on an origin in the top-left corner of the screen, with the positive *x*-axis going across the screen to the right and the positive *y*-axis going down the screen.

FIGURE 19-10

Movement

Moving the Sprite around the screen just requires the position of the ball to be updated so that the next time it is drawn it will be in a different location. Instead of hard-coding the position of the Sprite, the location will be determined by a state variable, `position`. During the `Update` method this value will be incremented according to the direction vector.

```
Vector2 direction = new Vector2(3,2);
Vector2 position = new Vector2(100, 100);

protected override void Update(GameTime gameTime) {
    // Allows the game to exit
    if (GamePad.GetState(PlayerIndex.One).Buttons.Back == ButtonState.Pressed)
        this.Exit();

    position = Vector2.Add(position, direction);
    if ((position.X + 100) > 480 || position.X < 0) direction.X *= -1;
    if ((position.Y + 100) > 800 || position.Y < 0) direction.Y *= -1;

    base.Update(gameTime);
}

protected override void Draw(GameTime gameTime) {
    GraphicsDevice.Clear(Color.CornflowerBlue);

    this.spriteBatch.Begin();
```

```
        this.spriteBatch.Draw(this.balanceBall,
                    new Rectangle((int)position.X,(int)position.Y,100,100),
                    Color.White);
        this.spriteBatch.End();
    }
```

Code snippet from DontTouchWallsGame.cs

When a wall is reached, the direction vector will be reversed, causing the ball to bounce off the wall, as illustrated in the sequence of screenshots in Figure 19-11.

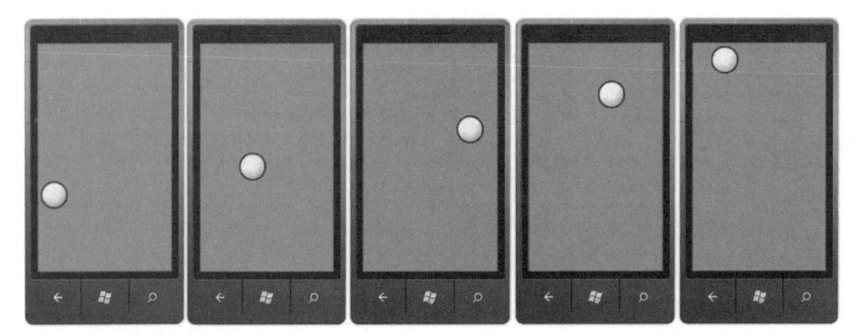

FIGURE 19-11

Text and Fonts

In this game the objective is to keep the ball from hitting the wall. If the ball does hit the wall, the game is over. To indicate this, you're going to write **Game Over** in the middle of the screen. The XNA Framework has built-in support for drawing text. However, in order to draw text, you need to tell the Framework which font to use. This is done by adding the font to the content project. Right-click the content project, DontTouchTheWallsContent, in the Solution Explorer window, and then select Add ⇨ New Item. This will display the "Add New Item" dialog shown in Figure 19-12 with the different types of content that you can add. In this case, you're going to add a Sprite Font called Lindsey.spritefont.

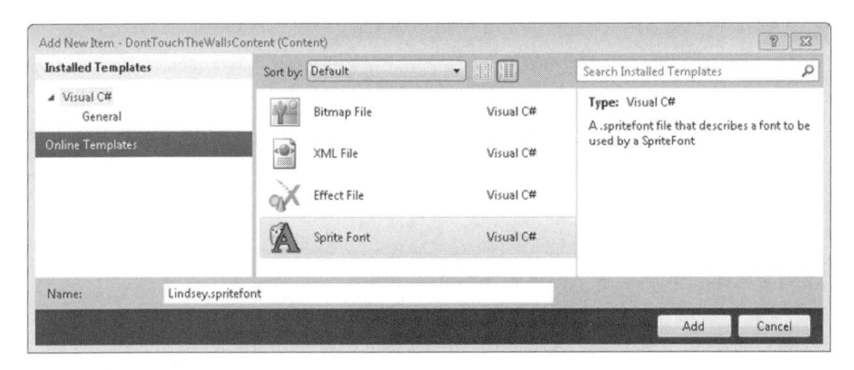

FIGURE 19-12

After clicking Add, Visual Studio will add a new spritefont file to the content project and will immediately open it for editing. You can think of the spritefont file as a declaration of the font that you want to be included when the content project is compiled. This means that the font family (specified in the `FontName` element in the following code snippet) has to correspond to a font that is installed on the computer that the project is going to be built on. This is important to remember if you have a build server that is responsible for generating release artifacts for your game. The spritefont file also specifies the size and style (i.e., Normal, Bold, or Italic) of the font that you wish to use. In this case, the code specifies the Lindsey font with size 45 and Bold style.

```
<?xml version="1.0" encoding="utf-8"?>
<XnaContent xmlns:Graphics="Microsoft.Xna.Framework.Content.Pipeline.Graphics">
  <Asset Type="Graphics:FontDescription">
    <FontName>Lindsey</FontName>
    <Size>45</Size>
    <Spacing>0</Spacing>
    <UseKerning>true</UseKerning>
    <Style>Bold</Style>
    <CharacterRegions>
      <CharacterRegion>
        <Start>&#32;</Start>
        <End>&#126;</End>
      </CharacterRegion>
    </CharacterRegions>
  </Asset>
</XnaContent>
```

Code snippet from Lindsey.spritefont

FONT LICENSING

As with other graphical content that you use within your game, you must ensure that you use fonts that you have permission to distribute with your game. The Lindsey font used in this example can be downloaded in the Redistributable Font Pack from the XNA Creators Club website (`http://creators.xna.com/en-US/contentpack/fontpack`). The license for these fonts is owned by Microsoft, who has made them freely available for you to use and distribute within your game. Make sure you check the licensing for any content that you use within your game.

To use the font in your game, you need to load it by calling the `LoadContent` method. In this case, since you're going to be displaying the text in the middle of the screen, you can pre-calculate its position:

```
SpriteFont gameOverFont;
Vector2 gameOverTextPosition;

protected override void LoadContent() {
    ...
```

```
        gameOverFont = Content.Load<SpriteFont>("Lindsey");
        gameOverTextPosition = new Vector2(graphics.GraphicsDevice.Viewport.Width / 2,
                                     graphics.GraphicsDevice.Viewport.Height / 2);
    }
```

Code snippet from DontTouchWallsGame.cs

The "Game Over" text should only be displayed after a wall has been hit. You can easily add a flag, `wallHit`, that is set to `true` when one of the walls is hit:

Available for download on Wrox.com

```
bool wallHit = false;

protected override void Update(GameTime gameTime) {
    if (GamePad.GetState(PlayerIndex.One).Buttons.Back == ButtonState.Pressed)
        this.Exit();

    position = Vector2.Add(position, direction);
    if ((position.X + 100) > 480 || position.X < 0) {
        wallHit = true;
        direction.X *= -1;
    }
    if ((position.Y + 100) > 800 || position.Y < 0) {
        direction.Y *= -1;
        wallHit = true;
    }

    base.Update(gameTime);
}
```

Code snippet from DontTouchWallsGame.cs

Lastly, when the flag is set to `true`, the text should be drawn using the `DrawString` method. The `DrawString` method allows you to make adjustments to the font such as translation, flipping (vertically or horizontally), or scaling:

Available for download on Wrox.com

```
private const string GameOverText = "Game Over";

protected override void Draw(GameTime gameTime) {
    GraphicsDevice.Clear(Color.CornflowerBlue);

    this.spriteBatch.Begin();
    this.spriteBatch.Draw(this.balanceBall,
                          new Rectangle((int)position.X,(int)position.Y,
                                        100,100), Color.White);
    if (wallHit == true) {
        // Find the center of the string
        Vector2 FontOrigin = gameOverFont.MeasureString(GameOverText) / 2;
        // Draw the string
        spriteBatch.DrawString(gameOverFont, GameOverText,
                               gameOverTextPosition, Color.DarkBlue, 0,
```

```
                                  FontOrigin, 1.0f, SpriteEffects.None, 0.5f);
        }
        this.spriteBatch.End();
    }
```

Code snippet from DontTouchWallsGame.cs

One thing to note is that although it is possible to scale the font using the `scale` parameter of the `DrawString` method, this can result in a font that is very fuzzy, blocky, or lacking detail. The font generated by a spritefont file consists of a series of bitmaps, that is, Sprites, one for each character in the font. This means that the font is rendered at a single font size. When a scale factor is involved, the Sprites are similarly stretched or shrunk as required. As an example, the following code magnifies the sprite font by using a scale factor of 3. This should generate text that is of a similar size to what you would get if you adjusted the Lidsey.spritefont file to specify a font size of 15 and used a scale of 1:3.

```
spriteBatch.DrawString(gameOverFont, GameOverText,
                                  gameOverTextPosition, Color.DarkBlue, 0,
                                  FontOrigin, 3.0f, SpriteEffects.None, 0.5f);
```

The left image of Figure 19-13 illustrates the effect of scaling a small font. Alternatively, in the right image, which is much clearer, the text has a font of 45 so is not scaled..

FIGURE 19-13

INPUT

Currently the game only displays a bouncing ball, and it doesn't accept any input from the user. In this section, you'll update the game to adjust the ball's speed and direction using the built-in accelerometer. You'll also allow the game to be restarted by the user touching the screen. Lastly, you'll learn how the user can enter his or her name using the keyboard when the game starts.

Accelerometer

In Chapter 10, you learned how to receive events from the accelerometer within your Silverlight application. The accelerometer works the same way within an XNA game. Since the accelerometer is specific to the Windows Phone version of the game, the logic for creating and working with the accelerometer needs to be conditional on the `WINDOWS_PHONE` compilation symbol:

Available for download on Wrox.com

```
#if WINDOWS_PHONE
        Accelerometer acc = new Accelerometer();
#endif
protected override void LoadContent() {
```

```
    ...
#if WINDOWS_PHONE
    acc.ReadingChanged += (s, e) => {
        if (e.X >= -1 && e.X <= 1){
            direction.X += (float)(e.X);
        }
        if (e.Y >= -1 && e.Y <= 1) {
            direction.Y -= (float)(e.Y);
        }
    };
    acc.Start();
#endif
}
```

Code snippet from DontTouchWallsGame.cs

The direction that the ball is traveling is incremented by the relevant accelerometer values, changing the speed that the ball is traveling. Now, as users tilt the device, the ball will speed up or slow down, allowing them to keep the ball away from the walls.

Touch

In a Silverlight application, when the user touches the screen, one or more mouse events are raised. In an XNA game, detecting user input is done within the `Update` method. The following code retrieves the current state of the `TouchPanel` as an array of `TouchLocation` objects. From this the `Position` property can be queried to determine the location of the touch point. Remember that all Windows Phone devices will support at least four touch points.

```
Random randomDirection = new Random((int)DateTime.Now.Ticks);
protected override void Update(GameTime gameTime) {
    ...
    if (wallHit) {
        var touchPoints = TouchPanel.GetState();
        if (touchPoints.Count > 0) {

            var t1 = touchPoints[0];
            position = t1.Position;
            direction = new Vector2(randomDirection.Next(0, 5),
                                    randomDirection.Next(0, 5));
            wallHit = false;
        }
    }

    base.Update(gameTime);
}
```

Code snippet from DontTouchWallsGame.cs

The `TouchLocation` object also has a `State` property that signals whether the point of contact has been Moved, Pressed, or Released. In the case of Moved, you can attempt to retrieve the previous location via the `TryGetPreviousLocation` method.

Keyboard

Capturing text from the keyboard can be done using the static `BeginShowKeyboardInput` method available on the `Guide` class. The following code invokes the keyboard input when the name variable is empty. Since this is done within the `Update` method, which will continue to be invoked in the background, it is important to first check that the keyboard is not already visible. The `IsVisible` property on the `Guide` class indicates if either the keyboard input or message box (call `BeginShowMessageBox` to display a message box) user interface (UI) elements are currently visible:

```
string name;
protected override void Update(GameTime gameTime) {
    if (string.IsNullOrEmpty(name)) {
        if (!Guide.IsVisible) {
            Guide.BeginShowKeyboardInput(PlayerIndex.One, "Enter your name",
                                    "You need to enter your name in order to play",
                                    "", NameCallback, null);
        }
    }
    else {
        ...
    }
    base.Update(gameTime);
}

private void NameCallback(IAsyncResult result)
{
    name = Guide.EndShowKeyboardInput(result);
}
```

Code snippet from DontTouchWallsGame.cs

When the user completes the text entry, the `NameCallback` method will be invoked. The text entered by the player can be retrieved from the `EndShowKeyboardInput` method. Figure 19-14 illustrates the keyboard input (left image) and the game in progress after entering a name.

FIGURE 19-14

3D RENDERING

In addition to working with 2D (two-dimensional) Sprites, the XNA Framework has support for loading, manipulating, and drawing 3D models. Building 3D games not only requires you to work in three dimensions (with the introduction of a *z*-axis for position and movement), but also introduces other complexities to do with lighting and other visual effects. This section provides a simple example of how you can load a 3D model and alter the lighting effects that are applied to it. You'll then learn how to create your own 3D shapes using drawing primitives and how to apply textures to them.

Before you jump in, let's cover a few basics. First, all 3D shapes are drawn in a World that starts at the origin, (0, 0, 0), and extends out in three dimensions (Figure 19-15, leftmost image). The first thing you need to do when drawing a 3D shape is define where it is positioned in the World (Figure 19-15, middle image) relative to the origin. The second thing you need to do is to determine where the camera or View of the World is located. Where the camera is located in the World, and the direction it is facing, will determine which objects, or more precisely which faces of the objects, can be seen (Figure 19-15, rightmost image).

To make working in 3D even more complex, you have to consider the number and type of lighting sources and the physical properties of the shapes you are drawing. You can also apply *textures* (in other words, part or all of an image) to the surfaces of the shapes to control how they look.

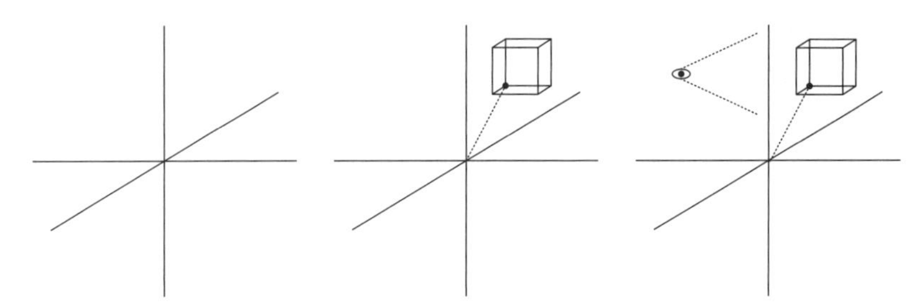

FIGURE 19-15

3D Model

One of the easiest ways to display 3D shapes is to load them from a 3D model that has been created using a 3D modeling tool. The example that you're going to use is a model of a sphere that has been created using Blender (`www.blender.org`) and exported as an FBX file (a format owned and developed by Autodesk that has built-in support within the XNA Framework Content Pipeline). This model is then added into the content project, DontTouchTheWallsContent, by right-clicking the project and selecting Add ➪ Existing Item. You can then browse and select the file, ball.fbx, which contains the exported 3D model. Once the model has been added to the content project, you may want to give it an easily identifiable name, in this case, *3dball*.

As you did with the 2D Sprite and the spritefont earlier, you need to load the 3D model via a call to the `LoadContent` method. You're also going to record the current position and rotation of the ball. The camera is going to be located at 500 along the *z*-axis and will look back toward the origin:

Available for download on Wrox.com

```
Model modelBall;
float aspectRatio;

Vector3 modelBallPosition = new Vector3(0, 50, 0);
float modelBallRotation = 0.0f;
Vector3 cameraPosition = new Vector3(0.0f, 0.0f, 500.0f);

protected override void LoadContent() {
    ...
    modelBall = this.Content.Load<Model>("3dball");
    aspectRatio = graphics.GraphicsDevice.Viewport.AspectRatio;
}
```

Code snippet from DontTouchWallsGame.cs

You're going to alter the rotation of the ball with each pass through the `Update` method:

Available for download on Wrox.com

```
protected override void Update(GameTime gameTime) {
    ...
    modelBallRotation += (float)gameTime.ElapsedGameTime.TotalMilliseconds *
                             MathHelper.ToRadians(0.05f);
    base.Update(gameTime);
}
```

Code snippet from DontTouchWallsGame.cs

Without getting into the depths of how a 3D model is structured, at a high level a Model is made up of a set of `ModelMesh` objects and a set of `ModelBone` objects. To render a 3D model, you need to iterate through the set of `ModelMesh` objects; set the World, View, and Projection of each `BasicEffect` object; and then draw the `ModelMesh`. The World position determines the location, rotation, and scaling that is applied to the part of the Model being rendered, whereas the View determines the location and direction of the camera. Lastly, the Projection defines the viewing range of the camera.

Available for download on Wrox.com

```
protected override void Draw(GameTime gameTime) {
    ...

    Matrix[] transforms = new Matrix[modelBall.Bones.Count];
    modelBall.CopyAbsoluteBoneTransformsTo(transforms);

    foreach (ModelMesh mesh in modelBall.Meshes) {
        foreach (BasicEffect effect in mesh.Effects) {
            effect.EnableDefaultLighting();

            effect.World = transforms[mesh.ParentBone.Index] *
                Matrix.CreateScale(0.1f) *
                Matrix.CreateRotationY(modelBallRotation) *
                Matrix.CreateTranslation(modelBallPosition);

            effect.View = Matrix.CreateLookAt(cameraPosition,
                                    Vector3.Zero, Vector3.Up);
```

```
            effect.Projection = Matrix.CreatePerspectiveFieldOfView(
                MathHelper.ToRadians(45.0f), aspectRatio,
                1.0f, 10000.0f);
        }
        mesh.Draw();
    }
    base.Draw(gameTime);
}
```

Code snippet from DontTouchWallsGame.cs

Color and Lighting

In the previous code snippet you may have noticed that you used the default lighting by calling `EnableDefaultLighting`. This sets up both a single directional lighting source and some ambient light so that you can see the models that are being rendered. If you want to control the lighting yourself, you can define up to three lighting sources. The following code enables lighting and then proceeds to set properties on the first lighting source, `DirectionalLight0`. For additional lighting you can enable `DirectionalLight1` and `DirectionalLight2`. In this case, the lighting is for a red light coming in from the left (i.e., in the positive *x*-direction). The colors are specified as a `Vector3` with the values being the `Red`, `Green`, and `Blue` values in the range 0 to 1:

Available for download on Wrox.com

```
effect.LightingEnabled = true;
effect.DirectionalLight0.DiffuseColor = new Vector3(0.5f, 0, 0);
effect.DirectionalLight0.Direction = new Vector3(1, 0, 0);
effect.DirectionalLight0.SpecularColor = new Vector3(0.9f, 0, 0);
effect.DirectionalLight0.Enabled = true;

effect.AmbientLightColor = new Vector3(0.2f, 0.2f, 0.2f);
effect.EmissiveColor = new Vector3(0.1f, 0, 0);
```

Code snippet from DontTouchWallsGame.cs

The difference between the default lighting and the single lighting source is illustrated in Figure 19-16. The left image uses the default lighting, while the right image has a single light source coming from the left. Notice how the color of the lighting has altered the appearance of the 3D object(s) within the scene.

FIGURE 19-16

Primitives

The other way to display 3D shapes in XNA is to compose your own 3D shapes out of basic geometric primitives, most commonly triangles. For each shape that you want to render, you need to determine a list of vertices (i.e., the corners) and a series of triangles between those points to make up the surface of the shape.

Each triangle will also have a normal that is a vector in a direction perpendicular to the face of the triangle.

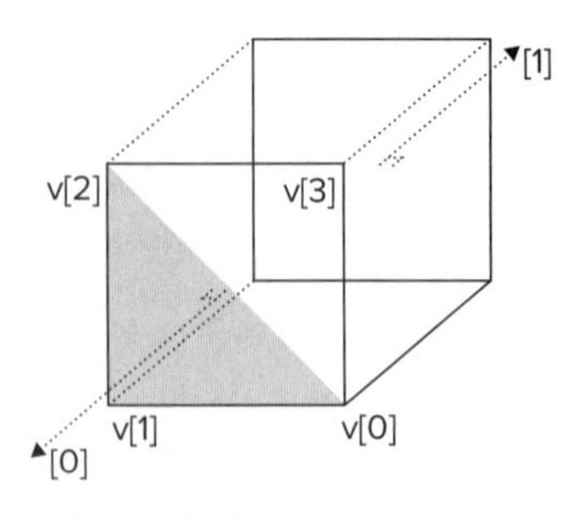

FIGURE 19-17

The following code defines a `Cube` class that has six faces. Each face is defined by a vector of size 1 pointing in a direction that is perpendicular to the face of the cube (i.e., the normal to the face). Figure 19-17 illustrates the first two normals pointing out from the page, [0], and into the page, [1]. These correspond to the positive z and negative z directions, respectively.

The vertices that make up the face are determined for each normal (these are annotated v[0]–v[3] in Figure 19-17). Each face is broken into two triangles. The shaded triangle in Figure 19-17 illustrates the first triangle defined in the `indices` list, defined by vertices 0, 1, and 2. The second triangle makes up the unshaded area, defined by vertices 0, 2, and 3.

Available for download on Wrox.com

```
public Cube(GraphicsDevice graphicsDevice, float size) {
    Vector3[] normals = {
        new Vector3(0, 0, 1),
        new Vector3(0, 0, -1),
        new Vector3(1, 0, 0),
        new Vector3(-1, 0, 0),
        new Vector3(0, 1, 0),
        new Vector3(0, -1, 0),
    };


    var vertices = new List<VertexPositionNormalTexture>();
    List<ushort> indices = new List<ushort>();
    // Create the 6 faces
    foreach (Vector3 normal in normals)
    {
        Vector3 side1 = new Vector3(normal.Y, normal.Z, normal.X);
        Vector3 side2 = Vector3.Cross(normal, side1);

        ushort firstVertex = (ushort)vertices.Count;

        // Break the square face into two triangles
        AddSideIndices(indices, firstVertex);

        // Four vertices per face.
        vertices.Add(new VertexPositionNormalTexture(
                                (normal - side1 - side2) * size / 2,
                                 normal, default(Vector2)));
        vertices.Add(new VertexPositionNormalTexture(
                                (normal - side1 + side2) * size / 2,
                                 normal, default(Vector2)));
        vertices.Add(new VertexPositionNormalTexture(
                                (normal + side1 + side2) * size / 2,
                                 normal, default(Vector2)));
        vertices.Add(new VertexPositionNormalTexture(
                                (normal + side1 - side2) * size / 2,
                                 normal, default(Vector2)));
    }
```

```
}

private void AddSideIndices(List<ushort> indices, ushort firstVertex) {
    // First triangle
    indices.Add((ushort)(firstVertex + 0));  // Bottom left corner
    indices.Add((ushort)(firstVertex + 1));  // Top left corner
    indices.Add((ushort)(firstVertex + 2));  // Top right corner

    // Second triangle
    indices.Add((ushort)(firstVertex + 0));  // Bottom left corner
    indices.Add((ushort)(firstVertex + 2));  // Top right corner
    indices.Add((ushort)(firstVertex + 3));  // Bottom right corner
}
```

Code snippet from Cube.cs

The list of vertices and indices is then used to generate a `VertexBuffer` and an `IndexBuffer`. These are provided by the XNA Framework to improve the performance of rendering 3D shapes using a large number of triangles. The following code also creates a `BasicEffect`, which will be used to determine the properties used to draw the shape, such as lighting:

Available for download on Wrox.com

```
// Vertex and Index buffers to make drawing more efficient
VertexBuffer vertexBuffer;
IndexBuffer indexBuffer;
BasicEffect basicEffect;

public Cube(GraphicsDevice graphicsDevice, float size) {
    ...
    // Create a vertex buffer, and copy our vertex data into it.
    vertexBuffer = new VertexBuffer(graphicsDevice,
                                    typeof(VertexPositionNormalTexture),
                                    vertices.Count, BufferUsage.None);
    vertexBuffer.SetData(vertices.ToArray());

    // Create an index buffer, and copy our index data into it.
    indexBuffer = new IndexBuffer(graphicsDevice, typeof(ushort),
                                  indices.Count, BufferUsage.None);
    indexBuffer.SetData(indices.ToArray());

    // Create a BasicEffect, which will be used to render the primitive.
    basicEffect = new BasicEffect(graphicsDevice);
    basicEffect.EnableDefaultLighting();
    basicEffect.PreferPerPixelLighting = true;
}
```

Code snippet from Cube.cs

To draw the cube, you'll actually define a `Draw` method on the `Cube` that will handle setting the location, rotation, and other properties. The `DrawIndexedPrimitives` method draws the shape using the buffered list of vertices and triangle indexes:

Available for download on Wrox.com

```
public void Draw(Matrix world, Matrix view, Matrix projection, Color color)
{
    basicEffect.World = world;
    basicEffect.View = view;
    basicEffect.Projection = projection;
    basicEffect.DiffuseColor = color.ToVector3();

    GraphicsDevice graphicsDevice = basicEffect.GraphicsDevice;
    graphicsDevice.SetVertexBuffer(vertexBuffer);
    graphicsDevice.Indices = indexBuffer;

    foreach (EffectPass effectPass in basicEffect.CurrentTechnique.Passes)
    {
        effectPass.Apply();

        int triangles = indexBuffer.IndexCount / 3;
        graphicsDevice.DrawIndexedPrimitives(PrimitiveType.TriangleList, 0, 0,
                                             vertexBuffer.VertexCount, 0, triangles);
    }
}
```

Code snippet from Cube.cs

The last thing to do is to invoke the `Cube`'s `Draw` method as part of the `Draw` method on the game. The location and orientation of the cube are defined by the `world` variable. The third parameter to the `Cube`'s `Draw` method is the color that you want the cube to appear, in this case, `DarkRed`.

Available for download on Wrox.com

```
protected override void Draw(GameTime gameTime)
{
    ...

    Matrix world = Matrix.CreateFromYawPitchRoll(modelBallRotation,
                                                 modelBallRotation,
                                                 modelBallRotation) *
                   Matrix.CreateTranslation(new Vector3(0,-100,0));
    Matrix view = Matrix.CreateLookAt(cameraPosition, Vector3.Zero, Vector3.Up);
    Matrix projection = Matrix.CreatePerspectiveFieldOfView(
                                                 MathHelper.ToRadians(45.0f),
                                                 aspectRatio, 1.0f, 10000.0f);
    cube.Draw(world, view, projection, Color.DarkRed);

    base.Draw(gameTime);
}
```

Code snippet from Cube.cs

Textures

In the previous section the cube was set to be a single color. An alternative is to paint the contents of a texture (in other words, an image) on each surface of the cube. You'll start by adding an image that you want to use to the content project, as you did earlier for the 2D Sprite. In this case, the name of the texture is *blurred*, and you again load this via a call to the `LoadContent` method. You also need to update the `Draw` method on the `Cube` to accept a fourth parameter that will be the texture to paint on each surface:

Available for download on Wrox.com

```
public void Draw(Matrix world, Matrix view, Matrix projection,
                 Color color, Texture2D texture) {
    basicEffect.World = world;
    basicEffect.View = view;
    basicEffect.Projection = projection;
    basicEffect.DiffuseColor = color.ToVector3();
    basicEffect.Texture = texture;
    basicEffect.TextureEnabled = true;
    ...
}
```

Code snippet from Cube.cs

Lastly, you need to add some additional information to each vertex in the Vertices list to specify the mapping between the vertex of the shape being rendered and the texture. This is done by specifying the coordinates on the texture that each vertex should be mapped to. The coordinates are on a scale from 0 to 1, as illustrated in the updated code for the generation of the Vertices list.

Available for download on Wrox.com

```
vertices.Add(new VertexPositionNormalTexture((normal - side1 - side2) * size / 2,
                                             normal, new Vector2(1, 1)));
vertices.Add(new VertexPositionNormalTexture((normal - side1 + side2) * size / 2,
                                             normal, new Vector2(1, 0)));
vertices.Add(new VertexPositionNormalTexture((normal + side1 + side2) * size / 2,
                                             normal, new Vector2(0, 0)));
vertices.Add(new VertexPositionNormalTexture((normal + side1 - side2) * size / 2,
                                             normal, new Vector2(0, 1)));
```

Code snippet from Cube.cs

Figure 19-18 illustrates the cube both without (left) and with (right) the texture applied to the surfaces.

FIGURE 19-18

SUMMARY

The XNA Framework provides an alternative way to build applications and games for Windows Phone. In this chapter, you've learned how to work with the Update–Draw game loop to display and move content around the screen. You've also seen how you can work with 3D models and shapes, which you can use to build more sophisticated games.

20

Where to Next?

WHAT'S IN THIS CHAPTER

- Getting your device set up for development and testing
- Migrating existing applications to Windows Phone
- How to monitor and improve performance
- Preparing your application to be published

In this final chapter, you will learn a few tips and tricks about building applications for Windows Phone. This includes how to add a splash screen to your application and an overview of some of the user interface (UI) toolkits that are available to help you build richer user experiences.

DEVICE DEBUGGING

Throughout this book, you have mostly focused on working with the Windows Phone emulator that ships with the Windows Phone development tools. However, it is important that at a minimum you perform a final set of tests, including some usability testing, on a real device. Unlike previous editions of Windows Mobile, this is not as simple as copying your application onto the device. With Windows Phone the only way to distribute an application to an end user is via the Windows Phone Marketplace. For development and testing purposes, there is an alternative that involves registering a real device for use in development. This will allow you to associate up to five devices with your Windows Live ID account.

Registering for Development

In order to deploy an application to a Windows Phone device without publishing it to the Windows Phone Marketplace, you need to register the device. The steps for registering a device for use in development and testing are relatively straightforward. Before you start, make sure that you have downloaded and installed the latest version of the Zune software from `www.zune.net`. The Zune software is the companion desktop experience for your Windows Phone, replacing both ActiveSync (Windows XP) and Windows Mobile Device Center (Windows Vista/Windows 7) that are used to connect and sync data with a legacy Windows Mobile device.

Connect your Windows Phone to your computer via the USB cable supplied with your device. This should launch the Zune software on your computer (if it doesn't, you may have to run it from the Start menu) and walk through the process of setting up a profile for your Windows Phone device. Once you have completed this process, you should see your Windows Phone device profile as shown in Figure 20-1.

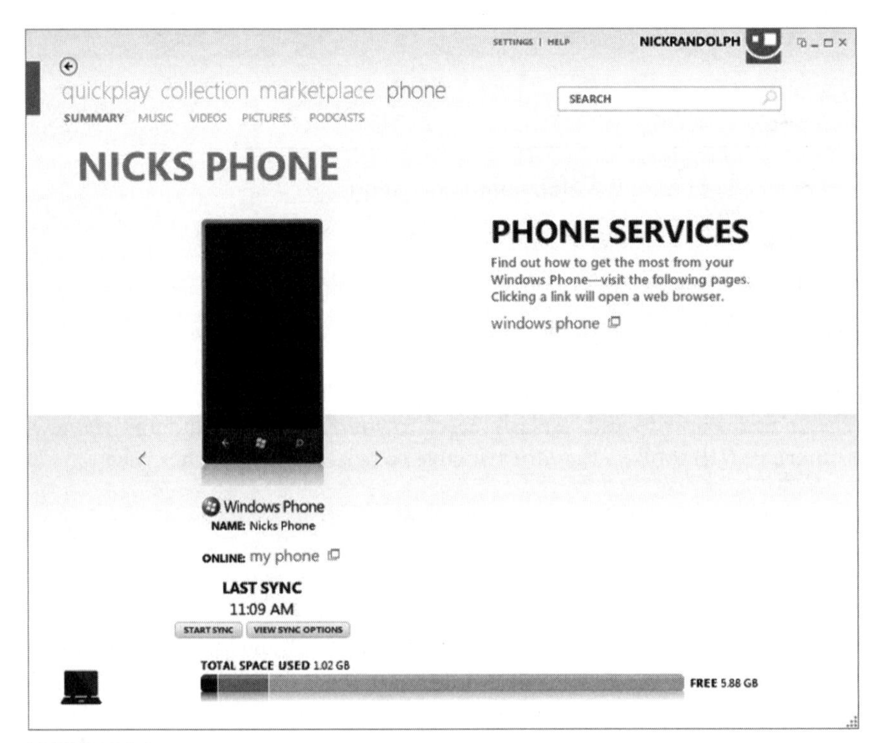

FIGURE 20-1

The next step is to run the Windows Phone Developer Registration tool, which can be found in the Start menu under All Programs ➪ Windows Phone Developer Tools. To register your device, all you need to do is enter your Windows Live ID and hit the Register button (left image of Figure 20-2). After a few seconds you should see a confirmation message indicating that your device has been registered (right image).

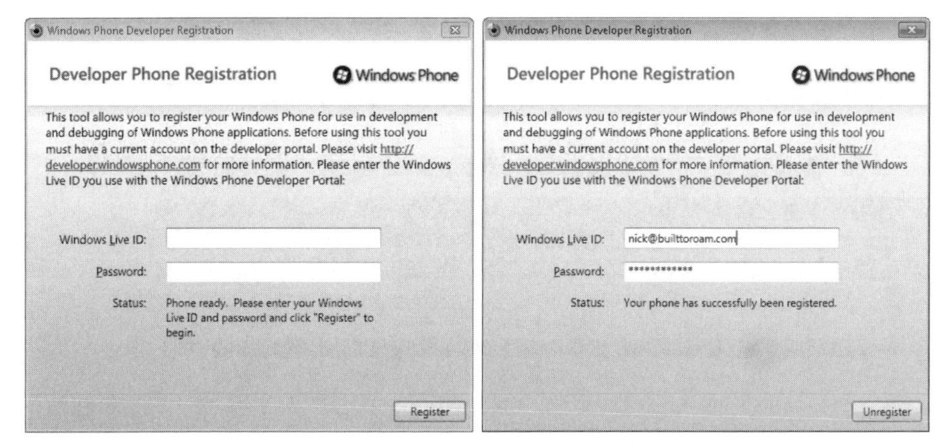

FIGURE 20-2

If for whatever reason you decide not to use a device for debugging (you might be selling the device or upgrading to a newer device), you should run the Windows Phone Developer Registration tool again and Unregister the device. As there is a limit to the number of devices you can register against your Windows Live account, it's recommended that you do this to ensure that you can continue to register new devices.

Debugging Applications

Once you have registered your Windows Phone, you can launch your application from either Visual Studio or Expression Blend. When you launch your Windows Phone application from Visual Studio, it will automatically attach the debugger, allowing you to step through code, hit break points, and inspect variables, the same as you would when the code is running in the emulator.

To start a debugging session on a real Windows Phone device, the only change you need to make is to select "Windows Phone 7 Device" from the Devices dropdown on the Standard toolbar in Visual Studio (shown in Figure 20-3). In Expression Blend there is a separate Device window where you again have the option as to whether you want to launch the application on the emulator or on a real device.

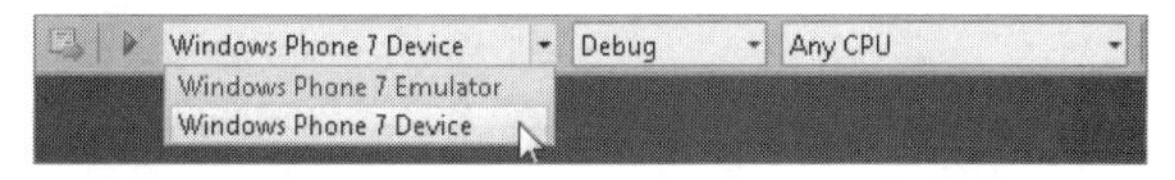

FIGURE 20-3

After selecting the "Windows Phone 7 Device" option when you press F5 to run your application, it will be deployed and launched on the real device instead of the emulator. One thing to be aware of is that if the device is in standby or the Lock screen is displayed, you may experience an "Access is denied" error (top image of Figure 20-4). Simply press the Power button to wake the device up out of standby, and dismiss the Lock screen (this may involve entering your PIN if you have enabled the Password feature). Now if you run your application, you should see that it is successfully deployed and that the application is launched (lower image of Figure 20-4).

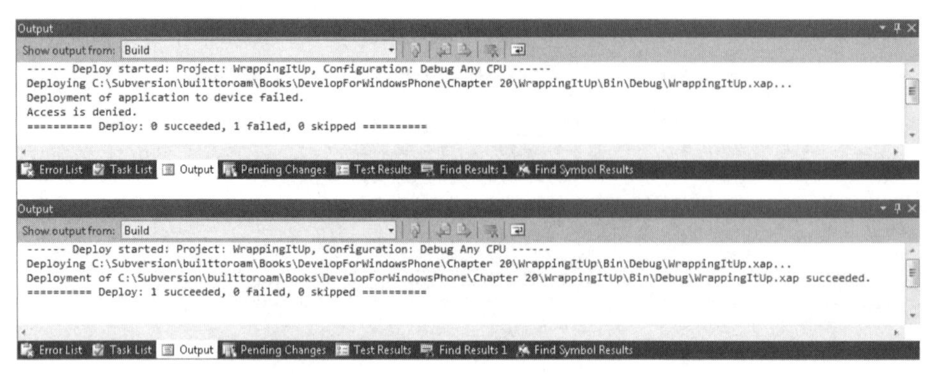

FIGURE 20-4

Deploying Applications

If you simply want to deploy an already compiled application to an emulator or a real Windows Phone device, you can use the Application Deployment tool that is installed in the Start menu at All Programs ➪ Windows Phone Developer Tools. Figure 20-5 illustrates the Application Deployment tool in action. Select the XAP file (the packaged form of your Windows Phone application, typically found in the Bin\Debug or Bin\Release subfolder of your Windows Phone application project) and which Target to deploy to, the emulator or a real device. When you click the Deploy button the Status will be updated, indicating that the application has been successfully deployed.

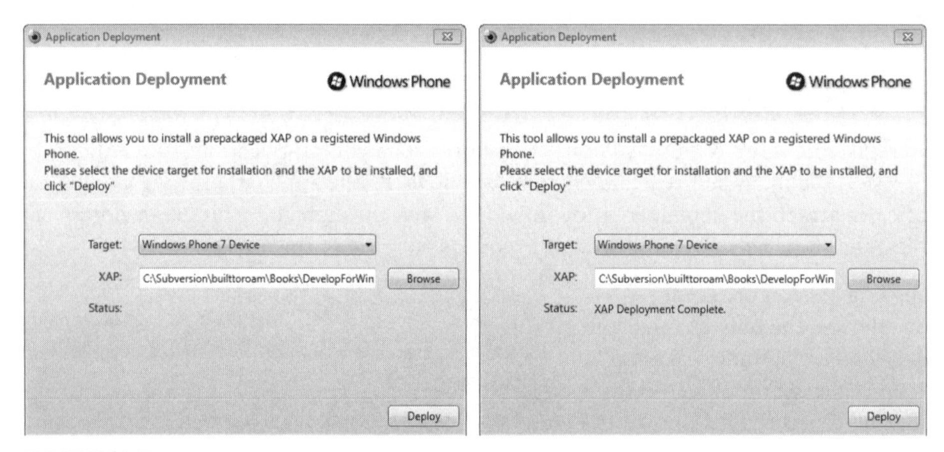

FIGURE 20-5

THIRD-PARTY COMPONENTS

There are only a limited number of controls and components that come with the Windows Phone SDK. Luckily there are a large number of vendors that have been building components for the Desktop version of Silverlight. Most of these can either be used directly, or require only minimal changes to work within a Windows Phone application.

Silverlight Toolkit

The Silverlight Toolkit (`http://silverlight.codeplex.com/`) literally includes a treasure trove full of controls, complete with source code, that can be reused within your application. Figure 20-6 illustrates the interactive sample viewer that allows you to see the list of controls available and how to work with them.

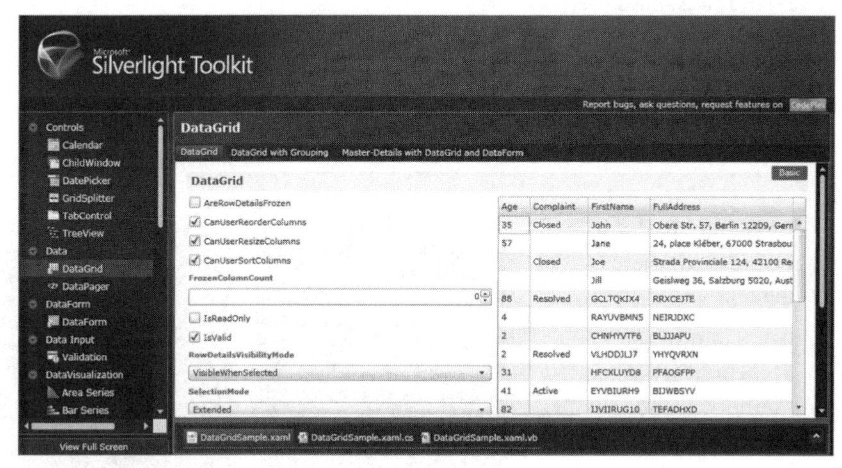

FIGURE 20-6

To work with the Silverlight Toolkit, you have a couple of options. Start by downloading the latest release, which includes the source code for Silverlight 3 and 4. Silverlight for Windows Phone is based on Silverlight 3, so you should work with the source code that is in the Silverlight 3 branch. You can either include the relevant source code for the controls you choose to use directly within your Windows Phone application, or you can simply add a reference to the Silverlight 3 class library project in Visual Studio. (The Toolkit is broken into a number of projects, so you will need to locate the right project that contains the control you want to use.)

Let's take a simple example of modifying a `ListBox` to use a `WrapPanel` from the Silverlight Toolkit to lay out list items in something other than a single vertical list. The `WrapPanel` is in the Controls.Toolkit class library project that can be found in the \Release\Silverlight3\Source\Controls.Toolkit folder. Open this project within Visual Studio. Build the project and copy the System.Windows.Controls.Toolkit.dll that is created into a subfolder of the solution that contains your Windows Phone application. Add a reference to that DLL from your Windows Phone application by right-clicking on the project node in Solution Explorer and selecting "Add Reference." Browse to where you copied the System.Windows. Controls.Toolkit.dll file, and select the assembly.

To make use of the `WrapPanel` within a `ListBox`, you need to start by adding a namespace reference to the `System.Windows.Controls.Toolkit` assembly. Then within the XAML, you can specify an `ItemsPanel` that will use the `WrapPanel` as the layout panel for the `ListBox`. You then need to specify the use of that template by setting the `ItemsPanel` property. This is all illustrated in the following code:

```
<phone:PhoneApplicationPage
    x:Class="WrappingItUp.MainPage"
    ...
    xmlns:toolkit="clr-namespace:System.Windows.Controls;
                   assembly=System.Windows.Controls.Toolkit">
```

```
<Grid x:Name="LayoutRoot" Background="Transparent">
    ...
    <Grid x:Name="ContentGrid" Grid.Row="1">
        <ListBox Margin="0,107,0,0">
         <ListBox.ItemsPanel>
                <toolkit:WrapPanel />
         </ListBox.ItemsPanel>
           <ListBoxItem Content="One" />
           <ListBoxItem Content="Two" />
           <ListBoxItem Content="Three" />
           ...
```

Code snippet from MainPage.xaml

Figure 20-7 illustrates the `ListBox` without the `WrapPanel` (left image) and with the use of the `WrapPanel` (right image).

There are many more controls within the Silverlight Toolkit that can be used without modification within your Windows Phone application.

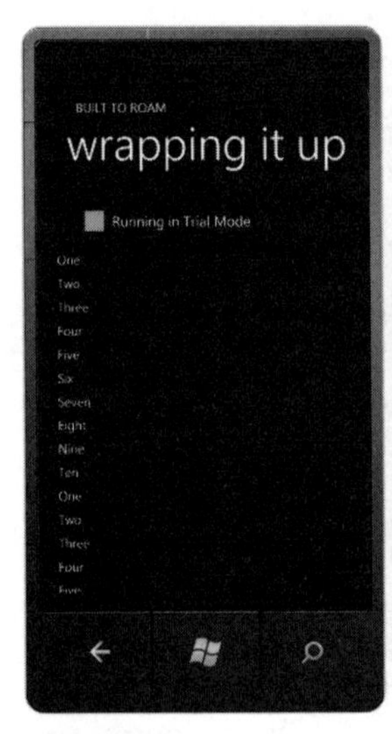

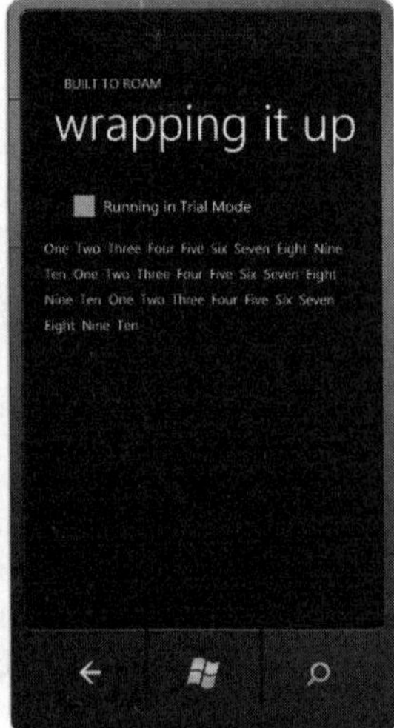

FIGURE 20-7

Database

There is currently no built-in database support for Windows Phone applications. However, there are several projects being worked on that can be used to persist data to Isolated Storage in a way that mimics the behavior of a traditional database. See Table 20-1.

TABLE 20-1: Database Alternatives

NAME	DESCRIPTION
Sterling; `http://sterling.codeplex.com`	Sterling is a lightweight object-oriented database implementation for Silverlight and Windows Phone 7 that works with your existing class structures. Sterling supports full LINQ to Object queries over keys and indexes for fast retrieval of information from large data sets.
Windows Phone 7 Database; `http://winphone7db.codeplex.com`	This project implements an Isolated Storage (IsolatedStorage)–based database for Windows Phone 7. The database consists of table objects, each one supporting any number of columns.

NAME	DESCRIPTION
WP7 SqliteClient Preview; http://sviluppomobile.blogspot.com/2010/07/wp7-sqliteclient-preview.html	This is an implementation of the Sqlite database engine to run under Windows Phone 7.
Perst; http://www.mcobject.com/perst	Perst is a commercial object-oriented database system from McObject that can be used to efficiently store and retrieve data within a Windows Phone or Silverlight application.

APPLICATION MIGRATION

There are two main scenarios to consider when discussing application migration. The first is existing desktop-based Silverlight or WPF applications that you may want to migrate to Windows Phone. Although it may be possible to migrate a large proportion of the application to run as a Windows Phone application, applications designed for the desktop don't lend themselves to being ported to a much smaller, touch-enabled form factor such as Windows Phone. To this end, it is recommended that you migrate as much of the business logic as you can but plan to rebuild the user interface (UI) layer, leveraging the page navigation model of Windows Phone. In doing this, you should also note that there are some significant differences between Silverlight for the desktop and Silverlight on Windows Phone. [The most significant ones are documented at `http://msdn.microsoft.com/en-us/library/ff426930(VS.96).aspx`.]

The second migration scenario is from an existing Windows Mobile application. These applications have a porting advantage in the fact that they typically are already designed to run on a similar form factor and a stylus-based touch interface, thus making them easier to port from an application flow and complexity perspective. The remainder of this section discusses some of the points to consider when doing such a migration.

User Interface

In order to convert an existing Windows Mobile application to Windows Phone, you will need to re-create the user interface. A Windows Mobile application written with the .NET Compact Framework will most likely use Windows Forms for its user interface, while Windows Phone uses Silverlight. There are some products emerging that can assist with converting from Windows Forms to XAML (such as the Windows Forms to XAML Converter by Ingenium; `www.ingeniumsoft.com/Products/WinForm2XAML/tabid/63/language/en-US/Default.aspx`).

As you go through your application there is a natural conversion from a Form into a PhoneApplicationPage. Most controls that you would have used in Windows Forms have an equivalent in Silverlight, and in the case in which there isn't a built-in control (e.g., the `DateTimePicker`), you can either build your own or leverage some controls that have been created for the Desktop version of Silverlight. Creating your own control in Silverlight is orders of magnitude easier than it is in Windows Forms with the designer support offered by Expression Blend.

There are some Windows Forms controls for which there isn't an actual Silverlight equivalent; however, there may be an alternative control that you can consider using that fits into the Windows

Phone style. For example, if you're converting a Form that has a `Tab` control, you might consider using the `Pivot` control on a `PhoneApplicationPage`.

You should also consider how controls that are specific to Windows Phone might be used to enhance your application. For example, you may want to use the `Panorama` control on the home page of your application, or you may want to integrate the `Map` and `Pivot` controls in order to build a richer user experience.

Services and Connectivity

As you saw in Chapter 12, there are some techniques that can be used to determine connectivity and make service requests on Windows Phone. These differ from what you may be used to coming from Windows Forms. Two of the notable differences are the absence of low-level networking such as sockets and the absence of synchronous methods. Within Silverlight all network communication is performed asynchronously.

If your Windows Mobile application relied on direct TCP or UDP socket access, you may also have to reconsider how your application is architected. Specifically, you may want to investigate if your usage of sockets could be replaced with one or more forms of Push Notification.

You should consider the absence of synchronous method calls for network operations as a forcing function to help you design your application in a way that ensures a good, responsive user experience. Whenever you have high latency operations, you should consider how you can do this in the background so as not to block the user from interacting with the application. This is particularly relevant when making service calls across the network.

Data

One of the most painful areas for application migration to Windows Phone will be how you handle data on the device. For relatively simple applications, reading and writing directly to Isolated Storage is the way to go. However, for applications that involve storing complex hierarchical or relational data offline, you might consider one of the database options mentioned in either Chapter 16 or the previous section.

Unfortunately, the lack of database support within Silverlight also means that the `System.Data` namespace isn't included. This means that classes such as `DataSet`, `DataTable`, and `DataRow` aren't available. When migrating your code you should consider creating classes that represent each data entity. For example, instead of having a `DataTable` containing rows of customer data, in your Windows Phone application you may have a list of `Customer` objects with properties such as `Name` and `Shipping Address` that represent the underlying data.

Device Capabilities

Another limitation of Windows Phone is the restricted access of device capabilities. Your Windows Mobile application may have made use of the State and Notification Broker, which allowed you to access and receive events when system properties were modified. You may have also utilized the ability to integrate into Pocket Outlook to send and intercept both e-mail and SMS (Short Message Service) messages. Windows Phone applications cannot access any of these device capabilities. In Chapter 8, you learned about using Windows Phone Tasks to access some of the device capabilities, and there are integration points for the Pictures and Music & Video hubs, as discussed in Chapter 11.

Background Processing

An important consideration when migrating your Windows Mobile application to Windows Phone is the application life-cycle model discussed in Chapter 6. Essentially this means that when your application goes into the background, it will be either suspended or terminated. As such, if your application does any long-running operations, you should consider passing the task off to a cloud-based service that can send a Push Notification to the application when it completes. This contrasts to a Windows Mobile application, which would continue to execute in the background.

USER INTERFACE PERFORMANCE

Silverlight encourages the development of rich user interfaces. Adding animation to your Windows Phone application can make your application both more exciting and intuitive to use. However, it can come at a cost, and it is important to understand how to monitor your application effectively to ensure that the user always has a smooth user experience that doesn't overly tax the hardware. In this section, you will learn about three different features that you can enable to help monitor, diagnose, and improve potential performance issues.

Performance Counters

The Silverlight for Windows Phone runtime maintains a set of performance counters that can be enabled to allow you to monitor rendering frame rates and memory usage. To enable the display of these counters, set the `EnableFrameRateCounter` property to `true` within the constructor of the `App` class:

Available for download on Wrox.com

```
public App(){
            UnhandledException += Application_UnhandledException;
            InitializeComponent();
            InitializePhoneApplication();

            Application.Current.Host.Settings.EnableFrameRateCounter = true;
}
```

Code snippet from App.xaml.cs

Figure 20-8 illustrates the various counters that are displayed at the top of the emulator when `EnableFrameRateCounter` is set to `true`. See Table 20-2 for a list of the counters.

FIGURE 20-8

If the `EnableFrameRateCounter` *property is set to* `true`*, the application will display the counters when it is run on either the emulator or on a real device. To avoid publishing an application with the counters visible, you should add a conditional compilation section where you set this property value.*

```
#if DEBUG
    Application.Current.Host.Settings.EnableCacheVisualization = true;
#endif
```

Make sure that you remember to publish using a Release build that doesn't have the `DEBUG` *compilation symbol declared.*

TABLE 20-2: Counters

COUNTER	DESCRIPTION
Compositor/Render Thread FPS	This is the frame rate (in frames per second) of the render thread.
User Interface Thread FPS	This is the frame rate of the user interface.
Texture Memory Usage	Indicates the amount of video memory used for application textures. As the number and complexity of controls, images, and media grow, so will this figure.
Surface Counter	Displays the number of surfaces that have been processed by the graphics chip. The more surfaces your application sends to the graphics chip, the more resources it will be consuming to render each refresh of the page.
Intermediate Texture Counter	A count of the number of intermediate textures that are created to render the page
Fill Rate Counter	This is the number of pixels that are painted for each frame, counted in terms of the number of full screens (i.e., 480 x 800 pixels).

The rendering engine of Silverlight for Windows Phone differs from that currently used by the Desktop version of Silverlight. Silverlight for Windows Phone uses a minimum of two threads to handle the user interface. The Render Thread handles the processing of simple animations, namely, double animations, and the built-in easing functions of simple properties, namely, render transforms, opacity, perspective transforms, and rectangular clipping. These functions can all be hardware-accelerated via the graphics processing unit (GPU). Note that this excludes scenes in which there is an opacity mask or a non-rectangular clip. The second thread is the User Input (UI) Thread, which handles processing of input, user callbacks, and custom control logic, visual layout, and more complex animations. It's important to monitor the frame rate of these two threads as they will provide early indication of any potential visual performance problems.

Redraw Regions

When you are designing your application, you should remember that the fewer items you have in the XAML visual tree, the less rendering code will need to be executed with each screen refresh, and hence you should see better performance.

While optimizing the rendering performance of your application, if you remember that "Less is More," you should be off to a great start. However, in case you have forgotten, there is another property, `EnableRedrawRegions`, that you can set so that you can see where redraw activity is occurring within your page:

```
Application.Current.Host.Settings.EnableRedrawRegions = true;
```

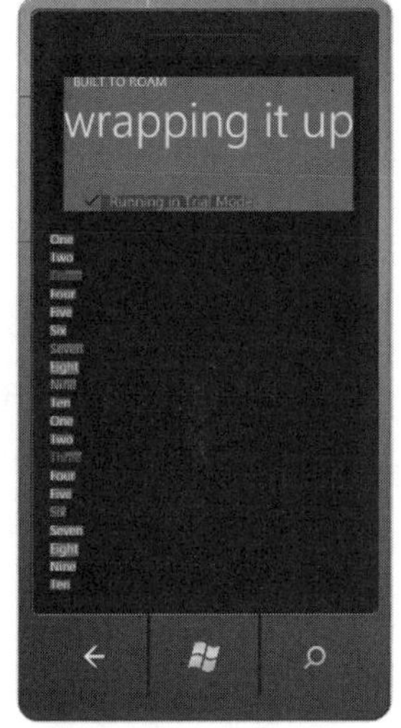

FIGURE 20-9

Figure 20-9 illustrates a page with two `TextBlocks`, a `CheckBox`, and a `ListBox`. Each time an area is redrawn, it is given a different background color (rotating between yellow, magenta, and blue). This allows you to easily see the parts of the screen that have been redrawn. The Framework will redraw any area that is considered *dirty*, which may be that an element's position has only moved by 0.01. Although this change isn't visible to the human eye, it would still result in the area being redrawn. This is something to consider if you are trying to improve the performance of your application.

Caching

The last option that you can enable to assist with monitoring performance is the `EnableCacheVisualization` property:

```
Application.Current.Host.Settings.EnableCacheVisualization = true;
```

This will tint any area that does not have its on-screen graphical representation cached. In other words, it highlights areas that may be slower to re-render when the screen requires a redraw. In the left image of Figure 20-10, the two `TextBlocks` and the `CheckBox` are not cached, whereas the `ListBox` is.

If you modify the XAML for the `TextBlock` to set the `CacheMode` to `BitmapCache`, you'll see that the `TextBlock` is now cached. This is illustrated in the right image of Figure 20-10, where the second `TextBlock` is now in white.

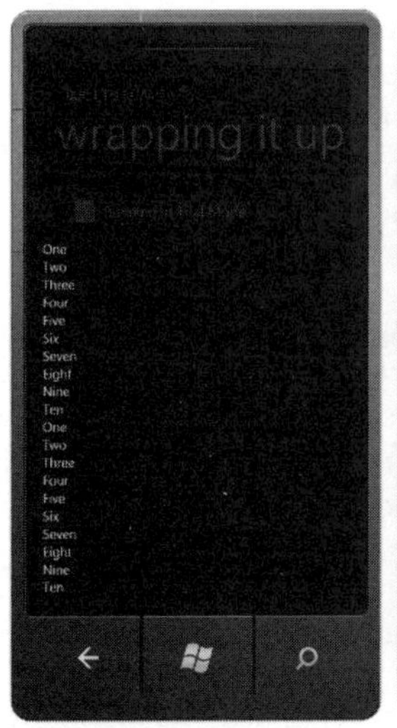

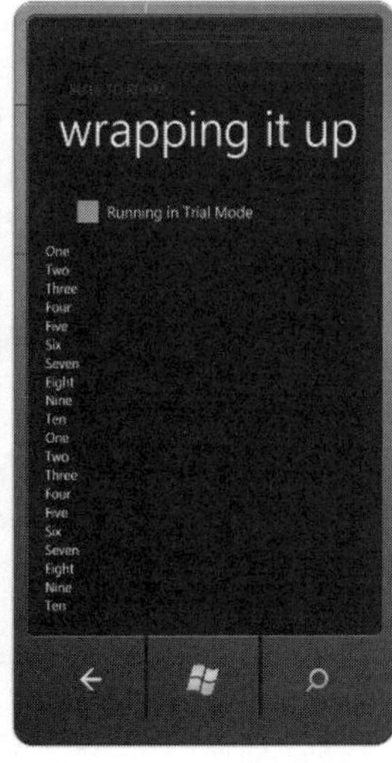

FIGURE 20-10

```
<TextBlock CacheMode="BitmapCache" x:Name="PageTitle" Text="wrapping it up"
          Margin="-3,-8,0,0" Style="{StaticResource PhoneTextTitle1Style}"/>
```

Throughout the design of your application, you should be aware of what can be hardware-accelerated. Typically this includes rendering of Render Transforms, Perspective Transforms, Opacity, and Rectangular Clips. JPEG and Media decoding is also hardware-accelerated. Leveraging this can significantly improve the performance of your application.

EXTERNAL SYSTEMS

One of the key tenants of Windows Phone applications is to make them *connected*. As discussed at the beginning of the book, this doesn't refer to the connectivity status of the device, but, rather, to the way that applications integrate with remote systems to deliver a connected and seamless experience to the end user. In this section, you'll look at two scenarios for referencing external systems.

Proxy Service (Exchange)

Quite often the external system that you want to integrate with doesn't expose an interface that is suited to being accessed directly from your Windows Phone application. Take, for example, the set of Web Services that are available for Exchange. If you look at the raw Web Services, they are overly complex and not the nicest to work with. An alternative is to use the Exchange Web Services Managed API, which is a .NET assembly that can be referenced from a desktop application or service, providing a much simpler API to work with. Unfortunately, you can't reference this assembly directly from your Windows Phone application.

Going back to the raw Web Services for a moment, the other thing that you will notice is that the payload being sent and received by the Exchange Server are large and verbose. In general, you should keep communications from your Windows Phone application as small as possible because this not only ensures a more responsive application, but can also reduce the costs associated with using your application.

One solution when faced with a Web Service such as the one offered by Exchange is to avoid communicating directly with the Exchange Server and, instead, introduce the concept of a *Proxy Service*. This is a much simpler WCF Service that has been specifically designed with your Windows Phone application in mind. All the communication with the Exchange Server will be done from within the WCF Service, with only the pertinent data being returned to the Windows Phone application.

To demonstrate this concept, create a new project called **ExchangeLink** based on the ASP.NET Empty Web Application template. Download and install the Exchange Web Services Managed API from the Microsoft downloads site (`www.microsoft.com/downloads`), and copy the Microsoft.Exchange .WebServices.dll from the installation directory, C:\Program Files\Microsoft\Exchange\Web Services\1.1, into a subfolder of the solution that contains the ExchangeLink project. Finally, add a reference to this assembly to the ExchangeLink project.

Within the ExchangeLink project, add a new item, based on the WCF Service item template, called **ExchangeLinkService.svc.** Add a new class called `MailItemSummary.cs` to the project and update it to include `From`, `Received`, and `Summary` properties:

Available for download on Wrox.com

```
public class MailItemSummary{
    public string From { get; set; }
    public string Subject { get; set; }
    public DateTime Received { get; set; }
}
```

Code snippet from MailItemSummary.cs

Next, update the WCF Service with the following code that will connect to the Exchange Server and retrieve the first 50 unread e-mail items from the Inbox:

Available for download on Wrox.com

```
[ServiceContract]
public interface IExchangeLinkService{
    [OperationContract]
    MailItemSummary[] RetrieveNumberOfUnreadItems(string emailAddress,
                                                  string password);
}
```

Code snippet from IExchangeLinkService.cs

```
public class ExchangeLinkService : IExchangeLinkService{
    private static bool ValidateRedirectionUrlCallback(string url){
        return url == "https://ex2010.myhostedservice.com/autodiscover" +
        "/autodiscover.xml";
    }

    public MailItemSummary[] RetrieveNumberOfUnreadItems(string emailAddress,
                                                         string password){
        ExchangeService service = new ExchangeService(ExchangeVersion.Exchange2010);
        service.Credentials = new NetworkCredential(emailAddress, password, "");
        service.AutodiscoverUrl(emailAddress, ValidateRedirectionUrlCallback);

        ItemView view = new ItemView(50);
        var filter = new SearchFilter.IsEqualTo(PostItemSchema.IsRead, false);
        FindItemsResults<Item> findResults =
                    service.FindItems(WellKnownFolderName.Inbox,filter, view);

        return (from item in findResults.Items.OfType<EmailMessage>()
                select new MailItemSummary(){
                    Received = item.DateTimeReceived,
                    From = item.From.Name,
                    Subject = item.Subject
                }).ToArray();
    }
}
```

Code snippet from ExchangeLinkService.svc.cs

To use this snippet to connect to your Exchange Server, simply change the URL specified in the `ValidateRedirectionUrlCallback` *method to the* `autodiscover` *URL of your Exchange Server.*

The next step is to add a reference to the ExchangeLinkService to your Windows Phone application. Right-click on the Windows Phone project node in the Solution Explorer window, and select "Add Service Reference." If the ExchangeLink project is in the same solution, then you can simply click the Discover button to locate the address of the Web Service; otherwise, you'll need to manually type the URL of the service into the `Address` field and click the Go button. Give the service a name, for example, **Exchange**, and click OK to complete the dialog.

The last step is to update the Windows Phone application to communicate with the ExchangeLinkService:

Available for download on Wrox.com

```
private void InboxButton_Click(object sender, RoutedEventArgs e)
{
    var client = new Exchange.ExchangeLinkServiceClient();
    client.RetrieveNumberOfUnreadItemsCompleted +=
                              new client_RetrieveNumberOfUnreadItemsCompleted;
    client.RetrieveNumberOfUnreadItemsAsync("<email address>", "<password>");
}

void client_RetrieveNumberOfUnreadItemsCompleted(object sender,
                          Exchange.RetrieveNumberOfUnreadItemsCompletedEventArgs e)
{
    this.UnreadItemsList.ItemsSource = e.Result;
}
```

Code snippet from ServicesPage.xaml.cs

Figure 20-11 illustrates this example in action.

The pattern used in this example can be used to connect to nearly any enterprise service such as Exchange or SharePoint in order to expose a minimal subset of data to your Windows Phone application. This not only reduces the amount of over-the-wire traffic, but also helps reduce the exposed service area of your internal services.

Since the credentials for the enterprise system are passed through in plaintext, you should ensure that all communications with your Proxy Service are done over SSL.

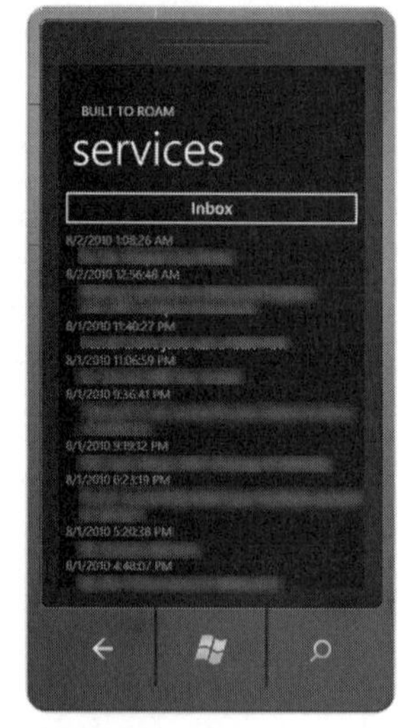

FIGURE 20-11

Shared Key Signatures (Windows Azure)

The second scenario that you're going to learn about is where you have an external system that requires an application to present a unique key in order to access resources. For example, in Windows Azure, the Blob

Storage allows you to save large blobs to a cloud-based storage. You can configure these blobs to be public, in which anyone can read them so long as they know the URL, or private, in which case you need to supply an access key as part of the URL. For most application data you are going to want to use the second option, where access to data is limited to applications that know the key.

To provide access to private blobs in Windows Azure storage from your Windows Phone application, you can either include the access key within the code of your application, or you could route all traffic through a Proxy Web Service. The first option is very bad practice as it involves distributing the access key, in other words a *secret*, inside your application as plaintext, making it easy for someone to disassemble your application in order to extract the access key. The second option is good from a security point of view since the access key is only known to the web application, which is presumably hosted within a trusted environment. However, as all traffic (including reading the blob data) would have to be routed through this service, it defeats one of the main purposes of having the data in Blob Storage. This is particularly true if you were planning on using a content distribution network to minimize download times and improve scalability.

The solution to this problem is what is known as a *Shared Key Signature*. Essentially this allows you to generate a time-limited signature that permits access to a specific blob URL. This signature would be generated by a cloud-hosted service for a blob URL that the Windows Phone application has requested access to. The generated signature is combined with the requested URL and passed back to the Windows Phone application. The Windows Phone application receives the blob URL (which contains the shared key signature) and uses that to download the blob directly from Windows Azure storage.

Let's see this in action by creating a WCF Service that will be hosted on Windows Azure. Before you begin, make sure that you download and install the latest version of the Windows Azure Tools for Visual Studio from the Microsoft downloads site (`www.microsoft.com/downloads`). Make sure that you are running Visual Studio as an Administrator and add a new project, called **SharedSignatures**, based on the Windows Azure Cloud Service project template. This will prompt you to specify which cloud projects you want to create. Figure 20-12 illustrates creating a project named **Signatures** based on the WCF Service Web Role.

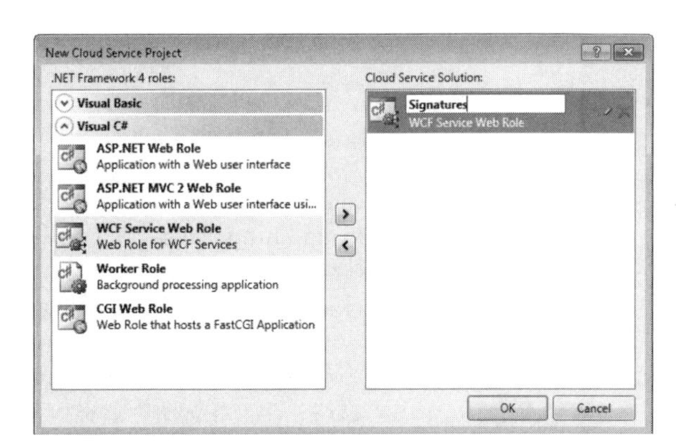

FIGURE 20-12

The WCF Service Web Role will create a default Service1.svc file, which you should delete, along with IService1.cs and Service1.svc.cs, rather than attempting to rename it. Instead, create a new item based on the WCF Service item template called **SharedSignatureService.svc**. Update the ISharedSignatureService interface and SharedSignatureService implementation as follows:

Available for download on Wrox.com

```
[ServiceContract]
public interface ISharedSignatureService{
    [OperationContract]
    string RetrieveContainerUrlWithSharedSignature(string containerName);
}
```

Code snippet from ISharedSignatureService.cs

```
public class SharedSignatureService : ISharedSignatureService{
    public string RetrieveContainerUrlWithSharedSignature(string containerName){
        var account =
      CloudStorageAccount.FromConfigurationSetting("DiagnosticsConnectionString");
        var blobs = account.CreateCloudBlobClient();

        var container = blobs.GetContainerReference(containerName.ToLower());
        container.CreateIfNotExist();

        var sas = container.GetSharedAccessSignature(new SharedAccessPolicy(){
                Permissions = SharedAccessPermissions.Write,
                SharedAccessExpiryTime = DateTime.UtcNow + TimeSpan.FromMinutes(10)
            });

        var builder = new UriBuilder(container.Uri) { Query = sas.TrimStart('?') };
        var ContainerUrl = builder.Uri.AbsoluteUri;

        return ContainerUrl;
    }
}
```

Code snippet from SharedSignatureService.svc.cs

The RetrieveContainerUrlWithSharedSignature method determines the URL for a storage container with a name matching the supplied containerName. It then generates the shared signature based on the access policy (in this case, Write) and scheduled expiry (in this case, after 10 minutes). The generated signature is combined with the container URL and returned.

WINDOWS AZURE CONFIGURATION

The RetrieveContainerUrlWithSharedSignature method relies on the configuration setting DiagnosticsConnectionString. For this to be accessible, you need to update the OnStart method in the WebRole class within the Signatures project.

```
public override bool OnStart(){
    DiagnosticMonitor.Start("DiagnosticsConnectionString");

    CloudStorageAccount.SetConfigurationSettingPublisher(
```

```
            (configName, configSetter) =>{
            // Provide the configSetter with the initial value
            configSetter(RoleEnvironment.GetConfigurationSettingValue
                                          (configName));

            RoleEnvironment.Changed += (sender, arg) =>{
                if (
          arg.Changes.OfType<RoleEnvironmentConfigurationSettingChange>()
                    .Any((change) =>
                        (change.ConfigurationSettingName == configName))){
                    if (!configSetter(
            RoleEnvironment.GetConfigurationSettingValue(configName))){
                        RoleEnvironment.RequestRecycle();
                    }
                }
            };
        });

        RoleEnvironment.Changing += RoleEnvironmentChanging;

        return base.OnStart();
    }
```

This adds a handler when the configuration information changes to allow the role to reload the configuration information and restart. As configuration information can be changed while the role is in operation, this allows the role to restart with the new configuration.

In your Windows Phone application add a reference to this new service by right-clicking on the project in Solution Explorer and selecting "Add Service Reference." Click the Discover button and then select the `SharedSignatureService`. Give the service a name, **Signatures**, and click OK. If you experience an error at this point, you may need to run the SharedSignatures cloud project in order for the WCF Service metadata end point to be accessible.

Now you're going to add functionality for the Windows Phone application to take a picture using the camera and then upload it to Windows Azure storage using a shared signature. To do this, you'll use the `CameraCaptureTask` that you saw in Chapter 8:

Available for download on Wrox.com

```
public partial class ServicesPage : PhoneApplicationPage{
    private CameraCaptureTask cameraTask = new CameraCaptureTask();
    private string filename;
    public ServicesPage(){
        InitializeComponent();
        cameraTask.Completed += new EventHandler<PhotoResult>(cameraTask_Completed);
    }

    void cameraTask_Completed(object sender, PhotoResult e){
```

```
            filename = e.OriginalFileName;
            filename = System.IO.Path.GetFileName(filename);

            CopyToIsolatedStorage(filename, e.ChosenPhoto);

            var bitmap = PictureDecoder.DecodeJpeg(e.ChosenPhoto);

            this.PhotoImage.Source = bitmap;
        }

        private void CopyToIsolatedStorage(string fileName, Stream dataStream){
            var position = dataStream.Position;
            using (var iso = new IsolatedStorageFileStream(filename,
                                        System.IO.FileMode.Create,
                                        IsolatedStorageFile.GetUserStoreForApplication())){
                var buffer = new byte[10000];
                var read = 0;
                while ((read = dataStream.Read(buffer, 0, buffer.Length)) > 0){
                    iso.Write(buffer, 0, read);
                }
            }
            dataStream.Seek(position, SeekOrigin.Begin);
        }

        private void TakePhotoButton_Click(object sender, RoutedEventArgs e){
            cameraTask.Show();
        }
    }
```

Code snippet from ServicesPage.xaml.cs

The image that is returned from the camera is copied into Isolated Storage and is displayed to the user via an `Image` control called `PhotoImage`. Uploading a blob to Windows Azure storage can involve multiple web requests, depending on the size of the blob. The following `BlobUploader` class handles splitting the file to be uploaded into appropriate blocks, uploading them, and then combining them. There is no need to call the `SharedSignatureService` because the shared signature will already be specified in the `uploadContainerUrl` parameter passed into the `BeginUpload` method.

Available for download on Wrox.com

```
    public class BlobUploader{
        public event PropertyChangedEventHandler PropertyChanged;
        private const long MaximumBlockSize = 4194304;

        public event EventHandler ProgressChanged;
        public event EventHandler UploadError;
        public event EventHandler UploadFinished;

        private Stream fileStream;
        private long totalBytesToSend;
        private long bytesSent;


        private string uploadUrl;
```

```
private bool isUsingBlocks;

private string currentBlock;
private List<string> blocks = new List<string>();

public void BeginUpload(string fileName, string uploadContainerUrl){

    // Open the file stream for the file to be sent
    fileStream = new IsolatedStorageFileStream(fileName, FileMode.Open,
                     IsolatedStorageFile.GetUserStoreForApplication());

    totalBytesToSend = fileStream.Length;
    bytesSent = 0;

    if (totalBytesToSend > MaximumBlockSize){
        isUsingBlocks = true;
    }
    else{
        isUsingBlocks = false;
    }

    // Inject the name of the file into the url
    var uriBuilder = new UriBuilder(uploadContainerUrl);
    uriBuilder.Path += string.Format("/{0}", fileName);
    uploadUrl = uriBuilder.Uri.AbsoluteUri;

    var sasBlobUri = uriBuilder.Uri;

    Upload();
}

private void Upload()
{
    long dataToSend = totalBytesToSend - bytesSent;

    var uriBuilder = new UriBuilder(uploadUrl);

    if (isUsingBlocks){
        // encode the block name and add it to the query string
        currentBlock = Convert.ToBase64String(
                          Encoding.UTF8.GetBytes(Guid.NewGuid().ToString()));
        uriBuilder.Query = uriBuilder.Query.TrimStart('?') +
            string.Format("&comp=block&blockid={0}", currentBlock);
    }

    // with or without using blocks, we'll make a PUT request with the data
    var webRequest = HttpWebRequest.Create(uriBuilder.Uri);
    webRequest.Method = "PUT";
    webRequest.BeginGetRequestStream(WriteToStreamCallback, webRequest);
}

private void WriteToStreamCallback(IAsyncResult asynchronousResult){
    var webRequest = asynchronousResult.AsyncState as HttpWebRequest;
    using (var requestStream = webRequest.EndGetRequestStream(asynchronousResult)){
```

```
            byte[] buffer = new Byte[4096];
            int bytesRead = 0;
            int tempTotal = 0;

            fileStream.Position = bytesSent;

            while ((bytesRead = fileStream.Read(buffer, 0, buffer.Length)) != 0 &&
                    tempTotal + bytesRead < MaximumBlockSize){
                requestStream.Write(buffer, 0, bytesRead);
                requestStream.Flush();

                bytesSent += bytesRead;
                tempTotal += bytesRead;

                RaiseProgressChanged();
            }
        }

        webRequest.BeginGetResponse(ReadHttpResponseCallback, webRequest);
    }

    private void ReadHttpResponseCallback(IAsyncResult asynchronousResult){
        try{
            var webRequest = asynchronousResult.AsyncState as HttpWebRequest;
            var webResponse = webRequest.EndGetResponse(asynchronousResult);
            using (var reader = new StreamReader(webResponse.GetResponseStream())){
                string responsestring = reader.ReadToEnd();
            }
        }
        catch{
            RaiseUploadError();
            return;
        }

        blocks.Add(currentBlock);

        // Check for more data
        if (bytesSent < totalBytesToSend){
            Upload();
        }
        else {
            fileStream.Close();
            fileStream.Dispose();

            if (isUsingBlocks){
                // One more request to connect all the blocks
                CombineBlocks();
            }
            else
            {
                RaiseUploadFinished();
            }
        }
    }

    private void CombineBlocks(){
```

```
        var webRequest = HttpWebRequest.Create(
                new Uri(string.Format("{0}&comp=blocklist", uploadUrl)));
        webRequest.Method = "PUT";
        webRequest.Headers["x-ms-version"] = "2010-08-03";
        webRequest.BeginGetRequestStream(WriteBlocksCallback, webRequest);
    }

    private void WriteBlocksCallback(IAsyncResult asynchronousResult){
        var webRequest = asynchronousResult.AsyncState as HttpWebRequest;
        using (var requestStream = webRequest.EndGetRequestStream(asynchronousResult)){
            var document = new XDocument(
                new XElement("BlockList",
                    from blockId in blocks
                    select new XElement("Uncommitted", blockId)));
            var writer = XmlWriter.Create(requestStream, new XmlWriterSettings()
                         { Encoding = Encoding.UTF8 });
            document.Save(writer);
            writer.Flush();
        }

        webRequest.BeginGetResponse(BlocksResponseCallback, webRequest);
    }

    private void BlocksResponseCallback(IAsyncResult asynchronousResult){
        try{
            var webRequest = asynchronousResult.AsyncState as HttpWebRequest;
            var webResponse = webRequest.EndGetResponse(asynchronousResult);
            using (var response = webResponse.GetResponseStream())
            using (var reader = new StreamReader(response)){

                string responsestring = reader.ReadToEnd();
            }
        }
        catch{
            RaiseUploadError();
            return;
        }

        RaiseUploadFinished();
    }

    private void RaiseProgressChanged(){
        if (ProgressChanged != null) {
            ProgressChanged(this, EventArgs.Empty);
            RaisePropertyChanged("PercentComplete");
        }
    }

    private void RaiseUploadError(){
        if (UploadError != null) {
            UploadError(this, EventArgs.Empty);
        }
    }

    private void RaiseUploadFinished(){
        if (UploadFinished != null) {
```

```
                UploadFinished(this, EventArgs.Empty);
                RaisePropertyChanged("PercentComplete");
            }
        }

        public double PercentComplete{
            get {
                if (totalBytesToSend > 0){
                    return ((double)bytesSent / (double)totalBytesToSend) * 100;
                }
                return 0;
            }
        }

        private void RaisePropertyChanged(string propertyName){
            if (PropertyChanged != null){
                PropertyChanged(this, new PropertyChangedEventArgs(propertyName));
            }
        }
    }
```

Code snippet from BlobUploader.cs

The last thing to do is to call the `SharedSignatureService` to retrieve the upload URL (which will contain the shared signature) and then to invoke the `BlobUploader`:

Available for download on Wrox.com

```
BlobUploader uploader;
private string filename;

public ServicesPage(){
    InitializeComponent();
    cameraTask.Completed += new EventHandler<PhotoResult>(cameraTask_Completed);

    Loaded += new RoutedEventHandler(ServicesPage_Loaded);
}

void ServicesPage_Loaded(object sender, RoutedEventArgs e){
    uploader = new BlobUploader();
    this.UploadProgress.DataContext = uploader;
}

private void TakePhotoButton_Click(object sender, RoutedEventArgs e){
    cameraTask.Show();
}

private void UploadPhotoButton_Click(object sender, RoutedEventArgs e){
    var client = new Signatures.SharedSignatureServiceClient();
    client.RetrieveContainerUrlWithSharedSignatureCompleted
                += client_RetrieveContainerUrlWithSharedSignatureCompleted;
    client.RetrieveContainerUrlWithSharedSignatureAsync("NicksPhotos");
}

void client_RetrieveContainerUrlWithSharedSignatureCompleted(object sender,
         Signatures.RetrieveContainerUrlWithSharedSignatureCompletedEventArgs e){
```

```
        var uploadUrlWithSignature = e.Result;

        uploader.BeginUpload(filename, uploadUrlWithSignature);
    }
```

Code snippet from ServicesPage.xaml.cs

Now you can run the application, take a photo, and upload it to the Windows Azure Blob Storage. At the moment, the SharedSignatures project is configured to run on the local development fabric, making use of the local development storage. This essentially mimics the behavior of the Windows Azure Services. Using the Server Explorer window in Visual Studio, you can browse the contents of the local development storage, as shown in Figure 20-13.

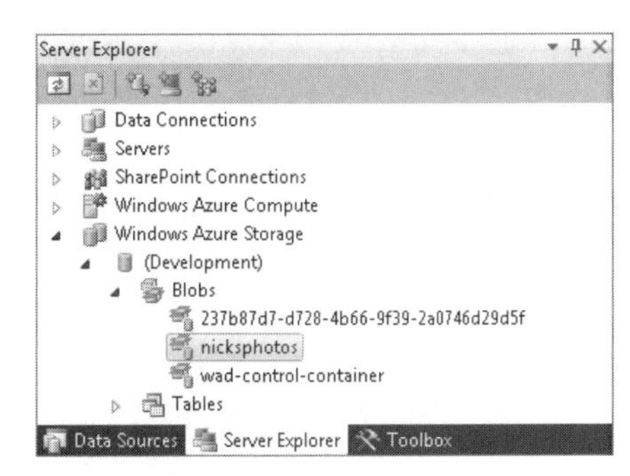

FIGURE 20-13

Double-clicking one of the container names under the Blobs node in Figure 20-13 will display a list of the files in that container, as shown at the top of Figure 20-14. If you select a file from the container, Visual Studio will attempt to display the contents of the file (middle of Figure 20-14) and the Activity Log (bottom of Figure 20-14) associated with that item.

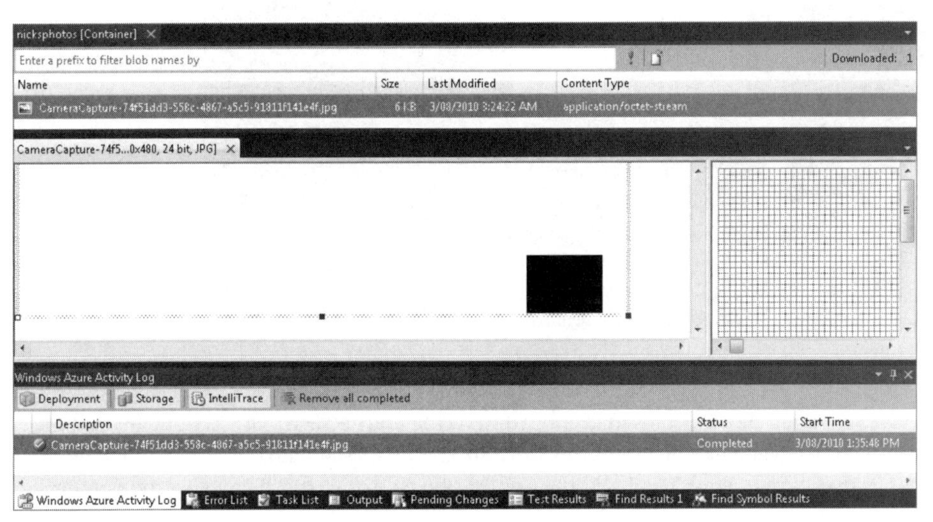

FIGURE 20-14

PUBLISHING

When it comes to distributing your application, there are a number of points that you should consider. Although this is the last section of this book, you should address these points as early as possible, if not at the beginning of the development process of your Windows Phone application. A good Windows Phone application typically involves many iterations of design, development, and end-user testing. Each iteration will involve publishing your application (if only via Visual Studio to test devices) so that you can get feedback from potential users. In the first iteration, if you set up your application so it's ready to be published, you will be in a position to quickly and easily take the last successful build and publish it in the Windows Phone Marketplace.

Application and Start Icons

In Chapter 2, you saw that in the project Properties page you can set both the `Icon` and the `Background` image properties. These correspond to the application icon, which is used in the Applications list and in the Marketplace, and the default Start screen image. By default, these properties are set to ApplicationIcon.png and Background.png, both of which are image files contained in the Windows Phone application project with their `Build Action` property set to `Content`. The ApplicationIcon.png is a 62 × 62 image, whereas the Background.png is 173 × 173. Figure 20-15 illustrates the Applications list with the WrappingItUp application listed with a 5-point star as its application icon. The rightmost image of Figure 20-15 illustrates the default tile for the WrappingItUp application.

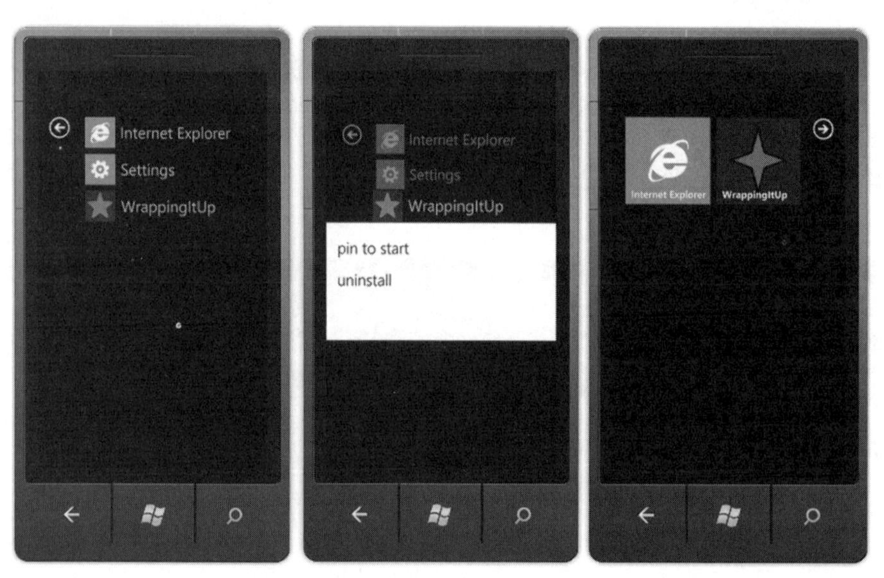

FIGURE 20-15

The default tile is used when the user pins an application to the Start screen. This is done by tapping and holding on an application in the Applications list until the dropdown appears (middle image of Figure 20-15). Selecting "pin to start" will add the application to the Start screen using the default image.

If you take a look at the WMAppManifest.xml file for the project, you will notice that the names of the application icon and default tile background are clearly identifiable:

```
<Deployment xmlns="http://schemas.microsoft.com/windowsphone/2009/deployment"
            AppPlatformVersion="7.0">
  <App xmlns="" ProductID="{64a63844-f6f6-4b77-b2be-4642e7c0a431}"
      Title="WrappingItUp" RuntimeType="Silverlight"
      Version="1.0.0.0" Genre="apps.normal"  Author="WrappingItUp author"
      Description="Sample description" Publisher="WrappingItUp">
```

```
    <IconPath IsRelative="true" IsResource="false">ApplicationIcon.png</IconPath>
    ...
    <Tokens>
      <PrimaryToken TokenID="WrappingItUpToken" TaskName="_default">
        <TemplateType5>
          <BackgroundImageURI IsRelative="true"
                              IsResource="false">Background.png</BackgroundImageURI>
          <Count>0</Count>
          <Title>WrappingItUp</Title>
        </TemplateType5>
      </PrimaryToken>
    </Tokens>
  </App>
</Deployment>
```

Code snippet from WMAppManifest.xml

If you look at the project Properties page, you will notice that there are also two `Title` fields. These correspond to the text that appears in the Applications list (and on Marketplace) and the default text that appears on the Start screen tile. The former is the `Title` attribute on the `App` element of the WMAppManifest.xml file, whereas the latter is the `Title` element later in the same file. The other thing you will notice is that you can also specify the default Count that will appear when the application is first pinned to the Start screen. In most cases, you will want to leave this as the default value of 0, in which case, the Count doesn't appear on the tile. However, you may want to set this to a number greater than 0 if you want the user to enter the application to perform some configuration of the tile or Push Notification.

Splash Screen

In Chapter 6, you saw that a splash screen can be displayed when the application is launched. This is done by creating a 480 × 800 image named **SplashScreenImage.jpg** and adding it to your Windows Phone application project with `Build Action` set to `Content`. One consideration here is that the splash screen is displayed whenever your application is launched. This means that if your application goes into the background (e.g., if the user presses the Start button) and is then terminated (i.e., tombstoned), when the user presses the Back button, your application will be restarted, displaying the splash screen. If you can reduce the startup time of your application so that you don't need a splash screen, it may be preferable to remove the splash screen to avoid users' confusion when they return to your application via the Back button.

Capabilities

Windows Phone introduces the concept of *capabilities*. In actual fact, this concept isn't new — if you go back to the Windows Mobile Marketplace, you can see that each application has a list of device requirements, for example, touch screen, camera, and so on. With Windows Phone the idea of a *capability* has been qualified to cover device or operating system resources for which there may be a privacy, security, cost, or other risk associated with their use. Take, for example, location information (acquired via the Location Services on Windows Phone) — there is a clear privacy concern regarding its use. In order for an application to access the Location Service, it needs to declare that it wants access to the `ID_CAP_LOCATION` capability in the WMAppManifest.xml. The following illustrates the list of capabilities that are requested by default by a new Windows Phone application created in Visual Studio:

```
<Deployment xmlns="http://schemas.microsoft.com/windowsphone/2009/deployment"
            AppPlatformVersion="7.0">
  <App ... >
    <Capabilities>
      <Capability Name="ID_CAP_NETWORKING" />
      <Capability Name="ID_CAP_LOCATION" />
      <Capability Name="ID_CAP_SENSORS" />
      <Capability Name="ID_CAP_MICROPHONE" />
      <Capability Name="ID_CAP_MEDIALIB" />
      <Capability Name="ID_CAP_GAMERSERVICES" />
      <Capability Name="ID_CAP_PHONEDIALER" />
      <Capability Name="ID_CAP_PUSH_NOTIFICATION" />
      <Capability Name="ID_CAP_WEBBROWSERCOMPONENT" />
    </Capabilities>
    ...
  </App>
</Deployment>
```

Code snippet from WMAppManifest.xml

There are three points at which a user may see the list of capabilities requested by an application. (See Table 20-3.) At a minimum, the list of required capabilities will be visible in the Application Details page within the Windows Phone Marketplace. Some capabilities will invoke a confirmation prompt when the application is purchased. These are typically capabilities for which there is a legal agreement regarding their use. Lastly, the user may be prompted prior to the capability being used by the application. For example, before the Location Service can be used, the user will be prompted to consent to the application using their current location. The Windows Phone capabilities are listed in Table 20-3, along with a description of what they control access to on the device.

TABLE 20-3: Windows Phone Capabilities

CAPABILITY	DESCRIPTION
ID_CAP_NETWORKING	Required for applications wanting to access data across the network. This includes access to parts of the `System.Net` namespace, the `WebBrowser` control, Smooth Streaming, and XNA GamerServices.
ID_CAP_LOCATION	Required in order to access the Location Services (System.Device.Location)
ID_CAP_MICROPHONE	Required in order to record sound via the microphone. Since this can be done without a visual prompt to the user, there is a potential privacy concern (`Microsoft.Xna.Framework.Audio.Microphone`).

CAPABILITY	DESCRIPTION
ID_CAP_MEDIALIB	Required for an application to access the media library. This covers access to the Radio, MediaLibrary, MediaSource, and MediaHistory classes.
ID_CAP_GAMERSERVICES	Required in order to access the Xbox Live APIs
ID_CAP_PHONEDIALER	Required in order for an application to initiate a phone call
ID_CAP_PUSH_NOTIFICATION	Required for an application to register for Push Notifications. There is a potential cost associated with network traffic generated by Push Notifications. This is also reliant on having the ID_CAP_NETWORKING capability.
ID_CAP_WEBBROWSERCOMPONENT	Required in order to use the WebBrowser control

Trial Mode and Marketplace

When you publish your application on the Windows Phone Marketplace, you can elect whether to include a Trial mode or not. If you enable Trial mode, users can download your application from the Marketplace without purchasing it. In this case, the Marketplace license associated with your application will have its IsTrial flag set to true. When they eventually purchase the application, this flag will change back to false. The following code illustrates how you can access the IsTrial flag within your application:

```
var license = new Microsoft.Phone.Marketplace.LicenseInformation();
var isInTrialMode = license.IsTrial();
```

The implementation of a Trial mode within your application is completely up to you. You could decide to limit the available functionality within the Trial mode, or you could enable all the functionality but limit the Trial to operate for a fixed period of time, after which all functions are disabled. When the application is running in Trial mode, you should provide visual clues to prompt the user to purchase the application. It would make sense to direct the user to the application page in the Marketplace via use of the MarketplaceDetailTask (as discussed in Chapter 8).

During development the IsTrial flag is always set to true. This can make it difficult to develop and test functionality for when the user has purchased the application. To get around this, you can use the DEBUG conditional compilation symbol to determine whether the application calls the IsTrial method or simply assumes a constant value:

Available for download on Wrox.com

```
public class ApplicationLicense{
    public bool IsInTrialMode{
        get {
#if !DEBUG
            var license = new Microsoft.Phone.Marketplace.LicenseInformation();
            return license.IsTrial();
#else
```

```
                return false;
#endif
            }
        }
    }
```

Code snippet from ApplicationLicense.cs

If you need to display or hide visual elements based on whether your application is running in Trial mode or not, you can create an instance of the `wrapper` class in the Resources dictionary of the application in the App.xaml file:

```
<Application
    x:Class="WrappingItUp.App"
    ...
    xmlns:local="clr-namespace:WrappingItUp">
    <Application.Resources>
        <local:ApplicationLicense x:Key="License" />
    </Application.Resources>
</Application>
```

Code snippet from App.xaml

You can then reference this static resource from elsewhere in your XAML:

```
<CheckBox Content="Running in Trial Mode" VerticalAlignment="Top"
          DataContext="{StaticResource License}" IsChecked="{Binding IsInTrialMode}"/>
```

Code snippet from MainPage.xaml

The left image of Figure 20-16 illustrates this checkbox with the application running in Trial mode (during debugging, the constant value is set to `true`), while the right image illustrates the application running in Full, or Purchased, mode (done by simply changing the constant value to return `false`).

You should ensure that when you publish your application you do so using the Release build configuration. This doesn't have the `DEBUG` symbol declared so the `IsTrial` method will be called on an instance of the `LicenseInformation` class.

FIGURE 20-16

In your XNA game you can determine whether you are running in Trial mode or not by querying the `Guide.IsTrialMode` *property. Rather than having to wrap this during development to simulate Trial and non-Trial modes, you can use the* `Guide.SimulateTrialMode` *property. The default value of the* `IsTrialMode` *property is* `false` *(i.e., the game is running in Full mode). Setting the* `SimulateTrialMode` *property to* `true` *will cause the* `IsTrialMode` *property to return* `true`*, allowing you to simulate Trial mode.*

SUMMARY

In this chapter, you've seen how to set up your Windows Phone so that you can debug applications to it. You took a quick look at some of the third-party frameworks you can reference within your application, as well as how you can integrate with existing enterprise systems. When you come to publish your application, you should consider leveraging Trial mode and providing as much detailed information to the user regarding the capabilities that your application requires as you can.

INDEX

Symbols

{} (curly brackets), Style, 33
"*", EntitySetsRights.All, 373

A

absolute positioning, 52
Accelerometer, 223, 224, 225
accelerometers, 221–240
  chassis design, 3
  emulator, 226–238
  games, 530–531
  orientation, 221–223
  page layout, 223
  random data, 229–230
  UI, 221
  vectors, 221–223
  Wii, 230–238
AccelerometerEventArgs, 226
AccelerometerReadingEventArgs, 225
AccelerometerStartFailedException, 226
Accept-Encoding header, 360, 362
access tokens
  Facebook, 513–514
  Twitter, 510–511
AccessTokenBaseUrl, 513
Activated, 141, 461
ActiveSync, 542
  Wi-Fi, 4
Add Existing Item, Solution Explorer, 374
Add Reference, Solution Explorer, 201
Add Service Reference, Solution Explorer, 367–368
AddressAccessDeniedException, 236
addressTask_Completed, 175
ADO.NET
  Entity Data Model, 372–377
  LastUpdated, 448
Advanced Encryption Standard (AES), 493
AES. *See* Advanced Encryption Standard
Alignment, 104
animation, 8, 79–88
  Expression Blend, 23
  navigation, 136–139
  storyboards, 138–139
  template transitions, 80–83
  transitions
    Back button, 134
    Expression Blend, 81–82
    states, 83–88
  Xbox, 7
Anonymous Authentication, 500
App, Launching, 140
Application Bar, 113–120
  EmailComposeTask, 187
  icons, 114–118
  menus, 118–119
  Opacity, 296
Application ID, 336
application templates, 65–66
Application Tile. *See* tiles
Application_Activated, 427
ApplicationBar
  properties, 116
  StateChanged, 120
ApplicationBarIconButton, 118
ApplicationBarMenuItem, 119
Application_Closing, 427
Application_Detactivated, 427
applicationGuid, 486
applicationHost.config, 362
ApplicationIdleDetectionMode, 149
Application_Launching, 427
ApplicationLifecycleObjects, App.xaml, 258
ApplicationLifetimeObjects, 468
Application.Resources, App.xaml, 65

ApplicationTitle, 55, 420
ApplicatonBarIconButton, 117
Apply Data Binding, 45
Apply3D, 263–267
App.xaml, 19, 55, 400
    ApplicationLifecycleObjects, 258
    Application.Resources, 65
    ExtendedButton, 71
    Resources window, 71–72
    Settings, 420
    UriMapping, 132
ASMX Web Service, 365–372
ASP.NET, 331, 341, 501–502
Assembly Information, 20
AssemblyCatalog, 461
assets, 516
Assets window
    Behavior tab, 75, 79
    controls, 30
    Expression Blend, 48, 251, 321
    GoToStateAction, 72
    Reset Pressed Status button, 73
AssociatedObject, 73, 74–75
AssociatedObject_Click, 75
Asynchronous, 479–480
attached properties, 52
audio
    playback, 272
    recording, 270–274
    saving, 272–274
AudioEmitter, 263, 264, 265
AudioListener, 263
AudioPlayer, 243
AudioStateConverter, 247–250
AudioStreamCount, 246
AudioStreamIndex, 246
Authenticate, 510–511
authentication, 497–514
    Anonymous Authentication, 500
    Basic Authentication, 500
        HTTP, 498
        SSL, 501
        WCF Service, 499
    connectivity, 328–347
    Delegated Authentication, 330–331, 337–341
        IIS, 331
    Forms-based Authentication, 501–507
    IIS, 498
    Live ID, 329–347
    personal and contact information, 341–347
    Web Authentication, 331–337
AuthenticationServiceClient, 506
AuthorizeUri, 509
autodiscover, 554
AutoPlay, 242, 243, 246–247
AutoResetEvent, 487
Avalon, 30

B

<b>, 40
Back button, 3–4
    emulator, 25
    MessageBox, 135
    navigation, 132–134
    no phantom back-stack, 142–143
    Silverlight, 139
Background Image, 152, 154
background processing, 140–149, 549
BackgroundColor, 116
BackKey, 135
BackKeyPress, 134, 472
Balance, 246
Base state, 66–70, 84, 104
Basic Authentication, 500
    HTTP, 498
    SSL, 501
    WCF Service, 499
BasicEffect, 537
basicHttpBinding, 365, 367, 499
battery, 5, 34
BeginExecute, 375
BeginGetRequestStream, 356
BeginGetResponse, 356
BeginInvoke, 79
BeginUpload, 558
behaviors
    motion, 71–79
    Tap gesture, 201–203
    TouchAndHold, 211
    Visual Studio, 74
Behavior tab, Assets window, 75, 79
Behavior<Button>, 74
Behavior<T>, 73, 74
BindableConnectivity, 318–320, 321

BindingMode, 388–392
BindToShellTile, 163, 170
BindToShellToast, 164–165
Bing button, 4
Bing Maps, 13, 295–307
credentials, 297
design, 296–297
events, 302–304
points of interest, 298–302
route calculator, 304–306
route display, 306–307
BitmapCache, 256, 551
BitmapImage, 43, 177, 179
BlobUploader, 562
Bluetooth, 231–232
Boolean, 247
Border
ContentGrid, 50
Grid, 51
MediaElement, 255
Pan gesture, 205
PlaneProjection, 224, 239
BorderBrush, 57, 255
BorderThickness, 57, 255
Bounce button, 88
Brush Resources tab, System Brushes, 61
BufferDuration, 271
Buffering, 244
BufferingProgress, 243
BufferingTime, 243
BufferReady, 271
Bug, 476–477
BuildAction, Content, 242, 243, 257, 565
Button, 35
Behavior<Button>, 74
Click, 49, 70
ContentGrid, 48–49
Focused state, 70
Grid, 53
Pressed state, 70
Properties window, 49
Style, 58
Tap gesture, 201
TemplateVisualState, 64
visual states, 64
WebBrowser, 321
ButtonBackground, 83
ButtonBackground Border, 57

C

cache
data, 422–442
performance, 551–552
CacheAction, 436, 438
CacheMode, 244, 441, 551
IStorageProvider, 438
ObjectCache, 437
Persist, 436
transforms, 256
CalculateRouteAsync, 304
Callback, 510–511
CameraCaptureTask, 174, 176–178, 557
CancelEventArgs, 129
CanClearTitle, 413
CanGoBack, 135–136
CanPause, 244, 246
CanSeek, 244
Canvas
Background, 209
Children, 35
ContentGrid, 51–52
Pan gesture, 205
PanAction, 208, 211
Panel, 35, 37
Silverlight, 51
TextBlock, 35
ZoomAction, 219
capabilities, 565–567
capacitance screens, 2
Central Processing Unit (CPU), 34
ChangedEntitiesFromServer, 446, 452
ChangeOrientationStateAction, 107
Channel URI, 156
ChannelErrorType, 170
ChannelOpenFailed, 170
ChannelUri, 158, 159, 160
ChannelUriUpdated, 157, 158
chassis design, 3–5
CheckBox, 36, 398, 551
Content, 251
IsChecked, 251
IsLooped, 262
Children, 35, 37
choosers, tasks, 174–175
ChosenPhoto, 177
chrome-free page layouts, 125

CID, 343
class library, 16
ClassCleanup, 480–481
ClassInitialize, 480–481
Cleanup, 423, 437
  IInitializeAndCleanup, 434
  ISyncProvider, 443
  TypeStorageContext, 435
  WCFSyncService, 450
ClearTitle, 412
Click
  Button, 49, 70
  ClickMode, 36
  MainPage, 73
  Tap gesture, 201
ClickMode, 36
ClientCredentials, 371
ClientCredentials.UserName, 500
clientCredentialType, 499
clipping, 256–257
Closed, CurrentState, 244
Closing, 140
cloud computing, 349–383
CLR. *See* Common Language Runtime
code-behind file, 30
Collapse Pane button, 19
Collapsed, 242, 296
Collection, 400
collection editor, 118
color, 6–7
  games, 535
Color, 55
Command Pattern, MVVM Light Toolkit, 411–413
Common Language Runtime (CLR), 30–31, 36–37
CommonStates, 66
compass, 3
CompleteCameraTask, 177
ComplexSilverlightInput, 326
composite WPF, 467
Composition Host, 461
Compositor/Render Thread FPS, 550
Compress-Data header, 363
compression, 360–365
Connected thread, 28, 29
ConnectionChangedEventArgs, 317
connectivity
  authentication, 328–347
  emulator, 312–315
  Internet, 309–347
  migration, 548
  network availability, 310
  service reachability, 310–311
  status, 309–320
Connectivity, 315–320
ConnectivityChanged, 317
ConsentToken, 341, 345
ConsentURL, 338
consistent start, 142
ConsoleAnalytics, 469
Contact Editor, 180
Contact, RetrieveContacts, 368
Contact Selector, 181
ContactServiceClient, 370
Content, 34–37, 49–52
  BuildAction, 242, 243, 257, 565
  CheckBox, 251
  MainPage, 223–224
  PasswordBox, 36
  TextBox, 36
Content Pipeline, 524
ContentControl, 34–35
ContentProperty, 36–37
contexts, SimpleStorageProvider, 431
context-sensitive search, 4
Control, 74
controls
  Assets window, 30
  layouts
    Expression Blend, 50
    Red Threads, 50–54
  media, 246–255
  Properties window, 30
  Red Threads, 29–50
  Silverlight, 30
  Tap gesture, 201
  templates, Red Threads, 56–59
  Toolbox, 29–30, 34–50
  WebBrowser, 320–326
ControlStoryboardAction, 88
ControlTemplate, 58
Convert, 247–248, 252, 393
ConvertBack, 247–248, 252, 393
ConvertRawSamplesToWav, 272
CookieContainer, 358–359, 506
cookies, 357–359, 505
Copy Always, 242, 257
Copy to Output Directory, 242, 257
CopyrightNotice, 326
Count, tiles, 152

Course, WCF Service, 292
CPU. *See* Central Processing Unit
Create Data Binding, 300
Create, HttpWebRequest, 507
Create Object Data Source, 405–406
CreateDirectory, 421
CreateFile, 421
CreateInstance, 260
Creates, Reads, Updates, and Deletes (CRUD), 372, 385
CreateTestPage, 484
credentials, 359, 371
Credentials, 499
CRM. *See* Customer Relationship Management
CRUD. *See* Creates, Reads, Updates, and Deletes
cubic Bézier curve, 47
current page
    hiding, 136–137
    states, 146–147
CurrentLocation, 287
CurrentState, 244
    IsEnabled, 249
    MediaElement, 247, 248
    MediaElementBinder, 249
    MediaElementState, 247
    WiiAccelerometer, 237
CurrentStateChanged, 248
CurrentStore, 435
Customer Relationship Management (CRM), 182

D

data
    cache, 422–442
    design with, 394–415
    design-time, 402–411
    encryption, 493–496
    migration, 548
    sample, 394–402
    storing, 417–441
    synchronization, 442–458
    visualization, 385–415
Data, 46, 47, 303
data binding, 33, 300, 385–386, 419–420
    Connectivity, 318–320
database, 546–547
DataContext, 386–388
    DesignTimeViewModel, 409
    Grid, 399
    INotifyPropertyChanged, 389–390
    MapData, 299–300
    PhoneApplicationPage, 299
    SoundInstancesList, 262
    ViewModel, 407
DataContract, 456
DataContractJsonSerializer, 381–382, 383, 431, 432, 452
DataMember, 456
DataRow, 548
DataServiceQuery, 375
DataSet, 548
DataSettings, 420
DataTable, 548
DataTemplate, 111
DateTime, 392
DateTime.MinValue, 447
DateToBoolConverter, 393
Deactivated, 141, 523
DEBUG, 567–568
debugging, 541–544
DecrementBusyServiceCounter, 484
Decrypt, 495–496
Deep Zoom, 326–328
DefaultTask, 127
Delegated Authentication, 330–331, 337–341
    IIS, 331
DelegationToken, 338
Delete, 430
    CacheMode, 441
    FileStream, 434
    ObjectCache, 445
    PropertyChanged, 439
    ShouldPersist, 438
Deleted
    Modified, 449
    SaveChangesToServer, 447
DeleteDirectory, 421
DeleteFile, 421
DeltaManipulation.Scale, 217
dependency properties
    MediaElementBinder, 248
    Silverlight, 77–79
DependencyObject, 78
DependencyProperty, 77–79
DependencyPropertyChangedEventArgs, 78
Description, 389, 476
design language, 5–9
design-time data, 402–411

DesignTimeViewModel, 402–405
    IRepository, 409–410
    LoadFromBook, 408
    UI, 406–407
DesiredAccuracy, 283
desktop gadgets, 152
destructive actions, 198
Details view, Back button, 4
developer landscape, 12–13
Developer Portal Services, 13
Developer Registration tool, 543
developer.windowsphone.com, 16
devices
    debugging, 541–544
    management, 492–493
    migration, 548
    security, 492
Devices window, Expression Blend, 105
DeviceType, 230, 314
DiagnosticsConnectionString, 556
dialogs, Back button, 134
Digital Rights Management (DRM), 244
DirectDraw, 12
DirectionalLight, 535
DirectoryExists, 421
DisableButtonAction, 76, 77, 78
DisabledTimeout, 77
DisabledTimeoutChanged, 77
Dispatcher, 79
Dispatcher.BeginInvoke, 356, 371
displaying new page, 137
DisplayName, 182
DoSomethingButton, 134
DoubleAnimation, 86
DoubleAnimationUsingKeyFrames, 86
Double-Tap gesture, 199, 203–214
    XAML, 204
DoubleTapAction, 203
DoubleTapTimeoutInMilliseconds, 203
DownloadProgress, 245
DownloadProgressChanged, 351
DownloadProgressOffset, 245
DownloadStringAsync, 350–352
Draw, 521, 537
DrawIndexedPrimitives, 537
DrawString, 529–530
drift effect, 209
DRM. *See* Digital Rights Management
Duration, 254, 255
dynamicTypes, 361

E

Edit in Visual Studio, 23, 74
element binding, 413–415
Eligible For Termination, 140, 141
Ellipse, 46–48
    Grid, 69
    SoundEffect, 264–266
    Visibility, 69
e-mail, 184
EmailAddressChooserTask, 174, 175–176, 186
EmailComposeTask, 174, 186–188
EmailResult, 186
emulator
    accelerometers, 226–238
    automation, 486–490
    Back button, 25
    connectivity, 312–315
    Toast Notification, 184
    WCF Service, 284
    Windows Phone Developer Tools, 24–26
EnableDefaultLighting, 535
EnableFrameRateCounter, 549–550
enableHttpCookieContainer, 505
Encrypt, 494–495
encryption, 493–496
endTestRun, 488, 489
endTestRunEvent, 488, 489
endTestRunEvent.WaitOne, 489
EnqueueConditional, 480
EnqueueTestComplete, 479
Entity Data Model, ADO.NET, 372–377
EntityCache, 425, 438, 446
EntityConfiguration, 441, 450
EntityExistsException, 441
EntityId, 433, 456
EntityKeyFunctions, 424
EntityKeyPropertyCache, 424
EntitySet, 452
EntitySetsRights.All, 373
EntityType, 423, 427
ErrorOccurred, 157, 169
errors, 169–170
EventArgs, 226
EventName, 72, 108
events
    Bing Maps, 302–304
    touch screens, 200–220
    TouchAndHold, 211
Excel, 190

Exchange Server, 184, 552–563
  WCF Service, 553
Exchange Web Services Managed API, 552
Exists, 426
Expand Pane button, 19
ExpectedException, 478–479
export, MEF, 460–463
Expression Blend, 13
  animation, 23
    transitions, 81–82
  Assets window, 48, 251, 321
  AudioStateConverter, 249
  Bing Maps, 296
  control layouts, 50
  Create Object Data Source, 405–406
  Devices window, 105
  free transport, 21–24
  keyframes, 82
  Load in WebBrowser Control, 56
  MapData, 298
  MediaElementBinder, 249
  Path, 47
  Search button, 23
  UI, 385–386
  windows, 22
  Windows Forms, 547
  workspaces, 22
expression.microsoft.com, 21
ExtendedButton, 72
  App.xaml, 71
eXtensible Application Markup Language (XAML), 15, 250, 251
  autoformatting, 18–21
  CLR, 30–31
  Converter, 547
  Double-Tap gesture, 204
  MainPage, 18–19
  ObjectKeyMap, 427
  Silverlight, 30
  StaticResources, 55
  Tap gesture, 201–202
  TargetType, 56
  Text Editor, 18
  user interface, 27
  ViewModel, 407
  Visual Studio, 18–21
  VisualStateManager, 69
  WCFSyncConfiguration, 450
  XML, 31
eXtensible Markup Language (XML)
  Extras, 193
  JSON, 380
  OData, 372
  Visualizer, 486
  XAML, 31
external systems, 552–563
Extras, 192–194

F

Facebook, 511–514
FakeAccelerometer, 229, 230
feedback
  touch screens, 7
  vibration, 239
Fiddler, 361
File menu, New Project, 16
FileExists, 421
FileMode, 421
FileStream, 432, 433, 434
FileUpload, 354
Fill, 43
Fill Rate Counter, 550
FinalVelocities, 209
Find, 158
FindByKey, 425–426
FindElementsInHostCoordinates, 207
FindName, 33
FirstOrDefault, 426
fixed orientation, 100–101
Flick gesture, 199, 209–211
FlickAction, 209
  PanAction, 211
  ZoomAction, 217
FlickDurationInMilliseconds, 209
FM tuner, 276–278
Focused state, Button, 70
FocusStates, 66
FontFamily, 36, 55–56
fonts, games, 527–530
FontSize, 36, 55–56
FontStretch, 36
FontStyle, 36, 40
FontWeight, 36, 40
Foreground, 56
ForegroundColor, 116
Forms-based Authentication, 501–507

FragmentNavigation, 131–132
FrameReported, 219–220
FrameworkElement, 55, 74
frameworks, 459–490
free transport
  Expression Blend, 21–24
  Visual Studio, 16–21
Frequency, 277
FromStream, 259
Full screen view, 125

G

gadgets, 152
games, 10–12
  accelerometers, 530–531
  capabilities, 567
  color, 535
  Deactivated, 523
  exiting, 52
  fonts, 527–530
  Initialize, 522
  keyboard, 532
  life cycle, 522–524
  lighting, 535
  LoadContent, 522
  primitives, 535–538
  rendering, 524–539
  text, 527–530
  textures, 538–539
  3D, 532–535
  touch screen, 531
  Trial mode, 569
  XNA, 15, 515–539
game loop, 521–524
  Draw, 521
  Update, 521
  variable-time-step, 522
GameTime, 522
Genre, 518
GeoCoordinateWatcher, 280–284
  Win7GeoPositionWatcher, 295
GeoEvent, 292
geo-location, 279–295
Geosense for Windows, 284
gestures, 2. *See also specific gestures*
  multi-touch, 200
  touch screen, 198–200
  Windows Phone Developer Tools, 200
Get Started tab, 16
GetContactsInformation, 343
GetContactsInformationAsync, 345
GetData, 271
GetIsNetworkAvailable, 310, 312
GetLanguageNames, 268
GetLanguagesForSpeak, 268
GetSensorsByTypeId, 286
GetUserInformation, 342
GetUserInformationAsync, 345
GetUserStoreForApplication, 421
GetValue, 77
GoBack, 135–136
GoogleAnalytics, 470
GoToConsentUrl, 339
GoToState, 70
GoToStateAction, 72, 79, 106
GPS, 3
GPU. *See* graphics processing unit
graphics acceleration, 2
graphics processing unit (GPU), 550
GraphicsDeviceManager, 518
Grid
  Border, 51
  Button, 53
  ContentGrid, 52
  DataContext, 399
  Ellipse, 69
  Panel, 37
  ScrollViewer, 102–103
Grid.Column, 53
Grid.Row, 53
Grid.RowDefinitions, 44
Grid.RowSpan, 43
GroupName, 36, 64
Guid, 433
Guide, IsVisible, 532

H

hardware
  buttons, 3–4
  requirements, 2
    RealAccelerometer, 230
HasBeenPressed, 66–67, 68, 71
HasNotBeenPressed, 66
Header, 90
Height, 20, 43
  Ellipse, 54

HidePage, 136–137
hiding current page, 136–137
Horizontal, 51
HorizontalAccuracy, 284
HorizontalAlignment, 20, 43, 54
HorizontalScrollbarVisibility, 51
Hover, ClickMode, 36
HTML tags, 40
HTTP
  Basic Authentication, 498
  headers, 371
  request, 349–350
HttpContext.Current.User.Identity, 506
HttpNotificationChannel, 156–157, 169
HttpNotificationReceived, 158, 166
HttpRequestMessageProperty, 371
HTTPS, 496
HttpWebRequest, 355–359, 496
  CookieContainer, 506
  cookies, 357–359
  Create, 507
  Credentials, 499
  WebClient, 355
hubs, 10–12
HVGA, 5

I

<i>, 40
IAccelerometer, 226, 229, 238
  UI, 227–228
  WCF Service, 236–237
IApplicationLifetime, 468
IApplicationService, 461, 468
icons
  Application Bar, 114–118
  library, 115
  Solution Explorer, 115
  Start screen, 564–565
IconUri, 117
Id
  EntityExistsException, 441
  KeyPropertyName, 423
ID_CAP_GAMERSERVICES, 567
ID_CAP_LOCATION, 566
ID_CAP_MEDIALIB, 567
ID_CAP_MICROPHONE, 566
ID_CAP_NETWORKING, 566
ID_CAP_PHONEDIALER, 567
ID_CAP_PUSH_NOTIFICATION, 567
ID_CAP_WEBBROWSERCOMPONENT, 567
identifiedChannelUri, 159
Identify, 159
IdentifyWithNoficationSource, 160
IEntityBase, 423–424
  LastUpdated, 445
  ObjectCache, 433
  SimpleEntityBase, 433
  UpdateFrom, 446
IEntityConfiguration, 430, 441, 443
IEntityTracking, 449
IGeoPositionWatcher, 284–295
Ignore, 478
IInitializeAndCleanup, 434
  IStorageProvider, 430, 443
  ISyncProvider, 450
  TypeStorageContext, 431
IIS
  Anonymous Authentication, 500
  authentication, 498
  Delegated Authentication, 331
ILogger, 462–463, 464
"I'm lost, take me to a known location" button, 4
Image, 43–46, 386–387
  Fill, 43
  PhotoImage, 558
  StackPanel, 40
IMAP. *See* Internet Message Access Protocol
Import and Export Settings, Visual Studio, 18
import, MEF, 460–463
ImportMany, 463–466
IndexBuffer, 537
indices, 536
INetworkInterface, 312–313
InheritedExport, 464
Initialize, 423, 424, 432, 434, 437, 461
  Composition Host, 461
  games, 522
  ISyncProvider, 443
  MEF, 461
  WCFSyncService, 450
InitializeComponent, 32
Inline, 37
InlineCollection, 37
InMemoryCache, 425
INotifyPropertyChanged, 389–391, 423–424, 427, 473, 474
  DataContext, 389–390

INotifyPropertyChanged (*continued*)
RaisePropertyChanged, 411
ViewModel, 403
InputScope, 111–112
Textbox, 110
Insert, 430, 441
PropertyChanged, 439
ShouldPersist, 438
SimpleEntityBase, 433
InstallAndLaunchApplication, 487
InstallApplication, 487
Instance.Pan, 262
InstanceWrapper, 262
IntelliSense, 49
interaction area, 197
Intermediate Texture Counter, 550
Internet
connectivity, 309–347
emulator, 25
security, 496–497
Internet Explorer, 25
RSS, 37
SIP, 109
WebBrowserTask, 189
Internet Message Access Protocol (IMAP), 184
Invoke, 76
OrientationChangedEventArgs, 107
IObjectCache, 423, 436
SyncService, 442
IObjectKeyMap, 423
IRepository, 409–410
IsChanged, 456
IsChecked, 36, 251
IsEnabled, 247, 249
ApplicationBarIconButton, 117
ApplicationBarMenuItem, 119
NetworkAvailable, 321
Target, 78
IsFixedTimeStep, 522
IsFullScreen, 518
IsInDesignTool, 314
IsLocked, 148
IsLooped, 262
IsMenuEnabled, 116
IsMuted, 245, 251
Isolated Storage, 417–422
database, 546
Guid, 433
LastUpdated, 444
Save, 420
Settings, 418–419
StorageFolder, 193
IsolatedStorageFile, 421
IsolatedStorageFileStream, 421–422
IsolatedStorageLogger, 464
IsolatedStorageSettings, 418
IsOpen, 303
IsOrientation, 97, 98
IsRunning, 230
IsRunningSlowly, 522
IsScriptEnabled, 322–323, 336
IStorageProvider
CacheMode, 438
DataContractJsonSerializer, 431
IInitializeAndCleanup, 430, 443
IObjectCache, 436
JSON, 431–436
ObjectCache, 436
Update, 445
WCFSyncService, 450
IsTrial, 567–568
IsTrialMode, 569
IsVisible
ApplicationBar, 116
Guide, 532
System Tray, 121
ISyncProvider, 443
ChangedEntitiesFromServer, 446
IInitializeAndCleanup, 450
RetrieveChangesFromServer, 446
SyncService, 442
WCF Data Service, 448–458
WCFSyncService, 450
ItemSource
ListBox, 111, 269
Loaded, 111
Pins, 301
ItemsPanel, 545–546
ItemTemplate, 111, 261, 302
IValueConverter, 247–248, 392–393
IWiimoteSensor, 233, 234

J

JSON, 379–383
IStorageProvider, 431–436
SaveChangesRequestWrapper, 452
WebBrowser, 325–326
XML, 380

K

keyboard, 2
    games, 532
    orientation, 100
    QWERTY, 110
keyframes, 82
KeyPropertyName, 423

L

Label, 36
Landscape, 5, 96
    accelerometers, 223
LandscapeLeft, 97, 98
LandscapeRight, 97, 98
Language, 269
LastUpdated
    ADO.NET, 448
    DateTime.MinValue, 447
    IEntityBase, 445
    Isolated Storage, 444
    SaveChanges, 449
    SaveChangesToServer, 447
Latest News tab, 16
Latitude, 298
launchers, 174–175, 188–192
LaunchersAndChoosers, 192
Launching, 140, 461
layout. *See also* page layout
    touch screens, 197–198
LayoutRoot, 43, 400
LayoutRoot Grid, 138
Life Maximizers, 8–9
light sensors, 3
Light Toolkit, MVVM, 411–415
lighting, games, 535
LineBreak, 37–42
LINQ
    Language, 269
    Rx, 238–239
List view, page layout, 124
ListBox
    Collection, 400
    Data, 303
    ItemSource, 111, 269
    ItemTemplate, 111, 261
    MainPage, 111
    StackPanel, 429
    WrapPanel, 545
List<IEntityBase>, 425
Live ID, 329–347
live tiles. *See* tiles
Load, 39
Load in WebBrowser Control, 50, 56
LoadAccelerometer, 230, 237–238
LoadCompleted, 341
LoadComponent, 32
LoadContent, 522, 528
Loaded, 39, 40–41, 111
LoadFromBook, 408
LoadUserInformation, 345
localhost, 25
LocalRedirectUrl, Navigated, 512
location, 4–5, 279–307
    Bing Maps, 295–307
    capabilities, 566
    geo-location, 279–295
    Silverlight, 52
    WCF Service, 284
    WPF, 52
Location
    Latitude, 298
    RoutePoints, 306
LocationCollection, 306
LocationGpsSensor, 286
LocationId, 338
    getUserInformation, 342
locator pattern, 407
Lock screen, 9–10, 148–149
LoggerImplementation, 461
Logical Tree, 207–208
login.live.com, 329, 334
loginUrl, 501
long running Web service, 168–169
Long Zheng, 284
Longtitude, 298
looping. *See also* game loop
    SoundEffect, 260–263

M

MailItemSummary.cs, 553
MainPage, 17
    Click, 73
    ContentGrid, 223–224
    ListBox, 111
    LoadComponent, 32
    Loaded, 39
    MEF, 462

MainPage (*continued*)
- PhoneApplicationFrame, 128
- PhoneApplicationPage, 127
- Red Threads, 31
- Source, 128
- Textbox, 111
- XAML, 18–19

MainPage.xaml, 19, 31, 57, 70
- DefaultTask, 127

Managed Extensibility Framework (MEF), 459–466
- import/export, 460–463
- Initialize, 461
- MainPage, 462

ManipulationCompleted.
- ManipulationStarted, 209

ManipulationDelta, 209
Map, 298
- MapItemsControl, 301

MapData
- DataContext, 299–300
- Expression Blend, 298
- LocationCollection, 306
- Pins, 301
- Pushpin, 301
- RoutePoints, 306
- SelectedPin, 303

MapItemsControl, 301
MapPolyline, 306
Margin, 20
- Map, 298
- orientation, 104

Marketplace, 10–12
- Developer Portal Services, 13
- registration, 541–543
- RSS, 16
- Trial mode, 567–569
- Windows Mobile, 3

MarketplaceDetailTask, 174, 191–192, 567
MarketplaceHubTask, 174, 191–192
MarketplaceReviewTask, 174, 191–192
MarketplaceSearchTask, 174, 191–192
Maximum
- Duration, 255
- ProgressBar, 44, 255
- PropertyBar, 45
- Ticks, 45

media
- controls, 246–255
- library, capabilities, 567
- playback, 241–257

Media Center, navigation, 7
MediaElement, 43–46, 190, 241–257
- AudioPlayer, 243
- BitmapCache, 256
- Border, 255
- clipping, 256–257
- CurrentState, 247, 248
- IsEnabled, 247
- IsMuted, 251
- NaturalDuration, 45
- Path tab, 45
- Position, 45
- properties and methods, 243–245
- SetSource, 242
- Slider, 45
- Speak, 270
- transforms, 255–256
- Visibility, 243
- WebClient, 242–243

MediaElementBinder
- CurrentState, 249
- dependency properties, 248
- Duration, 255
- Expression Blend, 249
- Position, 253
- XAML, 251

MediaElementState, 247, 250
MediaOpened, 44
MediaPlayerLauncher, 174, 190–191
MEF. *See* Managed Extensibility Framework
MEFApplicationService, 468
MemoryStream, 432–433
MessageBadContent, 170
MessageBox, 135
Microphone, 271, 272
- capabilities, 566
- XNA Audio Framework, 270

Microphone.Default, 270
Microsoft Silverlight Analytics Framework (MSAF), 463, 467–471
Microsoft Translator Service, 267–270
Microsoft.Devices, 240
Microsoft.Devices.Sensors, 224
Microsoft.Phone.Reactive, 238
Microsoft.SmartDevice.Connectivity, 486
MIME. *See* Multipurpose Internet Mail Extensions
Mobile World Congress, 2
Model View View Model (MVVM), 33, 402
- Light Toolkit, 411–415

Modified, 449

motion, 63–94
  animation, 8, 79–88
  behaviors, 71–79
  Panorama, 88–94
  Pivot, 88–94
  state management, 63–71
  Xbox, 7
MouseLeftButtonDown, 72, 201, 206, 302
MouseLeftButtonUp, 201, 206
MouseMove, 208, 266
MovementThreshold, 283
MSAF. *See* Microsoft Silverlight Analytics Framework
Multipurpose Internet Mail Extensions (MIME), 361
MultiScaleImage, 326–328
multi-touch, 2, 200, 214–220
Multi-Touch Vista project, 214–215
Music and Video, 10–12, 274–275
MVVM. *See* Model View View Model

N

NameCallback, 532
NaturalDuration, 245
  MediaElement, 45
  TimeSpan, 253
NaturalVideoHeight, 245
NaturalVideoWidth, 245
Navigate, 129, 321, 336
Navigate to Event Handler, 49
Navigated, 128, 512
NavigatedFrom, 146
NavigateToString, 50, 321
Navigating, 130
navigation, 123–149
  animation, 136–139
  Back button, 4, 132–134
  Media Center, 7
  Silverlight, 126–127
  WPF, 126
  Zune, 7
NavigationCancelEventArgs, 129, 135
NavigationContext, 132
NavigationFailed, 129
NavigationService, 129, 130–131, 135
NavigationStopped, 129
.NET Compact Framework, 12
network availability, 310
NetworkAddressChanged, 310, 312
NetworkAvailable, 321
NetworkChange, 310
NetworkInfo, 315
NetworkInterface, 310, 312
new page, displaying, 137
New Project, 16
no phantom back-stack, Back button, 142–143
non-destructive actions, 198
notifications
  errors, 169–170
  priority, 161–163
  Push Notification, 155–170, 565
    capabilities, 567
    Notification Service, 169
    Notification Source, 169
  Raw Notification, 165–166
    example, 168
    HttpNotificationReceived, 166
  Tile Notification, 163–164
    example, 166–167
  Toast Notification, 164–165
    emulator, 184
    example, 167–168
Notification Service, 156, 169
Notification Source, 156, 169
  WCF Service, 158
  WPF, 160
NotificationRateTooHigh, 170
notify, 324
NotImplementedException, 248
NotSupportedException, 248
nRoute, 467

O

OAuth 1.0, 507–511
OAuth 2.0, 511–514
object cache, 422–430
ObjectCache, 424, 426–427
  CacheMode, 437
  ChangedEntitiesFromServer, 452
  Delete, 445
  Exists, 426
  IEntityBase, 433
  IObjectCache, 436
  IStorageProvider, 436
  ISyncProvider, 443
  SaveChangesToServerCompleted, 456
ObjectKeyMap, 427

ObjectKeyMappings, 423
Objects and Timelines, 23, 50
Bounce button, 88
ContentGrid, 50
Record Keyframe button, 82, 108
Show Timeline button, 108
templates, 65–66
ToggleOpacityAction, 76
obscured screen, 148–149
observable sequence, 238
ObservableCollection, 404
ObservableReadingChanged, 239
Occasionally Connected, 310
OData. *See* Open Data Protocol
Office, 10–12
OnAttached, 73, 74, 75
OnBackKeyPress, 134, 135
OnDetaching, 73, 74, 75
OneTime, 388
OneWay, 388
OnFragmentNavigation, 130
OnNavigatedFrom, 130, 136–137
OnNavigatedTo, 130, 137, 193
OnNavigatingFrom, 130
OnStart, 556
OnTouchAndHoldCompleted, 214
OnTouchAndHoldStarted, 214
Opacity
Application Bar, 119, 296
ApplicationBar, 116
AssociatedObject_Click, 75
Open Data Protocol (OData), 372–379
Client Library, 374
OpenGL, 12
Opening, 244
OpenReadAsync, 351–352
OpenStream, 259
OpenWriteAsync, 352, 355
OpenWriteCompleted, 355
OperationContextScope, 371
OperationContract, 353
orientation, 5, 95–109
accelerometers, 221–223
Alignment, 104
auto-layout, 101–103
changes, 99
detection, 96–98
emulator, 25
fixed, 100–101
keyboard, 100
manual intervention, 103–104
Margin, 104
smooth transitions, 108–109
splitter bar, 19
states, 104–108
strategies, 99–109
VisualStateManager, 106
Orientation, 96
Slider, 251
StackPanel, 51
orientation, testOrientation, 97
OrientationChanged
EventName, 108
GoToStateAction, 106
OrientationChangedEventArgs, 107
OriginalFileName, 177

P

page layout, 123–126
accelerometers, 223
chrome-free, 125
Page_Load, 334
PageOrientation, 96, 98, 100
PageTitle, 19
Pan, 262
Pan gesture, 199, 205–209
Border, 205
Canvas, 205
MouseMove, 208
PanAction
Canvas, 208, 211
FlickAction, 211
MouseleftButtonDown, 206
MouseLeftButtonUp, 206
Panel
Canvas, 35, 37
Grid, 37
Panorama
motion, 88–94
Style, 90–92
PanoramaItem, 89, 90
Panoramic view, page layout, 124
Password, 500
PasswordBox, 36
PasswordChar, 36
Path, 46–48

    Data, 46, 47
    Expression Blend, 47
    Visual Studio, 47
Path tab, 45
Pause, 245
Paused, 244
PayloadFormatError, 170
Pen, 47
Pencil, 47
People, 10–12
performance counters, 549–550
Persist, 439–440
    CacheMode, 436
persistent storage, 430–441
persisting objects, 436–441
Personal thread, 28
Perst, 547
Phone Lock, 9
PhoneApplicationFrame, 128, 130
PhoneApplicationPage, 31
    BackKey, 135
    ChangeOrientationStateAction, 107
    DataContext, 299
    LayoutRoot Grid, 138
    MainPage, 127
    MSAF, 463
    NavigationService, 129, 130–131
    OnNavigatedTo, 193
    Pivot, 92, 548
    Properties window, 39
    SoundEffectInstance, 260
    State, 147
    TriggerAction<T>, 107
    virtual methods, 130
PhoneBackgroundBrush, 59–60
PhoneCallTask, 174, 182–183
PhoneNumber, 182
PhoneNumberChooserTask, 174, 181–182
PhoneNumberResult, 182
PhotoChooserTask, 174, 178–179
PhotoImage, 558
PhotoResult, 176, 177, 178
Photos, 90
photos, WCF Service, 347
PhotosExtrasApplication, 192
PickRuntimeNetworkInterface, 314
Picture Picker, 178–179
Pictures, 10–12
pinch, 2
Pinch and Stretch gesture, 200, 217–219
Pins
    ItemSource, 301
    ItemTemplate, 302
    MapData, 301
PipeTo, 364
Pitch, 260
Pivot
    motion, 88–94
    PhoneApplicationPage, 92, 548
Pivot view, page layout, 124
plain old CLR object (POCO), 474
PlaneProjection, 224, 239
Play, 245
    SoundEffect, 260, 272
Playing
    CurrentState, 244
    MediaElementState, 250
POCO. *See* plain old CLR object
points of interest, Bing Maps, 298–302
policy.html, 334
POP3. *See* Post Office Protocol
Popup, 303
Portrait, 95–96, 98
    accelerometers, 223
    Start button, 5
    SupportedOrientation, 101
PortraitDown, 96, 98
PortraitOrLandscape, 20, 100
PortraitUp, 98
Position, 245
    GeoCoordinateWatcher, 281–282
    MediaElement, 45
    MediaElementBinder, 253
    TimeSpan, 253
    Value, 255
PositionChanged, 281–282
PositionString, 284
Post Office Protocol (POP3), 184
PowerLevelChanged, 170
PowerPoint, 190
PreferredBackBufferHeight, 518
PreferredBackBufferWidth, 518
Press, ClickMode, 36
Pressed state, button, 70
PressMeButton, 70
previous application, Back button, 134
previous page, Back button, 134
primitives, games, 535–538
Priority, notification priority, 161
Prism, 467

privacy, location information, 4–5
ProceedDialog, 135
ProgressBar, 43–46
 Maximum, 255
Projection, 255–256
Projects window, 23, 74
Properties window, 19
 Button, 49
 controls, 30
 IsEnabled, 249
 PhoneApplicationPage, 39
 RenderTransform, 23
PropertyBar, 45
PropertyChanged, 439, 474–475, 476
PropertyInfo, 424
proximity sensors, 3
Proxy Service, 552–563
PublishedDate, 392, 398–399
publishing, 563–569
Push Notification, 155–170, 565
 capabilities, 567
 Notification Service, 169
 Notification Source, 169
Pushpin
 ItemTemplate, 302
 MapData, 301
 MouseLeftButtonDown, 302

Q

QueryString, 131–132
QWERTY keyboard, 110

R

RadioBinder, 277
RadioButtons, 36
RaisePropertyChanged, 411
random data, accelerometers, 229–230
Raw Notification, 165–166
 example, 168
 HttpNotificationReceived, 166
Reactive Extensions for .NET (Rx), 238–239
Read, 432–433
ReadingChanged
 Accelerometer, 224
 IAccelerometer, 226, 229, 238
 Rx, 239
Real Time, notification priority, 161
RealAccelerometer, 228–229, 230
Really Simple Syndication (RSS)
 Internet Explorer, 37
 Latest News tab, 16
 Marketplace, 16
 TextBlock, 37–42
ReceiveChangesFromServer, 443
Record Keyframe button, Objects and Timeline, 82, 108
Rectangle, 46–48, 256
Red Threads, 8, 27–61
 controls, 29–50
  layouts, 50–54
  templates, 56–59
 MainPage, 31
 themes, 59–61
redraw regions, 551
RedThreadLayout, 31, 43
RegisterChannelButton_Click, 157
registration, Marketplace, 541–543
Regular Expressions, HTML tags, 40
Regular, notification priority, 161
RelayCommand, 412–413
Release, ClickMode, 36
Relevant thread, 28–29
RemoteImageUrl, 155
RenderedFramesPerSecond, 246
RenderTransform, 255–256
 ButtonBackground, 83
 Properties window, 23
ReportFinalResult, 489
Representational State Transfer (REST), 341–342, 349, 353
 FileUpload, 354
 GetContactsInformation, 343
request token, Twitter, 508–509
RequestUri, 508
Reset Pressed Status button, 73, 76
resilience, 156
resource binding, 413–415
resource dictionary, 55
Resources window, 71–72
REST. *See* Representational State Transfer
RetrieveChangesFromServer, 446, 452
RetrieveContacts, 368, 370–371
RetrieveContainerUrlWithSharedSignature, 556
RetrieveRequestToken, 509
RetrieveSetting, 418

Return URL, 331
ReturnToken, 334, 338
Rivera, Rafael, 284
RootPage, 463
RotateBounceButton, 88
Route Service, Bing Maps, 304–307
RouteHelper, 304–305
RoutePoints, 306
RouteServiceClient, 304
RowDefinition, 34
RSS. *See* Really Simple Syndication
Run, 40
  IsRunning, 230
  Start, 229
  TextBlock, 37–42
  while, 230
Runtimes, 13
Rx. *See* Reactive Extensions for .NET

S

SafeDictionaryValue, 147
sample data, 394–402
SampleRate, 272
Save, 420
SaveChanges, 449
SaveChangesRequestWrapper, 452
SaveChangesToServer, 443, 447
  WCF Data Services, 452
SaveChangesToServerCompleted, 456
SaveContact, 370
SaveEmailAddressTask, 174, 185–186
SavePhoneNumberTask, 174, 179–181
SaveReportFile, 482, 485, 489
SaveReportFileAsync, 484
ScaleVisibility, 296
ScaleX, 82–83
ScaleY, 82–83
screen resolution, 5
script, 341
ScriptNotify, 323, 330
  ConsentToken, 345
  WebBrowser, 340
  window.external.Notify, 337
ScrollViewer, 38
  ContentGrid, 38
  Grid, 102–103
  HorizontalScrollbarVisibility, 51
  StackPanel, 48, 51
SeaDragon, 326–328
Search button, 3
  context-sensitive search, 4
  emulator, 25
  Expression Blend, 23
SearchButton, 23, 24
SearchTask, 174, 188–189
SearchTextBox, 20, 25
Secure Sockets Layer (SSL), 359, 496
  Basic Authentication, 501
  WCF Authentication Service, 507
security, 491–514
  device, 492
  Internet, 496–497
  WCF, 496
Select, 425, 430
  FileStream, 433
Select Device, Visual Studio, 24
SelectedElement, 266
SelectedItem, 111–112
SelectedPin, 303
SendNotification, 168
sensor_ReadingChanged, 224
SensorState, 225
SensorText, 224
Serializer, 432
ServerChangesRequestWrapper, 452
ServiceContract, 353
ServiceReachable, 321
ServiceReportingProvider, 484, 489
services
  migration, 548
  reachability, 310–311
ServiceTestUrl, 315
ServiceTimeout, 315
SetSource, 242, 245
Settings
  App.xaml, 420
  DataSettings, 420
  Isolated Storage, 418–419
Settings, Phone Lock, 9
SetValue, 77
shared key signatures, 554–563
SharedSignatureService, 557, 562
SharpZipLib, 363
shell:Application, 118
ShellTileSchedule, 155
ShellToastNotificationReceived, 158
Short Message Service (SMS), 548

ShouldPersist, 438
Show
- CameraCaptureTask, 176
- SavePhoneNumberTask, 180
- SMSComposeTask, 183
- tasks, 175
- WebBrowserTask, 190

Show Timeline button, 108
ShowCamera, 179
Sign In button, 336
Sign Out button, 336
SignalStrength, 276
SignIn, 357
Silverlight, 13, 15
- Back button, 139
- behaviors, 71, 76
- Canvas, 51
- controls, 30
- dependency properties, 77–79
- IValueConverter, 247
- location, 52
- navigation, 126–127
- Toolkit, 545–546
- UserControl, 126
- vectors, 223
- windows, 126
- Windows Forms, 547–548
- XAML, 30
- XNA games, 518

Silverlight for Windows Phone, 17
Silverlight Unit Test Framework (SUTF), 471–474
silverlightContent, 324
SilverlightInput, 324
Simple Mail Transfer Protocol (SMTP), 184
Simple Object Access Protocol (SOAP), 349, 353
SimpleEntityBase, 433, 456
SimpleLogger, 462, 464
SimpleStorageProvider, 440
- contexts, 431
- IInitializeAndCleanup, 434
- StoreName, 431

SIP. *See* Soft Input Panel
Slider
- MediaElement, 45
- Orientation, 251
- Pan, 262
- ProgressBar, 45
- Value, 262

smartphones, 6
smooth transitions, 108–109
SMS. *See* Short Message Service
SmsComposeTask, 174, 183–184
SMTP. *See* Simple Mail Transfer Protocol
snapping, 82
Snippet Editor, 381
SOAP. *See* Simple Object Access Protocol
Soft Input Panel (SIP), 109–113
- default layouts, 112
- Internet Explorer, 109
- TextBox, 110

SolidColorBrush, 55
Solution Explorer, 19, 20
- Add Existing Item, 374
- Add Reference, 201
- Add Service Reference, 367–368
- icons, 115
- Update Service Reference, 369
- WCF Service, 159

Souce, BitmapImage, 43
SoundEffect
- balance, 260–263
- CreateInstance, 260
- Ellipse, 264–266
- FromStream, 259
- looping, 260–263
- Pitch, 260
- Play, 260, 272
- SoundEffectInstances, 263
- SoundsList ListBox, 262
- Speak, 270
- 3D, 263–267
- Volume, 260–263
- XNA, 257–267

SoundEffectInstance, 260, 263, 264
SoundInstancesList, 262
SoundsList ListBox, 262
SoundUtility, 272
Source, 245, 386–387
- AutoPlay, 242
- MainPage, 128
- Navigate, 129
- URI, 242

source, UI, 385
Speak
- GetLanguagesForSpeak, 268
- MediaElement, 270
- SoundEffect, 270

Speech API, 267
Speed, WCF Service, 292
splash screen, 128, 565

splitter bar, 19
sprites, 525–526
spritesheets, 525
SSL. *See* Secure Sockets Layer
StackPanel, 38
  ContentGrid, 38
  Horizontal, 51
  Image, 40
  ListBox, 429
  Orientation, 51
  PanoramaItem, 89
  ScrollViewer, 48, 51
  TextBlock, 39
Start
  Accelerometer, 223
  GeoEvent, 292
  IAccelerometer, 226
  Run, 229
  TimeSpan, 240
  Try-Catch, 226
Start button, 3
  Eligible For Termination, 140, 141
  emulator, 25
  Portrait mode, 5
Start screen, 9–10, 151
  icons, 564–565
  New Project, 16
  states, 146
State
  PhoneApplicationPage, 147
  TryGetPreviousLocation, 531
states
  animation transitions, 83–88
  current page, 146–147
  management, 63–71
  orientation, 104–108
  saving, 146–148
  Start screen, 146
State Recording mode, 67
StateChanged, ApplicationBar, 120
StateName, VisualStateManager, 72
States window, 65–66
StaticResource, Style, 55
staticTypes, 361
Status, Connectivity, 317
Sterling, 546
Stop
  IAccelerometer, 226
  MediaElement, 245
  Microphone, 272
Stopped
  CurrentState, 244
  MediaElementState, 250
StorageFolder, 193
StoreName
  Initialize, 434
  SimpleStorageProvider, 431
storyboards, animation, 138–139
stream of data, 238
StreamReader, 421
StreamWriter, 421
string-binding, 401
Style
  {} (curly brackets), 33
  ApplicationTitle, 55
  Button, 58
  FontFamily, 55–56
  FontSize, 55–56
  Foreground, 56
  FrameworkElement, 55
  Panorama, 90–92
  StaticResource, 55
  Template, 58
  templates, 72
  TextBlock, 55
stylus, 2
Subscribe, 238–239
SupportedOrientations, 20, 100, 101
  accelerometers, 223
SupportedPageOrientation, 100
Surface Counter, 550
SUTF. *See* Silverlight Unit Test Framework
swipe, 2, 205
Sync, 442
  RetrieveChangesFromServer, 446
SyncCallback, 452, 455
synchronization
  data, 442–458
  WCF Data Services, 442
SyncService, 442
SyndicationFeed, 39
System Brushes, 61
System Tray, 120–121
System.Data, 548
System.Environment.DeviceType, 314
System.Net, 566
System.Observable, 238
System.Runtime.Serialization, 383
System.ServiceModel.Syndication, 39
System.ServiceModel.Web, 381

System.Windows.Controls.Grid, 33
System.Windows.Controls.Toolkit, 545
System.Windows.Interactivity, 468
System.Windows.Media, 250

T

Tag, 477–478
tap, 2
Tap gesture, 199, 201–203
- behaviors, 201–203
- Button, 201
- Click, 201
- controls, 201
- XAML, 201–202

TapAction, 203
Target, 78
TargetedTriggerAction<Button>, 76
TargetedTriggerAction<T>, 73, 79
TargetElapsedTime, 521–522
TargetName, 73
TargetType, 56
targetType, 392
tasks, 173–194
- choosers, 174–175
- e-mail, 184
- launchers, 174–175
- Show, 175

tasks, 20
TCP. *See* Transmission Control Protocol
TDD. *See* test-driven development
Team Foundation Server (TFS), 477
Template
- ControlTemplate, 58
- Style, 58

templates
- animation transitions, 80–83
- Objects and Timelines, 65–66
- Red Threads, 56–59
- Style, 72

TemplateVisualState, 64
TestClass, 473
TestCleanup, 480
test-driven development (TDD), 471
testing, 471–486
- Visual Studio, 481–486

TestInitialize, 480
TestMethod, 473
testOrientation, 97
TestReportingProvider, 481, 484
TestReportService, 483
TestReportServiceClient, 484
Text, 20, 386–387
- ApplicationBarIconButton, 117
- ApplicationBarMenuItem, 119

Text Editor, 18
text, games, 527–530
Text Visualizer, 486
TextBlock, 19, 36, 386–387, 551
- <b>, 40
- Canvas, 35
- ContentProperty, 37
- HTML tags, 40
- <i>, 40
- InputScope, 111–112
- LineBreak, 37–42
- PanoramaItem, 89
- RSS, 37–42
- Run, 37–42
- SensorText, 224
- StackPanel, 39
- Style, 55
- Wiimote, 238

TextBlock.Text, 33
TextBox
- ApplicationTitle, 420
- Content, 36
- InputScope, 110
- MainPage, 111
- SIP, 110
- Title, 389
- WebBrowser, 321

Texture Memory Usage, 550
textures, XNA games, 538–539
TFS. *See* Team Foundation Server
themes
- Red Threads, 59–61
- Toolbox, 59

third-party components, 544–547
this.FindName, 32
3D
- games, 532–535
- graphics, 15
- SoundEffect, 263–267
- vectors, 221

3G+ networks, 4

Ticks
    Duration, 254
    Maximum, 45
    TimeSpan, 254
tiles, 9–10, 151–155
    RemoteImageUrl, 155
Tile Notification, 163–164
    example, 166–167
Timeout, 480
TimeoutInMilliseconds, 290
TimerGeoPositionWatcher, 293
    GeoEvent, 292
TimeSpan, 243, 253
    Start, 240
    Ticks, 254
TimeSpanConverter, 255
Title, 565
    TextBox, 389
    tiles, 152
TitleContainer, 259
Toast Notification, 164–165
    emulator, 184
    example, 167–168
ToggleControlSwitch, 36
ToggleOpacityAction, 76
Tombstone, 141, 144
Toolbox
    controls, 29–30, 34–50
    themes, 59
Tools and Support, 13
Touch, 219
Touch and Hold gesture, 200, 211–214
Touch Gesture Reference Guide (Wroblewski), 200
touch screens, 195–220
    events, 200–220
    feedback, 7
    games, 531
    gestures, 198–200
    guidelines, 196–200
    layout, 197–198
    multi-touch, 214–220
    user experience, 195–200
touch target, 197
TouchAndHold, 211
TouchAndHoldAction, 211
TouchLocation, 531
TouchPanel, 531
TrackAction, 469–470
transforms
    CacheMode, 256
    MediaElement, 255–256
transitions, 7–8
    animation, 134
TranslateX, 105, 108
TranslateY, 105, 108
Transmission Control Protocol (TCP), 349
    migration, 548
Trial mode, Marketplace, 567–569
TriggerAction<T>, 73
    PhoneApplicationPage, 107
    ToggleOpacityAction, 76
try-catch
    Start, 226
    window.external.Notify, 324
TryConnect, 234
TryGetPreviousLocation, 531
TryStart, 282
Turn off Timeline snapping button, 82
Twitter, 507–511
    access tokens, 510–511
    request token, 508–509
TwoWay, 389
Type, 423, 425
    EntityCache, 438, 446
    TypeStorageContext, 431, 435
TypeConverter, 427
TypeStorageContext, 431–432
    Cleanup, 435
    CurrentStore, 435
    IInitializeAndCleanup, 431
    Type, 431, 435

U

UDP. *See* User Datagram Protocol
UI. *See* user interface
UIElement, 74
    FrameReported, 219–220
    TouchAndHoldAction, 211
Uniform, 43
Uniform Resource Identifier (URI), 233
    Source, 242
UniformFile, 43
unit testing, 471–486
Update, 430

Update (*continued*)
CacheMode, 441
FileStream, 434
game loop, 521
GameTime, 522
IStorageProvider, 445
Update Service Reference, Solution Explorer, 369
UpdateAsync, 479
UpdateFrom, 446
UpdateSounds, 265
uploadContainerUrl, 558
UploadStringAsync, 355
URI. *See* Uniform Resource Identifier
UriMapping, 132–133
App.xaml, 132
User Datagram Protocol (UDP), 349
migration, 548
user experience, 8–9
Music and Video, 274
SIP, 109
touch screens, 195–200
user interface (UI)
accelerometers, 221
DesignTimeViewModel, 406–407
Expression Blend, 385–386
IAccelerometer, 227–228
migration, 547–548
performance, 549–552
smartphones, 6
source, 385
XAML, 27
User Interface Thread FPS, 550
UserControl, 126, 139
UserName, 500
UserToken, 341
Utilities, 110

V

ValidateRedirectionUrlCallback, 554
Value
Position, 255
ProgressBar, 45
Slider, 262
value converters, 392–394
variable-time-step, game loop, 522
vectors
accelerometers, 221–223
Silverlight, 223
VertexBuffer, 537
VerticalAlignment, 20, 43
Ellipse, 54
VibrateController, 240
vibration, 239–240
chassis design, 3
feedback, 239
VideoPlayButton
IsEnabled, 249
XAML, 251
VideoPlayer
AutoPlay, 246–247
Volume, 252
XAML, 251
VideoStopButton, 251
ViewChangeEnd, 302
ViewChangeOnFrame, 302
ViewChangeStart, 302
ViewModel, 402–405
DataContext, 407
INotifyPropertyChanged, 403
XAML, 407
virtual methods, 130
Visibility, 69, 243
Collapsed, 242
Ellipse, 69
HasBeenPressed, 68
MediaElement, 243
Visible, 69, 243
visual states, 64–71
Button, 64
GroupName, 64
Visual Studio, 13
behaviors, 74
free transport, 16–21
Import and Export Settings, 18
Path, 47
Select Device, 24
testing, 481–486
TouchAndHold, 211
WCF Test Client, 369–371
Windows Mobile, 12
XAML, 18–21
Visual Tree, 207–208
VisualStateManager, 104
GoToState, 70
HasBeenPressed, 71
orientation, 106
StateName, 72
XAML, 69

VisualStudioLogProvider, 481
VisualTreeHelper, 207
Volume, 245
  SoundEffect, 260–263
  VideoPlayer, 252
VolumeConverter, 252–253

W

WCF. *See* Windows Communication Foundation
WCFSyncConfiguration, 450
WCFSyncService, 450
Web Authentication, 331–337
Web Platform Installer, 360, 498
Web Services, Bing Maps, 304–307
WebAnalytics, 469
WebAnalytics.Behaviors, 469
WebAnalyticsService, 468
WebBrowser, 48–50
  Button, 321
  capabilities, 567
  ContentGrid, 48–49
  controls, 320–326
  IsScriptEnabled, 322–323, 336
  JSON, 325–326
  LoadCompleted, 341
  Navigate, 321, 336
  notify, 324
  RetrieveRequestToken, 509
  ScriptNotify, 340
  TextBox, 321
WebBrowserTask, 174, 189–190
  Internet Explorer, 189
  Show, 190
WebClient, 350–355, 496
  Credentials, 499
  HttpWebRequest, 355
  MediaElement, 242–243
WebGet, 357
WebInvoke, 357
WebRole, 556
while, Run, 230
widgets, 152
Width, Ellipse, 54
Wi-Fi
  ActiveSync, 4
  chassis design, 3
  Zune, 4
Wii, accelerometers, 230–238
WiiAccelerometer
  CurrentState, 237
  LoadAccelerometer, 237–238
WiiDataService, 235–236
Wiimote, 231–232
  TextBlock, 238
  WCF Service, 233–235
WiimoteChanged, 234
WiimoteLib, 231
WiimoteSensor, 234
WiiWrapper, 233–234
windows, 8
  Expression Blend, 22
  Silverlight, 126
window.external.Notify, 323, 330, 337
  ReturnToken, 334, 338
  try-catch, 324
Windows Azure, 13, 554–563
Windows Communication Foundation (WCF)
  Authentication Services, 501–507
    ASP.NET, 501–502
    cookies, 505
    SSL, 507
  Data Services, 372–379
    custom methods and stored procedures, 376–378
    EntityConfiguration, 450
    ISyncProvider, 448–458
    OData, 372–379
    queries, 374–376
    RetrieveChangesFromServer, 452
    SaveChangesToServer, 452
    synchronization, 442
    updates, inserts, and deletes, 378–379
  security, 496
  Service, 349, 365–372
    ASP.NET, 341
    Basic Authentication, 499
    Course, 292
    emulator, 284
    example, 168–169
    Exchange Server, 553
    IAccelerometer, 236–237
    location, 284
    Notification Source, 158
    photos, 347
    Solution Explorer, 159
    Speed, 292
    testing, 369–371
    Wiimote, 233–235
  Test Client, Visual Studio, 369–371

Windows Forms, 30
Expression Blend, 547
Label, 36
Silverlight, 547–548
Windows Mobile, 1
Marketplace, 3
problems with, 2
Visual Studio, 12
Windows Phone Developer Tools
emulator, 24–26
gestures, 200
Windows Phone Game and Game Library, 16
Windows Phone Location Service, 280
Windows Presentation Foundation (WPF), 27
Avalon, 30
composite, 467
location, 52
navigation, 126
Notification Source, 160
Windows Workflow Foundation, 30
WindowsLiveIdServiceClient, 345
WindowsLiveLogin.User, 334
Win7GeoPositionWatcher, 289–290
GeoCoordinateWatcher, 295
wizards, 139
Back button, 134
Word, 190
WorkItem, 477
workspaces, Expression Blend, 22
WP7 SqlliteClientPreview, 547
WPF. *See* Windows Presentation Foundation
wptestsite.com, 331–332
WrapPanel, 545
Write, 432
WriteableBitmap, 177, 179
WriteLog, 484
Wroblewski, Luke, 200
WVGA, 5

X

XAML. *See* eXtensible Application Markup Language
XAP, 544
Xbox
animation, 7
motion, 7
Xbox Live, 13
x:Key, 55
XML. *See* eXtensible Markup Language
XNA, 13, 16
Audio Framework, Microphone, 270
Creators Club, 528
Framework, 515–516
Game Studio 4.0, 16
games, 15, 515–539
accelerometers, 530–531
color, 535
Deactivated, 523
exiting, 52
fonts, 527–530
Initialize, 522
keyboard, 532
life cycle, 522–524
lighting, 535
LoadContent, 522
primitives, 535–538
rendering, 524–539
text, 527–530
textures, 538–539
3D, 532–535
touch screen, 531
Trial mode, 569
SoundEffect, 257–267
XNADispatcher, 257

Z

ZoomAction
Canvas, 219
FlickAction, 217
ZoomBarVisibility, 296
Zune, 542
FM tuner, 278
navigation, 7
Wi-Fi, 4
ZuneHD, 7